Queueing Networks

WILEY-INTERSCIENCE SERIES IN SYSTEMS AND OPTIMIZATION

CHAO/MIYAZAWA/PINEDO—Queueing Networks: Customers, Signals and Product Form Solutions

GITTINS—Multi-armed Bandit Allocation Indices

KALL/WALLACE—Stochastic Programming

KAMP/HASLER—Recursive Neural Networks for Associative Memory

KIBZUN/KAN—Stochastic Programming Problems with Probability and Quantile Functions

RUSTEM—Algorithms for Nonlinear Programming and Multiple-objective Decisions

VAN DIJK—Queueing Networks and Product Forms: A Systems Approach

WHITTLE—Optimal Control: Basics and Beyond

WHITTLE—Risk-sensitive Optimal Control

WHITTLE— Neural Nets and Chaotic Carriers

The concept of a system as an entity in its own right has emerged with increasing force in the past few decades in, for example, the areas of electrical and control engineering, economics, ecology, urban structures, automation theory, operational research and industry. The more definite concept of a large-scale system is implicit in these applications, but is particularly evident in such fields as the study of communication networks, computer networks, and neural networks. The Wiley-Interscience Series in Systems and Optimization has been established to serve the needs of researchers in these rapidly developing fields. It is intended for works concerned with developments in quantitative systems theory, applications of such theory in areas of interest, or associated methodology.

Queueing Networks

Customers, Signals and Product Form Solutions

Xiuli Chao
New Jersey Institute of Technology, USA

Masakiyo Miyazawa
Science University of Tokyo, Japan

Michael Pinedo
New York University, USA

JOHN WILEY & SONS, LTD
Chichester • New York • Weinheim • Brisbane • Singapore • Toronto

Baffins Lane, Chichester,
West Sussex, PO19 1UD, England

National 01243 779777
International (+44) 1243 779777

e-mail (for orders and customer service enquiries): cs-books@wiley.co.uk

Visit our Home Page on http://www.wiley.co.uk or http://www.wiley.com

Other Wiley Editorial Offices

John Wiley & Sons, Inc., 605 Third Avenue,
New York, NY 10158-0012, USA

Weinheim • Brisbane • Singapore • Toronto

Library of Congress Cataloging-in-Publication Data

Chao, Xiuli
Queueing networks : customers, signals, and product form solutions / Xiuli Chao, Masakiyo Miyazawa, Michael Pinedo.
p. cm. — (Wiley-Interscience series in systems and optimization)
Includes bibliographical references and index.
ISBN 0-471-98309-8 (hardcover : alk. paper)
1. Production scheduling. 2. Queueing theory. I. Miyazawa, Masakiyo. II. Pinedo, Michael. III. Title. IV. Series.
TK157.5.C453 1999 99-26511
658.5′3—dc21 CIP

British Library Cataloguing in Publication Data

A catalogue record for this book is available from the British Library

ISBN 0 471 98750 6

Produced PostScript files supplied by the authors
Printed and bound in Great Britain by Antony Rowe, Chippenham
This book is printed on acid-free paper responsibly manufactured from sustainable forestry in which at least two trees are planted for each one used for paper production.

To our families

Contents

Preface

Over the last four decades many books have been written on queueing theory with a number of these focusing on queueing networks. So the reader may ask what the motivation behind this book is. This book basically contains material that is relatively new. The content is mostly based on results that have appeared in the literature during the last ten years.

The focus of the book is mainly theoretical; we establish a broad framework and try to present results that are as general as possible. The modeling framework adopted is, we believe, more general than the frameworks used in most other queueing network books. We assume throughout the book that a queueing system not only has customers moving around, but also signals instructing customers what to do. With a customer one usually associates a "body"; with a signal one usually does not.

Since the structure of the models considered is more general than that in any existing text, we believe that the application domain is also wider than that of any other text. We do refer regularly to examples that motivate the theoretical models. These examples come mainly from areas such as manufacturing, distribution, computers and telecommunications. They include production models with quality control stations that scrap multiple items at the same time, computer virus models, circuit switched telecommunication networks, and predator–prey models.

The book is meant for Masters and Ph.D. students in Engineering, Operations Research, and Probability and Statistics. At the end of each chapter there are exercises, where the student is asked either to prove results that are presented in the text without proofs or to consider models that require some extension of the theory. The book can be used for a Masters/Ph.D. level course in, for example, electrical engineering with students that are somewhat theoretically inclined. Students do *not* need a course in elementary queueing as a prerequisite. However, they do need a course in stochastic processes. Parts of the book have been used in Masters/Ph.D. level courses at Columbia University and at Science University of Tokyo.

In the references we include only those that are directly related to the material covered in the text. We believe that references on topics that are similar but not directly related can be found in other queueing network books; including all of them would detract from the thrust of the book.

We have developed and will continue to maintain a website which contains information that could be of interest to readers of this book. The address is

http://www.is.noda.sut.ac.jp/~qnet

This website includes material that could be useful to an instructor who teaches a course in queueing networks. It provides hints to the exercises in the book as well as some additional exercise problems. It also has links to other related websites.

This book has benefited greatly from joint work and discussions with our research collaborators, including Richard Boucherie, Issei Kino, Hideo Ōsawa, Richard F. Serfozo, Karl Sigman, Hiroyuki Takada, Peter G. Taylor, Nico M. van Dijk, Ronald W. Wolff, Hideaki Yamashita, and Shaohui Zheng. We are indebted to our students at Columbia University and Science University of Tokyo for their detailed comments and suggestions. We are also grateful for the helpful advice from Richard Freedman and Kanji Yoneyama. Finally, we acknowledge Yasutaka Hattori for drawing many of the figures in the book.

XIULI CHAO
MASAKIYO MIYAZAWA
MICHAEL PINEDO

1

Introduction

1.1 INTRODUCTION

Waiting lines have confounded humanity since the dawn of civilization. However, the formal study of queueing only began at the outset of the twentieth century with the work of A.K. Erlang (1917). This Danish engineer provided the first mathematical analysis of a queueing model for a telephone switch. Since then, emerging technologies have steadily fostered the need for increasing research on queueing phenomena. The developments in manufacturing, computer and communication systems during the second half of the twentieth century motivated, in particular, the study of interconnected systems in the form of queueing networks.

A queueing network is a system consisting of a finite number of *stations* that provide services to *customers*. The service stations in the network are typically referred to as *nodes*. Examples of queueing networks include computer systems, manufacturing systems, job shops, airport terminals, railway or highway systems, and telephone systems. In these settings, customers (data, parts or subassemblies, jobs, planes, vehicles, phone calls, etc.) arrive at the system and require some form of service (operation executions, assembly processes, machining, airplane take-offs, bridge or toll booth passings, phone conversations, etc.). Queueing network models have been successfully applied in the performance evaluation and optimization of computer systems, communication systems, manufacturing systems and logistic systems. System design and optimization issues include dynamic routing control of customers (packets), trunk designs, resource allocation, load balancing, and throughput maximization. Typical performance measures of practical interest are sojourn time, congestion level, blocking probability, and throughput.

A number of methods have been developed for analyzing queueing networks, each with its own limitations. For example, rapid advances in computers have made it possible to perform large-scale simulations. Still, in addition to the high costs because of the extensive use of computer time and memory, the results of large-scale simulations tend to be application-specific, and are not always useful for detecting general trends in performance measures. Another method for evaluating a queueing system is approximation. However, a theoretical basis is often required to guarantee that an approximation is not far from the real solution. A theoretical analysis of a queueing model is therefore not only important per se; it is also important to complement simulation results and approximations. Such a theoretical analysis involves determining the stationary distribution of the network states, e.g., the customer populations at all nodes, from which various performance measures can be derived.

Clearly, a closed form solution for the stationary distribution, if obtainable, is the most preferred. It is the class of queueing networks with tractable solutions that we will consider in this book. Of the networks with tractable solutions, networks with product form stationary distributions are the ones most researchers have focused on and most applications are based on. In this class of networks, in spite of the high level of interaction between the nodes, the joint distribution of all the nodes is the product of the marginal distributions of the individual nodes. This joint distribution is referred to as a *product form solution.* Due to this property, the analysis of a queueing network reduces to the analyses of single node queues, simplifying the applications tremendously. Nevertheless, we shall see that the class of product form queueing networks is, from a theoretical as well as from a practical point of view, not as narrow as it seems. Indeed, were it not because of Jackson's celebrated product form result and its extensions, applications of queueing networks would most likely not have been as widespread as they are today.

The conventional queueing networks that are typically considered in textbooks have limitations. In such networks, the arrival of a customer increases the population at a node while the departure of a customer decreases the population. With the advances in information technology, many networks have, in addition to conventional customers or jobs, also signals or messages circulating through the system. A signal carries information and instructions and may trigger simultaneous events at one or more nodes. The arrival of a signal at a node may, for example, decrease the population or induce more general transactions at that node. Queueing networks that include these features are referred to as networks with signals. The signals add a new dimension to classical queueing networks and enlarge their modeling capabilities. Clearly, they cover conventional networks (without signals) as special cases.

The objective of this book is to present, in a unified framework, the latest developments in queueing networks with signals that have tractable stationary distributions.

1.2 QUEUEING NETWORKS WITH SIGNALS

In the traditional view of queues and queueing networks, customers or jobs arrive at a service facility, bring a certain amount of work to the system, and leave after the services have been provided; a customer may have to wait for a service provider. In a mathematical model, a customer's arrival increases the queue length and a customer's departure decreases the queue length.

In the networks of queues considered in this book there are, besides the regular customers, also signals that carry commands and move around from node to node. A signal may or may not have a physical "body" associated with it. When a signal arrives at a node, it causes a certain event to occur. This event may imply either the deletion of one or more customers or the addition of one or more customers. It may be even more general.

A major feature of networks with signals, which distinguishes them from conventional models, is that signals can trigger simultaneous events throughout the network. For instance, if an arriving signal induces a customer to move, then the customer may either leave the network or go to another node; when it goes to another node, it may arrive either as a customer or as a signal. If it arrives as a signal it may again induce customers to move, and consequently the cascading effect may result in a sequence of simultaneous events. When a signal deletes a customer or triggers a customer to move, it is called a negative signal. If its effect is the addition of customers, it is called a positive signal. In particular, if an arriving signal simply kills a customer at a node and does not trigger any other event, it may be referred to as a negative customer.

An example of a queueing network with signals is a computer system with a virus. When there are no viruses, computer networks have been modeled and analyzed using conventional queueing networks without signals. The customers represent the execution of algorithmic and logic operations, and the nodes represent CPUs, I/O devices, etc. However, when the effects of computer viruses are included, a conventional queueing network is no longer an appropriate model.

Example 1.1 (Computer viruses) Consider a computer network with each node subject to infections from viruses. When a virus enters a node, one or more files may be infected, and the system manager may have to go through a number of backups to recover the infected files. In some cases, they may not be recoverable. A virus may originate from outside the network, e.g., through a floppy disk, or may come from another node in the network, e.g., by an electronic mail. A virus cannot be modeled as a customer, since it does not require a service. However, a virus can change the status of a node.

Assume the system has N nodes. Exogenous non-infected jobs (denoted by c) arrive at node j according to a Poisson process with rate λ_j, and infected jobs (denoted by s) arrive at node j according to a Poisson process with rate λ_j^-. When a job's processing at node j is completed, it goes to node k as a non-infected job with probability $r_{jc,kc}$, and as an infected job with probability $r_{jc,ks}$. Whenever an infected job arrives at node j, either from outside the network or from another node, a random number of jobs at node j will be infected and will carry the virus. Assume that the number of infected jobs at node j is an arbitrarily distributed random variable with probability mass function $\{b_j(k); k = 1, 2, \ldots\}$. These infected jobs are then moved to a backup drive and are repaired. After a random amount of time the repair is complete and with probability ϑ_j the batch is recovered and all jobs go back to node j for further processing. However, with probability $1 - \vartheta_j$ these jobs are found nonrecoverable and are scrapped. □

In the example above the signals are the viruses or the infected jobs and the effects of the signals are the infection of batches of other jobs. So the arrival of an infected job immediately triggers an event at a node: a batch of jobs becomes infected. This cannot be captured by a conventional queueing network model. However, it fits well within the framework of queueing networks with signals, since the virus is nothing more than a signal. This problem will be studied in more detail in Chapters 7 and 8.

Queueing networks with signals also allow for concurrent movements. A typical example is the following problem, which is referred to as the bathroom problem.

Example 1.2 (The bathroom problem) In a park there is a men's room and a ladies' room, referred to as nodes 1 and 2, respectively. Men arrive at node 1 and women at node 2 according to random arrival processes. There are also arrivals of couples and each couple consists of a man and a woman, see Figure 1.1.

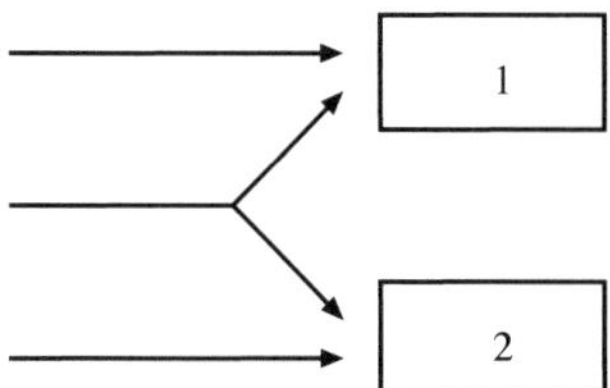

Figure 1.1 The bathroom problem

This problem can be modeled as a queueing network with signals. In addition to the two regular arrival processes at nodes 1 and 2, we introduce an arrival process of positive signals. Assume that positive signals arrive at node 1. Each signal adds one customer to node 1, immediately leaves for node 2 as a positive signal, adds one customer there, and then leaves the network. Clearly, the effect of a positive signal is equivalent to the arrival of a couple. □

The bathroom problem and the effects of positive signals are discussed in more detail in Chapters 4 and 8.

The idea of signals originated with Henderson and Taylor (1989) and Gelenbe (1991). The terms *interruption* and *negative customer* were then used since the effect of a signal was defined as the interruption of a service and the deletion of a customer, respectively. The signals considered in this book are more general. It is clear that signals add a new dimension to classical queueing networks and enlarge their modeling capabilities.

1.3 APPLICATIONS

The main objective of this book is to investigate analytically tractable solutions for queueing networks with signals and to identify networks with such solutions. In this section we illustrate through several examples the modeling capabilities of networks with signals relative to those of conventional networks.

There are many well-known applications of conventional queueing networks. The best known application areas are computer networks, communication networks and manufacturing systems. The usefulness of queueing networks became clear with the work of Scherr in the 1960s, who studied the delays and waiting times in computer networks. In the 1970s many researchers started to use queueing networks for the modeling of flexible manufacturing systems. In the 1980s queueing network models started being used for the analysis of communication networks. During the last decade a number of textbooks appeared focusing on the applications of queueing networks.

It is clear that the models discussed in this book are at least as useful as the conventional models without signals, since the signals enlarge the application domain. Because of the signals, we can, for example, model sequential and simultaneous activities at one or more nodes. This makes it possible to build strong dependencies between the behavior of the different nodes in a network. Such dependencies occur often in practice, e.g., in manufacturing as well as in communication systems.

Example 1.3 (A production and distribution system) Consider a manufacturing facility, denoted by f. Raw materials arrive at the facility according to a Poisson process with rate λ. After a unit of the product has been manufactured at the facility, it is transferred to a distribution station, denoted by s. Station s can be considered as a queue with processing rate 0. Orders for the finished product come in according to a Poisson process with rate λ^- and they queue up at an order processing station, denoted by d. The order processing times are also random variables. The order quantities are random, and the order size is k with probability $b(k)$, $k = 1, 2, \ldots$. When an order of size k is processed at station d, the queue at station d, which represents the number of orders, goes down by one, and simultaneously, the queue at station s goes down by k, provided k or more are present. Otherwise the station is emptied out (partial order fulfillment). See Figure 1.2.

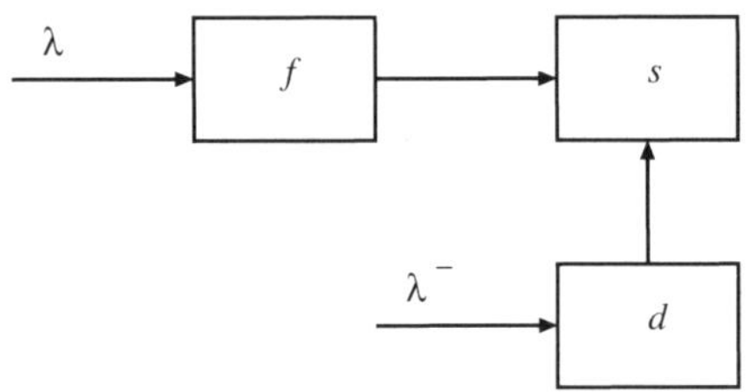

Figure 1.2 A production and distribution system

This network cannot be modeled as a conventional queueing network since a departure at station d triggers a batch departure at station s. However, this is a typical feature in queueing networks with signals. □

We will study this problem in more detail in Chapter 7.

Example 1.4 (A production and inventory control model) A company manufactures K classes of products. Assume that the manufacturing process of product j can be modeled as a single-server facility, which produces units at rate μ_j, $j = 1, \ldots, K$. There are I classes of demands, and demands of class i arrive according to a Poisson process with rate λ_i, $i = 1, 2, \ldots, I$. Each demand of class i consists of an order for a_{ij} units of product j, $j = 1, \ldots, K$. A demand is immediately satisfied if there is enough inventory of each class of product in stock; otherwise it is backordered. The company uses a base stock policy, which places an order for the manufacturing of a product j whenever its inventory position drops below R_j. The inventory position is defined as

$$\begin{aligned}\text{Inventory position} = \; & \text{Inventory on hand} \\ & + \text{Outstanding manufacturing orders} \\ & - \text{Demand back orders,}\end{aligned}$$

where Outstanding manufacturing orders is the number of products that have been ordered from manufacturing and not yet been produced.

Let $I_j(t)$ and $B_j(t)$ be the inventory on hand and the backordered demand at time t. Note that $B_j(t)$ can be positive only when $I_j(t) = 0$. The state space for $I_j(t) - B_j(t)$ is

$$\{\ldots, -2, -1, 0, 1, 2, \ldots, R_j - 1, R_j\}.$$

Every time a class i demand comes in, $I_j(t) - B_j(t)$ decreases by a_{ij}, $j = 1, 2, \ldots, K$. To study the performance of the inventory control policy, let

$$X_j(t) = R_j - (I_j(t) - B_j(t)).$$

Clearly, $X_j(t)$ has state space

$$\{0, 1, 2, \ldots\}.$$

The process $\{X_j(t); t \geq 0\}$ can be regarded as an individual queue with service rate μ_j and batch arrivals: every time a demand of class i arrives, $X_j(t)$ goes up by a_{ij}, $j = 1, 2, \ldots, K$. Hence $(X_1(t), \ldots, X_K(t))$ represents a queueing network with concurrent batch arrivals at the K queues.

A queueing system with concurrent batch arrivals can be modeled as a queueing network with positive signals, as described in Example 1.2. This production and inventory control problem will be studied in Chapter 8. □

One particularly useful feature that can be modeled with a queueing network with signals is *cause-dependent feedback*. In networks with cause-dependent feedback the routing probabilities are allowed to be functions of the cause of departure, e.g., whether it is due to an actual service completion or due to a trigger by a signal of a particular class. Hence if a customer is induced to move by a signal that represents a request for rework at a specific node, then the routing probabilities can force the customer to proceed to that node. Thus dependencies can be constructed throughout the network between events such as node failures or detection of defects, as the following example shows.

Example 1.5 (Quality control in manufacturing) Consider a quality control station in a flexible manufacturing process. There are a number of stages in series. Each job has its own route through the system. The final stage is a quality control station. At this station each part is checked for defects. If a part is defective, then three things can happen: First, it is possible that the part only needs rework, i.e., it has to go to an upstream station and then follow a new route through the flow shop. Such a situation can be handled within the framework of conventional queueing networks with feedback. Second, suppose the part has to be discarded. If the part has to be discarded and it is the only part affected by such an event, then the situation can be modeled as a conventional network also. Third, suppose the quality control station determines the cause of the defect and it is clear that the part under consideration is not the only one that has to be scrapped. The cause of the defect, which may be attributed to a particular upstream station, indicates that several other parts in the system have to be discarded as well. This situation cannot be handled by a conventional queueing network, but it can be taken care of by introducing signals with feedback that remove discarded parts throughout the network. □

A similar phenomenon may occur in communication networks. For example, consider a communication system with nodes subject to breakdowns. A breakdown at a node may happen due to exogenous reasons; it may also be caused by other nodes in the transmission process. Furthermore, if a node breaks down, all the messages residing at that node are lost. This cannot be modeled as a conventional queueing network, but it fits perfectly well within the framework of queueing networks with signals. We simply introduce a signal that induces a catastrophe at that node (see Chapter 7).

One important application area of conventional queueing networks is population dynamics. However, the following predator–prey example, due to Henderson, Northcote and Taylor (1994*a*), provides an instance that cannot be modeled as a conventional queueing network, but again fits well within the framework of networks with signals.

Example 1.6 (A predator–prey model) Consider an ecosystem of N species contained in a finite area. Assume that each species has a fixed diet, i.e., a particular species is a predator of a fixed subset of the N species in the ecosystem, and will prey on a particular species in that subset for a fixed proportion of the time when it is looking for food. In addition to possibly being killed by predators from a subset of the N species, the members of each species arrive at the ecosystem due to migration or birth, and depart due to migration or death, where death may occur due to starvation (there may not be enough of a particular species present in the ecosystem) or other natural causes. Let node 0 represent the world outside the area.

To model this problem each species is represented by a single node in an N-node network. A transition from node j to node 0 represents the death or migration of species j. Assume that there are also transitions from node j to node k. Its effect is that, whenever it occurs, it triggers an individual from node k to transfer to node j, provided there are individuals present at node k; otherwise it leaves the system. This is because when an individual switches from node j to node k, it reduces the population size at node k by one, and it immediately returns to node j and increases the population size at node j by one. This implies that a member of species j eliminates a member of species k from the ecosystem and returns to its base node. The case where an arrival of species j at node k disappears when node k is empty represents the death of species j due to starvation.

This problem cannot be modeled as a conventional queueing network since a departure at one node triggers movements of species at another. However, it is again easily handled by queueing networks with signals. The predator–prey model will be studied in Chapter 9. □

1.4 DEFINITIONS OF THE BASIC CONCEPTS

This book is about networks of queues with customers and signals. In this section we elaborate on the terminology that we will use throughout the book.

Definition 1.7 (Queue) *A queue* is a system driven by two types of events: arrivals and departures (Figure 1.3).

Figure 1.3 A queue

The effects of arrivals and departures on the system are not specified in the definition and they can be very general. The effect of an arrival is characterized by the *arrival effect probability* that specifies changes in the state of the queue. For example, when the state of the queue is the number of entities, an arrival at the queue may lead to a decrease in the number of entities in the system or other types of changes in the state. Similarly, a departure may lead to an increase in the number of entities in the system. This is in contrast to a conventional queue where an arrival increases the number of entities and a departure decreases the number of entities. Furthermore, we do not make any assumptions implying that an arrival will eventually depart the system, since arrival and departure events do not necessarily represent real entities. Moreover, since a queue can be considered as a black box, there can be *internal transitions* within the system that are not triggered by arrivals or departures.

Definition 1.8 (Network of queues) A *network of queues* is a collection of queues that interact with one another through departure and arrival events, e.g., a departure from one queue becomes an arrival at another queue. We often refer to the queues in a network as *nodes*.

We describe arrivals and departures by movements of entities. As we explained, entities are not necessarily customers and may not have any physical bodies.

Definition 1.9 (Class) A class consists of entities that have similar characteristics with regard to the routing specifications, the effects on each node they visit (at their arrival as well as at their departure), and the effects they have on the network dynamics. In other words, entities of the same class are governed by the same probability laws.

The state space of each queue as well as of the network can also be very general. In the dynamics of a queueing network, upon arrival at a queue some entities change the state of the queue but do not induce an immediate departure, while other entities upon arrival at a queue may change the state of the queue and at the same time also induce an immediate departure. If, upon arrival at a queue, an entity has a positive probability of triggering an immediate departure, it is called a *signal*.

Definition 1.10 (Signal) A *signal* is an entity which, upon arrival at a queue, may change the state of the queue and may trigger an immediate departure. In other words, a signal may cause instantaneous movements (see Figure 1.4).

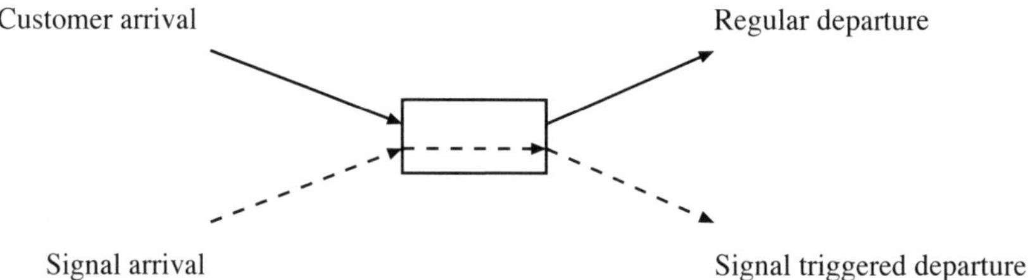

Figure 1.4 Customers and signals

Clearly, the cascading effects of signals may generate an arbitrary number of arrivals and departures simultaneously throughout the network.

Remark 1.11 It follows from the definitions above that only an arrival has to be classified as to whether it is a signal, and there is no need to specify a departure. However, when a departure is routed to another node as an arrival, it has to be specified whether it is a signal since its effect on the node where it arrives depends on such a classification. Nevertheless, for convenience we shall often refer to a departure as a signal in order to reduce the number of different classes of entities in the network.

In contrast to these general entities, in a queueing network model with an arbitrary state space we use the terminology "customer" in the conventional sense when the state of the queue is the number of entities. More specifically, when the state space of the network is based on its population, we shall further classify the entities according to their effects on the queue, as *(regular) customers, negative customers, positive signals*, and *negative signals*. The remaining part of this section focuses on such special classes of networks.

Definition 1.12 (Customer) A *customer* is an entity which, upon arrival at a queue, may change the number of entities in the queue but does not cause an instantaneous departure.

In particular, if the effect of an arrival is an increase in the number of entities and does not cause an instantaneous departure, it is called a regular customer (or simply a customer). On the other hand, if the effect of an arrival is a decrease in the number of entities in the queue, it is called a *negative customer*.

Definition 1.13 (Positive signals and negative signals) If the arrival of a signal increases the number of entities at the queue and then triggers an immediate departure, it is called a *positive signal*; on the other hand, if the arrival of a signal decreases the number of entities at the queue and then triggers an immediate departure, it is called a *negative signal*.

Remark 1.14 Since the arrival of a negative customer decreases the number of entities, its effect is equivalent to the arrival of a negative signal that triggers an entity to leave the system. For this reason, a negative customer is also often regarded as a negative signal.

Regular and negative customers as well as positive and negative signals can be considered as different classes of entities. We may have more than one class of regular customers and negative customers and more than one class of positive and negative signals since the size of the increase or the decrease in the number of entities may differ. Furthermore, even when customers and signals cause the same changes in the states of the nodes where they arrive or from where they depart, they may belong to different classes because of their different routing and service characteristics.

1.5 OUTLINE OF THE BOOK

This book requires some basic probability theory as a prerequisite, but it does *not* require any knowledge of queueing theory. Even though the book focuses on networks with signals, it can be used as a textbook for a course in standard queueing network theory and applications. The results for conventional queueing network models without signals can be obtained by simply ignoring the signals.

Almost all networks considered have an *outside source*, i.e., customers arrive at the system from the outside and leave the system for the outside. Even though many results hold for closed networks, in this book we focus mainly on open networks. However, a network becomes closed when the outside source is considered a regular node and all customers that are removed by signals return to the outside node.

Chapters 2, 3, and 4 develop the foundations and some basic tools used throughout the book. Chapters 5 to 8 cover a number of different queueing network models with various features. Chapters 9, 10, and 11 focus on more fundamental structures of queueing networks, and Chapter 12 presents a framework for discrete time queueing network models.

Chapter 2 introduces some fundamental results that form a basis for the rest of the book. In addition to some of the basics of Markov processes, the three most commonly used techniques for showing product form solutions, i.e., local balance, the Kelly lemma, and reversed processes, are discussed. Single server queues with batch arrivals and batch services, which constitute the building blocks for a number of network models in later chapters, are explored in some detail. Other important concepts, i.e., Markov driven counting processes, reversibility, product form solutions, and decomposability, are also discussed in Chapter 2.

Chapters 3 and 4 develop a framework for quasi-reversibility of queues with signals. The readers mainly interested in learning about the various models and their solutions can skip these two chapters in the first reading of the book. A probability decomposition approach is used in this development. The definitions of quasi-reversibility are given in Chapter 3 and illustrated with several examples. The product form result for networks of quasi-reversible nodes is shown in Chapter 4. As an illustration we present a class of network models with positive and negative signals. Many queueing networks with concurrent batch movements are, in one way or another, special cases of this simple network.

Chapter 5 focuses on networks with exponential service times. These include the conventional Jackson networks as well as the Kelly networks. Chapter 6 focuses on networks with arbitrary service times. These networks include the Baskett–Chandy–Muntz–Palacios (BCMP) networks and Kelly networks with so-called symmetric service disciplines.

Chapters 7 and 8 describe results for networks with batch movements. The models presented in Chapter 7 have simultaneous batch departures throughout the network, while the models of Chapter 8 have simultaneous batch departures as well as simultaneous batch arrivals. A key difference between these models is that, while the models of Chapter 7 always have product form stationary distributions, the models in Chapter 8 do not have stationary distributions that are product form. Some extra boundary conditions are introduced and it is shown that with these modifications the networks in Chapter 8 do have product form solutions. Furthermore, it is shown that in many cases the product form solutions thus obtained can actually serve as stochastic upper bounds for the networks without these extra conditions.

The signals in Chapter 7 have the features that their arrivals either delete a batch of customers or trigger a batch of customers to move. So, in a sense, they may be considered as negative signals or negative customers. On the other hand, the effects of the signals in Chapter 8 can be either the removal of a batch of customers, or the addition of a batch of customers. Thus the models studied in Chapter 8 can be regarded as networks with positive and negative signals. Indeed, the results of both Chapters 7 and 8 can be considered as special cases of the networks with positive and negative signals discussed in Chapter 4.

Chapter 9 extends the results of earlier chapters to allow state-dependent transitions and stage-dependent transitions. It is shown that all the models in Chapters 5 to 8 have state-dependent versions in which the transition rates depend on the state of the entire network.

Chapter 10 continues with the theoretical development of Chapter 4. It is shown that local balance plays a central role in extending results for networks with state-independent transition rates to networks with state-dependent transition rates. Important relationships between local balance, product form, and quasi-reversibility are explored.

The main approach for proving product form stationary distributions, up to Chapter 8, is quasi-reversibility. As shown in Chapter 4, quasi-reversibility is a sufficient condition for product form. In some cases, quasi-reversibility is also a necessary condition for a network to have a product form distribution. This is true for a conventional queueing network with Poisson arrivals when all the nodes are internally balanced (their arrival and departure rates are identical) and each node has an additional property when it is in an empty state. Such results have led in the literature to the conjecture that quasi-reversibility is a necessary condition for a product form in a queueing network. However, this is not true, as shown in Chapter

11. A necessary and sufficient condition for product form is given in Chapter 11 in terms of the marginal distributions. This condition yields a general procedure to check whether the network has a product form stationary distribution and to compute the distribution when it exists. It is shown that the nodes of the network are quasi-reversible if and only if it has a product form distribution that satisfies the *biased local balance* condition discussed in Chapter 10. The other topic discussed in Chapter 11 is stability. Several queueing network models are examined in detail and their stability conditions are presented.

The final chapter deals with discrete time queueing network models. Quasi-reversibility is extended to the discrete time framework. A number of discrete time queueing network models are discussed in detail. The readers interested only in discrete time models may go directly from Chapter 2 to Chapter 12.

The three appendices contain some basic material. Appendix A presents some elementary probability theories. Appendix B focuses on Markov processes and Poisson processes, and Appendix C describes Brouwer's fixed point theorem, a result frequently used in Chapter 11 for proving the existence of solutions to traffic equations.

1.6 REFERENCE NOTES

The research on queueing networks began with the work of Jackson (1957) and has gained momentum ever since. Several texts have appeared that discuss product form solutions, including Kelly (1979), Whittle (1986), Walrand (1988), Van Dijk (1993), Gelenbe and Pujolle (1998) and Serfozo (1999). In addition, at least a dozen books have been published on the applications of queueing networks, dealing with modeling, analysis, and optimization of manufacturing and logistics systems, as well as computer and communication networks.

2

Fundamentals

This chapter covers some preliminary results. Section 2.1 reviews some basic aspects of Markov chains. Section 2.2 discusses counting processes driven by Markov chains. Section 2.3 studies local balance, while Section 2.4 focuses on reversed processes and reversibility. Section 2.5 introduces several notions of product form solutions. Sections 2.6 and 2.7 analyze two simple single-server queues, i.e., $M/M/1$ queues with batch services and those with batch arrivals and with batch services, which constitute the basic building blocks for several of the networks considered in later chapters.

2.1 MARKOV CHAINS

Let $\{X(t); t \geq 0\}$ be a stochastic process defined on a countable state space $\mathcal{S}$. The stochastic process $X(t)$ is called a Markov chain if, for all $s \geq 0$,

$$\begin{aligned} &P\big(X(t+s) = x' | X(t) = x, X(u) = x(u), 0 \leq u \leq t\big) \\ &= P\big(X(t+s) = x' | X(t) = x\big), \end{aligned} \tag{1.1}$$

whenever the conditioning event $(X(t) = x, X(u) = x(u), 0 \leq u \leq t)$ has a positive probability. Intuitively, a Markov chain is a process with the property that the conditional distribution of the future state $X(t+s)$, given the current state $X(t)$ and the past history $X(u), 0 \leq u \leq t$, depends only on the current state and is independent of the past. This is called the *Markov property*. It can be shown that this property is equivalent to the fact that, given the current state, the past and the future are independent. It implies that the current state contains all the necessary information for predicting the future. In addition, if

$$P(X(t+s) = x' \mid X(t) = x) = p_s(x, x') \tag{1.2}$$

is independent of t, then the Markov chain is said to be time homogeneous. We assume time homogeneity throughout this chapter.

The time in a Markov chain can be either discrete or continuous. When the time is continuous it is denoted by t and $\{X(t); t \geq 0\}$ is referred to as a continuous time Markov chain. When the time is discrete it is denoted by k, the process $\{X(k); k = 0, 1, 2, \ldots\}$ is also written as $\{X_k; k = 0, 1, 2, \ldots\}$ and it is referred to as a discrete time Markov chain. The probability

$$p(x, x') = P(X_{k+1} = x' | X_k = x), \qquad x, x' \in \mathcal{S},$$

is called the transition probability from x to x'. The transition probabilities satisfy

$$\sum_{x' \in \mathcal{S}} p(x, x') = 1, \qquad x \in \mathcal{S}.$$

The value $p(x, x')$ represents the probability that the process, when in state x, makes its next transition to state x'.

For a discrete time Markov chain $\{X_k; k \geq 0\}$ and an integer $n \geq 1$, the probability

$$p^{(n)}(x, x') = P(X_{k+n} = x' | X_k = x), \qquad x, x' \in \mathcal{S},$$

is called the n-step transition probability from x to x'. State x' is said to be accessible from x if $p^{(n)}(x, x') > 0$ for some $n \geq 1$. Two states x and x' that are accessible from one another are said to *communicate*. It is easily seen that communication is an *equivalence* relationship, which implies that the states in $\mathcal{S}$ can be partitioned into classes in such a way that any two states in the same class communicate, and any two states in different classes do not communicate. If a Markov chain contains a single class, i.e., any two states communicate, the Markov chain is called *irreducible*.

Suppose the Markov chain is in state x, say at time 0. If the process will revisit state x for sure at some time in the future (i.e., with probability 1), then state x is said to be *recurrent*; otherwise it is said to be *transient*. If x is a recurrent state, and if in addition the expected amount of time to revisit state x is finite, then x is said to be *positive recurrent*; otherwise it is said to be *null recurrent*.

A Markov chain is termed *periodic* if there exists an integer $\delta \geq 2$ such that

$$P(X_{k+m} = x' | X_k = x) = 0,$$

unless m is an integer multiple of δ. Otherwise the Markov chain is called *aperiodic*.

It can be shown that positive recurrence, null recurrence and transience are class properties. That is, if one state is positive recurrent (null recurrent, or transient, respectively), then all the states in that same class are also positive recurrent (null recurrent, or transient, respectively). If a Markov chain is irreducible and its states are aperiodic and positive recurrent, then the Markov chain is called ergodic.

For any ergodic Markov chain $\{X_k; k = 0, 1, \ldots\}$ there exists a set of positive numbers $\pi(x)$, $x \in \mathcal{S}$, such that

$$\pi(x) = \lim_{n\to\infty} p^{(n)}(x', x), \qquad x, x' \in \mathcal{S}.$$

The collection $\{\pi(x), x \in \mathcal{S}\}$ represents the stationary distribution of the process in state space $\mathcal{S}$. It is also often referred to as the limiting distribution, equilibrium distribution or steady state probability of the Markov chain. By the ergodic theorem, the $\pi(x)$ can also be interpreted as the relative frequency of the Markov chain taking value x if the Markov chain has been under observation for a long time. A stationary distribution exists if we can find a collection of positive numbers $\pi(x)$, $x \in \mathcal{S}$, such that

$$\pi(x) = \sum_{x'\in\mathcal{S}} \pi(x')p(x', x), \qquad x \in \mathcal{S}, \tag{1.3}$$

and

$$\sum_{x\in\mathcal{S}} \pi(x) < \infty. \tag{1.4}$$

Any $\pi(x)$ satisfying (1.3) is called a stationary measure. Under condition (1.4) the stationary measure $\pi(x)$ can be normalized to obtain the stationary distribution. The equations (1.3) are known as the *global balance equations*. If there exists no π satisfying (1.3) and (1.4), the process does not have a stationary distribution.

For a continuous time Markov chain $\{X(t); t \geq 0\}$, the transition rate from state x to state x' is defined as

$$q(x, x') = \lim_{h\downarrow 0} \frac{P(X(t+h) = x' | X(t) = x)}{h}, \qquad x \neq x'. \tag{1.5}$$

Note that $q(x, x)$ is not defined. In the literature $q(x, x)$ is often defined as

$$q(x, x) = -\sum_{x'\neq x} q(x, x').$$

However, in this book we use a different definition, which we will discuss in the next chapter. In this chapter, $q(x, x)$ is assumed to be zero unless otherwise specified.

The Markov property (1.1) leads to an alternative definition for a continuous time Markov chain. Namely, a continuous time Markov chain is a stochastic process having the properties that

(i) each time it enters a state (say) x, the amount of time it spends in that state before making a transition into a different state is exponentially distributed with mean $1/q(x)$, i.e., its distribution function $F(t)$ is given by

$$F(t) = 1 - e^{-q(x)t}, \qquad t \geq 0;$$

(ii) when the process leaves state x, it enters state x' with some probability, say $p(x, x')$, called the transition probability. Of course these $p(x, x')$ must satisfy

$$\sum_{x' \in \mathcal{S}} p(x, x') = 1, \qquad x \in \mathcal{S}.$$

The proof of (i) is technical, and is given in Appendix B, while (ii) is a direct consequence of the Markov property (1.1). Theorem B.6 of Appendix B shows that the parameter $q(x)$ is related to $q(x, x')$ by

$$q(x) = \sum_{x' \in \mathcal{S}} q(x, x'), \qquad x \in \mathcal{S}, \tag{1.6}$$

while $q(x)p(x, x')$ is the transition rate from state x to state x'. Hence

$$p(x, x') = \frac{q(x, x')}{q(x)}, \qquad x, x' \in \mathcal{S}. \tag{1.7}$$

If X_k is the state just after the kth state change of the continuous time Markov chain, then $\{X_k; k = 0, 1, \ldots\}$ is a discrete time Markov chain. The two properties (i) and (ii) can be used to construct the continuous time Markov chain as follows. The process starts in state $X_0 = X(0)$ and stays in this state for an exponentially distributed amount of time with parameter $q(X_0)$, then jumps to state X_1 according to transition probability (1.7). Then the process stays in X_1 for an exponentially distributed amount of time with parameter $q(X_1)$ before it jumps to another state, and so on. This procedure continues and generates the entire sample path of the stochastic process $X(t), t \geq 0$.

This argument relates a continuous time Markov chain to a discrete time Markov chain. It states that the behavior of a continuous time Markov chain resembles that of a discrete time chain with transition probabilities $p(x, x')$. However, whenever it enters a state, say x, it stays there for a random amount of time (instead of one time unit) before making the next transition. This random amount of time is exponentially distributed and its parameter depends on the current state of the process, i.e., $q(x)$.

We can construct continuous time Markov chains with strange behaviors, e.g., a process that visits an infinite number of states within a finite period of time. Such processes will not be considered in this book. Throughout this book we consider only well-behaved stochastic processes that are time homogeneous, irreducible, and incapable of passing through an infinite number of states in any finite time. Furthermore, we typically assume that the process is ergodic, i.e.,

$$\lim_{t\to\infty} P(X(t) = x | X(0) = x') = \pi(x), \qquad x \in \mathcal{S},$$

for some positive number $\pi(x)$. This $\pi(x)$ can also be interpreted as the relative frequency of the process $X(t)$ visiting x. Suppose the stationary distribution π_0 exists for the discrete time Markov chain defined by (ii). From (i), the relative frequency that $X(t)$ resides at x is equal to that of the discrete time Markov chain multiplied by the mean sojourn time in x, i.e., $1/q(x)$. Thus we have

$$\pi(x) = \pi_0(x)\frac{1}{q(x)}. \tag{1.8}$$

Substituting (1.8) and (1.7) into (1.3) yields

$$\pi(x)\sum_{x'} q(x, x') = \sum_{x'} \pi(x')q(x', x), \qquad x \in \mathcal{S}. \tag{1.9}$$

These are the global balance equations for the continuous time Markov chain $X(t)$. Again, $\{\pi(x); x \in \mathcal{S}\}$ is called the stationary distribution of the continuous time Markov chain. Similar to the case of a discrete time Markov chain, a stationary distribution exists if we can find a collection of positive numbers $\pi(x)$, $x \in \mathcal{S}$, such that (1.9) is satisfied and

$$\sum_{x\in\mathcal{S}} \pi(x) < \infty.$$

In this case the π's can be normalized to yield the stationary probability. Any π's that satisfy (1.9) is called a stationary measure.

The global balance equations simply state that in steady state, the transition rate going out of any state is equal to the transition rate going into that state, i.e., they represent a simple rate conservation law. Using this property we conclude that in steady state, the rate going out of any subset of states is equal to the rate going into that subset. In other words, the probability flow each way across a cut balances, as shown in Figure 2.1. Mathematically, it means that for any subset $A \subset \mathcal{S}$,

$$\sum_{x \in A} \sum_{x' \in \mathcal{S} \setminus A} \pi(x) q(x, x') = \sum_{x' \in \mathcal{S} \setminus A} \sum_{x \in A} \pi(x') q(x', x). \tag{1.10}$$

We refer to (1.10) as *a cross balance equation*. Its formal verification is left to the reader as an exercise. Note that if $A = \{x\}$, then (1.10) reduces to a global balance equation.

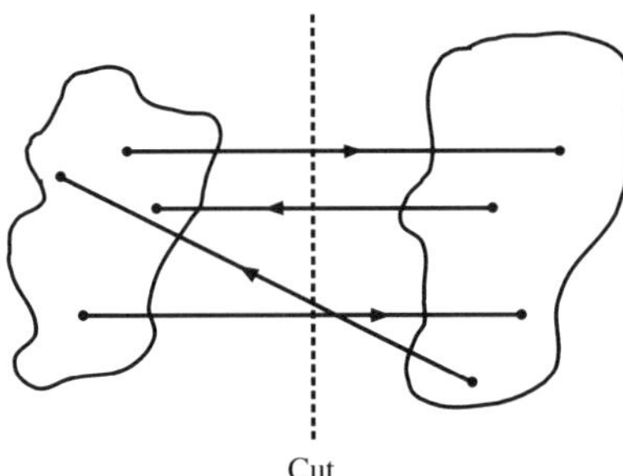

Figure 2.1 A cross balance

A stochastic process $\{X(t); t \geq 0\}$ is called stationary if $(X(t_1), \ldots, X(t_n))$ has the same distribution as $(X(t_1 + s), \ldots, X(t_n + s))$ for all n and any $t_1, \ldots, t_n, s$. In particular, if a process is stationary, then $X(t)$ has the same distribution as $X(0)$ for all t. For a Markov chain with stationary distribution π, it can be shown that if

$$P(X(0) = x) = \pi(x), \qquad x \in \mathcal{S},$$

then

$$P(X(t) = x) = \pi(x), \qquad x \in \mathcal{S}, \quad \text{for all } t.$$

It is for this reason that π is referred to as the stationary distribution.

We conclude this section with two classical examples of continuous time Markov chains: the $M/M/1$ queue and the birth and death process.

Example 2.1 (The $M/M/1$ queue) Consider a service system, e.g., an ATM machine in a bank, with a single-server. At any time at most one customer can be served. Customers arrive at this service facility one at a time, and for any t, the amount of time from t to the next arrival is exponentially distributed with mean $1/\lambda$. The service times are independent and exponentially distributed with mean $1/\mu$.

Define $X(t)$ as the number of customers in the system at time t; $\{X(t); t \geq 0\}$ is a stochastic process. Actually, it is a continuous time Markov chain. To see this, we apply the second definition. Note that whenever the state of the system is $X(t) = n \geq 1$, the time until the next arrival is exponentially distributed with mean $1/\lambda$, and the time until the next departure, i.e., the remaining service time, is also exponentially distributed with mean $1/\mu$. Since these two random variables are independent, the amount of time the process stays in state n is the minimum of the two, and by the properties of exponential random variables, this minimum is exponentially distributed with mean $1/(\lambda + \mu)$, i.e.,

$$q(n) = \lambda + \mu.$$

Furthermore, given that the process leaves state n, it is due to an arrival with probability $\lambda/(\lambda + \mu)$ and due to a departure with probability $\mu/(\lambda + \mu)$ (see Exercise 2.2). Hence

$$p(n, n+1) = \frac{\lambda}{\lambda + \mu}, \qquad n \geq 0,$$
$$p(n, n-1) = \frac{\mu}{\lambda + \mu}, \qquad n \geq 1.$$

Similarly, we can argue that when $n = 0$,

$$q(0) = \lambda, \quad p(0, 1) = 1.$$

Hence $\{X(t); t \geq 0\}$ is a continuous time Markov chain. Its transition rates are given by

$$q(n, n+1) = \lambda, \qquad n \geq 0,$$
$$q(n, n-1) = \mu, \qquad n \geq 1.$$

Clearly, n and $n + 1$ communicate, hence the Markov chain is irreducible if and only if $\lambda > 0$ and $\mu > 0$. It can be shown that the system is ergodic if and only if $0 < \lambda < \mu$. This is quite intuitive since when $\lambda \geq \mu$, there are more arrivals than the system can handle, and consequently the system would explode, i.e., the queue size would increase to infinity. □

An example that is more general than the $M/M/1$ queue is the so-called birth and death process.

Example 2.2 (The birth and death process) Consider a stochastic process $\{X(t); t \geq 0\}$ that represents the size of a population at time t. It is assumed that whenever the population size is n, the time until the next birth is exponentially distributed with mean $1/\lambda(n)$ and the time until the next death is exponentially distributed with mean $1/\mu(n)$, and these two events are independent of each other. The process $\{X(t); t \geq 0\}$ is called a birth and death process with birth rates $\lambda(n), n \geq 0$, and death rates $\mu(n), n \geq 1$.

Using the same argument as in the previous example, we can show that $\{X(t); t \geq 0\}$ is a continuous time Markov chain with transition rates

$$q(n, n+1) = \lambda(n), \qquad n \geq 0,$$
$$q(n, n-1) = \mu(n), \qquad n \geq 1.$$

As in Example 2.1, the process is irreducible if $\lambda(n) > 0$ for all $n \geq 0$ and $\mu(n) > 0$ for all $n \geq 1$. It can be shown that the process is ergodic if and only if

$$\sum_{n=1}^{\infty} \frac{1}{\lambda(n)} = \infty,$$
$$\sum_{n=1}^{\infty} \prod_{\ell=1}^{n} \frac{\lambda(\ell - 1)}{\mu(\ell)} < \infty.$$

The first condition ensures that there cannot be an infinite number of jumps in any finite period of time, and the second condition guarantees that the stationary measure can be normalized to a stationary distribution. □

2.2 COUNTING PROCESSES DRIVEN BY MARKOV CHAINS

As discussed in Section 1.4, queueing systems are driven by arrivals and departures, which constitute the basic events for describing a queueing model. There are two ways of defining these discrete events in a queueing system. One is by counting the number of events up to time t, and the other is by considering the times at which events occur. Clearly, these two types of data can be obtained from one another, provided not more than one event can occur at one point in time (see Figure 2.2). The stochastic process describing these discrete events is called a counting process.

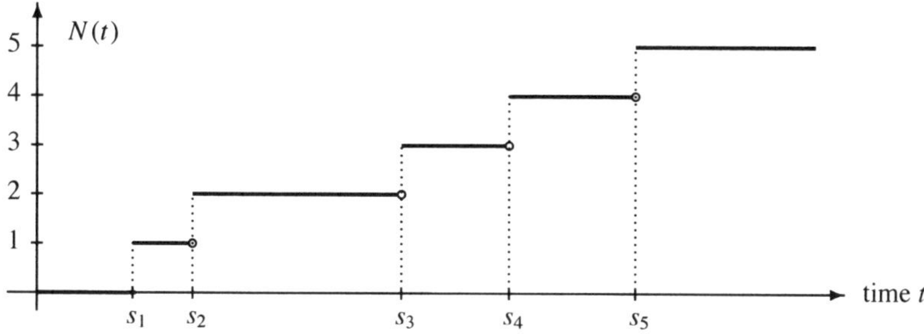

Figure 2.2 A counting process $\{N(t)\}$ with increasing epochs s_i

Definition 2.3 (Counting process) A nonnegative integer-valued stochastic process $\{N(t); t \geq 0\}$ is called a *counting process* if $N(0) = 0$ and $N(t)$ is non-decreasing in t for each sample path. In particular, if not more than one event occurs at any time t, i.e.,

$$N(t+) - N(t-) \leq 1, \qquad \text{for all } t > 0,$$

where $N(t+) = \lim_{\epsilon \downarrow 0} N(t + \epsilon)$ and $N(t-) = \lim_{\epsilon \downarrow 0} N(t - \epsilon)$, then the counting process $\{N(t); t \geq 0\}$ is called *simple*.

Since we are concerned with a continuous time Markov chain, counting processes driven by a Markov chain are of special interest (see Figure 2.3).

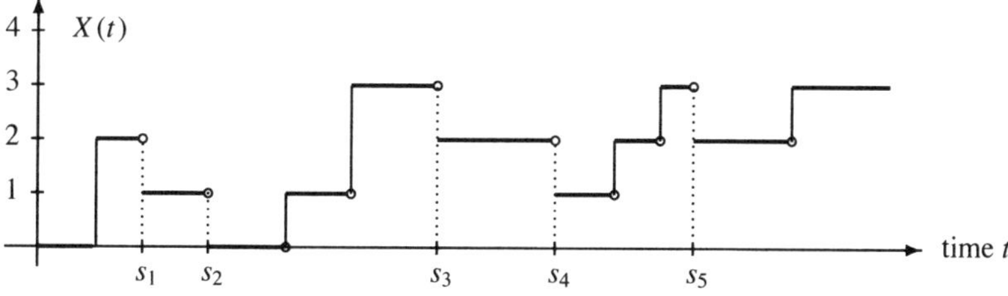

Figure 2.3 The counting process $\{N(t)\}$ of Figure 2.2, driven by $\{X(t)\}$ with respect to downward jumps

Definition 2.4 (Markov-driven counting process) Let $\{X(t); t \geq 0\}$ be a continuous time Markov chain with state space $\mathcal{S}$. A counting process $\{N(t); t \geq 0\}$ is said to be *driven by* the Markov chain $X(t)$ if there exists a set A of pairs of states (x, y), $x, y \in \mathcal{S}$, such that $N(t)$ is the number of transitions the process X makes from x to y, for $(x, y) \in A$, during the interval $[0, t]$. That is,

$$N(t) = \int_0^t 1[(X(s-), X(s)) \in A]ds,$$

where $1[.]$ is the indicator function which assumes the value 1 if the statement in the square brackets is true and 0 otherwise.

A very important counting process is the Poisson process.

Definition 2.5 (Poisson process) A counting process $\{N(t); t \geq 0\}$ is a Poisson process with rate $\lambda > 0$ if it is driven by all the jump epochs of a pure birth process with constant birth rate λ, i.e.,

$$q(n, n+1) = \lambda, \qquad n \geq 0.$$

The λ is called the arrival *rate* (or *intensity*) of the Poisson process.

In view of the properties (i) and (ii) in Section 2.1, the duration from any point in time in the pure birth process until the next event is exponentially distributed with mean $1/\lambda$. Thus we have the following result.

Lemma 2.6 *A Poisson process is a counting process such that the time intervals between adjacent events are independent and exponentially distributed random variables with the same mean.*

This result provides an alternative definition for a Poisson process, since the independent time intervals and their common distribution uniquely determine a point process. It follows from Lemma 2.6 that the mean interarrival time is $1/\lambda$, so the mean number of events in a unit interval equals λ, i.e.,

$$E(N(t+1) - N(t)) = \lambda.$$

Some additional properties of Poisson processes are summarized in Appendix B. In particular, we quote the following fact.

Lemma 2.7 *For a Markov chain with transition rate q and a subset A of $\mathcal{S} \times \mathcal{S}$, if there exists a positive constant λ such that*

$$\sum_{x':(x,x')\in A} q(x, x') = \lambda, \qquad \text{for all } x \in \mathcal{S}, \tag{2.1}$$

then the counting process driven by the Markov chain with respect to A is a Poisson process with rate λ. Furthermore, the future arrivals of the counting process are independent of the current state and the past history of the Markov chain.

The Poisson process derived in Lemma 2.7 is said to be *embedded* in the continuous time Markov chain. In general, any Markov-driven counting process can be regarded as a process embedded in the original Markov chain. Such an embedded counting process is useful in deriving the probability distribution just before (or after) the epochs when these events occur, e.g., customer arrivals. This distribution is referred to as the *embedded* or *Palm distribution* just before (or after) the counting epochs. For a stationary Markov chain, this distribution is in general not equal to the stationary distribution. To see this, let $\{N_A(t)\}$ be a counting process driven by a stationary Markov chain $X(t)$ with transition rate q that counts the number of jumps from state x to state y with $(x, y) \in A$, i.e.,

$$N_A(t) = \sum_{i=1}^{N(t)} 1[(X(s_i-), X(s_i)) \in A],$$

where s_i is the ith jump epoch of $\{N(u)\}$. Since the rate of jumps from state x to x' is $\pi(x)q(x, x')$, the embedded distribution π_A just before a counting epoch is

$$\pi_A(x) = \frac{\pi(x) \sum_{(x,x')\in A} q(x, x')}{\sum_{(x',x'')\in A} \pi(x')q(x', x'')}, \qquad x \in \mathcal{S}. \tag{2.2}$$

In particular, if N_A is the Poisson process, i.e., (2.1) holds, then we have

$$\begin{aligned}\sum_{(x',x'')\in A} \pi(x')q(x', x'') &= \sum_{x'} \pi(x') \sum_{x'':(x',x'')\in A} q(x', x'') \\ &= \sum_{x'} \pi(x')\lambda \\ &= \lambda.\end{aligned}$$

Hence, (2.2) is simplified to $\pi(x)$. We arrive at the following well-known result for stationary continuous time Markov chains due to Wolff (1982).

Theorem 2.8 (PASTA) *If an embedded counting process in a continuous time Markov chain is Poisson, then the embedded distribution just before a counting epoch is identical to the stationary distribution of the Markov chain.*

This important property is known as PASTA, which stands for "Poisson Arrivals See Time Averages". The time average refers to the stationary distribution and the event average refers to the embedded distribution. When the embedded counting process is Poisson, the time average is equal to the event average.

Example 2.9 (The Jackson network) In many applications individual queues are elements of a larger system and there are interactions between the queues. For instance, in a job shop, a part, after having been processed at one station, may have to go through several other stations before becoming a finished product. In such applications we have to study the stochastic behavior and the performance of the entire system (see Figure 2.4).

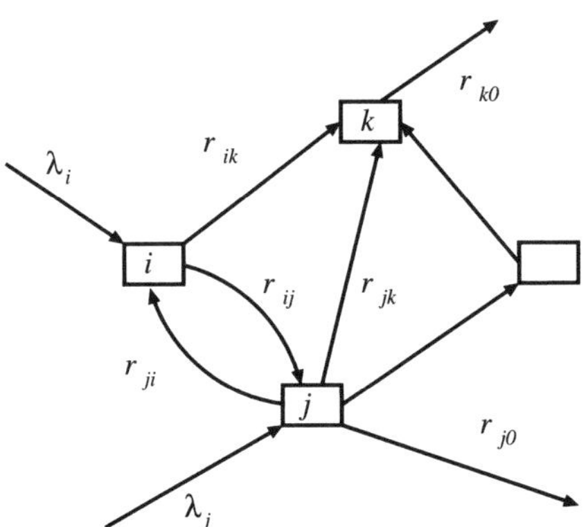

Figure 2.4 A Jackson network

Consider a network that consists of N processing stations. There is a single-server at each node. Jobs arrive at node j from the outside according to a Poisson process with rate λ_j. The processing times at node j are exponentially distributed with mean $1/\mu_j$. After a job is processed at node j, it joins node k for further processing with probability r_{jk} and it leaves the network as a finished product with probability r_{j0}, where

$$\sum_{k=0}^{N} r_{jk} = 1, \qquad j = 1, \ldots, N. \tag{2.3}$$

The processing discipline is assumed to be First–Come First–Served.

To study this system, let $X_j(t)$ be the number of jobs at node j at time t, and let

$$\boldsymbol{X}(t) = (X_1(t), \ldots, X_N(t))$$

denote the state of the network. Define $\boldsymbol{n} = (n_1, \ldots, n_N)$, and let $\mathbf{e}_j$ be the N-dimensional vector with a 1 as its jth entry and a 0 everywhere else. It is easily shown that $\boldsymbol{X}(t)$ is a continuous time Markov chain with transition rates

$$\begin{aligned} q(\boldsymbol{n}, \boldsymbol{n} + \mathbf{e}_j) &= \lambda_j, && j = 1, \ldots, N, \\ q(\boldsymbol{n}, \boldsymbol{n} - \mathbf{e}_j + \mathbf{e}_k) &= \mu_j r_{jk}, && n_j > 0 \text{ and } j, k = 1, \ldots, N, \\ q(\boldsymbol{n}, \boldsymbol{n} - \mathbf{e}_j) &= \mu_j r_{j0}, && n_j > 0 \text{ and } j = 1, \ldots, N. \end{aligned}$$

All other transition rates are zero. This is the so-called Jackson network.

In this example, the external arrivals at node j are Markov driven with

$$A = \{(\boldsymbol{n}, \boldsymbol{n} + \mathbf{e}_j); \boldsymbol{n} \geq \mathbf{0}\}.$$

That the external arrivals at node j are Poisson with rate λ_j can be seen from the fact that

$$\sum_{\boldsymbol{n}':(\boldsymbol{n},\boldsymbol{n}')\in A} q(\boldsymbol{n}, \boldsymbol{n}') = \lambda_j, \qquad \text{for all } \boldsymbol{n} \in \mathcal{S}.$$

Let π be the stationary distribution of this Markov chain when it exists. From (1.9), π is determined by the global balance equations

$$\begin{aligned} \pi(\boldsymbol{n}) \sum_{j=1}^{N} (\lambda_j + \mu_j 1(n_j > 0)) &= \sum_{j=1}^{N} \pi(\boldsymbol{n} - \mathbf{e}_j)\lambda_j + \sum_{j=1}^{N} \pi(\boldsymbol{n} + \mathbf{e}_j)\mu_j r_{j0} \\ &+ \sum_{j=1}^{N} \sum_{k=1}^{N} \pi(\boldsymbol{n} + \mathbf{e}_k - \mathbf{e}_j)\mu_k r_{kj}. \end{aligned} \tag{2.4}$$

In the next section we determine the stationary distribution π. □

A distinct feature of this book is that different types of events may result in the same state change. This implies that events, e.g., epochs in a counting process, cannot be defined in terms of a set A of $\mathcal{S} \times \mathcal{S}$. For instance, if an event of a counting process changes the state of a Markov chain from n to $n + 1$, then there may be other events that also trigger transitions from n to $n+1$. The more general embedded counting process defined below in terms of rate decompositions (or thinned transitions) plays a key role in subsequent chapters.

Definition 2.10 (Embedded counting process) Consider a continuous time Markov chain $\{X(t)\}$ with transition rate q. Let q_* be a thinned transition rate of q, i.e.,

$$q_*(x, x') \leq q(x, x'), \qquad x, x' \in \mathcal{S}.$$

Let $N(t)$ be the thinned counting process generated by q_*, i.e., it counts the number of state transitions in process $X(t)$, but a transition from x to x' is counted with probability $q_*(x, x')/q(x, x')$ independently of other events. The process $N(t)$ is called the *embedded counting process* generated by q_* with respect to q.

The following is an extension of Lemma 2.7 to embedded counting processes. See Appendix B for its proof.

Lemma 2.11 *Consider a Markov chain with transition rate q and thinned transition rate q_*. If there exists a positive constant λ such that*

$$\sum_{x' \in \mathcal{S}} q_*(x, x') = \lambda, \qquad \text{for all } x \in \mathcal{S}, \tag{2.5}$$

then the counting process generated by q_* with respect to q is a Poisson process with arrival rate λ. Furthermore, the future epochs of the embedded counting process are independent of the current state and the past history of the Markov chain.

2.3 LOCAL BALANCE EQUATIONS

The global balance equation (1.9) states that in steady state, the rate going out of each state x balances with the rate going into that same state x. In queueing applications, it often happens that the rate going out of any given state due to a specific event balances with the rate going into that same state due to a corresponding event. When this is the case the Markov chain is said to be *locally balanced* with respect to these events. If q_* and q'_* are the transition rates of the events in question, which may be considered as thinned transition rates of q, i.e.,

$$q_*(x, x') \leq q(x, x'), \qquad q'_*(x, x') \leq q(x, x'), \qquad x,\ x' \in \mathcal{S},$$

then the local balance can be defined as follows.

Definition 2.12 (Local balance) A continuous time Markov chain with transition rates q is said to satisfy the *local balance* with respect to thinned transitions q_* and q'_* if

$$\pi(x) \sum_{x' \in \mathcal{S}} q_*(x, x') = \sum_{x' \in \mathcal{S}} \pi(x') q'_*(x', x), \qquad x \in \mathcal{S}. \tag{3.1}$$

Remark 2.13 Definition 2.12 is very general since no restriction is imposed on the thinned transitions q_* and q'_*. In applications, q_* and q'_* have to be clearly specified and they usually represent transition rates of particular events. In the literature, local balance is defined in a more restrictive form. Namely, for a subset A of $\mathcal{S} \times \mathcal{S}$, it is defined as

$$\pi(x) \sum_{x' : (x, x') \in A} q(x, x') = \sum_{x' : (x, x') \in A} \pi(x') q(x', x), \qquad x \in \mathcal{S}. \tag{3.2}$$

In this case, the rate going out of state x through jumps in A is balanced with the rate going into the same state x through the corresponding jumps under time reversal. However, as noted in the last section, such a definition does not apply to situations where different events result in the same state change, which is the case in this book. Note that the local balance (3.2) is a special case of the local balance (3.1). It also shows that q_* and q'_* are different in general (see (3.4) below).

In applications we often have a set of local balance equations indexed by k in a countable set $\mathcal{J}$, i.e., there are two sets of thinned transition rates $\{q_k(x, x'); x, x' \in \mathcal{S}\}$ and $\{q'_k(x, x'); x, x' \in \mathcal{S}\}$, $k \in \mathcal{J}$, such that

$$\pi(x) \sum_{x' \in \mathcal{S}} q_k(x, x') = \sum_{x' \in \mathcal{S}} \pi(x') q'_k(x', x), \qquad x \in \mathcal{S}, \tag{3.3}$$

for each $k \in \mathcal{J}$. For instance, for jump sets $A_k \in \mathcal{S} \times \mathcal{S}$, if the thinned transitions are defined as

$$\begin{aligned} q_k(x, x') &= q(x, x') 1[(x, x') \in A_k], \\ q'_k(x, x') &= q(x, x') 1[(x', x) \in A_k], \end{aligned} \tag{3.4}$$

then the set of local balance equations is

$$\pi(x) \sum_{x' : (x, x') \in A_k} q(x, x') = \sum_{x' : (x, x') \in A_k} \pi(x') q(x', x), \qquad x \in \mathcal{S}, k \in \mathcal{J}.$$

This is the case discussed in Remark 2.13. In particular, if these local balance equations imply the global balance equations, then they are called a *full set of local balance equations*.

Definition 2.14 (Full set of local balance equations) If the transition rates $q(x, x')$ of a continuous time Markov chain are decomposed into two sets of thinned transition rates $q_k(x, x'),\ q'_k(x, x'),\ x, x' \in \mathcal{S}$ indexed by $k \in \mathcal{J}$, i.e.,

$$q(x, x') = \sum_{k \in \mathcal{J}} q_k(x, x') = \sum_{k \in \mathcal{J}} q'_k(x, x'), \qquad x, x' \in \mathcal{S},$$

and the local balance equations (3.3) are satisfied for each $k \in \mathcal{J}$, then these equations are referred to as a full set of local balance equations.

In the special case that $q_* = q'_* = q$, the local balance becomes the global balance. On the other hand, if the local balance equations are defined for each jump $k \equiv (x, x') \in \mathcal{S} \times \mathcal{S}$ with respect to transitions (3.4) with $A_{(x,x')} = \{(x, x')\}$, i.e.,

$$q_{(x,x')}(y, y') = q(y, y')1[(y, y') = (x, x')],$$
$$q'_{(x,x')}(y, y') = q(y, y')1[(y', y) = (x, x')],$$

then (3.3) becomes

$$\pi(x)q(x, x') = \pi(x')q(x', x), \qquad x, x' \in \mathcal{S}. \tag{3.5}$$

Equations (3.5) are called the *detailed balance equations*, which clearly constitute a full set of local balances. They imply that for any pair of states x and x', the rate going from state x to state x' is equal to the rate going from state x' to state x. Detailed balance equations are equivalent to the reversibility of the Markov chain, which we will discuss in the next section.

Local balance equations are useful for computing the stationary distributions of queueing systems. In what follows we illustrate this through several examples. Since in these examples different events result in different state transitions, it suffices to specify the subsets A_k of $\mathcal{S} \times \mathcal{S}$ in (3.3), for which thinned transition rates are defined by (3.4). More general cases will be discussed in Chapter 10. The first example is the birth and death process introduced in Example 2.2.

Example 2.15 (Local balance in birth and death process) Consider the birth and death process of Example 2.2. Let $\pi(n)$ be its stationary distribution. The global balance equations are

$$\begin{aligned}&\pi(n)\lambda(n) + \pi(n)\mu(n)1[n > 0]\\ &\quad = \pi(n+1)\mu(n+1) + \pi(n-1)\lambda(n-1)1[n > 0], \qquad n \geq 0,\end{aligned}$$

where $\mu(0)$ and $\lambda(-1)$ are 0. Clearly, global balance is obtained by summing up the following two sets of equations:

$$\pi(n)\lambda(n) = \pi(n+1)\mu(n+1), \tag{3.6}$$

$$\pi(n)\mu(n) = \pi(n-1)\lambda(n-1). \tag{3.7}$$

Thus, these two sets of local balance equations constitute a full set. Define

$$A_1 = \{(n, n+1); \quad n = 0, 1, \ldots\},$$

$$A_{-1} = \{(n, n-1); \quad n = 1, 2, \ldots\},$$

where A_1 represents the upward jumps, and A_{-1} represents the downward jumps. The thinned transition rates for A_1 and A_{-1} are defined, for $k = 1, -1$, by (3.4), i.e., for $n, n', \geq 0$,

$$q_1(n, n') = \lambda(n)1[(n, n') \in A_1],$$

$$q_1'(n, n') = \mu(n)1[(n', n) \in A_1],$$

$$q_{-1}(n, n') = \mu(n)1[(n, n') \in A_{-1}],$$

$$q_{-1}'(n, n') = \lambda(n)1[(n', n) \in A_{-1}].$$

These thinned transition rates correspond to arrival and departure events. The local balance equations with respect to q_1 and q_1' (q_{-1} and q_{-1}') are (3.6) (respectively, (3.7)). Note that (3.6) and (3.7) are equivalent, which basically states that the rate going into state n due to an arrival event balances with the rate going out of state n due to a departure event. Thus it suffices to compute π using only one of them. Solving (3.6) inductively yields

$$\pi(n) = \pi(0) \prod_{\ell=1}^{n} \frac{\lambda(\ell-1)}{\mu(\ell)}, \qquad n \geq 0,$$

where

$$\pi(0) = \left(1 + \sum_{n=1}^{\infty} \prod_{\ell=1}^{n} \frac{\lambda(\ell-1)}{\mu(\ell)}\right)^{-1}.$$

In particular, when the birth rates $\lambda(n) \equiv \lambda$ and the death rates $\mu(n) \equiv \mu$ are constants, we obtain the following stationary distribution for the $M/M/1$ queue discussed in Example 2.1:

$$\pi(n) = \left(1 - \frac{\lambda}{\mu}\right)\left(1 - \frac{\lambda}{\mu}\right)^n, \quad n = 0, 1, \ldots. \qquad \square$$

The next example shows how local balance equations can be applied to obtain the stationary distribution for a tandem queue.

Example 2.16 (Local balance in a tandem queue) Consider a special case of the Jackson network described in Example 2.9 with two nodes in series. Such a system is called a tandem queue. There is a single-server at each node. Customers arrive at node 1 according to a Poisson process with rate λ, and the service times at the two nodes are exponentially distributed with rates μ_1 and μ_2, respectively. Customers are first served at node 1, and then at node 2. After completing service at node 2, the customers leave the system. The service disciplines at both nodes are First–Come First–Served. See Figure 2.5.

Figure 2.5 A two-node tandem queue

Let $\pi(n_1, n_2)$ be the stationary probability of having n_1 customers at node 1 and n_2 customers at node 2. The stationary probabilities satisfy the global balance equations

$$\begin{aligned}
&\pi(n_1, n_2)\mu_1 1[n_1 \geq 1] + \pi(n_1, n_2)\mu_2 1[n_2 \geq 1] + \pi(n_1, n_2)\lambda \\
&\quad = \pi(n_1 - 1, n_2)\lambda 1[n_1 \geq 1] + \pi(n_1 + 1, n_2 - 1)\mu_1 1[n_2 \geq 1] \\
&\quad +\pi(n_1, n_2 + 1)\mu_2.
\end{aligned}$$

A local balance for this system could be that the terms on the left-hand side are equal to the corresponding terms on the right-hand side. If this is the case we conclude that

$$\pi(n_1, n_2)\mu_1 1[n_1 \geq 1] = \pi(n_1 - 1, n_2)\lambda 1[n_1 \geq 1], \quad (3.8)$$

$$\pi(n_1, n_2)\mu_2 1[n_2 \geq 1] = \pi(n_1 + 1, n_2 - 1)\mu_1 1[n_2 \geq 1], \quad (3.9)$$

$$\pi(n_1, n_2)\lambda = \pi(n_1, n_2 + 1)\mu_2. \quad (3.10)$$

These local balances are with respect to arrival and departure events at nodes 1, 2 and the outside, and they correspond to

$$A_1 = \{((n_1, n_2), (n_1 - 1, n_2 + 1)), ((n_1, n_2), (n_1 - 1, n_2)); n_1 \geq 1, n_2 \geq 0\},$$

$$A_2 = \{((n_1, n_2), (n_1, n_2 - 1)), ((n_1, n_2), (n_1 + 1, n_2 - 1)); n_1 \geq 0, n_2 \geq 1\},$$

$$A_3 = \{((n_1, n_2), (n_1 + 1, n_2)), ((n_1, n_2), (n_1, n_2 + 1)); n_1, n_2 \geq 0\}.$$

Using (3.8) we find recursively that

$$\pi(n_1, n_2) = \left(\frac{\lambda}{\mu_1}\right)^{n_1} \pi(0, n_2),$$

and using (3.9), we obtain

$$\pi(0, n_2) = \left(\frac{\lambda}{\mu_2}\right)^{n_2} \pi(0, 0).$$

Hence

$$\pi(n_1, n_2) = \left(\frac{\lambda}{\mu_1}\right)^{n_1} \left(\frac{\lambda}{\mu_2}\right)^{n_2} \pi(0, 0).$$

From the fact that

$$\sum_{n_1=0}^{\infty} \sum_{n_2=0}^{\infty} \pi(n_1, n_2) = 1,$$

we obtain

$$\pi(0, 0) = \left(1 - \frac{\lambda}{\mu_1}\right)\left(1 - \frac{\lambda}{\mu_2}\right).$$

Therefore,

$$\pi(n_1, n_2) = \prod_{j=1}^{2} \left(1 - \frac{\lambda}{\mu_j}\right)\left(\frac{\lambda}{\mu_j}\right)^{n_j}. \tag{3.11}$$

Note that (3.10) is a direct consequence of (3.8) and (3.9), i.e.,

$$\pi(n_1, n_2)\lambda = \pi(n_1 + 1, n_2)\mu_1 = \pi(n_1, n_2 + 1)\mu_2,$$

where the first equality follows from (3.8) and the second equality from (3.9). Hence $\pi(n_1, n_2)$ also satisfies the local balance equations (3.10) and is therefore indeed the stationary distribution of the tandem queue.

There is a physical interpretation for these three local balance equations. The first one simply states that in steady state the rate going out of state (n_1, n_2) due to a service completion at node 1 is equal to the rate going into that same state due to an arrival at node 1. Similarly, the second equation states that the rate going out of state (n_1, n_2) due to a service completion at node 2 is the same as the rate going into that same state due to an arrival at node 2. If we consider the outside world also as a node (node 0, say) with service rate λ, then the third local balance states that the rate going out of state (n_1, n_2) due to a service completion at node 0 is the same as the rate going into the same state due to an arrival at node 0. This shows that these local balance equations imply that, for each node j, the rate going out of a state due to a departure from node j balances with the rate going into that same state due to an arrival at node j. We will see in later chapters that such a result holds for more general queueing networks also. □

The same arguments can be extended to Jackson networks.

Example 2.17 (Local balances in Jackson networks) Consider the Jackson network introduced in Example 2.9. We now derive its stationary distribution $\pi(\boldsymbol{n})$ using local balance equations. Let α_j be the average number of visits to node j per unit time, $j = 1, 2, \ldots, N$. The α_j are the solution to the traffic equations

$$\alpha_j = \lambda_j + \sum_{k=1}^{N} \alpha_k r_{kj}, \qquad j = 1, \ldots, N. \tag{3.12}$$

Similar to the tandem queue of Example 2.16, we decompose the global balance equations (2.4) according to the rate out of state $\boldsymbol{n}$ due to a departure from node j and the rate into state $\boldsymbol{n}$ due to an arrival at node j:

$$\pi(\boldsymbol{n}) \sum_{j=1}^{N} \lambda_j = \sum_{k=1}^{N} \pi(\boldsymbol{n} + \mathbf{e}_k)\mu_k r_{k0}, \tag{3.13}$$

$$\pi(\boldsymbol{n})\mu_j = \pi(\boldsymbol{n} - \mathbf{e}_j)\lambda_j + \sum_{k=1}^{N} \pi(\boldsymbol{n} + \mathbf{e}_k - \mathbf{e}_j)\mu_k r_{kj}, \quad n_j > 0, \tag{3.14}$$

These local balance equations correspond to the jump sets

$$A_0 = \{(\boldsymbol{n}, \boldsymbol{n} + \mathbf{e}_k);\, \boldsymbol{n} \in \mathcal{S},\, 1 \le k \le N\},$$

$$A_j = \{(\boldsymbol{n}, \boldsymbol{n} + \mathbf{e}_k - \mathbf{e}_j);\, \boldsymbol{n} \in \mathcal{S},\, n_j \ge 1, 0 \le k \le N\}, \qquad j = 1, 2, \ldots, N,$$

where $\mathbf{e}_0 = (0, 0, \ldots, 0)$ is the N-dimensional null vector. For obvious reasons, we often use $\mathbf{0}$ to denote a vector of all zeros. A comparison of (3.14) with (3.12) shows that, if multiplying both sides of (3.12) by

$$\left(\frac{\alpha_j}{\mu_j}\right)^{-1} \prod_{\ell=1}^{N} \left(\frac{\alpha_\ell}{\mu_\ell}\right)^{n_\ell},$$

we have (3.14) with

$$\pi(\boldsymbol{n}) = \pi(\mathbf{0}) \prod_{j=1}^{N} \left(\frac{\alpha_j}{\mu_j}\right)^{n_j}. \tag{3.15}$$

On the other hand, summing (3.12) over $j = 1, \ldots, N$ and applying (2.3), we obtain

$$\sum_{j=1}^{N} \alpha_j r_{j0} = \sum_{j=1}^{N} \lambda_j. \tag{3.16}$$

It is easily seen that (3.13) and (3.15) reduce to (3.16). Thus, π satisfies the local balance equations and consequently also the global balance equations (2.4). Clearly, π is a stationary distribution if and only if the total sum of (3.15) is finite, which is equivalent to

$$\alpha_j < \mu_j, \qquad j = 1, \ldots, N. \qquad \square$$

The local balance equations (3.14) and (3.13) imply that the rate going out of state $\boldsymbol{n}$ due to a service completion at node j balances with the rate going into $\boldsymbol{n}$ due to an arrival at node j, $j = 0, 1, \ldots, N$. In many queueing models, however, such a balance equation is not satisfied. This is the case in networks with signals, since the arrival of a signal at a node may have a similar effect as a regular service completion. Thus some new classes of local balance equations, namely biased local balance and cross local balance equations, are needed. These new forms of local balance equations will be discussed in Chapter 10.

2.4 TIME REVERSED PROCESSES AND REVERSIBILITY

The time reversed process, or simply the reversed process, of a stochastic process $X(t)$ on $t \in (-\infty, \infty)$, is defined as

$$\tilde{X}(t) = X(-t).$$

If a stationary process is originally defined on the half line $[0, +\infty)$, it can be extended to $(-\infty, +\infty)$.

Assume that $X(t)$ is a stationary continuous time Markov chain with state space $\mathcal{S}$, transition rates $q(x, x')$, $x, x' \in \mathcal{S}$, and stationary distribution $\pi(x)$, $x \in \mathcal{S}$. To obtain the probability structure of its reversed process, we first compute the following conditional probability. For each positive integer $n \geq 1$ and nonnegative numbers s and t, let $x_i \in \mathcal{S}$ and $u_i < t$ for $i = 1, \ldots, n$. Now

$$\begin{aligned}
&P(\tilde{X}(t+s) = x' \mid \tilde{X}(u_i) = x_i, u_i < t, i = 1, \ldots, n, \tilde{X}(t) = x) \\
&= P(X(-t-s) = x' \mid X(-u_i) = x_i, -u_i > -t, i = 1, \ldots, n, X(-t) = x)
\end{aligned}$$

$$
\begin{aligned}
&= \frac{P\big(X(-t-s)=x',\, X(-u_i)=x_i,\, -u_i > -t,\, i=1,\dots,n,\, X(-t)=x\big)}{P(X(-u_i)=x_i,\, -u_i > -t,\, i=1,\dots,n,\, X(-t)=x)} \\
&= \frac{P\big(X(-t-s)=x',\, X(-t)=x\big)}{P(X(-t)=x)} \\
&\quad \times \frac{P\big(X(-u_i)=x_i,\, -u_i > -t,\, i=1,\dots,n \mid X(-t-s)=x',\, X(-t)=x\big)}{P(X(-u_i)=x_i,\, -u_i > -t,\, i=1,\dots,n \mid X(-t)=x)} \\
&= P(X(-t-s)=x' \mid X(-t)=x) \\
&= P\big(\tilde{X}(t+s)=x' \mid \tilde{X}(t)=x\big),
\end{aligned}
$$

where the fourth equality follows from the Markov property (1.1) (see Appendix B for its formal verification). Thus the reversed process $\{\tilde{X}(t)\}$ is also a continuous time Markov chain. The transition rate from state x to state x' in the reversed process is, by (1.5),

$$
\begin{aligned}
\tilde{q}(x,x') &= \lim_{h\downarrow 0} \frac{P\big(\tilde{X}(t+h)=x' \mid \tilde{X}(t)=x\big)}{h} \\
&= \lim_{h\downarrow 0} \frac{P\big(\tilde{X}(t+h)=x',\, \tilde{X}(t)=x\big)}{hP\big(\tilde{X}(t)=x\big)} \\
&= \lim_{h\downarrow 0} \frac{P\big(X(-t-h)=x',\, X(-t)=x\big)}{hP\big(X(-t)=x\big)} \\
&= \lim_{h\downarrow 0} \frac{P\big(X(-t-h)=x'\big)}{P\big(X(-t)=x\big)} \frac{P\big(X(-t)=x \mid X(-t-h)=x'\big)}{h} \\
&= \frac{\pi(x')q(x',x)}{\pi(x)}.
\end{aligned}
$$

Note that the transition intensity $\tilde{q}(x)$ from x in the reversed process is

$$
\begin{aligned}
\tilde{q}(x) &= \sum_{x'\in\mathcal{S}} \tilde{q}(x,x') \\
&= \sum_{x'\in\mathcal{S}} \frac{\pi(x')q(x',x)}{\pi(x)} \\
&= \sum_{x'\in\mathcal{S}} q(x,x') \\
&= q(x),
\end{aligned}
$$

where the third equality follows from the global balance equations (1.9).

We summarize these results in the following theorem.

Theorem 2.18 *If $X(t)$ is a stationary continuous time Markov chain with transition rates $q(x, x')$, $x, x' \in \mathcal{S}$, then the reversed process $X(-t)$ is also a stationary Markov chain with transition rates*

$$\tilde{q}(x, x') = \frac{\pi(x')q(x', x)}{\pi(x)}, \qquad x, x' \in \mathcal{S}. \tag{4.1}$$

The reversed process has the same stationary distribution $\pi(x)$, $x \in \mathcal{S}$, and the same sojourn time distribution for each state x as the original process.

The following example illustrates the theorem.

Example 2.19 (The reversed process of a tandem queue) Consider the tandem queue of Example 2.16. Its transition rates are

$$\begin{aligned} q((n_1, n_2), (n_1 + 1, n_2)) &= \lambda, & n_1 \geq 0, n_2 \geq 0, \\ q((n_1, n_2), (n_1 - 1, n_2 + 1)) &= \mu_1, & n_1 \geq 1, n_2 \geq 0, \\ q((n_1, n_2), (n_1, n_2 - 1)) &= \mu_2, & n_1 \geq 0, n_2 \geq 1, \end{aligned}$$

and all other transition rates are zero. In Example 2.16 we showed that its stationary distribution is given by (3.11). By (4.1), the reversed process has the transition rates

$$\begin{aligned} \tilde{q}((n_1, n_2), (n_1, n_2 + 1)) &= \lambda, & n_1 \geq 0, n_2 \geq 0, \\ \tilde{q}((n_1, n_2), (n_1 + 1, n_2 - 1)) &= \mu_2, & n_1 \geq 0, n_2 \geq 1, \\ \tilde{q}((n_1, n_2), (n_1 - 1, n_2)) &= \mu_1, & n_1 \geq 1, n_2 \geq 0, \end{aligned}$$

and all other transition rates are zero. These transition rates correspond to the transition rates of another tandem queue with two nodes in series. The arrival rate at the first queue is λ and the service rates are μ_2 and μ_1, respectively, i.e., the service rates are interchanged. In fact, it does not take much effort to realize that the reversed process has to be another tandem queue, where customers arrive at the second node, first receive service at the second node and then proceed to the first node (see Figure 2.6). □

Figure 2.6 A time reversed tandem queue

A study of the reversed process often leads to a better understanding of the original process, as the following example shows.

Example 2.20 (An $M/M/1$ queue with state-dependent service) Consider an $M/M(n)/1$ queue with state dependent service rates. The arrival process is Poisson with rate λ, and the service rate is $\mu(n)$ when there are n customers present. For instance, if

$$\mu(n) = \min\{n, C\}\mu, \qquad n = 0, 1, \ldots,$$

for some fixed integer $C \geq 1$ and positive number μ, then the queue is an $M/M/C$ queue with mean service times μ^{-1}.

Let $X(t)$ be the number of customers in the system at time t. It is clear that this process is a special case of the birth and death process discussed in Example 2.2. The transition rates are

$$\begin{aligned} q(n, n+1) &= \lambda, & n &\geq 0, \\ q(n, n-1) &= \mu(n), & n &\geq 1. \end{aligned}$$

It follows from Example 2.15 that the stationary distribution is

$$\pi(n) = \pi(0)\frac{\lambda^n}{\prod_{\ell=1}^{n}\mu(\ell)},$$

where

$$\pi(0) = \left(1 + \sum_{n=1}^{\infty}\frac{\lambda^n}{\prod_{\ell=1}^{n}\mu(\ell)}\right)^{-1}.$$

From (4.1) it follows that the transition rates of the reversed process are

$$\begin{aligned} \tilde{q}(n, n+1) &= \lambda, & n &\geq 0, \\ \tilde{q}(n, n-1) &= \mu(n), & n &\geq 1. \end{aligned}$$

These transition rates are exactly the same as the transition rates of the forward process. Therefore, the reversed process is also an $M/M(n)/1$ queue with customers arriving according to a Poisson process with rate λ, and state dependent service rates $\mu(n)$. On the other hand, an arrival in the reversed process corresponds to a departure in the forward process. Since in the reversed process the arrivals are Poisson, the future arrivals in the reversed process are independent of the current number of customers in the system, i.e., $X(t)$. This implies that the departure process in the original process is also Poisson with rate λ, and that past departures, which correspond to future arrivals in the reversed process, are independent of the current state of the system. This property is known as quasi-reversibility, which will be studied in more detail in Chapters 3 and 4. □

The reversed process not only enables us to gain a better insight into the original process, it may also be helpful in finding the stationary distributions of complicated systems, which may be otherwise hard to obtain. This approach is known as the Kelly lemma, which is very useful in proving product form solutions for networks of queues.

Lemma 2.21 (Kelly lemma) *Let $X(t)$ be a stationary continuous time Markov chain with transition rates $q(x, x')$, $x, x' \in \mathcal{S}$. If we can find a collection of non-negative numbers $\tilde{q}(x, x')$, $x, x' \in \mathcal{S}$, and a collection of positive numbers $\pi(x)$, $x \in \mathcal{S}$, summing to unity, such that*

$$\sum_{x' \in \mathcal{S}} q(x, x') = \sum_{x' \in \mathcal{S}} \tilde{q}(x, x'), \qquad x \in \mathcal{S}, \tag{4.2}$$

and

$$\pi(x')q(x', x) = \pi(x)\tilde{q}(x, x'), \qquad x, x' \in \mathcal{S}, \tag{4.3}$$

then $\tilde{q}(x, x')$, $x, x' \in \mathcal{S}$ are the transition rates of the reversed process, and $\pi(x)$, $x \in \mathcal{S}$, is the stationary distribution of both processes.

PROOF. Applying (4.2) and summing (4.3) over all $x' \in \mathcal{S}$ yields

$$\begin{aligned} \pi(x) \sum_{x' \in \mathcal{S}} q(x, x') &= \pi(x) \sum_{x' \in \mathcal{S}} \tilde{q}(x, x') \\ &= \sum_{x' \in \mathcal{S}} \pi(x')q(x', x). \end{aligned}$$

This is the global balance for q. Hence, by the assumption that $\pi(x)$, $x \in \mathcal{S}$, sum to unity we conclude that $\pi(x)$, $x \in \mathcal{S}$, is the stationary distribution for the continuous time Markov chain $X(t)$. In addition, it follows from (4.3) that

$$\tilde{q}(x, x') = \frac{\pi(x')q(x', x)}{\pi(x)}, \qquad x, x' \in \mathcal{S}.$$

By Theorem 2.18, $\tilde{q}(x, x')$ is indeed the transition rate of the reversed process, and the two processes have the same stationary distribution. □

The following example illustrates the usefulness of the Kelly lemma.

Example 2.22 (Application of the Kelly lemma to tandem queues) We again consider the two-node tandem queue of Example 2.19. The transition rates are presented in that example. As we observed, a natural candidate for the reversed process is also a tandem queue with two nodes. Customers arrive at node 2 according to a Poisson process, and leave the system after being served at node 1, as depicted in Figure 2.6. Since any customer that enters the system in the forward process will eventually leave, the arrival rate at queue 2 in the reversed process should also be λ. Thus its transition rates are conjectured to be the $\tilde{q}$ of Example 2.19. If a product form solution holds, the stationary distribution should be

$$\pi(n_1, n_2) = \left(1 - \frac{\lambda}{\mu_1}\right)\left(\frac{\lambda}{\mu_1}\right)^{n_1}\left(1 - \frac{\lambda}{\mu_2}\right)\left(\frac{\lambda}{\mu_2}\right)^{n_2}.$$

To show that $\pi(n_1, n_2)$ is indeed the stationary distribution and the reversed process is indeed characterized by the $\tilde{q}$ of Example 2.19, we have to verify the Kelly lemma. This is straightforward since it follows immediately from the transition rates of both processes that

$$\begin{aligned}\sum_{(n_1', n_2')} q((n_1, n_2), (n_1', n_2')) &= \sum_{(n_1', n_2')} \tilde{q}((n_1, n_2), (n_1', n_2')) \\ &= \lambda + \mu_1 1[n_1 \geq 1] + \mu_2 1[n_2 \geq 1],\end{aligned}$$

and (4.3) can be verified using the $\tilde{q}$. For example, consider

$$\begin{aligned}&\pi(n_1, n_2) q((n_1, n_2), (n_1 - 1, n_2 + 1)) \\ &\quad = \pi(n_1 - 1, n_2 + 1)\tilde{q}((n_1 - 1, n_2 + 1), (n_1, n_2)).\end{aligned}$$

If $n_1 = 0$, both sides are zero. If $n_1 \geq 1$, the left-hand side is $\pi(n_1, n_2)\mu_1$, and the right-hand side is

$$\pi(n_1 - 1, n_2 + 1)\mu_2 = \pi(n_1, n_2)\frac{\mu_1}{\lambda}\frac{\lambda}{\mu_2}\mu_2,$$

which is the same as the left-hand side. Other cases can be verified in the same way and are left to the reader. □

The main advantage of this approach lies in the fact that it does not require any summation to verify the global balance equations. Moreover, this approach gives as a by-product the structure of the reversed process. This structure, as we have seen in Example 2.20, often reveals important structural properties of the forward process, e.g., quasi-reversibility.

For the applications of this book, the following detailed form of the Kelly lemma is more convenient to use.

Lemma 2.23 (Detailed Kelly lemma) *Let $X(t)$ be a stationary continuous time Markov chain with transition rates $q(x, x')$, $x, x' \in \mathcal{S}$. Assume that $q(x, x')$ is decomposed into the transition rates q_σ, indexed by σ in a countable set $\mathcal{U}$, i.e.,*

$$q(x, x') = \sum_{\sigma \in \mathcal{U}} q_\sigma(x, x'), \qquad x \in \mathcal{S}.$$

If we can find a collection of nonnegative numbers $\tilde{q}_{\tilde{\sigma}}(x, x')$, $\tilde{\sigma} \in \mathcal{U}$, where there is a one-to-one correspondence between σ and $\tilde{\sigma}$, and a collection of positive numbers $\pi(x)$, $x \in \mathcal{S}$, summing to unity, such that

$$\sum_{\sigma \in \mathcal{U}} \sum_{x'} q_\sigma(x, x') = \sum_{\tilde{\sigma} \in \mathcal{U}} \sum_{x'} \tilde{q}_{\tilde{\sigma}}(x, x'), \qquad x \in \mathcal{S}, \tag{4.4}$$

and

$$\pi(x') q_\sigma(x', x) = \pi(x) \tilde{q}_{\tilde{\sigma}}(x, x'), \qquad x, x' \in \mathcal{S},\ \sigma \in \mathcal{U}, \tag{4.5}$$

then

$$\tilde{q}(x, x') = \sum_{\tilde{\sigma} \in \mathcal{U}} \tilde{q}_{\tilde{\sigma}}(x, x'), \quad x, x' \in \mathcal{S},$$

is the transition rate of the reversed process, and $\pi(x), x \in \mathcal{S}$, is the stationary distribution of both processes. In this case, for each σ, the embedded counting process generated by $\tilde{q}_{\tilde{\sigma}}$ with respect to $\tilde{q}$ follows the same probability law as the time reversal of the embedded counting process generated by q_σ with respect to q in the original process.

PROOF. Define $\tilde{q}$ as in the lemma. Summing (4.5) over all σ leads to (4.3). Since (4.4) is equivalent to (4.2), the first half follows from the Kelly lemma. Dividing the two sides of (4.5) by the two sides of (4.3), which are assumed to be positive, we obtain

$$\frac{q_\sigma(x', x)}{q(x', x)} = \frac{\tilde{q}_{\tilde{\sigma}}(x, x')}{\tilde{q}(x, x')}.$$

Hence, by Definition 2.10, the probability of choosing the embedded transitions q_σ

from transitions (x', x) in the original Markov chain is identical to the probability of choosing the corresponding $\tilde{q}_{\tilde{\sigma}}$ from transitions (x, x') in the reversed Markov chain. Since these selections are independent of one another, we can use the embedded counting process generated by q_σ with respect to q as the time reversal of the counting process generated by $\tilde{q}_{\tilde{\sigma}}$ with respect to $\tilde{q}$. This proves the second half of the lemma. □

Remark 2.24 If q_σ and $q_{\sigma'}$ (or equivalently, $\tilde{q}_{\tilde{\sigma}}$ and $\tilde{q}_{\tilde{\sigma}'}$) do not overlap when $\sigma \neq \sigma'$, i.e., $q_\sigma(x, x') > 0$ only if $q_{\sigma'}(x, x') = 0$, then the first half of this lemma reduces to Lemma 2.21. This, however, is not true in most applications in this book. For example, a service completion may result in a state change that is identical to the state change caused by the arrival of a negative customer at the node. The second half is crucial in the interpretation of the embedded counting process of the reversed process. In general, this embedded counting process is not necessarily identical to the one embedded in the original process, because the embedded transitions are randomly chosen.

In Example 2.20 we observe that the reversed process has exactly the same structure as the forward process, i.e., $q = \tilde{q}$. A process with this property is called reversible. Its mathematical definition is given below.

Definition 2.25 (Reversibility) A stochastic process is reversible if $(X(t_1), X(t_2), \ldots, X(t_n))$ has the same distribution as $(X(s-t_1), X(s-t_2), \ldots, X(s-t_n))$ for all n and any $t_1, \ldots, t_n$, and s.

Intuitively, reversibility implies that when a process is viewed reversed in time, as if we were to film such a process and then run the film backwards, it is probabilistically indistinguishable from the process when it is viewed forward in time.

Lemma 2.26 *A reversible process is stationary.*

PROOF. By setting $s = 0$, it follows that $(X(t_1), X(t_2), \ldots, X(t_n))$ has the same distribution as $(X(-t_1), X(-t_2), \ldots, X(-t_n))$. By setting $s = t$ and replacing t_i by $-t_i$ it follows that $(X(-t_1), X(-t_2), \ldots, X(-t_n))$ has the same distribution as $(X(t+t_1), X(t+t_2), \ldots, X(t+t_n))$. Hence $(X(t_1), X(t_2), \ldots, X(t_n))$ has the same distribution as $(X(t+t_1), X(t+t_2), \ldots, X(t+t_n))$ for any n and $t, t_1, \ldots, t_n$. This shows that the process is stationary. □

At the outset of this section we computed the reversed process of a continuous time stationary Markov chain. The transition rates of the reversed process are given by (4.1). Apparently, when

$$q(x, x') = \tilde{q}(x, x'), \qquad x, x' \in \mathcal{S},$$

the transition rates of the reversed process are the same as those of the forward process. By (4.1), $q(x, x') = \tilde{q}(x, x')$ can be expressed as

$$\pi(x)q(x, x') = \pi(x')q(x', x), \qquad x, x' \in \mathcal{S}. \tag{4.6}$$

These equations are identical to (3.5), and are called the detailed balance equations.

Theorem 2.27 *The detailed balance equations are satisfied if and only if the Markov chain is reversible.*

PROOF. If $\{X(t); t \geq 0\}$ is reversible, then it follows from Definition 2.25 that, if we let $n = 2$, $t_1 = t$, $t_2 = t + h$ and $s = 2t + h$, then

$$P\big(X(t) = x, X(t+h) = x'\big) = P\big(X(t) = x', X(t+h) = x\big). \tag{4.7}$$

The process is stationary by Lemma 2.26, so let $\pi(x) = P(X(t) = x)$. From (4.7) it follows that

$$\pi(x)P\big(X(t+h) = x' | X(t) = x\big) = \pi(x')P\big(X(t+h) = x | X(t) = x'\big). \tag{4.8}$$

Divide both sides by h and let $h \downarrow 0$, equations (4.6) immediately follow. □

The detailed balance equations provide a necessary and sufficient condition for a Markov chain to be reversible, and it gives a simple criterion to check whether a process is reversible: If we can find a pair of states x and x' such that

$$q(x, x') \neq 0, \quad \text{and} \quad q(x', x) = 0,$$

then the process is not reversible.

Example 2.28 (Non-reversibility of tandem queues) Consider the same tandem queue as in Example 2.16. Customers arrive at node 1 with rate λ, and the service times at the two nodes are exponentially distributed with rates μ_1 and μ_2. This process cannot be reversible because, for $x = (n_1, n_2 + 1)$ and $x' = (n_1, n_2)$,

$$q(x, x') = \mu_2 > 0, \quad \text{but} \quad q(x', x) = 0.$$ □

Let G be the transition diagram of a Markov chain, with the set of vertices being $\mathcal{S}$ (the state space), and with an arc connecting vertices x and x' if either $q(x, x')$ or $q(x', x)$ is positive. For example, the transition diagram of a birth and death process is a single chain, see Figure 2.7.

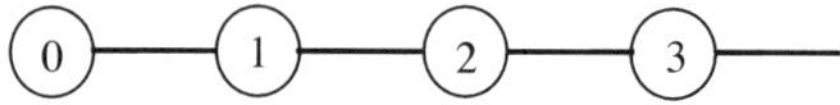

Figure 2.7 The transition diagram of a birth and death process

The following result gives a sufficient condition for a process to be reversible.

Lemma 2.29 *If the transition diagram G of an ergodic Markov chain is a tree, then the process is reversible.*

PROOF. Suppose G is a tree (see Figure 2.8). For any two states x and x', if there is an edge connecting x and x', the removal of this edge will cut the graph into two disconnected subgraphs. If we use $\mathcal{A}$ to denote the first subgraph containing state x, then $\mathcal{A}$ and $\mathcal{S} \setminus \mathcal{A}$ are connected in G only by the edge joining x and x'. Applying the cross balance equations (1.10) yields the detailed balance equations and the reversibility of the Markov chain follows. □

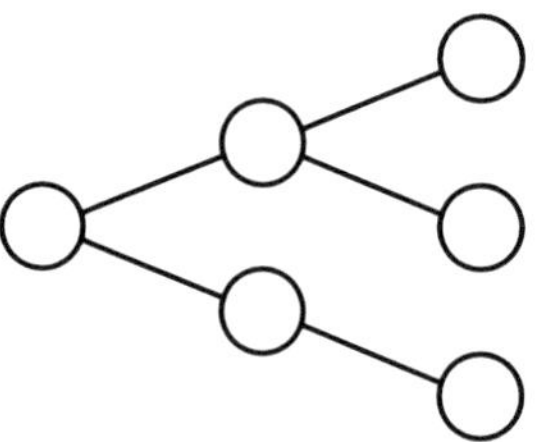

Figure 2.8 A tree structure

By Lemma 2.29, all birth and death processes are reversible, since their transition diagram constitutes a single chain. In particular, the $M/M(n)/1$ queue with Poisson arrivals and state dependent service rates discussed in Example 2.20 is reversible, since it is a birth and death process.

The most important property of a reversible process is that when the process is truncated to a subspace, its stationary distribution is also truncated to this subspace. A process with state space $\mathcal{S}$ and transition rate $q(x, x')$ is said to be truncated to subspace $\mathcal{A}$ if we focus on its transitions within $\mathcal{A}$, and ignore all the transitions within $\mathcal{S} \setminus \mathcal{A}$ and transitions between $\mathcal{A}$ and $\mathcal{S} \setminus \mathcal{A}$. The resulting continuous time Markov chain with state space $\mathcal{A}$ is referred to as the truncated Markov chain. For instance, an $M/M/1$ queue with its state space $\{0, 1, 2, \ldots\}$ truncated to the state space $\{0, 1, \ldots, N\}$ becomes an $M/M/1/N$ queue.

Theorem 2.30 *If a stationary reversible process $X(t)$ with state space $\mathcal{S}$ and stationary distribution π is truncated to a subspace $\mathcal{A}$, then the truncated process is also reversible and has the stationary distribution*

$$\pi_{\mathcal{A}}(x) = \frac{\pi(x)}{\sum_{x' \in \mathcal{A}} \pi(x')}, \qquad x \in \mathcal{A}. \tag{4.9}$$

Note that (4.9) is the conditional distribution given that the process is in set $\mathcal{A}$.

PROOF. Since

$$\pi_{\mathcal{A}}(x) q(x, x') = \pi_{\mathcal{A}}(x') q(x', x), \qquad x, x' \in \mathcal{A},$$

and

$$\sum_{x \in \mathcal{A}} \pi_{\mathcal{A}}(x) = 1,$$

the $\pi_{\mathcal{A}}$ satisfies the detailed balance equations of the truncated Markov chain, and is therefore the stationary distribution. □

Reversibility is a sufficient condition for the truncation property, but it is by no means necessary. It turns out that a necessary and sufficient condition for the truncation property is local balance.

Theorem 2.31 *Suppose a continuous time Markov chain with stationary distribution $\pi(x)$ and state space $\mathcal{S}$ is truncated to a subset $\mathcal{A}$. Then the stationary distribution for the truncated process is the truncated distribution (4.9) if and only if the local balance equations*

$$\pi(x) \sum_{x' \in \mathcal{A}} q(x, x') = \sum_{x' \in \mathcal{A}} \pi(x') q(x', x), \qquad x \in \mathcal{A},$$

are satisfied.

PROOF. If $\pi_{\mathcal{A}}$ is the stationary distribution for the truncated process, then it satisfies the global balance equations

$$\pi_{\mathcal{A}}(x) \sum_{x' \in \mathcal{A}} q(x, x') = \sum_{x' \in \mathcal{A}} \pi_{\mathcal{A}}(x') q(x', x), \qquad x \in \mathcal{A}. \tag{4.10}$$

Substituting (4.9) into (4.10) yields the local balance equations. On the other hand, if (4.10) is satisfied, (4.9) will also be satisfied, so $\pi_{\mathcal{A}}$ is the stationary distribution of the truncated Markov chain. □

2.5 PRODUCT FORMS AND DECOMPOSABILITY

The stationary distributions obtained in Examples 2.16, 2.17 and 2.22 are the product of their marginal distributions. This type of product form stationary distribution is typical for the networks discussed in this book. Product forms are also important from the application point of view, since they simplify the analysis of complicated systems.

A distribution π on a product space $\mathcal{S} \equiv \mathcal{S}_1 \times \mathcal{S}_2 \times \cdots \times \mathcal{S}_N$ is said to have a product form or to be a product form solution if there exist marginal distributions π_j, $j = 1, \ldots, N$, such that

$$\pi(\boldsymbol{x}) = \prod_{j=1}^{N} \pi_j(x_j), \qquad \boldsymbol{x} = (x_1, x_2, \ldots, x_N) \in \mathcal{S}. \tag{5.1}$$

This product form can be relaxed to the case where $\mathcal{S}$ is a subspace of the product space $\mathcal{S}_1 \times \mathcal{S}_2 \times \cdots \times \mathcal{S}_N$, where we may not be able to choose the x_j independently. In this case, we require a normalizing constant C such that

$$\pi(\boldsymbol{x}) = C \prod_{j=1}^{N} \pi_j(x_j), \qquad \boldsymbol{x} = (x_1, x_2, \ldots, x_N) \in \mathcal{S},$$

where

$$C^{-1} = \sum_{\boldsymbol{x} \in \mathcal{S}} \prod_{j=1}^{N} \pi_j(x_j).$$

A typical example of this product form is the stationary distribution of a closed Jackson network, in which the total number of customers in the system is fixed.

There is another type of product form in the queueing literature. For some queueing network models, the system dynamics, i.e., the state transition rates, can be decomposed into two parts, namely the service part and the routing part. The stationary distribution π of the state of the network can often be expressed as

$$\pi(\boldsymbol{x}) = \Phi(\boldsymbol{x}) f(\boldsymbol{x}), \qquad \boldsymbol{x} \in \mathcal{S}, \tag{5.2}$$

where Φ and f are only related to the service and routing parts, respectively. When we have to make a distinction between the two types of product forms, we call (5.1) the *distributional product form*, and (5.2) the *functional product form*. However, we may refer to any one of the two simply as the *product form* when the meaning is clear from the context.

For certain queueing networks, the state includes not only the populations of the different classes of customers, but also supplementary information such as the remaining or attained service times of the customers. In such a case, the system's state $\boldsymbol{x}$ has two components, a main part $\boldsymbol{n}$ and a supplementary part $\boldsymbol{y}$. Suppose $\boldsymbol{y}$ is a $k(\boldsymbol{n})$ dimensional vector, which is a function of state $\boldsymbol{n}$. Often $\boldsymbol{n}$ also includes information about classes of customers $u \in T$, where T is a set of entity classes. For example, the ℓth component y_ℓ may be the remaining service phase of a type u customer in position ℓ when the service time distribution is of the phase type. If distribution π has the form

$$\pi(\boldsymbol{x}) = \pi_{\mathcal{N}}(\boldsymbol{n}) \prod_{\ell=1}^{k(\boldsymbol{n})} h_{c_\ell(\boldsymbol{n})}(y_\ell),$$

where $c_\ell(\boldsymbol{n})$ is the class of the ℓth component, and $\pi_{\mathcal{N}}$ and $h_{c_\ell(\boldsymbol{n})}$ are probability distributions, then π is said to be *decomposable* with respect to the supplementary information. Clearly, this is yet another type of product form, which we refer to in what follows as *decomposable* to distinguish it from those described above.

2.6 THE $M/M^Y/1$ QUEUE: BATCH SERVICES

This section presents an elementary result for the $M/M/1$ queue with batch services. The result will be used in subsequent chapters in the analysis of networks of queues with batch processing features.

Assume that customers arrive according to a Poisson process with rate λ. Customers are served in batches, and the processing time of each batch is exponentially distributed with rate μ. The batch sizes are arbitrarily distributed integer random variables with distribution

$$P(Y = n) = b(n), \qquad n \geq 1,$$

where Y is a generic batch size. Assume that if the number of customers in the system upon a service completion is less than the batch size of the server, then the queue size reduces to zero. Thus the transition rates for such a process are

$$q(n, n+1) = \lambda, \qquad n \geq 0,$$

$$q(n, n-\ell) = \mu b(\ell), \qquad n > \ell \geq 1,$$

$$q(n, 0) = \mu \bar{b}(n), \qquad n \geq 1,$$

where $\bar{b}(n)$ is the tail function, i.e., $\bar{b}(n) = \sum_{\ell=n}^{\infty} b(\ell)$. This batch arrival queue is referred to as $M/M^Y/1$.

Lemma 2.32 *The stationary distribution of the number of customers in the system has the geometric distribution*

$$\pi(n) = (1-\rho)\rho^n, \qquad n = 0, 1, \ldots, \tag{6.1}$$

where ρ is the solution of the equation

$$\lambda = \mu \sum_{n=1}^{\infty} \bar{b}(n)\rho^n,$$

provided $\rho < 1$. Moreover, this equation has a unique solution ρ in (0,1) if and only if $\lambda < \mu E(Y)$.

PROOF. The balance equations of the Markov chain are

$$(\lambda + \bar{b}(1)\mu)\pi(n) = \lambda\pi(n-1) + \mu \sum_{\ell=1}^{\infty} \pi(n+\ell)b(\ell), \qquad n \geq 1, \tag{6.2}$$

$$\lambda\pi(0) = \mu \sum_{\ell=1}^{\infty} \pi(\ell)\bar{b}(\ell). \tag{6.3}$$

Since the Markov chain is irreducible, e.g., state 0 is reachable from any state, its stationary distribution is unique if it exists. If the distribution is of the form $\pi(n) = C\rho^n$, $n \geq 0$, then (6.2) and (6.3) are reduced to

$$\lambda + \mu = \lambda/\rho + \mu \sum_{\ell=1}^{\infty} \rho^\ell b(\ell), \tag{6.4}$$

$$\lambda = \mu \sum_{\ell=1}^{\infty} \rho^\ell \bar{b}(\ell). \tag{6.5}$$

It is not hard to verify that (6.4) and (6.5) are equivalent. To see this, observe that

$$\bar{b}(n) = b(n) + \bar{b}(n+1).$$

Hence (6.5) can be written as

$$\begin{aligned}\lambda &= \mu \sum_{\ell=1}^{\infty} \rho^{\ell} b(\ell) + \mu \sum_{\ell=1}^{\infty} \rho^{\ell} \bar{b}(\ell+1) \\ &= \mu \sum_{\ell=1}^{\infty} \rho^{\ell} b(\ell) + \mu \left(\rho^{-1} \sum_{\ell=1}^{\infty} \rho^{\ell} \bar{b}(\ell) - \bar{b}(1) \right) \\ &= \mu \sum_{\ell=1}^{\infty} \rho^{\ell} b(\ell) + \frac{\lambda}{\rho} - \mu,\end{aligned}$$

where the last equality follows from (6.5). Thus, (6.5) implies (6.4).

To prove the second part of the lemma, let

$$f(\rho) = \mu \sum_{\ell=1}^{\infty} \rho^{\ell} \bar{b}(\ell) - \lambda.$$

So $f(0) = -\lambda < 0$, and, since $\sum_{\ell=1}^{\infty} \bar{b}(\ell) = E(Y)$, $f(1) = \mu E(Y) - \lambda$. Furthermore, $f(\rho)$ is a strictly increasing function. Hence $f(\cdot)$ has a root in (0,1) if and only if $f(1) = \mu E(Y) - \lambda > 0$. □

The stability condition $\lambda < \mu E(Y)$ is quite intuitive. In order for the system to be stable the arrival rate has to be less than the total service capacity. Since each service completion reduces the queue size by Y (assuming the number of customers in the system is sufficiently large), the total service capacity is $\mu E(Y)$ per unit time. The stability condition thus follows.

2.7 THE $M^X/M^Y/1$ QUEUE: BATCH ARRIVALS AND BATCH SERVICES

We now consider a single-server queue with both batch arrivals and batch services. Batches of customers arrive according to a Poisson process with rate λ, and each batch contains ℓ customers with probability $a(\ell)$, $\ell = 1, \ldots$, where

$$\sum_{\ell=1}^{\infty} a(\ell) = 1.$$

The customers are also served in batches, and the batch size is ℓ with probability $b(\ell)$, $\ell = 1, \ldots$, where

$$\sum_{\ell=1}^{\infty} b(\ell) = 1.$$

If a service batch is ℓ and there are at least ℓ customers in the system at the service completion, then the queue length will be reduced by ℓ; otherwise the queue will be emptied out, in the same way as in Section 2.6. We assume that the service rate is constant, μ, regardless of how many customers are in the system, and customers who arrive when the batch in service is not full join the batch being served immediately. This model is referred to as $M^X/M^Y/1$.

We now assume that there is an additional batch Poisson arrival process with rate λ^* that is activated only when the queue is empty. The batch sizes of this additional arrival process are i.i.d. random variables with probability mass function $a^*(\ell)$, $\ell = 1, \ldots$, where

$$\sum_{\ell=1}^{\infty} a^*(\ell) = 1.$$

All the arrival processes and service processes are independent of each other.

Denote the moment generating functions for $\{a(\ell); \ell \geq 1\}$, $\{b(\ell); \ell \geq 1\}$, and $\{a^*(\ell); \ell \geq 1\}$ by $\hat{A}$, $\hat{B}$, and $\hat{A}^*$, respectively. Furthermore, let X, Y, and X^* be generic batch sizes for an arrival batch, a service batch, and an additional arrival batch. We assume that the convergence radius of $\hat{A}$ is greater than 1, i.e.,

$$\sup\{z \geq 0, \hat{A}(z) < \infty\} > 1, \tag{7.1}$$

which implies that $\{a(\ell); \ell \geq 1\}$ has a finite mean, i.e., $E(X) < \infty$.

Let $X(t)$ be the number of customers in the system at time t. It is a continuous time Markov chain with state space $\{0, 1, 2, \ldots\}$ and transition rates

$$\begin{aligned} q(0, \ell) &= \lambda a(\ell) + \lambda^* a^*(\ell), && \ell \geq 1, \\ q(n, n+\ell) &= \lambda a(\ell), && n \geq 1, \ell \geq 1, \\ q(n, n-\ell) &= \mu b(\ell), && n > \ell \geq 1, \\ q(n, 0) &= \mu \bar{b}(n), && n \geq 1, \end{aligned}$$

where $\bar{b}(n) = \sum_{\ell=n}^{\infty} b(\ell)$.

It follows from the Kelly lemma that a probability distribution π is the stationary distribution of $X(t)$ if and only if there exists a transition rate function $\tilde{q}$

satisfying (4.2) and (4.3). So suppose $X(t)$ has a stationary distribution π of a geometric form, i.e.,

$$\pi(n) = (1-\rho)\rho^n, \qquad n \geq 0,$$

for some $\rho \in (0, 1)$. Substituting the q given above and the geometric distribution into (4.3), we obtain

$$\begin{aligned}
\tilde{q}(0,n) &= \mu\bar{b}(n)\rho^n, && n \geq 1,\\
\tilde{q}(n,n+\ell) &= \mu b(\ell)\rho^\ell, && \ell \geq 1, n \geq 1,\\
\tilde{q}(n,n-\ell) &= \lambda a(\ell)\rho^{-\ell}, && n > \ell \geq 1,\\
\tilde{q}(n,0) &= (\lambda a(n) + \lambda^* a^*(n))\rho^{-n}, && n \geq 1.
\end{aligned}$$

Substituting the $\tilde{q}$ into (4.2) with $n = 0$ yields

$$\mu \sum_{\ell=1}^{\infty} \bar{b}(\ell)\rho^\ell = \lambda + \lambda^*,$$

and from (4.3) it follows that, for $n \geq 1$,

$$\mu \sum_{\ell=1}^{\infty} b(\ell)\rho^\ell + \lambda \sum_{\ell=1}^{n} a(\ell)\rho^{-\ell} + \lambda^* a^*(\ell)\rho^{-n} = \lambda + \mu. \tag{7.2}$$

For (7.2) to be true for all $n \geq 1$, it is necessary and sufficient that there exists some $h \geq 0$ such that

$$\mu \sum_{\ell=1}^{\infty} b(\ell)\rho^\ell + \lambda \sum_{\ell=1}^{\infty} a(\ell)\rho^{-\ell} + h = \lambda + \mu, \tag{7.3}$$

or

$$\lambda\hat{A}(\rho^{-1}) + \mu\hat{B}(\rho) + h = \lambda + \mu.$$

Subtracting (7.2) from (7.3) yields

$$\lambda^* a^*(n) = \lambda \sum_{\ell=n+1}^{\infty} a(\ell)\rho^{-(\ell-n)} + h\rho^n. \tag{7.4}$$

Let

$$\delta = \sup\{z \geq 0;\ \hat{A}(z) < \infty\},$$

which is greater than 1 because of (7.1). Then, the right-hand side of (7.4) is finite for $\rho \in (1/\delta, 1)$. Clearly, (7.4) depends on h and the minimal additional arrival process is obtained by setting $h = 0$. Set $h = 0$ and let

$$f(\rho) \equiv \lambda(\hat{A}(\rho^{-1}) - 1) - \mu(1 - \hat{B}(\rho)). \tag{7.5}$$

Then (7.3) can be written as $f(\rho) = 0$. It can be verified that $f(\rho)$ is a strictly convex function, unless all batch sizes are equal to one (see Exercise 2.8). Furthermore, since $f(1) = 0$, $f(\rho) = 0$ has a solution in (0, 1) only if $f'(1) > 0$, i.e., f is strictly increasing at 1. This is equivalent to

$$\lambda E(X) < \mu E(Y). \tag{7.6}$$

This is a typical stability condition for a queueing system. By (7.4), λ^* should be selected as

$$\lambda^* = \lambda \sum_{n=1}^{\infty} \sum_{\ell=n+1}^{\infty} a(\ell)\rho^{-(\ell-n)} = \frac{\lambda\rho\hat{A}(\rho^{-1}) - \lambda}{1-\rho}. \tag{7.7}$$

Summarizing, we have the following result.

Theorem 2.33 *Under the regularity condition (7.1), the single node queue with batch arrivals and batch services has a geometric solution for the queue length if the stability condition (7.6) holds and the additional arrival process is given by (7.4).*

The following theorem shows how the modified model can be used to evaluate the original system.

Theorem 2.34 *Let π^0 denote the stationary distribution of the single node queue with no additional batch arrivals. Then, under conditions (7.1) and (7.6), π^0 is stochastically less than or equal to π, i.e.,*

$$\sum_{\ell=n}^{\infty} \pi^0(\ell) \le \rho^n, \qquad n = 0, 1, \ldots, \tag{7.8}$$

where ρ is the solution of the equation

$$\lambda(\hat{A}(\rho^{-1}) - 1) = \mu(1 - \hat{B}(\rho)).$$

The equality in (7.8) holds if and only if $a(1) = 1$. *Furthermore, this* ρ *gives the best possible geometric bound, where* ρ *is said to be a best geometric bound for a function* $g(n)$ *if* ρ *is the smallest positive number such that there is a constant* c *and* $g(n) \le c\rho^n$ *for all* n.

PROOF. The proof of the first part of the theorem is straightforward using coupling (see Exercise 2.4), since additional arrivals can only make the system more congested. Construct two systems, one with an additional arrival process and one without. For convenience, call these two systems 1 and 2. Since we are only comparing the number of customers in the two systems, the order of service of the customers is immaterial. Hence let us assume that in system 1, customers from the additional arrival process have lower priority, and the arrival of a regular customer can preempt a lower priority customer if a lower priority customer is being served upon its arrival. Since the services of regular customers in system 1 are not affected by those from the additional arrival stream, the numbers of regular customers in both systems are clearly the same, provided we use the same service times for the regular customers in the two systems. However, there may be customers of lower priority in system 1, so the number of customers in system 1 is at least as large as that in system 2. By the exponential service time assumption, the preempted customers still have exponentially distributed service times with the same rate. This shows that in the sample path sense, the number of customers in system 1 dominates that in system 2. Hence the number of customers in the two systems are stochastically ordered.

For the second part, note that the number of customers in the system without additional arrivals is affected only by regular arrival epochs or departure epochs. Order the arrival and departure epochs in increasing order, and let $N^0(n)$ be the number of customers in the system after the nth epoch. Note that if the $(n+1)$st epoch is an arrival epoch, then

$$N^0(n+1) = N^0(n) + X,$$

where X is the arrival batch size, and

$$N^0(n+1) = \max\{0, N^0(n) - Y\}$$

if the $(n+1)$st epoch is a departure, where Y is the service batch size. If N^0 denotes the stationary number of customers immediately after a transition (which has distribution π^0), then by conditioning on whether the transition is an arrival or a departure we obtain

$$N^0 \cong \max\{0, N^0 + U^0\},$$

where "$\cong$" denotes the equality in distribution, and the distribution of the random variable U^0 is given by

$$P(U^0 = \ell) = \begin{cases} \frac{\lambda a(\ell)}{\lambda+\mu}, & \text{if } \ell \geq 0, \\ \frac{\mu b(-\ell)}{\lambda+\mu}, & \text{if } \ell < 0, \\ 0, & \text{if } \ell = 0. \end{cases}$$

Thus N^0 satisfies the so-called Lindley's equation for $GI/G/1$ queues. Using Kingman's exponential bound for N^0, we conclude that the best exponential bound for $P(N^0 \geq x)$ is $e^{-\theta_0 x}$, where θ_0 is the solution of the equation

$$E(e^{\theta_0 U}) = \frac{\lambda \hat{A}(e^{\theta_0}) + \mu \hat{B}(e^{-\theta_0})}{\lambda + \mu} = 1. \tag{7.9}$$

Comparing (7.9) with (7.3) we obtain that $\rho = e^{-\theta_0}$ is a solution for the queue with additional arrivals (7.4). So (7.8) has to be the best geometric bound. □

When $a(1) = 1$, the single queue is reduced to the model discussed in Section 2.6, and in this case, by (7.4), $a^*(n) = 0$ for all n. On the other hand, if $a(1) < 1$, then $\lambda^* > 0$. From Theorem 2.34 we conclude that in this case the single queue without additional batch arrivals cannot have a geometric stationary distribution.

Corollary 2.35 *The single queue with batch arrivals and batch services has a geometric stationary distribution if and only if the arrival batches are of size one.*

2.8 REFERENCE NOTES

A clear and elementary treatment of Markov chains can be found in the textbooks of Ross (1983, 1997). More in-depth studies can be found in Kemeny, Snell and Knapp (1966), Chung (1967) and Asmussen (1987). Counting processes are also called point processes in the literature. A general formulation of point processes can be found in Bremaud (1981), Daley and Vere-Jones (1988), while their applications to queues are discussed by Franken, König, Arndt and Schmidt (1982) and Baccelli and Bremaud (1994). The PASTA property is due to Wolff (1982). The Jackson network is first studied in Jackson (1957) (see also Jackson (1963)). Local balance has been one of the major tools for showing product form solutions and the insensitivity of queueing networks (and more generally, of generalized semi-Markov processes); Whittle (1967) (see also Whittle (1968)) is apparently the first one to apply it to queueing networks. The reversed process approach enabled Kelly to generalize the class of product form queueing networks, see Kelly (1975, 1976). An excellent reference is his monograph, Kelly (1979). In this book we refer to his reversed process argument as the Kelly lemma. Discussions of various product form distributions can be found in Kelly (1979), Whittle (1986) and Van Dijk (1993). The terminology *decomposability* is taken from Miyazawa (1993). The geometric stationary distribution of $M/M/1$ queue with batch services goes back to Kleinrock (1975) and Gross and Harris (1985). The material of Section 2.7 is taken from Miyazawa and Taylor (1997). For a comprehensive discussion of coupling approach to stochastic ordering relationships for queueing and other stochastic systems, see Ross (1983) and Stoyan (1983).

EXERCISES

2.1 Show that (1.10) is equivalent to the global balance (1.9).

2.2 Let X and Y be independent exponential random variables with means $1/\lambda$ and $1/\mu$.

(a) Show that the minimum of the two independent exponential random variables, i.e., $\min\{X, Y\}$, is an exponential random variable with mean $1/(\lambda + \mu)$.

(b) Show that

$$P(X < Y) = \frac{\lambda}{\lambda + \mu}.$$

2.3 Show that a Jackson network is irreducible if and only if the Markov chain with state space $\{0, 1, 2, \ldots, N\}$ and transition probabilities $\hat{p}_{jk}$, $j, k = 0, 1, \ldots, N$, is irreducible, where

$$\hat{p}_{jk} = p_{jk}, \qquad j = 1, \ldots, N, \text{ and } k = 0, 1, \ldots, N,$$
$$\hat{p}_{0j} = \frac{\lambda_j}{\sum_{k=1}^{N} \lambda_k}, \qquad j = 1, 2, \ldots, N.$$

2.4 A random variable X is said to be stochastically less than a random variable Y, denoted by $X \leq_{st} Y$, if

$$P(X > x) \leq P(Y > x)$$

for all x. Show that $X \leq_{st} Y$ if and only if two random variables $\hat{X}$ and $\hat{Y}$ can be constructed on the same probability space such that

$$P(\hat{X} \leq x) = P(X \leq x), \qquad \text{for all } x,$$
$$P(\hat{Y} \leq x) = P(Y \leq x), \qquad \text{for all } x,$$
$$\hat{X} \leq \hat{Y}, \qquad \text{almost surely.}$$

Thus X is stochastically less than Y if and only if these two random variables can be constructed on the same probability space such that they are ordered almost surely. This process of constructing two random variables on the same probability space is referred to as *sample path analysis* or *coupling*, and it can be extended to the stochastic comparison of two processes. The stochastic ordering result in this chapter as well as those of later chapters can all be shown using coupling.

2.5 Consider a network of N single-server queues with exogenous Poisson arrivals with rate $\lambda_1 > 0$ at node 1. Furthermore, let r_{jk} be the probability that after a customer completes its service at node j it proceeds to node k, $k = 0, 1, \ldots, N$, with 0 representing the outside. Show that if $r_{10} = 0$ then the network process cannot be reversible.

2.6 Show that (3.11) is the stationary distribution of the network in Example 2.16 by verifying the global balance equations.

2.7 Show that if $\{X_k; k = \ldots, -2, -1, 0, 1, 2, \ldots\}$ is a stationary Markov chain, then the reversed process $\{X_{-k}; k = \ldots, -2, -1, 0, 1, 2, \ldots\}$ is also a stationary Markov chain, i.e.,

$$P(X_{-(k+1)} = x'|X_{-\ell}, \ell < k, X_{-k} = x) = P(X_{-(k+1)} = x'|X_{-k} = x).$$

2.8 Consider a Markov chain with state space $\mathcal{S}$, stationary distribution $\pi(x)$, and transition rates $q(x, x')$, $x, x' \in \mathcal{S}$. Assume that the transition rates are changed in such a way that $q(x, x')$ becomes $cq(x, x')$, for $x \in \mathcal{A}$, $x' \in \mathcal{S} \setminus \mathcal{A}$ and some $c \neq 0, 1$. Show that the resulting Markov chain has a stationary distribution of the form

$$\begin{array}{ll} B\pi(x), & x \in \mathcal{A}, \\ Bc\pi(x), & x \in \mathcal{S} \setminus \mathcal{A}, \end{array}$$

if and only if its distribution $\pi(x)$ satisfies the local balance equations

$$\pi(x) \sum_{x' \in \mathcal{A}} q(x, x') = \sum_{x' \in \mathcal{A}} \pi(x') q(x', x), \qquad x \in \mathcal{A}.$$

2.9 Suppose a Markov chain with transition matrix P is irreducible and has a stationary distribution. Prove that the Markov chain is reversible if and only if

$$P = AD,$$

where A is a symmetric matrix with nonnegative entries and D is a diagonal matrix with positive diagonal entries.

2.10 Show that $f(\rho)$ of (7.5) is a strictly convex function unless all the batch sizes are equal to 1.

2.11 Show that under the stability condition (7.6), (7.3) has a solution $\rho \in (0, 1)$ for any $h \geq 0$. Thus it follows from (7.4) that λ^* can also be chosen as

$$\lambda^* = \lambda \sum_{n=1}^{\infty} \left(\sum_{\ell=n+1}^{\infty} a(\ell)\rho^{-(\ell-n)} + h\rho^n \right) = \frac{\lambda(\rho A(\rho^{-1}) + h) - \lambda}{1 - \rho}.$$

This implies that the modified queue has a geometric stationary distribution if and only if ρ is the solution of (7.3), and the additional arrival process is given by (7.4). Show that under condition (7.1) and stability condition (7.6), there exists a unique $\tau \in (0, 1)$ such that

$$\lambda A'(\tau^{-1}) = \mu\tau^2 B'(\tau).$$

Furthermore, (7.3) has a solution $\rho \in (0, 1)$ if and only if

$$0 \leq h \leq \mu(1 - B(\tau)) - \lambda(A(\tau^{-1}) - 1),$$

where the right-hand side is always positive. This implies that for each h satisfying

$$0 < h < \mu(1 - B(\tau)) - \lambda(A(\tau^{-1}) - 1),$$

there are two sets of parameters a^* and λ^* that result in two different geometric stationary distributions (since there are two solutions ρ for (7.3)). These modified queues differ only in the additional arrival processes when the queue is empty.

3

Quasi-Reversibility

Quasi-reversibility is an input–output property of queues. It implies that when the system is in stochastic equilibrium, the future arrival processes, the current state of the system, and the past departure processes are independent. In this chapter we present a formal theory of quasi-reversibility for queues with and without signals. In the first section we modify the definition of continuous time Markov chains by including a transition rate from each state back to itself. In Sections 3.2 and 3.3 we discuss quasi-reversibility of queues with and without signals, respectively.

3.1 CONTINUOUS TIME MARKOV CHAINS REVISITED

We modify the definition of a continuous time Markov chain with state space $\mathcal{S}$ so that its transition rate function $q(x, x')$ includes transitions from state x back to itself for each $x \in \mathcal{S}$. Such transitions are usually excluded in the theory of continuous time Markov chains. Clearly, such modifications do not change the stationary distribution of the Markov chain. However, these modifications are important and convenient for defining quasi-reversibility since, in the framework of this book, the arrival of an entity may not result in any change of state.

To see the need for such a modification, consider the following example. In this example as well as in the rest of this book, when transition rates are only partially defined, the unspecified rates are assumed to be zero.

Example 3.1 (Effect of arrivals) Consider an $M/M/1$ queue with two classes of arrivals, denoted by $\{c, c^-\}$. Class c refers to the regular customer and class c^- refers to an entity, called *negative customer*. Assume that both regular customers and negative customers arrive according to independent Poisson processes with rates α and α^-, and the effect of an arriving negative customer is the deletion of the customer in service. However, when a negative customer arrives at an empty node, nothing happens.

The state of the queue is the number of customers in the system, denoted by n. When the state is 0, there are two types of events that can occur: one is the arrival of a regular customer, and the other is the arrival of a negative customer. The first event changes the state from 0 to 1, whereas the second event does not change the state. Mathematically, the statement that regular customers arrive according to a Poisson process is expressed as

$$q(n, n+1) = \alpha, \qquad n = 0, 1, \ldots.$$

The arrival of a negative customer when the state is not 0 can be reflected in a change of state from n to $n-1$. However, when the state is 0, such an arrival does not cause a change of state. It is for this reason that in a Markov chain we need a rate $q(x, x)$ that represents the occurrence of an event that does not change the state of the system. □

We briefly introduce the modification technique. As discussed in Section 2.1, a continuous time Markov chain with transition rates $\{q(x, y); x, y \in \mathcal{S}\}$ can also be formulated as a process with a jump kernel p defined by

$$p(x, x') = \frac{q(x, x')}{q(x)}, \qquad x \neq x', x, x' \in \mathcal{S},$$

and with sojourn times in state x exponentially distributed with mean $1/q(x)$ for each $x \in \mathcal{S}$, where

$$q(x) = \sum_{x' \in \mathcal{S} \setminus \{x\}} q(x, x').$$

In the theory of Markov chains, $q(x, x)$ is defined as $-q(x)$, but we use it to denote the rate of those events in state x that do not cause a change of state. For a number $q(x, x) \geq 0$, we modify the Markov chain with the jump kernel p so that its mean sojourn time in state x is $1/q^*(x)$ and its transition kernel is p^*, where

$$q^*(x) = q(x) + q(x, x),$$
$$p^*(x, x') = \frac{q(x, x')}{q^*(x)}.$$

It is not hard to see that the mean sojourn time in state x up to a jump into another state equals $1/q(x)$, and when this occurs, the process jumps to state $x' \neq x$

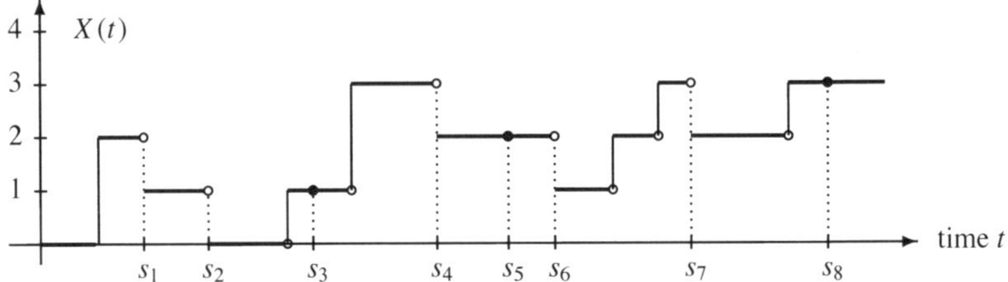

Figure 3.1 The counting process $\{s_i\}$ driven by $\{X(t)\}$ with respect to downward jumps, where s_3, s_5 and s_8 are generated by $q(x, x)$.

with probability $p(x, x')$. Hence, the sample path of the modified process can be chosen to be identical to the one in the original process. Furthermore, if $\mathcal{G}_t$ denotes the history of all jump epochs up to time t, then the conditional probabilities of events after time t, given $X(s)$ $(s \le t)$ and $\mathcal{G}_t$, only depend on $X(t)$. In other words, introducing $q(x, x)$ does not change the sample path of the original continuous time Markov chain. Thus, the modified process may be regarded as identical to the original continuous time Markov chain as far as the process $\{X(t); t \ge 0\}$ is concerned (see Figure 3.1).

Remark 3.2 Note that after $q(x, x)$ (the jump rate from x to itself) is introduced, $\{X(s); s \le t\}$ *does not* contain all the information about the jump process because there are jumps even when $X(t)$ does not change. In other words, the modification does not change the sample path of the process $X(t)$, but it does change the jump process. However, $\{X(t)\}$ is still a continuous time Markov chain with respect to $\{\mathcal{F}_t\}$, where $\mathcal{F}_t$ is the information up to time t including all the jump epochs, i.e., the behavior of $X(s)$ after time t given $\mathcal{F}_t$ depends only on $X(t)$ (see Appendix B for a formal definition). The standard Markov chain typically assumes

$$\mathcal{F}_t = \sigma\{X(s); \quad s \le t\}$$

to be the σ-field generated by $X(t)$, but for the modified Markov chain, we define $\mathcal{F}_t$ to be the smallest σ-field containing

$$\sigma\{X(s); s \le t\} \cup \mathcal{G}_t.$$

Throughout the book, we tacitly assume this σ-field for every Markov chain unless otherwise specified. See Appendix A for a formal description of σ-fields.

Clearly, a different set of nonnegative numbers $q(x, x)$, $x \in \mathcal{S}$, yields a different modified continuous time Markov chain. Even though the actual values of $q(x, x)$ do not affect the stationary distribution of the Markov chain, we will see in the next section that they do affect the quasi-reversibility of the modified continuous time Markov chain, because the arrival and departure epochs include transitions from x to x.

3.2 QUASI-REVERSIBILITY OF QUEUES

In Example 3.1 both the arrival of a negative customer and the service completion of a regular customer result in a transition from $n + 1$ to n. Hence, different events may result in the same transition from x to x'. For this reason, we introduce the following notation. For each pair of states (x, x'), we decompose the transition rate function $q(x, x')$ of the queue into three types of rates, namely

$$\begin{aligned} &q_u^{\text{A}}(x, x'), \qquad && u \in T, \\ &q_v^{\text{D}}(x, x'), \qquad && v \in T, \\ &q^{\text{I}}(x, x'), \end{aligned}$$

where T is the set of the classes of arrivals and departures, which is countable. Even though in many queueing systems the classes of arrivals are different from the classes of departures, we use a single index set T because we can take T as the union of both arrival and departure classes. Thus the transition rate of the queue can be written as

$$q(x, x') = \sum_{u \in T} q_u^{\text{A}}(x, x') + \sum_{v \in T} q_v^{\text{D}}(x, x') + q^{\text{I}}(x, x'), \qquad x, x' \in \mathcal{S}. \quad (2.1)$$

In view of Definition 2.10, these thinned transition rate functions q_u^{A}, q_v^{D} and q^{I} generate the embedded point processes corresponding to class u arrivals, class v departures and the internal transitions, respectively. The first two embedded point processes are often referred to as the *arrival process of class u entities* and the *departure process of class v entities*. The superscripts "A", "D", and "I" stand for "arrival", "departure" and "internal". An internal transition typically represents a change of status of the customers such as a decrease in their remaining service times, or, in the case when the node contains multiple service stations, a movement of a customer between different stations of the node.

If the supports of the rate functions in (2.1) are disjoint, the decomposition above would have only one term. However, we do not make any restriction with regard to their supports. They are distinguished only by the probabilities, i.e., it is a rate decomposition. It should be noted that, even though $q(x, x')$ is said to be decomposed into three types of components (arrival, departure, and internal transition rates) which result in the same transition from x to x', the opposite occurs in applications: one is usually given the arrival, departure and internal rates that result in the same transition, and they have to be aggregated in order to obtain $q(x, x')$.

Example 3.3 (An $M/M/1$ queue with negative customers) Consider the $M/M/1$ queue of Example 3.1 with two classes of arrivals. Let $T = \{c, c^-\}$ with c and c^- representing regular and negative customers. If a service completion is classified as class c, then it follows from the assumption of exponential service times with mean $1/\mu$ that

$$q_c^{\mathrm{D}}(n, n-1) = \mu, \qquad n = 1, 2, \ldots.$$

Customers arrive according to a Poisson process with rate α, so

$$q_c^{\mathrm{A}}(n, n+1) = \alpha, \qquad n = 0, 1, \ldots.$$

Since negative customers arrive according to a Poisson process with rate α^- and reduce the number of customers by 1, we have

$$q_{c^-}^{\mathrm{A}}(n, n-1) = \alpha^-, \qquad n = 1, 2, \ldots.$$

Finally, since a negative customer that arrives at an empty node simply disappears, we have

$$q_{c^-}^{\mathrm{A}}(0, 0) = \alpha^-.$$

Let $q(n, n')$ denote the transition rate of the queue. The nonzero transition rates are

$$\begin{aligned} q(n, n+1) &= q_c^{\mathrm{A}}(n, n+1), & n \geq 0, \\ q(n, n-1) &= q_c^{\mathrm{D}}(n, n-1) + q_{c^-}^{\mathrm{A}}(n, n-1), & n \geq 1, \\ q(0, 0) &= q_{c^-}^{\mathrm{A}}(0, 0). \end{aligned}$$

□

Now we are ready to give the first definition of quasi-reversibility. As we discussed at the beginning of this chapter, quasi-reversibility is an input-output property of queueing systems. The property is presented using the arrival and departure transition rates.

Definition 3.4 (First definition of quasi-reversibility) The continuous time Markov chain with transition rate q is called *quasi-reversible with respect to* $\{q_u^{\mathrm{A}}(x, x'); u \in T\}$, $\{q_u^{\mathrm{D}}(x, x'); u \in T\}$ *and* $q^{\mathrm{I}}(x, x')$ if there exist two sets of nonnegative numbers $\{\alpha_u; u \in T\}$ and $\{\beta_u; u \in T\}$ such that

$$\sum_{x' \in \mathcal{S}} q_u^{\mathrm{A}}(x, x') = \alpha_u, \qquad x \in \mathcal{S}, u \in T, \tag{2.2}$$

$$\sum_{x' \in \mathcal{S}} \pi(x') q_u^{\mathrm{D}}(x', x) = \beta_u \pi(x), \qquad x \in \mathcal{S}, u \in T, \tag{2.3}$$

where π is the stationary distribution of the Markov chain q.

The nonnegative numbers α_u and β_u are often called the arrival rate and departure rate of class u entities. We illustrate this concept with the queueing system of Example 3.3.

Example 3.5 (Quasi-reversibility of $M/M/1$ with negative customers) Consider the $M/M/1$ queue of Example 3.3. This queue is quasi-reversible with respect to $\{q_c^{\mathrm{A}}, q_{c^-}^{\mathrm{A}}\}$, $\{q_c^{\mathrm{D}}\}$ and q^{I} (set equal to zero). To see this, note that as far as the stationary probability is concerned, this system behaves in exactly the same way as a conventional $M/M/1$ queue with regular customers arriving at rate α and a modified service rate $\mu + \alpha^-$. The stationary probability is thus given by

$$\pi(n) = \left(1 - \frac{\alpha}{\mu + \alpha^-}\right)\left(\frac{\alpha}{\mu + \alpha^-}\right)^n, \qquad n \geq 0. \tag{2.4}$$

It follows from Example 3.3 that

$$\sum_{n'} q_c^{\mathrm{A}}(n, n') = q_c^{\mathrm{A}}(n, n+1) = \alpha, \qquad n \geq 0,$$

$$\sum_{n'} q_{c^-}^{\mathrm{A}}(n, n') = q_{c^-}^{\mathrm{A}}(n, n-1) = \alpha^-, \qquad n \geq 1.$$

Hence condition (2.2) is satisfied since, when $n = 0$,

$$\sum_{n'} q_{c^-}^{\mathrm{A}}(0, n') = q_{c^-}^{\mathrm{A}}(0, 0) = \alpha^-.$$

By (2.4), we obtain, for all $n \geq 0$,

$$\begin{aligned}\sum_{n'} \pi(n') q_c^{\mathrm{D}}(n', n) &= \pi(n+1) q_c^{\mathrm{D}}(n+1, n) \\ &= \pi(n) \frac{\alpha}{\mu + \alpha^-} \mu.\end{aligned}$$

Hence (2.3) is satisfied with departure rate

$$\beta = \frac{\alpha}{\mu + \alpha^-}\mu.$$

This shows that the queue is quasi-reversible. □

This simple example shows the importance of using $q(0,0)$. It is clear that quasi-reversibility will not be satisfied without introducing $q(0,0)$, even though the stationary distribution of the queue is not affected at all.

Quasi-reversibility is a property concerning the arrival and departure processes. In fact, it is often useful to study this property with regard to only a *portion* of the arrival and departure processes. This is particularly true in networks of queues where only some of the departures from one node join another node, while the rest are either absorbed at the node or exit the network. The following example illustrates this.

Example 3.6 (An $M/M/1$ queue with services in batches of two) Consider an $M/M/1$ queue with only regular customers. The arrival rate is α, and services are provided in batches of two customers. However, when there is only one customer in the system, the server will work on that one customer; if a new customer arrives before this customer's service is completed, the arriving customer joins the customer being served. Assume that the service times are exponentially distributed with parameter μ, regardless of the number of customers in service. This model is a special case of the $M/M/1$ queue with batch departures described in Section 2.6. The transition rates for this model are

$$q(n, n+1) = \alpha, \qquad n = 0, 1, \ldots,$$

$$q(n, n-2) = \mu, \qquad n = 2, 3, \ldots,$$

$$q(1, 0) = \mu.$$

Let $\pi(n)$ denote the stationary distribution. The global balance equations are

$$(\alpha + \mu)\pi(n) = \alpha\pi(n-1) + \mu\pi(n+2), \qquad n = 1, 2, \ldots,$$

$$\alpha\pi(0) = \mu\pi(1) + \mu\pi(2).$$

It follows from Lemma 2.32 that

$$\pi(n) = (1-\rho)\rho^n, \qquad n = 0, 1, 2, \ldots,$$

provided $\rho < 1$, where ρ is a positive solution of the equation

$$\alpha = \mu(\rho + \rho^2).$$

Solving this equation yields

$$\rho = \frac{\sqrt{\mu^2 + 4\alpha\mu} - \mu}{2\mu}.$$

First, let us include all the service completions as departure epochs. The decomposed rates for such a classification are

$$\begin{aligned} q_c^{\mathrm{A}}(n, n+1) &= \alpha, \qquad n = 0, 1, \ldots, \\ q_c^{\mathrm{D}}(n, n-2) &= \mu, \qquad n = 2, 3, \ldots, \\ q_c^{\mathrm{D}}(1, 0) &= \mu. \end{aligned}$$

Hence for $n \geq 1$,

$$\begin{aligned} \sum_{n'} \pi(n') q_c^{\mathrm{D}}(n', n) &= \pi(n+2) q_c^{\mathrm{D}}(n+2, n) \\ &= \pi(n)\rho^2\mu. \end{aligned}$$

While for $n = 0$,

$$\begin{aligned} \sum_{n'} \pi(n') q_c^{\mathrm{D}}(n', 0) &= \pi(2) q_c^{\mathrm{D}}(2, 0) + \pi(1) q_c^{\mathrm{D}}(1, 0) \\ &= \pi(0)(\rho^2 + \rho)\mu. \end{aligned}$$

So, condition (2.3) is not satisfied, and the queue is not quasi-reversible.

Let us now only include the service completions of pairs of customers as departures; the transitions from 1 to 0 are classified as internal. Thus q_c^{D} is changed to

$$\begin{aligned} q_c^{\mathrm{D}}(n, n-2) &= \mu, \qquad n = 2, 3, \ldots, \\ q^{\mathrm{I}}(1, 0) &= \mu. \end{aligned}$$

In this case it is easily verified that, for all $n \geq 0$,

$$\sum_{n'} \pi(n') q_c^{\mathrm{D}}(n', n) = \pi(n)\rho^2\mu.$$

Thus condition (2.3) is satisfied for all $n \geq 0$. This shows that the queue is quasi-reversible with departure rate $\beta = \rho^2\mu$. □

An alternative definition for quasi-reversibility is the following.

Definition 3.7 (Second definition of quasi-reversibility) A stationary continuous time Markov chain $\{X(t); t \geq 0\}$ with transition rate q of (2.1) is quasi-reversible if the following two conditions hold.

(i) The $X(t)$ is independent of the arrival process of class u entities subsequent to time t for all $u \in T$.

(ii) The $X(t)$ is independent of the departure process of class u entities prior to time t for $u \in T$.

Quasi-reversibility is closely related to Poisson flows, as is shown in the following theorem.

Theorem 3.8 *Definitions 3.4 and 3.7 are equivalent, and each one of the two implies that*

(a) the arrival process of class $u \in T$ entities are Poisson and the arrival processes of different classes of entities are independent, and

(b) the departure process of class $u \in T$ entities are Poisson and the departure processes of different classes of entities are independent.

PROOF. By Lemma 2.11, the first quasi-reversibility condition (2.2) implies condition (i) of Definition 3.7 and part (a) of the theorem. On the other hand, condition (i) implies that the left-hand side of condition (2.2) is a constant. Denoting this constant by α_u, we obtain (2.2). We next show that condition (2.3) of the first definition implies condition (ii) of the second definition and part (b) of the theorem. To this end, consider the time reversed process $X(-t)$. Define $\tilde{q}_u^{\mathrm{A}}$, $\tilde{q}_u^{\mathrm{D}}$ and $\tilde{q}^{\mathrm{I}}$ as

$$\begin{aligned}
\tilde{q}_u^{\mathrm{A}}(x, x') &= \frac{\pi(x')}{\pi(x)} q_u^{\mathrm{D}}(x', x), \\
\tilde{q}_u^{\mathrm{D}}(x, x') &= \frac{\pi(x')}{\pi(x)} q_u^{\mathrm{A}}(x', x), \\
\tilde{q}^{\mathrm{I}}(x, x') &= \frac{\pi(x')}{\pi(x)} q^{\mathrm{I}}(x', x).
\end{aligned}$$

Let $\tilde{q}$ be the transition rate function of $X(-t)$. From Theorem 2.18, we have

$$\tilde{q}(x, x') = \sum_u \tilde{q}_u^{\mathrm{A}}(x, x') + \sum_u \tilde{q}_u^{\mathrm{D}}(x, x') + \tilde{q}^{\mathrm{I}}(x, x').$$

Thus, the reversed process also represents a queueing model with thinned transitions $\tilde{q}_v^{\mathrm{A}}$, $\tilde{q}_u^{\mathrm{D}}$ and $\tilde{q}^{\mathrm{I}}$. From condition (2.3), we have

$$\sum_{x'} \tilde{q}_u^{\mathrm{A}}(x, x') = \frac{1}{\pi(x)} \sum_{x'} \pi(x') q_u^{\mathrm{D}}(x', x) = \beta_u .$$

Hence, Lemma 2.11 implies that the class u arrival process in $X(-t)$ is Poisson with rate β_u. However, from the definition of $\tilde{q}_u^{\mathrm{A}}$ and the detailed Kelly lemma (Lemma 2.23), a class u arrival in the reversed process $X(-t)$ corresponds to a class u departure in the process $X(t)$. Thus we obtain condition (ii) of the second definition. Since a Poisson process reversed in time is Poisson with the same rate, we have (b). Finally, condition (ii) implies that the arrival epochs in the reversed process $X(-t)$ are generated at a constant rate independent of the current state, which gives (2.3). □

The quasi-reversibility of Example 3.6 can be extended to the case of arbitrary service batch size distributions. The following theorem shows that the queue is quasi-reversible even with arbitrarily distributed batch arrivals, as long as the modification of Section 2.7 is made.

Theorem 3.9 *Consider the $M^X/M^Y/1$ queue with batch arrivals and batch services of Section 2.7. With the additional arrivals given by (7.4), the queue is quasi-reversible with respect to batch arrivals and full batch departures of any batch size. Recall that a departing batch is full if its size is equal to the requested service batch size.*

PROOF. Define a class ℓ arrival (departure) as a batch arrival (full batch departure) of size ℓ, $\ell = 1, 2, \ldots$. Then the set T of the classes of arrivals and departures is

$$T = \{1, 2, \ldots\}.$$

The modified queue is described by the following thinned transition rate functions:

$$q_\ell^{\mathrm{A}}(n, n+\ell) = \begin{cases} \lambda a(\ell) + \lambda^* a^*(\ell), & n = 0, \ell \geq 1, \\ \lambda a(\ell), & n \geq 1, \ell \geq 1, \end{cases}$$

$$q_\ell^{\mathrm{D}}(n, n-\ell) = \mu b(\ell), \qquad n \geq \ell \geq 1,$$

$$q^{\mathrm{I}}(n, 0) = \mu \bar{b}(n+1), \qquad n \geq 1.$$

From Theorem 2.33, it follows that the stationary distribution is

$$\pi(n) = (1-\rho)\rho^n, \qquad n \geq 0,$$

where ρ is determined by (7.2). Clearly, the first condition (2.2) of quasi-reversibility is satisfied. The second condition (2.3) is verified as follows:

$$\begin{aligned}\sum_{n'=0}^{\infty} \pi(n') q_\ell^{\mathrm{D}}(n', n) &= \pi(n+\ell) q_\ell^{\mathrm{D}}(n+\ell, n) \\ &= \pi(n)\mu b(\ell)\rho^\ell.\end{aligned}$$

Thus the queue is quasi-reversible with the departure rate of class ℓ entities being $\beta_\ell = \mu b(\ell)\rho^\ell$. This completes the proof. □

If the batch sizes of the arrivals are all equal to 1, then, as discussed in Chapter 2, no additional arrivals are needed. Hence, we immediately obtain the following result.

Corollary 3.10 *If the $M/M^Y/1$ queue with batch services described in Section 2.6 satisfies the stability condition $\lambda < \mu E(Y)$, then it is quasi-reversible with respect to the arrivals of customers and the full batch departures of any batch size ℓ, $\ell = 1, 2, \ldots$.*

3.3 QUASI-REVERSIBILITY OF QUEUES WITH SIGNALS

The quasi-reversibility concept defined in the previous section is concerned with arrival and departure epochs. One prominent restriction is that an arrival cannot occur at the same time as a departure. In queueing systems with signals, however, the arrival of a signal may immediately trigger a departure. Thus the definition of quasi-reversibility is then not applicable. In this section we extend the notion of quasi-reversibility to include such simultaneous events.

As before, let q be the transition rate of the node and let it be decomposed into the components $\{q_u^{\mathrm{A}}; u \in T\}$, $\{q_u^{\mathrm{D}}; u \in T\}$ and q^{I} of (2.1). Assume that q admits the stationary distribution π. Furthermore, assume that when a class u entity arrives and induces the state of the node to change from x to x', it instantaneously triggers a class v departure with probability $f_{u,v}(x, x')$, where

$$\sum_{v \in T} f_{u,v}(x, x') \leq 1, \qquad u \in T, x, x' \in \mathcal{S}.$$

With probability

$$1 - \sum_{v \in T} f_{u,v}(x, x')$$

the class u arrival does not trigger any departure.

We refer to $f_{u,v}(x, x')$ as the *triggering probability*. Note that when $f_{u,v} \equiv 0$ for all $u, v \in T$, the system reduces to the queue of the previous section without signals.

Definition 3.11 (Quasi-reversibility of queues with signals) If there exist two sets of nonnegative numbers $\{\alpha_u; u \in T\}$ and $\{\beta_u; u \in T\}$ such that

$$\sum_{x' \in \mathcal{S}} q_u^{\mathrm{A}}(x, x') = \alpha_u, \qquad x \in \mathcal{S}, u \in T, \quad (3.1)$$

$$\sum_{x' \in \mathcal{S}} \pi(x') \left(q_u^{\mathrm{D}}(x', x) + \sum_{v \in T} q_v^{\mathrm{A}}(x', x) f_{v,u}(x', x) \right) = \beta_u \pi(x), \quad x \in \mathcal{S}, u \in T, \quad (3.2)$$

then the queue with signals is said to be quasi-reversible with respect to $\{q_u^{\mathrm{A}}, f_{u,v}; u \in T, v \in T\}$, $\{q_u^{\mathrm{D}}; u \in T\}$, and q^{I}.

As in the previous section, quasi-reversibility for queues with signals implies that the arrivals of different classes of entities form independent Poisson processes, and the departures of different classes of entities, including both triggered and non-triggered departures, also form independent Poisson processes. Moreover, future arrivals and past departures are independent of the current state of the system.

In many applications, triggered and non-triggered departures belong to different classes, i.e.,

$$q_v^{\mathrm{A}}(x, x') f_{v,u}(x, x') q_u^{\mathrm{D}}(x, x') = 0, \qquad \text{for all } x, x' \text{ and } u, v \in T.$$

Let T' and T'' be the sets of the non-triggered and triggered departure classes, respectively, such that

$$T = T' \cup T'' \quad \text{and} \quad T' \cap T'' = \emptyset.$$

Then (3.2) is reduced to

$$\sum_{x' \in \mathcal{S}} \pi(x') q_u^{\mathrm{D}}(x', x) = \beta_u \pi(x), \qquad x \in \mathcal{S}, u \in T',$$

$$\sum_{x' \in \mathcal{S}} \pi(x') \sum_{v \in T'} q_v^{\mathrm{A}}(x', x) f_{v,u}(x', x) = \beta_u \pi(x), \qquad x \in \mathcal{S}, u \in T''.$$

This is equivalent to saying that both the triggered and non-triggered departure processes are independent Poisson with rate β_u for class $u \in T$.

The arrivals that trigger immediate departures are referred to as signals since they pass through a node and change its state instantaneously. The following example illustrates this.

Example 3.12 (An $M/M/1$ queue with signals) Consider an $M/M/1$ queue with two classes of arrivals, denoted by c and s. Class c refers to the regular customers, and class s refers to signals. When a signal arrives at the node, it triggers a customer to depart immediately as a class s departure, provided the queue is not empty upon its arrival. If a signal arrives at an empty queue, nothing occurs and no departure is triggered. The customer departures generated by regular service completions are still classified as class c departures. The decomposed transition rates are

$$\begin{aligned} q_c^{\mathrm{A}}(n, n+1) &= \alpha, && n \geq 0, \\ q_s^{\mathrm{A}}(n, n-1) &= \alpha^-, && n \geq 1, \\ q_s^{\mathrm{A}}(0, 0) &= \alpha^-, && \\ q_c^{\mathrm{D}}(n, n-1) &= \mu, && n \geq 1. \end{aligned}$$

All other transition rates are zero. By the triggering mechanism, we have

$$\begin{aligned} f_{c,c}(n, n') = f_{c,s}(n, n') &= 0, && n, n' \geq 0, \\ f_{s,s}(n, n-1) &= 1, && n \geq 1. \end{aligned}$$

Since the dynamics of this queue are the same as those of a regular $M/M/1$ queue with arrival rate α and service rate $\mu + \alpha^-$, its stationary distribution π is given by

$$\pi(n) = \left(1 - \frac{\alpha}{\mu + \alpha^-}\right)\left(\frac{\alpha}{\mu + \alpha^-}\right)^n, \qquad n \geq 0.$$

If we set

$$\beta = \frac{\alpha\mu}{\mu + \alpha^-}, \qquad \beta^- = \frac{\alpha\alpha^-}{\mu + \alpha^-},$$

then this system is quasi-reversible with departure rates β and β^-, since

$$\sum_{n'} q_c^{\mathrm{A}}(n, n') = \alpha, \qquad n \geq 0,$$

$$\sum_{n'} q_s^{\mathrm{A}}(n, n') = \alpha^-, \qquad n \geq 0,$$

$$\sum_{n'} \pi(n')\left(q_c^{\mathrm{D}}(n', n) + \sum_{u=c,s} q_u^{\mathrm{A}}(n', n) f_{u,c}(n', n)\right)$$
$$= \sum_{n'} \pi(n') q_c^{\mathrm{D}}(n', n) = \beta\pi(n), \qquad n \geq 0,$$

$$\sum_{n'} \pi(n')\left(q_s^{\mathrm{D}}(n', n) + \sum_{u=c,s} q_u^{\mathrm{A}}(n', n) f_{u,s}(n', n)\right)$$
$$= \sum_{n'} \pi(n') q_s^{\mathrm{A}}(n', n) f_{s,s}(n', n) = \beta^-\pi(n), \qquad n \geq 0.$$

This is a very simple system, but many queueing networks with negative signals are generated by this model, as we will see in subsequent chapters. □

In the above example, triggered and non-triggered departures are distinguished by classes c and s. The next example allows them to be of the same class. This example also shows that a queue may be quasi-reversible with respect to one classification, but not with respect to another.

Example 3.13 (The $M/M(n)/1$ queue with signals) Consider the same model as in Example 3.12 except that now the service rates are state dependent, i.e., the $M/M(n)/1$ queue. Again there are two classes of arrivals $T = \{c, s\}$, with respective arrival rates α and α^-. When there are n customers in the system, the service rate is $\mu(n)$. As in the previous examples, the arrival of a class c entity increases the number of customers by 1, and the arrival of a class s entity triggers a customer to depart immediately, provided the queue is not empty. If we classify a regular service completion as a class c departure, and a class s triggered departure as a class s departure, then the decomposed transition rates and the triggering probabilities are the same as those in Example 3.5 except that $q_c^{\mathrm{D}}(n, n-1)$ is now

$$q_c^{\mathrm{D}}(n, n-1) = \mu(n), \qquad n = 1, 2, \ldots.$$

Since the transition rates of the queue are

$$q(n, n+1) = \alpha, \qquad n \geq 0,$$
$$q(n, n-1) = \mu(n) + \alpha^-, \qquad n \geq 1,$$

the queue is equivalent to a birth and death process with stationary distribution

$$\pi(n) = \pi(0)\frac{\alpha^n}{\prod_{\ell=1}^{n}(\alpha^- + \mu(\ell))}, \qquad n = 0, 1, \ldots,$$

with

$$\pi(0) = \left(1 + \sum_{n=1}^{\infty} \frac{\alpha^n}{\prod_{\ell=1}^{n}(\alpha^- + \mu(\ell))}\right)^{-1}.$$

According to such a classification, the queue is not quasi-reversible since

$$\begin{aligned}\sum_{n'} \pi(n')\left(q_s^{\mathrm{D}}(n', n) + \sum_{u=c,s} q_u^{\mathrm{A}}(n', n) f_{u,s}(n', n)\right) &= \pi(n+1) q_s^{\mathrm{A}}(n+1, n)\\ &= \frac{\alpha\alpha^-}{\alpha^- + \mu(n+1)}\pi(n).\end{aligned}$$

So we cannot find any nonnegative number β^- such that (3.2) is satisfied for all $n \geq 0$, unless $\mu(n)$ is independent of n, i.e., the $M/M/1$ queue.

Let us now classify the departures triggered by signals also as class c, i.e.,

$$f_{s,c}(n, n-1) = 1, \qquad n \geq 1.$$

So the departures from the queue (triggered as well as non-triggered) are all of the same class. In this case

$$\begin{aligned}\sum_{n'} \pi(n')\left(q_c^{\mathrm{D}}(n', n) + \sum_{u=c,s} q_u^{\mathrm{A}}(n', n) f_{u,c}(n', n)\right) &= \pi(n+1)(\mu(n+1) + \alpha^-)\\ &= \alpha\pi(n),\\ \sum_{n'} \pi(n')\left(q_s^{\mathrm{D}}(n', n) + \sum_{u=c,s} q_u^{\mathrm{A}}(n', n) f_{u,s}(n', n)\right) &= 0 \times \pi(n) = 0.\end{aligned}$$

Hence (3.2) is satisfied with $\beta = \alpha$ and $\beta^- = 0$. This shows that the queue is now quasi-reversible. □

In the previous two examples all the arrivals and departures are taken into account. The following is an example where the system is quasi-reversible with respect to some of the departures, but is not quasi-reversible with respect to all departures. This example runs in parallel with Example 3.6.

Example 3.14 (An $M/M/1$ queue with batch services and signals) Consider the $M/M/1$ queue discussed in Example 3.6, except that now there are also Poisson arrivals of signals with rate α^-, denoted by s. The service mechanism is the same as that of Example 3.6. Assume that when a signal arrives, two customers are induced to move immediately, provided that there are at least two customers present. In the case when there is only one customer present when the signal arrives, the system is emptied out. Signals that arrive at an empty system will, as in previous examples, disappear. The transition rates for this model are

$$\begin{aligned} q(n, n+1) &= \alpha, && n = 0, 1, \ldots, \\ q(n, n-2) &= \mu + \alpha^-, && n = 2, 3, \ldots, \\ q(1, 0) &= \mu + \alpha^-. \end{aligned}$$

Note that these transition rates are identical to those in Example 3.6 provided μ in that example is replaced by $\mu + \alpha^-$. Hence, the stationary probabilities for these equations are

$$\pi(n) = (1 - \rho)\rho^n, \qquad n \geq 0,$$

where

$$\rho = \frac{\sqrt{(\mu + \alpha^-)^2 + 4\alpha(\mu + \alpha^-)} - (\mu + \alpha^-)}{2(\mu + \alpha^-)}.$$

If we include all the departure epochs, with regular service completions being class c departures, and triggered departures being class s departures, then the transition rates are decomposed as follows:

$$\begin{aligned} q_c^{\mathrm{A}}(n, n+1) &= \alpha, && n \geq 1, \\ q_c^{\mathrm{D}}(n, n-2) &= \mu, && n \geq 2, \\ q_c^{\mathrm{D}}(1, 0) &= \mu, \\ q_s^{\mathrm{A}}(n, n-2) &= \alpha^-, && n \geq 2, \\ q_s^{\mathrm{A}}(1, 0) &= q_s^{\mathrm{A}}(0, 0) = \alpha^-. \end{aligned}$$

The triggering probabilities are

$$\begin{aligned} f_{s,s}(n, n-2) &= 1, && n \geq 2, \\ f_{s,s}(1, 0) &= 1. \end{aligned}$$

Condition (3.1) is easily verified for $u = c, s$. With regard to condition (3.2), note that for $u = c$,

$$\sum_{n'} \pi(n') \left(q_c^{\mathrm{D}}(n', n) + \sum_{u=c,s} q_u^{\mathrm{A}}(n', n) f_{u,c}(n', n) \right) = \pi(n+2)\mu$$
$$= \pi(n)\rho^2\mu, \qquad n \geq 1;$$

and

$$\sum_{n'} \pi(n') \left(q_c^{\mathrm{D}}(n', 0) + \sum_{u=c,s} q_u^{\mathrm{A}}(n', 0) f_{u,c}(n', 0) \right) = \pi(1)\mu + \pi(2)\mu$$
$$= \pi(0)(\rho + \rho^2)\mu.$$

This shows that (3.2) is not satisfied for $u = c$ since the left-hand side of (3.2) is not $\pi(x)$ multiplied by a constant. Similarly, we can verify that (3.2) is not satisfied for $u = s$. Therefore the node is not quasi-reversible.

If we only consider part of the departure process, then the system becomes quasi-reversible. Let us only count the departures of customers in batches of two, and classify the departures of single customers, i.e., the state transitions from 1 to 0, as internal transitions. So q_c^{D}, q^{I}, and $f_{s,s}$ are now

$$q_c^{\mathrm{D}}(1, 0) = 0,$$
$$q^{\mathrm{I}}(1, 0) = \mu + \alpha^-,$$
$$f_{s,s}(1, 0) = 0.$$

Since

$$\sum_{n'} \pi(n') \Big(q_c^{\mathrm{D}}(n', n) + \sum_{u=c,s} q_u^{\mathrm{A}}(n', n) f_{u,c}(n', n) \Big) = \pi(n+2)\mu$$
$$= \pi(n)\rho^2\mu, \qquad n \geq 0,$$

condition (3.2) is satisfied for $u = c$ and for all $n \geq 0$. For $u = s$,

$$\sum_{n'} \pi(n') \Big(q_s^{\mathrm{D}}(n', n) + \sum_{u=c,s} q_u^{\mathrm{A}}(n', n) f_{u,s}(n', n) \Big)$$
$$= \pi(n+2)\alpha^- f_{s,s}(n+2, n)$$
$$= \pi(n)\rho^2\alpha^-, \qquad n \geq 0.$$

Thus condition (3.2) is satisfied with departure rates

$$\beta = \rho^2 \mu, \qquad \beta^- = \rho^2 \alpha^-,$$

and the queue is quasi-reversible. □

Example 3.15 (A finite buffer queue with blocking) Consider a finite buffer queue $M/M(n)/1/K$ with buffer size K, including the customer in service. Customers arrive according to a Poisson process with rate α, and when n customers are present, the service rate is $\mu(n)$. The state space is $\{0, 1, \ldots, K\}$. Let the customers be denoted by c.

If we only include regular service completions as departures, then the queue is characterized by

$$\begin{aligned} q_c^{\mathrm{A}}(n, n+1) &= \alpha, & 0 \le n < K, \\ q_c^{\mathrm{A}}(K, K) &= \alpha, & \\ q_c^{\mathrm{D}}(n, n-1) &= \mu(n), & n \ge 1. \end{aligned}$$

This is a birth and death process with stationary distribution

$$\pi(n) = \pi(0) \frac{\alpha^n}{\prod_{\ell=1}^n \mu(\ell)}, \qquad n = 0, 1, \ldots, K,$$

where

$$\pi(0) = \left(1 + \sum_{n=1}^{K} \frac{\alpha^n}{\prod_{\ell=1}^n \mu(\ell)}\right)^{-1}.$$

Since

$$\begin{aligned} \sum_{n'} \pi(n') q_c^{\mathrm{D}}(n', n) &= \pi(n+1)\mu(n+1) \\ &= \alpha \pi(n), \qquad n = 0, 1, \ldots, K-1, \qquad (3.3) \\ \sum_{n'} \pi(n') q_c^{\mathrm{D}}(n', K) &= 0, \end{aligned}$$

the queue is not quasi-reversible.

If we consider both service completions and blocked customers as departures, then the blocked customers can be considered as triggered departures, i.e.,

$$f_{c,c}(K, K) = 1,$$

and all other triggering probabilities are 0. Now we have, in addition to (3.3), for $n = K$,

$$\sum_{n'} \pi(n')\Big(q_c^{\mathrm{D}}(n', K) + q_c^{\mathrm{A}}(n', K) f_{c,c}(n', K)\Big) = \alpha\pi(K).$$

Thus the queue is quasi-reversible with departure rate α. □

This example (and Exercise 3.4) constitutes a building block for queueing networks with finite buffers and blocking.

3.4 REFERENCE NOTES

Burke (1956) is apparently the first to note the Poisson departure process in a queueing system, see also Burke (1968) and Reich (1957, 1963). The notion of quasi-reversibility was first introduced by Muntz (1972), and it was further developed by Kelly (1975, 1976); see also Walrand and Varaiya (1980), Whittle (1986), Serfozo (1989*b*) and Chao and Pinedo (1993). The earlier definitions of quasi-reversibility were given in terms of transition sets, e.g., an arrival event is defined by a set of pairs of states (x, x') such that there is an arrival if and only if a transition from x to x' occurs in the set. These definitions cannot be applied to networks with signals since there may be different types of events that result in the same transition. The decomposition approach employed in this chapter is taken from Chao and Miyazawa (1996*a*) (see also Henderson, Pearce, Taylor and Schassberger (1995)), and the quasi-reversibility of queues with triggering was first studied by Chao and Miyazawa (1996*b*), from which the content of Section 3.3 is taken.

EXERCISES

3.1 Use the arguments in the proof of Theorem 3.8 to find the time reversed process of the batch departure queue in Example 3.6.

3.2 Suppose a single server queue has a Poisson arrival process with rate λ, and infinite buffers. Customers are served one by one according to First–Come First–Served. The service times are i.i.d., and have k ($k \geq 1$) phases. Each phase is exponentially distributed with mean $1/(k\mu)$. The service of a customer is completed when all k phases are processed. This service time distribution is called the *Erlang-k* distribution. Let $L(t)$ and $J(t)$ be the number of customers in the system and the service phase the customer in service has reached, where $J(t) = 0$ if $L(t) = 0$.

(a) Explain that $(L(t), J(t))$ is a continuous time Markov chain, and find its transition rate q.

(b) Find q^{A}, q^{D} and q^{I} so that they represent the arrivals, the departures and the service phase transitions, respectively.

(c) Find the stationary distribution of q for $k = 2$ when $\lambda < \mu$, and show that the queue is not quasi-reversible.

3.3 Consider an $M/M/1$ queue where customers arrive according to a Poisson process with rate λ. The service times are exponentially distributed with rate μ. However, upon a service completion, the customer leaves the system with probability $1 - \bar{b}(2)/\bar{b}(1)$, where $\bar{b}(k) = \sum_{\ell=k}^{\infty} b(\ell)$, and with probability $\bar{b}(2)/\bar{b}(1)$ it returns as a class 1 negative signal, where $\{b(\ell);\ \ell \geq 1\}$ is a probability mass function. A class 1 negative signal removes one customer from the node, and joins the queue again as a class 2 negative signal with probability $\bar{b}(3)/\bar{b}(2)$, etc. In general, a class k negative signal removes one customer from the system, and returns with probability $\bar{b}(k+1)/\bar{b}(k)$. A negative signal arriving at an empty system disappears.

(a) Show that this system is equivalent to a batch service queue with batch size distribution $\{b(\ell);\ \ell \geq 1\}$.

(b) Calculate the average arrival rate α_k^- of class k negative signals assuming the stationary distribution of the queue is $\pi(n) = (1-\rho)\rho^n$, $n = 0, 1, \ldots$.

(c) Because the effect of a negative signal is the same as a regular service completion, its stationary distribution should be the same as that of an $M/M/1$ queue with arrival rate λ and service rate $\mu + \sum_{k=1}^{\infty} \alpha_k^-$. Thus its stationary distribution should be of a geometric form with

$$\rho = \frac{\lambda}{\mu + \sum_{k=1}^{\infty} \alpha_k^-}.$$

Show that this analysis yields the same result as that obtained in Section 2.6.

3.4 Consider an $M/M/1/K$ queue with a finite buffer and blocking. There are two classes of arrivals, called regular customers c, and negative signals s, and their arrival rates are α and α^-. The service rate is μ. An arrival of class c increases the number of customers in the system by 1, as long as the number of customers in the system is less than K, and a class c arrival is blocked when it arrives and finds K customers in the system. When a signal arrives at the system, it induces a customer to depart, provided the system is not empty. The arrival of a signal at an empty system has no effect. Show that the queue is quasi-reversible when the blocked customers are considered as triggered departures, with triggering probabilities

$$f_{c,c}(K, K) = \frac{\mu}{\mu + \alpha^-},$$

$$f_{c,s}(K, K) = \frac{\alpha^-}{\mu + \alpha^-}.$$

4

Networks of Quasi-Reversible Queues

In this chapter we connect N quasi-reversible nodes which may have signals and instantaneous triggerings into a queueing network with Markovian routing mechanisms. The main result is that such a network has a product form solution, i.e., the stationary distribution of the network factorizes into the product of the marginal distributions of the individual nodes. To illustrate the essence of this result, we present in Section 4.1 an intuitive approach to quasi-reversibility. In Section 4.2 we describe a general network process without signals, and in Section 4.3 we show how the reversed process approach can be used to prove product form results for this class of networks. These results are extended to networks with signals in Section 4.4. Finally, the results of Section 4.4 are used in Section 4.5 to analyze a network with positive and negative signals.

4.1 AN INTUITIVE APPROACH TO QUASI-REVERSIBILITY

In this section we use intuitive arguments to illustrate how product form results of queueing networks can be obtained as a consequence of the quasi-reversibility of the individual nodes. The goal is to provide some insights into this important concept. This is done via two simple examples.

Consider a tandem queue with two nodes in series. Customers and signals arrive at node 1 according to independent Poisson processes with rates λ and λ^-, respectively, and the service times at the two nodes are exponentially distributed with rates μ_1 and μ_2. A customer, after completing service at node 1, goes to node 2, and after being served at node 2, leaves the network. The arrival of a signal at node 1 induces the customer that is being served, if there is one, to depart for node

2. If, however, a signal arrives at node 1 when it is empty, then nothing happens. A customer induced to leave node 1 joins node 2 as a signal and removes one customer from node 2, if at least one is present, before leaving the network. Again, when a signal arrives at an empty node 2, nothing happens. As a result, the arrival of a signal at node 1 may simultaneously remove one customer from node 1 and one customer from node 2. See Figure 4.1.

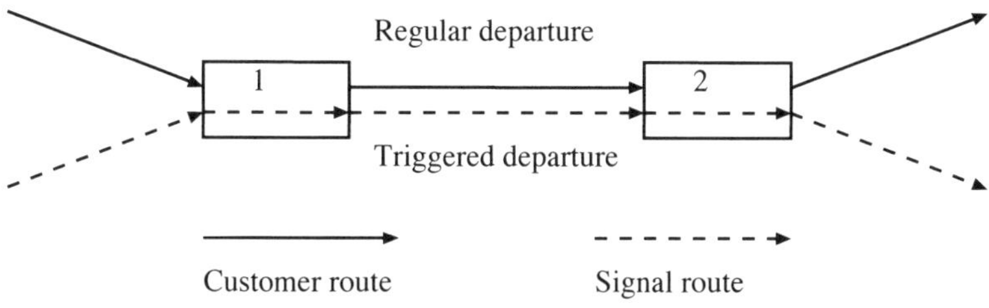

Figure 4.1 Triggered departure

Let $X_j(t)$ be the number of customers at node j at time t for $j = 1, 2$, and let

$$\boldsymbol{X}(t) = (X_1(t), X_2(t))$$

denote the state of the tandem queue. Assume

$$\lambda < \lambda^- + \mu_1.$$

Since node 1 behaves like an $M/M/1$ queue with arrival rate λ and service rate $\mu_1 + \lambda^-$, $X_1(t)$ has the stationary distribution π_1:

$$\pi_1(n_1) = (1 - \rho_1)\rho_1^{n_1}, \qquad n_1 = 0, 1, \ldots,$$

where

$$\rho_1 = \frac{\lambda}{\mu_1 + \lambda^-}.$$

Since

$$\pi_1(n_1 + 1)\mu_1 = \frac{\lambda\mu_1}{\mu_1 + \lambda^-}(1 - \rho_1)\rho_1^{n_1} = \frac{\lambda\mu_1}{\mu_1 + \lambda^-}\pi_1(n_1),$$
$$\pi_1(n_1 + 1)\lambda^- = \frac{\lambda\lambda^-}{\mu_1 + \lambda^-}(1 - \rho_1)\rho_1^{n_1} = \frac{\lambda\lambda^-}{\mu_1 + \lambda^-}\pi_1(n_1),$$

node 1 is a quasi-reversible queue with signals. Thus the departure processes of customers and signals from node 1 are independent Poisson with rates $\lambda\mu_1/(\mu_1 + \lambda^-)$ and $\lambda\lambda^-/(\mu_1 + \lambda^-)$. They become the arrivals of customers and signals, respectively, at node 2. This shows that node 2 also behaves as an $M/M/1$ queue with service rate μ_2, customer arrival rate $\lambda\mu_1/(\mu_1+\lambda^-)$, and signal arrival rate $\lambda\lambda^-/(\mu_1+\lambda^-)$. Thus if

$$\frac{\lambda\mu_1}{\mu_1 + \lambda^-} < \frac{\lambda\lambda^-}{\mu_1 + \lambda^-} + \mu_2$$

is satisfied, then node 2 has the stationary distribution π_2 given by

$$\pi_2(n_2) = (1 - \rho_2)\rho_2^{n_2}, \qquad n_2 = 0, 1, \ldots,$$

where

$$\rho_2 = \frac{\lambda\mu_1}{\lambda^-\lambda + \mu_2(\lambda^- + \mu_1)}.$$

Furthermore, because of the quasi-reversibility of node 1, the past departures of customers and signals from node 1 prior to time t are independent of the current state of the node, $X_1(t)$. On the other hand, the state of node 2, i.e., $X_2(t)$, only depends on the departure processes of customers and signals from node 1 up to time t, thus $X_1(t)$ and $X_2(t)$ are independent. As a result, the joint distribution π for $\boldsymbol{X}(t)$ is given by

$$\pi(n_1, n_2) = (1 - \rho_1)(1 - \rho_2)\rho_1^{n_1}\rho_2^{n_2}, \qquad n_1 \geq 0,\ n_2 \geq 0.$$

This shows that the stationary distribution of the tandem network is the product of the marginal distributions of the individual nodes.

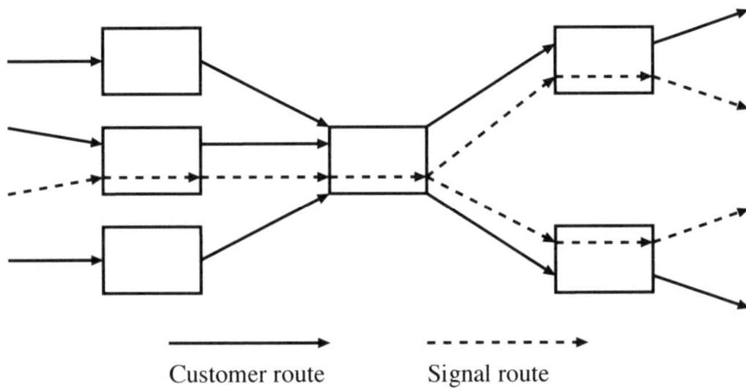

Figure 4.2 A network with feedforward routing

It is not hard to see that this argument can be applied to tandem queues with more than two nodes in series. Actually, it easily extends to more general queueing networks with feedforward routing. See Figure 4.2.

However, the argument above fails when there is feedback. In order to see this, modify the tandem queue by adding a feedback loop from node 2 to node 1, and disregard all signals. That is, customers arrive at node 1 according to a Poisson process with rate λ. A customer, after being served at node 1, proceeds to node 2, and after being served at node 2, it is fed back to node 1 with probability p and it leaves the network with probability $q = 1 - p$. See Figure 4.3.

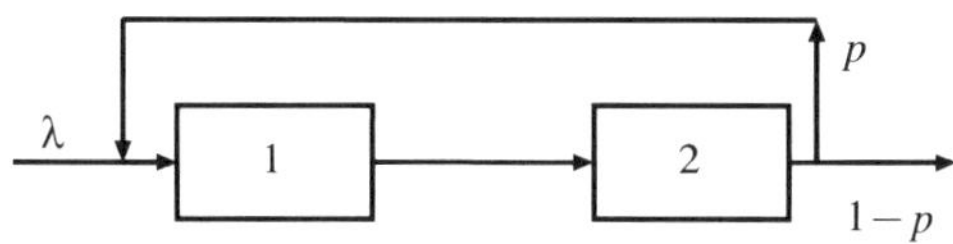

Figure 4.3 A tandem queue with feedback

Each node in isolation is an $M/M/1$ queue, and is thus quasi-reversible. However, the argument used above breaks down because the customer flows in the network are no longer Poisson. To see this, consider the case of light traffic with μ_1 and μ_2 large (i.e., the mean processing times are small). Assume that the process is stationary, and at time 0, one customer arrives at node 2. This information indicates that at time 0 node 2 is not empty. Hence with probability p this customer, after completing service at node 2, joins node 1, and after completing service at node 1 it joins node 2. However, since μ_1 and μ_2 are large, this customer will arrive at node 2 again in a very short time. This implies that an arrival at node 2 will result in a cluster of arrivals at node 2; a typical sample path of the arrival epochs at node 2 is like the one presented in Figure 4.4. This implies that the information that there is an arrival at node 2 is not independent of the future arrivals at node 2. This violates the properties of a Poisson process and the flow process from node 1 to node 2 is therefore not Poisson. One can similarly argue that the overall arrival process at node 1 is not Poisson either.

Figure 4.4 A sample path of a non-Poisson process

In order to show that this network still has a product form solution, we modify the network by introducing a delay of d time units for each customer that goes from one node to another, including those arriving at node 1 from the outside, as shown in Figure 4.5.

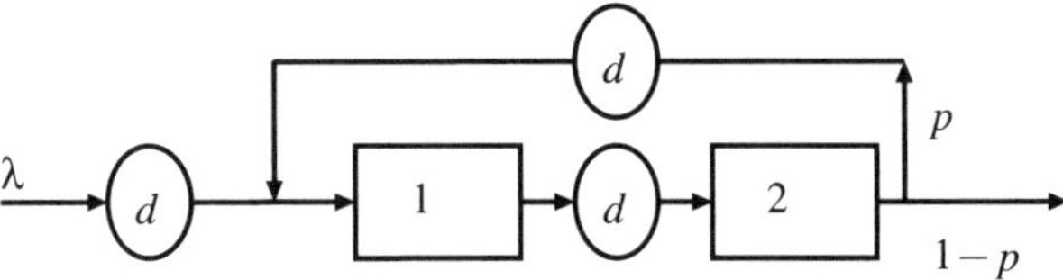

Figure 4.5 Delay lines

Assume that at time 0, $(X_1(0), X_2(0))$ has the distribution

$$\pi(n_1, n_2) = \prod_{j=1}^{2} \left(1 - \frac{\alpha_j}{\mu_j}\right) \left(\frac{\alpha_j}{\mu_j}\right)^{n_j}, \tag{1.1}$$

where

$$\alpha_1 = \alpha_2 = \alpha = \lambda/q.$$

Because of the delay d, the customers from the outside and the customers from node 1 (2) to node 2 (1) will not arrive before time d. To initiate the process, we assume that in the initial time interval $[0, d]$, there is a Poisson arrival process at each of the two queues with rate α, and these two processes are independent. That is,

(a) at the beginning of the interval $[0, d]$ the distribution of $(X_1(0), X_2(0))$ is given by (1.1), and

(b) in the time interval $[0, d]$ the arrivals at the two nodes are independent Poisson with rates $(\alpha_1, \alpha_2) = (\alpha, \alpha)$.

In what follows we show that (a) and (b) imply that $(X_1(t), X_2(t))$ has distribution (1.1) for all $t \geq 0$.

From the construction above, the two nodes are independent of each other during the time interval $[0, d]$. Since both nodes started in a stationary state, the probability of having n_j customers at node j at time $t \in [0, d]$ is also

$$\left(1 - \frac{\alpha_j}{\mu_j}\right) \left(\frac{\alpha_j}{\mu_j}\right)^{n_j},$$

and the joint probability in $[0, d]$ is

$$P(X_1(t) = n_1, X_2(t) = n_2) = \prod_{j=1}^{2} \left(1 - \frac{\alpha_j}{\mu_j}\right)\left(\frac{\alpha_j}{\mu_j}\right)^{n_j}, \qquad 0 \le t \le d.$$

Because of the quasi-reversibility of the two nodes, the departure processes from the two nodes during $[0, d]$ are independent Poisson with rate α, and the departure processes in $[0, d]$ are independent of the states of the two nodes at time d, i.e., $(X_1(d), X_2(d))$. The departures from node 1 in $[0, d]$ arrive at node 2 during $[d, 2d]$. Hence the arrival process at node 2 during $[d, 2d]$ is also Poisson with rate α, and it is independent of $(X_1(d), X_2(d))$.

Similarly, the departure process from node 2 during $[0, d]$ is Poisson, and each departure is fed back to node 1 with probability p. Hence the arrival process from node 2 to node 1, which arrives in $[d, 2d]$ because of the delay, is Poisson with rate αp, and it is independent of $(X_1(d), X_2(d))$ because of the quasi-reversibility. Since there is also a Poisson arrival process with rate λ from the outside in $[0, d]$, which arrive at node 1 in $[d, 2d]$ because of the delay, the composite arrival process at node 1 in $[d, 2d]$ is Poisson with rate

$$\lambda + \alpha p = \alpha,$$

and this arrival process in $[d, 2d]$ is independent of $(X_1(d), X_2(d))$. Moreover, the arrival processes at the two nodes in $[d, 2d]$ are also independent of each other. Summarizing:

(a′) At the beginning of the interval $[d, 2d]$ the distribution of $(X_1(d), X_2(d))$ is given by (1.1).

(b′) The arrivals at the two nodes in the time interval $[d, 2d]$ are independent Poisson with rates $(\alpha_1, \alpha_2) = (\alpha, \alpha)$.

The situation during the interval $[d, 2d]$ is therefore identical to the situation during the interval $[0, d]$. So the argument can be repeated and continued, and as a result, we conclude that the process $(X_1(t), X_2(t))$ has distribution (1.1) for all $t \ge 0$.

This proves that in the presence of a delay d, the delay network has distribution π for all $d > 0$. We can show that the result is continuous in the sense that the distribution of the delay network converges to that of the original network when d goes to zero. This then proves that the original network also has the product form stationary distribution π.

Remark 4.1 Properties (b) and (b′) do not imply that the arrival processes at nodes 1 and 2 are Poisson, although they are Poisson for a time interval of length d. For instance, the departures from node 2 in $[d, 2d]$, which depend on the arrivals at node 2 in $[d, 2d]$, become with probability p the arrivals at node 1 in $[2d, 3d]$. Thus the arrivals at node 1 in $[2d, 3d]$ depend on the arrivals at node 2 in $[d, 2d]$. On the other hand, a similar argument shows that the arrivals at node 2 in $[d, 2d]$ depend on the arrivals at node 1 in $[0, d]$. This implies that the arrivals at node 1 in $[2d, 3d]$ depend on the arrivals at node 1 in $[0, d]$, which shows that the arrival process at node 1 is not Poisson. Similarly, it can be shown that the arrival process at node 2 is not Poisson either.

These intuitive arguments suggest that a network of quasi-reversible nodes with general entity classes and arbitrary Markovian routing mechanisms has a stationary distribution of product form. This will be formally proved in Sections 4.3 and 4.4.

4.2 CONNECTING NODES INTO NETWORKS

We consider a queueing network with an arbitrary Markovian routing mechanism and multiple classes of entities. As discussed in Chapter 3, the arrival effects of entities on the nodes can be quite general. For instance, the arrival of an entity may decrease the number of customers or trigger other actions before instantaneously moving to another node. In this section we consider a network without signals, i.e., an arrival does not trigger any instantaneous departures. This enables us to give an explicit expression for the network transition rates. The model in this section forms a basis for the network with signals that is discussed in Section 4.4. However, when signals are present, the model becomes more involved, and a mathematical expression for the network transition rates is complicated without the use of matrix operators. For this reason, a mathematical expression for the transition rates of networks with signals is deferred until Chapter 10.

Suppose the network has N nodes. Each node represents a single service station, or a cluster of stations (subnetwork). In addition to these nodes we have node 0, which represents the outside world. In this chapter, the state space $\mathcal{S}_0$ of node 0 is the singleton, i.e., $\mathcal{S}_0 = \{0\}$. Node 0 is a Poisson source, i.e., exogenous entities arrive at the network according to a Poisson process. In Chapter 11 this assumption will be relaxed. Even though our main concern is an open network, the arguments can be applied to closed networks as well by simply removing node 0. However, throughout this chapter we keep node 0 for consistency. Let T_j denote the class of arrival and departure entities at node j, $j = 0, 1, \ldots, N$. As discussed in Chapter 3, a common set T_j is used for arrival classes and departure classes, even though they may be different. In the case when they are different we simply let T_j be the union of the arrival and departure classes. For instance, in Example 3.1, $T_j = \{c, c^-\}$, even though the arrival classes are $\{c, c^-\}$ and the departure class is $\{c\}$.

Let x_j be the state of node j with state space $\mathcal{S}_j$. For node 0, apparently $x_0 \equiv 0$. When node j contains a single station, x_j may represent, for instance, the number of each class of customers as well as their positions in the queue. Since node j may also be a subnetwork, x_j can be more general, e.g., it may represent the number of each class of customers present at each station as well as their position in the queue at their stations within the node. It may also include the remaining service times at the node when the service times are not exponentially distributed. Furthermore, since each node may be a subnetwork, there may be internal transitions within x_j, e.g., customer movements between different stations within the same node, or between positions in the same station of node j.

To describe the system dynamics of a queueing network, we need to specify two types of information: node (or local) information, i.e., the manner in which each node operates and reacts to arrivals from other nodes; and inter-node (or global) information, i.e., the mechanism in which the nodes are connected.

With regard to the node information, we first note that the arrival process at each node of the network is not known before the network is put together. Therefore the arrival transition rate of each node, i.e., the q_j^{A} of Chapter 3, is not known, nor is it needed for characterizing the queueing network. What we do need to know is what will happen at the node when an arrival occurs. Thus, for each node we need to specify the following information.

(i) Arrival effects: the rules according to which the node changes state with the arrival of an entity.

(ii) Departure transition rates: the rate at which the state of the node changes, possibly affecting the states of other nodes as well.

(iii) Internal transition rates: the rate at which the state of a node changes without affecting the states of other nodes and not due to an arrival or a departure.

For these reasons, we specify each node by a transition *probability* function that describes the changes in the state with arrivals, and transition *rate* functions that describe changes in the state due to departures and internal transitions. Thus, for node j and an entity of class u, we introduce the functions p_{ju}^{A}, q_{ju}^{D} and q_j^{I} on the state space $\mathcal{S}_j$:

$p_{ju}^{\mathrm{A}}(x_j, x_j') =$ the probability that a class u arrival at node j changes the state from x_j to x_j', where it is assumed that

$$\sum_{x_j' \in \mathcal{S}_j} p_{ju}^{\mathrm{A}}(x_j, x_j') = 1, \qquad x_j \in \mathcal{S}_j;$$

$q_{ju}^{\mathrm{D}}(x_j, x_j') =$ the rate at which class u departures change the state of node j from x_j to x_j';

$q_j^{\text{I}}(x_j, x_j')$ = the rate at which internal transitions change the state of node j from x_j to x_j'.

For node 0, we set $p_{0,u}^{\text{A}}(0,0) = 1$, $q_{0,u}^{\text{D}}(0,0) = \beta_{0u}$, and $q_j^{\text{I}}(0,0) = 0$. This implies that exogenous class u entities, i.e., class u departures from node 0, arrive at the network from the outside according to a Poisson process with rate β_{0u}. That is, node 0 is the Poisson source. Note that $p_{ju}^{\text{A}}(x_j, x_j)$ may be positive, i.e., an arrival may, with a positive probability, not cause a change of state. We refer to p_{ju}^{A} as the *arrival effect function.*

In Chapter 3 we described each queue by the three rates q_u^{A}, q_u^{D} and q^{I}. If a queue in the network is initially characterized by q_u^{A}, q_u^{D} and q^{I}, then the arrival effect function may be defined as

$$p_u^{\text{A}}(x, x') = \frac{q_u^{\text{A}}(x, x')}{\sum_y q_u^{\text{A}}(x, y)}, \tag{2.1}$$

and q_u^{D} and q^{I} are the departure and internal transition functions. However, unless a node is an isolated queue, we assume that it is characterized by p_u^{A}, q_u^{D} and q^{I} because, as discussed earlier, the arrival process at a node of a network depends on the structure of the entire network.

Example 4.2 (A queue with negative customers) Assume that node j of the network is a queue with negative customers, i.e., Example 3.3. It has two classes of arrivals and a single class of departures, $T_j = \{c, c^-\}$, and is characterized by the following arrival and departure functions:

$$p_{jc}^{\text{A}}(n_j, n_j') = \begin{cases} 1, & n_j' = n_j + 1, n_j \geq 0, \\ 0, & \text{otherwise,} \end{cases}$$

$$p_{jc^-}^{\text{A}}(n_j, n_j') = \begin{cases} 1, & n_j' = n_j - 1, n_j \geq 1, \\ 0, & \text{otherwise,} \end{cases}$$

$$p_{jc^-}^{\text{A}}(0, 0) = 1,$$

$$q_{jc}^{\text{D}}(n_j, n_j') = \begin{cases} \mu_j, & n_j' = n_j - 1, n_j \geq 1, \\ 0, & \text{otherwise.} \end{cases}$$

There are no class c^- departures and there are no internal transitions, so

$$q_{jc^-}^{\text{D}}(n_j, n_j') = q_j^{\text{I}}(n_j, n_j') = 0, \qquad n_j, n_j' \geq 0.$$

Thus node j has a single server with service rate μ_j. When a regular customer arrives, the number of customers in the node increases by 1, and when a negative customer arrives, the number of customer decreases by 1, provided the node is not empty. When a negative customer arrives at an empty node, the state does not change. Note that the arrival process at node j, which was denoted in Chapter 3 by q^{A}_{ju}, is not given. It is characterized by p^{A}_{ju}, which describes what happens when a class u ($u = c, c^-$) entity arrives at the node when it is in state n_j. Also note that p^{A}_{ju} and q^{A}_{ju} in Example 3.3 satisfy (2.1). □

The interactions between the nodes are defined as follows. A class u departure from node j enters node k as a class v arrival with probability $r_{ju,kv}$, and an exogenous class u arrival is routed to node k as a class v arrival with probability $r_{0u,kv}$. It is assumed that

$$\sum_{k=0}^{N} \sum_{v \in T_k} r_{ju,kv} = 1, \qquad j = 0, 1, \dots, N, \ u \in T_j. \tag{2.2}$$

It is clear that class u departures from node j leave the network with probability $\sum_{v \in T_0} r_{ju,0v}$. This probability is often denoted by $r_{ju,0}$. Similarly, $r_{0u,kv}$ is often written as $r_{0,kv}$, since there is no need to distinguish between the different classes of departures from node 0. In this way, we associate the departures from one node with the arrivals at another.

Let

$$\mathcal{S} = \mathcal{S}_1 \times \mathcal{S}_2 \times \cdots \times \mathcal{S}_N$$

be the product state space, and let

$$\boldsymbol{x} = (x_1, x_2, \dots, x_N) \in \mathcal{S}$$

denote the state of the network. This network is a continuous time Markov chain with state space $\mathcal{S}$ and transition rate function q, where

$$\begin{aligned} q(\boldsymbol{x}, \boldsymbol{x}') = {} & \sum_{j=0}^{N} \sum_{k \neq j} \sum_{u \in T_j} \sum_{v \in T_k} q^{\mathrm{D}}_{ju}(x_j, x'_j)\, r_{ju,kv}\, p^{\mathrm{A}}_{kv}(x_k, x'_k)\; 1[x_\ell = x'_\ell, \ell \neq j, k] \\ & + \sum_{j=0}^{N} \sum_{u,v \in T_j} \sum_{y_j \in \mathcal{S}_j} q^{\mathrm{D}}_{ju}(x_j, y_j)\, r_{ju,jv}\, p^{\mathrm{A}}_{jv}(y_j, x'_j) 1[x_\ell = x'_\ell, \ell \neq j] \\ & + \sum_{j=0}^{N} q^{\mathrm{I}}_j(x_j, x'_j)\; 1[x_\ell = x'_\ell, \ell \neq j], \end{aligned} \tag{2.3}$$

for $\boldsymbol{x} = (x_1, x_2, \ldots, x_N) \in \mathcal{S}$ and $\boldsymbol{x}' = (x_1', x_2', \ldots, x_N') \in \mathcal{S}$. The first summation on the right-hand side of (2.3) represents the state changes due to customer transfers from one node to another, the second summation represents the feedback, and the third summation represents internal state changes. If

$$q_{ju}^{\mathrm{D}}(x_j, x_j) = p_{ju}^{\mathrm{A}}(x_j, x_j) = q_j^{\mathrm{I}}(x_j, x_j) = 0,$$

then the transition rate function (2.3) can be partitioned into disjoint sets:

$$q(\boldsymbol{x}, \boldsymbol{x}') = \begin{cases} \sum_{u \in T} \sum_{v \in T} q_{ju}^{\mathrm{D}}(x_j, x_j')\, r_{ju,kv}\, p_{kv}^{\mathrm{A}}(x_k, x_k'), & x_\ell = x_\ell' \text{ for all } \ell \neq j, k, \\ \sum_{u,v \in T_j} \sum_{y_j \in \mathcal{S}_j} q_{ju}^{\mathrm{D}}(x_j, y_j)\, r_{ju,jv}\, p_{jv}^{\mathrm{A}}(y_j, x_j') + q_j^{\mathrm{I}}(x_j, x_j'), & \\ & x_\ell = x_\ell' \text{ for all } \ell \neq j, \\ 0, & \text{otherwise.} \end{cases}$$

This is a typical situation in a conventional queueing network such as a Jackson network. However, it is not true in general. For instance, if a departing customer transforms itself into a negative signal and there is no customer present at the node where it arrives, then the signal does not have any effect. Only the state of the node from where the customer departs changes. So, the transition caused by a signal may result in a change of state that is similar to an internal transition.

4.3 NETWORKS OF QUASI-REVERSIBLE NODES

In this section we derive the stationary distribution for the queueing network defined in the previous section. Assuming quasi-reversibility for each node in isolation, which will be defined below, we show that the stationary distribution of the network process is product form.

Consider for each node j the following auxiliary process:

$$q_j^{(\boldsymbol{\alpha}_j)}(x_j, x_j') = \sum_{u \in T_j} \left(\alpha_{ju} p_{ju}^{\mathrm{A}}(x_j, x_j') + q_{ju}^{\mathrm{D}}(x_j, x_j') \right) + q_j^{\mathrm{I}}(x_j, x_j'), \qquad (3.1)$$

for $x_j, x_j' \in \mathcal{S}_j$, $j = 0, 1, \ldots, N$. Clearly, $q_j^{(\boldsymbol{\alpha}_j)}(x_j, x_j')$ can be viewed as node j being in isolation, with class $u \in T_j$ entities arriving according to a Poisson process with rate α_{ju}. In general, p_{ju}^{A}, q_{ju}^{D}, and q_j^{I} are allowed to be functions of $\boldsymbol{\alpha}_j = \{\alpha_{ju}; u \in T_j\}$. However, this dependency on $\boldsymbol{\alpha}_j$ is, for simplicity, made implicit.

Suppose $q_j^{(\boldsymbol{\alpha}_j)}$ has a stationary distribution $\pi_j^{(\boldsymbol{\alpha}_j)}$, i.e.,

$$\pi_j^{(\boldsymbol{\alpha}_j)}(x_j) \Bigg(\sum_{u \in T_j} \Bigg(\alpha_{ju} + \sum_{x_j' \in \mathcal{S}_j} q_{ju}^{\mathrm{D}}(x_j, x_j') \Bigg) + \sum_{x_j' \in \mathcal{S}_j} q_j^{\mathrm{I}}(x_j, x_j') \Bigg)$$

$$= \sum_{u \in T_j} \sum_{x_j' \in \mathcal{S}_j} \pi_j^{(\boldsymbol{\alpha}_j)}(x_j') \Big(\alpha_{ju} p_{ju}^{\text{A}}(x_j', x_j) + q_{ju}^{\text{D}}(x_j', x_j) \Big) + \sum_{x_j' \in \mathcal{S}_j} \pi_j^{(\boldsymbol{\alpha}_j)}(x_j') q_j^{\text{I}}(x_j', x_j),$$

$$j = 0, 1, \ldots, N, \; x_j, x_j' \in \mathcal{S}_j. \tag{3.2}$$

For $j = 0$, (3.2) is obviously satisfied by

$$\pi_0^{(\boldsymbol{\alpha}_0)}(0) = 1,$$

which is independent of $\boldsymbol{\alpha}_0$. For $j \neq 0$, the $\pi_j^{(\boldsymbol{\alpha}_j)}$ is expected to be the marginal distribution of node j for some parameters $\boldsymbol{\alpha}_j = (\alpha_{ju}; u \in T_j)$. However, the exact values of $\boldsymbol{\alpha}_1, \ldots, \boldsymbol{\alpha}_N$ are not yet known. Thus, for the time being, the $\boldsymbol{\alpha}_j$ may be regarded as dummy parameters, and their values will be determined later by the traffic equations.

First, note that we always have

$$\sum_{x_j' \in S_j} \alpha_{ju} p_{ju}^{\text{A}}(x_j, x_j') = \alpha_{ju}, \qquad j = 1, \ldots, N, \; u \in T_j.$$

Hence, quasi-reversibility of $q_j^{(\boldsymbol{\alpha}_j)}$ for $j = 1, \ldots, N$ is equivalent to the existence of a set of nonnegative numbers $\{\beta_{ju}; u \in T_j\}$ such that

$$\sum_{x_j' \in \mathcal{S}} \pi_j^{(\boldsymbol{\alpha}_j)}(x_j') q_{ju}^{\text{D}}(x_j', x_j) = \beta_{ju} \pi_j^{(\boldsymbol{\alpha}_j)}(x_j), \qquad x_j \in \mathcal{S}_j, \tag{3.3}$$

for all $j = 1, 2, \ldots, N$ and $u \in T_j$. By (3.3), β_{ju} is determined by

$$\beta_{ju} = \sum_{x_j, x_j' \in \mathcal{S}} \pi_j^{(\boldsymbol{\alpha}_j)}(x_j) q_{ju}^{\text{D}}(x_j, x_j'), \qquad j = 1, \ldots, N. \tag{3.4}$$

Node j in isolation is said to be quasi-reversible with $\boldsymbol{\alpha}_j$ if (3.3) is satisfied. For convenience we often drop "in isolation" when it is clear from the context.

Consider now the network generated by connecting nodes $1, 2, \ldots, N$ through the routing probability matrix $R = \{r_{ju,kv}\}$, where node j is defined by p_{ju}^{A}, q_{ju}^{D}, and q_j^{I}. Assume that class $u \in T_0$ entities arrive at the network from the outside (node 0) at rate β_{0u}, which is given, and that each entity joins node k as a class v entity with probability $r_{0u,kv}$. Let β_{kv} be the average departure rate of class v entities from node k. The average arrival rate of class u entities at node j satisfies

$$\alpha_{ju} = \sum_{k=0}^{N} \sum_{v \in T_k} \beta_{kv} r_{kv,ju}, \qquad j = 0, 1, \ldots, N, \; u \in T_j. \tag{3.5}$$

These equations are referred to as the *traffic equations*. Note that β_{jv} is a nonlinear function of $\boldsymbol{\alpha}_j$, which is determined by (3.4). So the traffic equations are, in general, nonlinear in the α_{ju}'s. Finding solutions of (3.5) can be considered a fixed point problem concerning the vector $\boldsymbol{\alpha} = \{\alpha_{ju}; j = 0, 1, \ldots, N, u \in T_j\}$.

Theorem 4.3 *If each node j, $j = 1, \ldots, N$, of the network with transition rate $q(\boldsymbol{x}, \boldsymbol{x}')$ is in isolation quasi-reversible with $(\boldsymbol{\alpha}_0, \boldsymbol{\alpha}_1, \ldots, \boldsymbol{\alpha}_N)$ that satisfy (3.4) and (3.5), then the stationary distribution of the network is*

$$\pi(\boldsymbol{x}) = \prod_{j=1}^{N} \pi_j^{(\boldsymbol{\alpha}_j)}(x_j), \qquad \boldsymbol{x} \equiv (x_1, x_2, \ldots, x_N) \in \mathcal{S}, \tag{3.6}$$

where $\pi_j^{(\boldsymbol{\alpha}_j)}$ is the stationary distribution of $q_j^{(\boldsymbol{\alpha}_j)}$, $j = 1, \ldots, N$.

PROOF. Define the distribution π by (3.6). We apply the detailed Kelly lemma (Lemma 2.23) to verify that this π is indeed the stationary distribution of the network process q. For convenience, we drop the superscript $(\boldsymbol{\alpha}_j)$. Assume that the time reversed process corresponds to another network with a similar structure. Let

$$\begin{aligned}
\tilde{q}_{ju}^{\mathrm{D}}(x_j', x_j) &= \frac{\pi_j(x_j)\alpha_{ju} p_{ju}^{\mathrm{A}}(x_j, x_j')}{\pi_j(x_j')}, & & j = 0, 1, \ldots, N,\ u \in T_j, \\
\tilde{q}_j^{\mathrm{I}}(x_j', x_j) &= \frac{\pi_j(x_j) q_j^{\mathrm{I}}(x_j, x_j')}{\pi_j(x_j')}, & & j = 0, 1, \ldots, N, \\
\tilde{p}_{ju}^{\mathrm{A}}(x_j', x_j) &= \frac{\pi_j(x_j) q_{ju}^{\mathrm{D}}(x_j, x_j')}{\pi_j(x_j')\beta_{ju}}, & & j = 0, 1, \ldots, N,\ u \in T_j, \\
\tilde{r}_{ju,kv} &= \frac{\beta_{kv} r_{kv,ju}}{\alpha_{ju}}, & & j, k = 0, 1, \ldots, N,\ u \in T_j,\ v \in T_k,
\end{aligned}$$

where $\pi_0(0) = 1$. Note that

$$\sum_{x_j' \in \mathcal{S}_j} \tilde{p}_{ju}^{\mathrm{A}}(x_j, x_j') = 1, \qquad j = 0, 1, , \ldots, N,\ u \in T_j,$$

and

$$\sum_{k=0}^{N} \sum_{v \in T_j} \tilde{r}_{ju,kv} = 1, \qquad j = 0, 1, , \ldots, N,\ u \in T_j.$$

So we can define a transition rate $\tilde{q}$ for a queueing network process with arrival probability function $\tilde{p}_{ju}^{\mathrm{A}}$, departure rate function $\tilde{q}_{ju}^{\mathrm{D}}$, internal transition rate $\tilde{q}_j^{\mathrm{I}}$ and routing probabilities $\tilde{r}_{ju,kv}$.

The detailed Kelly lemma requires the verification of two conditions. To check the first condition, i.e.,

$$\sum_{\boldsymbol{x}'} q(\boldsymbol{x}, \boldsymbol{x}') = \sum_{\boldsymbol{x}'} \tilde{q}(\boldsymbol{x}, \boldsymbol{x}'),$$

note that the state of the network changes only when there is a departure or when there is an internal transition. Thus

$$\begin{aligned}
\sum_{\boldsymbol{x}'} \tilde{q}(\boldsymbol{x}, \boldsymbol{x}') &= \sum_{j=0}^{N} \Big[\sum_{u \in T_j} \sum_{x'_j} \tilde{q}^{\mathrm{D}}_{ju}(x_j, x'_j) + \sum_{x'_j} \tilde{q}^{\mathrm{I}}_j(x_j, x'_j) \Big] \\
&= \sum_{j=0}^{N} \Bigg[\sum_{u \in T_j} \sum_{x'_j} \frac{\pi_j(x'_j)\alpha_{ju} p^{\mathrm{A}}_{ju}(x'_j, x_j)}{\pi_j(x_j)} + \sum_{x'_j} \frac{\pi_j(x'_j) q^{\mathrm{I}}_j(x'_j, x_j)}{\pi_j(x_j)} \Bigg] \\
&= \sum_{j=0}^{N} \frac{1}{\pi_j(x_j)} \Bigg[\sum_{u \in T_j} \sum_{x'_j} \pi_j(x'_j) \Big(\alpha_{ju} p^{\mathrm{A}}_{ju}(x'_j, x_j) + q^{\mathrm{D}}_{ju}(x'_j, x_j) \Big) \\
&\qquad + \sum_{x'_j} \pi_j(x'_j) q^{\mathrm{I}}_j(x'_j, x_j) - \sum_{u \in T_j} \beta_{ju} \Bigg] \\
&= \sum_{j=0}^{N} \Big[\sum_{u \in T_j} \sum_{x'_j} q^{\mathrm{D}}_{ju}(x_j, x'_j) + \sum_{x'_j} q^{\mathrm{I}}_j(x_j, x'_j) + \sum_{u \in T_j} (\alpha_{ju} - \beta_{ju}) \Big] \\
&= \sum_{j=0}^{N} \Big[\sum_{u \in T_j} \sum_{x'_j} q^{\mathrm{D}}_{ju}(x_j, x'_j) + \sum_{x'_j} q^{\mathrm{I}}_j(x_j, x'_j) \Big] \\
&= \sum_{\boldsymbol{x}'} q(\boldsymbol{x}, \boldsymbol{x}'),
\end{aligned}$$

where the third equality follows from (3.3), the fourth equality follows from the fact that π_j is the stationary distribution of q_j, i.e., (3.2), and the last equality follows from the total balance

$$\sum_{j=0}^{N} \sum_{u \in T_j} \alpha_{ju} = \sum_{j=0}^{N} \sum_{u \in T_j} \beta_{ju}, \tag{3.7}$$

which is immediate from the traffic equation.

We next verify the second condition of the detailed Kelly lemma. We decompose $q(\boldsymbol{x}, \boldsymbol{x}')$ and $\tilde{q}(\boldsymbol{x}, \boldsymbol{x}')$ according to the types of transitions. Let $\boldsymbol{x}_j(x'_j)$ be the vector $\boldsymbol{x}$ with its jth element x_j replaced by x'_j, and denote $(\boldsymbol{x}_j(x'_j))_k(x'_k)$ by $\boldsymbol{x}_{jk}(x'_j, x'_k)$. Note that the types of transitions out of state $\boldsymbol{x}$ under q are $(\boldsymbol{x}, \boldsymbol{x}_j(x'_j))$ and $(\boldsymbol{x}, \boldsymbol{x}_{jk}(x'_j, x'_k))$. The first represents an internal transition at node j and the second a departure from node j that triggers an arrival at node k. In the latter transition, j may be equal to k, representing a feedback. When this is the case the sequence of transitions is

$$\boldsymbol{x} \xrightarrow{u} \boldsymbol{x}_j(y_j) \xrightarrow{v} \boldsymbol{x}_j(x'_j), \tag{3.8}$$

where u and v are the classes of entities that cause the corresponding transitions. Denote the transition rate of this sequence by

$$q_{ju(y_j),jv}(\boldsymbol{x}, \boldsymbol{x}') = \begin{cases} q^{\mathrm{D}}_{ju}(x_j, y_j) r_{ju,jv} p^{\mathrm{A}}_{jv}(y_j, x'_j), & \text{for } \boldsymbol{x}' = \boldsymbol{x}_j(x'_j), \\ 0, & \text{otherwise.} \end{cases}$$

Similarly, denote the transition rate of the sequence of transitions

$$\boldsymbol{x} \xrightarrow{u} \boldsymbol{x}_j(x'_j) \xrightarrow{v} \boldsymbol{x}_{jk}(x'_j, x'_k), \qquad j \neq k, \tag{3.9}$$

by

$$q_{ju,kv}(\boldsymbol{x}, \boldsymbol{x}') = \begin{cases} q^{\mathrm{D}}_{ju}(x_j, x'_j) r_{ju,kv} p^{\mathrm{A}}_{kv}(x_k, x'_k), & \text{for } \boldsymbol{x}' = \boldsymbol{x}_{jk}(x'_j, x'_k), \\ 0, & \text{otherwise.} \end{cases}$$

We define the corresponding rates in the reversed process in a similar way. Let $\tilde{q}_{ju(y_j),jv}$ denote the rate of the transitions (3.8), and, for $j \neq k$, let $\tilde{q}_{ju,kv}$ be the rate of the transitions (3.9). Then, we have

$$\begin{aligned} q(\boldsymbol{x}, \boldsymbol{x}') = \sum_{j=0}^{N} \Bigg(\sum_{u \in T_j} \Bigg(\sum_{k=0}^{N} \sum_{v \in T_k} q_{ju,kv}(\boldsymbol{x}, \boldsymbol{x}') + \sum_{u \in T_j} \sum_{y_j \in \mathcal{S}_j} q_{ju(y_j),jv}(\boldsymbol{x}, \boldsymbol{x}') \Bigg) \\ + q^{\mathrm{I}}_j(\boldsymbol{x}, \boldsymbol{x}') \Bigg), \end{aligned}$$

$$\begin{aligned} \tilde{q}(\boldsymbol{x}, \boldsymbol{x}') = \sum_{j=0}^{N} \Bigg(\sum_{u \in T_j} \Bigg(\sum_{k=0}^{N} \sum_{v \in T_k} \tilde{q}_{ju,kv}(\boldsymbol{x}, \boldsymbol{x}') + \sum_{u \in T_j} \sum_{y_j \in \mathcal{S}_j} \tilde{q}_{ju(y_j),jv}(\boldsymbol{x}, \boldsymbol{x}') \Bigg) \\ + \tilde{q}^{\mathrm{I}}_j(\boldsymbol{x}, \boldsymbol{x}') \Bigg). \end{aligned}$$

We now verify the second condition of the detailed Kelly lemma for each sequence of state transitions and for each internal transition. The latter is immediate from

$$\begin{aligned} \pi(\boldsymbol{x}_j(x'_j)) \tilde{q}^{\mathrm{I}}_j(x'_j, x_j) &= \pi(\boldsymbol{x}_j(x'_j)) \frac{\pi_j(x_j) q^{\mathrm{I}}_j(x_j, x'_j)}{\pi_j(x'_j)} \\ &= \pi(\boldsymbol{x}) q^{\mathrm{I}}_j(x_j, x'_j), \end{aligned}$$

where we have used the fact that $\pi(\boldsymbol{x})$ is the product of the $\pi_j(x_j)$. Similarly,

$$\begin{aligned} \pi(\boldsymbol{x}_j(x'_j)) \tilde{q}_{ju(y_j),jv}(\boldsymbol{x}_j(x'_j), \boldsymbol{x}) &= \pi(\boldsymbol{x}_j(x'_j)) \tilde{q}^{\mathrm{D}}_{ju}(x'_j, y_j) \tilde{r}_{ju,jv} \tilde{p}^{\mathrm{A}}_{jv}(y_j, x_j) \\ &= \pi(\boldsymbol{x}_j(x'_j)) \frac{\pi_j(y_j) \alpha_{ju} p^{\mathrm{A}}_{ju}(y_j, x'_j)}{\pi_j(x'_j)} \frac{\beta_{jv} r_{jv,ju}}{\alpha_{ju}} \frac{\pi_j(x_j) q^{\mathrm{D}}_{jv}(x_j, y_j)}{\pi_j(y_j) \beta_{jv}} \\ &= \pi(\boldsymbol{x}_j(x'_j)) \frac{\pi_j(x_j)}{\pi_j(x'_j)} q^{\mathrm{D}}_{jv}(x_j, y_j) r_{jv,ju} p^{\mathrm{A}}_{ju}(y_j, x'_j) \\ &= \pi(\boldsymbol{x}) q_{jv,ju(y_j)}(\boldsymbol{x}, \boldsymbol{x}_j(x'_j)). \end{aligned}$$

A similar argument yields, for $j \neq k$,

$$\pi(\boldsymbol{x}_{jk}(x'_j, x'_k))\tilde{q}_{ju,kv}(\boldsymbol{x}_{jk}(x'_j, x'_k), \boldsymbol{x}) = \pi(\boldsymbol{x})q_{kv,ju}(\boldsymbol{x}, \boldsymbol{x}_{kj}(x'_k, x'_j)).$$

Thus the second condition of Lemma 2.23 is also satisfied. This completes the proof of the theorem. □

Remark 4.4 Theorem 4.3 assumes that node 0, i.e., the outside of the network, has a single state. However, it can be easily seen from the proof of Theorem 4.3 that node 0 can be an arbitrary quasi-reversible node. In this case node 0 has a non-trivial stationary distribution π_0, and the product form solution (3.6) is changed to

$$\pi(\boldsymbol{x}) = b \prod_{j=0}^{N} \pi_j^{(\boldsymbol{\alpha}_j)}(x_j),$$

where b is the normalizing constant. The b may not be one since $x_0, \ldots, x_N$ may not be independently chosen. This observation can be used to consider closed queueing networks. See Example 9.3 for more details.

In many queueing systems (3.3) holds for a range of $\boldsymbol{\alpha}_j$. If this is the case, then the queue is called uniformly quasi-reversible.

Definition 4.5 (Uniform quasi-reversibility) Node j, characterized by $\{p^{\mathrm{A}}_{ju}; u \in T\}$, $\{q^{\mathrm{A}}_{ju}; u \in T\}$ and q^{I}_j, is called *uniformly quasi-reversible* if it is quasi-reversible for all $\boldsymbol{\alpha}_j$ for which the stationary distribution $\pi_j^{(\boldsymbol{\alpha}_j)}$ exists.

If node j is uniformly quasi-reversible, then the departure rate β_{ju} is well defined over the range of $\boldsymbol{\alpha}_j$ for which $\pi_j^{(\boldsymbol{\alpha}_j)}$ exists. Thus β_{ju} can be considered a function of $\boldsymbol{\alpha}_j$, and the right hand side of traffic equations (3.5) are functions of vector $(\boldsymbol{\alpha}_1, \ldots, \boldsymbol{\alpha}_N)$. This implies that uniform quasi-reversibility enables us to view $\boldsymbol{\alpha}_j$ as solutions to nonlinear equations. The remaining problem is whether the traffic equations have a solution and how they can be calculated. As we stated before, this is a fixed point problem, and it will be addressed in Chapter 11.

We refer to a network of quasi-reversible nodes as a *quasi-reversible network*. Indeed, as can be seen from the following corollary, when the entire network is viewed as a single queue, it is quasi-reversible.

Corollary 4.6 *In quasi-reversible queueing networks, the class u departure process from node j to the outside is Poisson with rate $\beta_{ju} \sum_{v \in T_0} r_{ju,0v}$. The network is quasi-reversible with respect to arrivals from the outside and departures to the outside.*

PROOF. From the proof of Theorem 4.3, the time reversed network process also represents a quasi-reversible network characterized by $\tilde{p}^{\mathrm{A}}_{ju}$, $\tilde{q}^{\mathrm{D}}_{ju}$ and $\tilde{q}^{\mathrm{I}}_{j}$, and routing probabilities $\tilde{r}_{ju,kv}$. Hence, in the time reversed network, the class u entities arrive at node j from the outside according to a Poisson process with rate

$$\begin{aligned}\sum_{v \in T_0} \alpha_{0v} \tilde{r}_{0v,ju} &= \sum_{v \in T_0} \alpha_{0v} \frac{\beta_{ju}}{\alpha_{0v}} r_{ju,0v} \\ &= \beta_{ju} \sum_{v \in T_0} r_{ju,0v} \,.\end{aligned}$$

The corollary follows from the fact that the arrivals from the outside in the time reversed process are the departures from the network to the outside in the original network. □

However, as discussed in Section 4.1, this does not imply that the departure and arrival processes at each node of the network are Poisson (see also Remark 4.1 and Exercise 4.1). Actually, it can be shown that the flow on a link is Poisson if and only if it is not part of a cycle. For instance, the flows in between any two nodes in a feedforward network are Poisson.

Example 4.7 (A network with negative customers) Consider a network with N single-server nodes. Each node has exponentially distributed service times and two classes of entities: regular customers and negative customers, i.e., the type of node discussed in Example 4.2. Regular as well as negative customers arrive at node j from the outside according to independent Poisson processes with rates λ_j and λ_j^-. Upon a service completion at node j, a customer joins node k as a regular customer with probability $r_{jc,kc}$ and as a negative customer with probability r_{jc,kc^-}, $k = 0, 1, \ldots, N$. A negative customer is a signal that removes a customer from the node. The state of the network is represented by a vector $\boldsymbol{n} = (n_1, \ldots, n_N)$, where n_j is the number of regular customers at node j, $n_j \in \mathcal{S}_j = \{0, 1, \ldots\}$.

Node j, characterized by (3.1), is subject to Poisson arrivals of regular and negative customers with rates α_j and α_j^-. From Example 3.3, it follows that the stationary distribution π_j of node j is of a geometric form, that the node is uniformly quasi-reversible, and that the π_j exists if and only if

$$\alpha_j < \mu_j + \alpha_j^- \,. \tag{3.10}$$

The departure rate from node j is

$$\beta_j = \frac{\alpha_j \mu_j}{\mu_j + \alpha_j^-}.$$

The traffic equations are

$$\alpha_j = \lambda_j + \sum_{k=1}^{N} \frac{\alpha_k \mu_k}{\mu_k + \alpha_k^-} r_{kc,jc}, \qquad j = 1, \ldots, N,$$

$$\alpha_j^- = \lambda_j^- + \sum_{k=1}^{N} \frac{\alpha_k \mu_k}{\mu_k + \alpha_k^-} r_{kc,jc^-}, \qquad j = 1, \ldots, N.$$

Thus, if the stability condition (3.10) is satisfied, then, by Theorem 4.3, the stationary distribution of the network, π, is the product of the π_j, i.e.,

$$\pi(\boldsymbol{n}) = \prod_{j=1}^{N} \left(1 - \frac{\alpha_j}{\mu_j + \alpha_j^-}\right) \left(\frac{\alpha_j}{\mu_j + \alpha_j^-}\right)^{n_j}. \qquad \square$$

4.4 NETWORKS WITH SIGNALS AND INSTANTANEOUS MOVEMENTS

We extend the results of the previous two sections to networks with signals and instantaneous movements. In these networks the arrival of an entity may immediately trigger a departure. The cascading effects of these entities may create instantaneous movements throughout the network.

Consider a network with N nodes. Each node is a quasi-reversible queue with signals as described in Section 3.3. Let $\mathcal{S}_j$ be the state space of node j, and let T_j be the set of arrival and departure entity classes, $j = 1, 2, \ldots, N$. As discussed in Section 4.2, we need to specify for each node the transition *probability* functions that describe state changes due to arrivals, and the transition *rate* functions that describe departures and internal state changes. For these, we use the same notation as in Section 4.2, i.e., p_{ju}^{A}, q_{ju}^{D} and q_j^{I}. However, since there are signals in the network, we also have to specify the probability functions with regard to the arrivals that induce departures, i.e., the *triggering probability* functions $f_{ju,v}(x_j, x_j')$. Assume that when a class u entity arrives at node j and the state changes from x_j to x_j', it simultaneously induces a class v departure with triggering probability $f_{ju,v}(x_j, x_j')$. These probabilities satisfy

$$\sum_{v \in T_j} f_{ju,v}(x_j, x_j') \le 1, \qquad u \in T_j,\ x_j, x_j' \in \mathcal{S}_j,\ j = 1, 2, \ldots, N.$$

Let $f_{0u,v} \equiv 0$ for all $u, v \in T_0$, i.e., we assume that there are no triggerings at node 0. As in Section 4.2 we allow p^{A}_{ju}, $f_{ju,v}$, q^{D}_{ju}, and q^{I}_{j} to be functions of a nonnegative vector $\boldsymbol{\alpha}_j = \{\alpha_{ju}; u \in T_j\}$, even though this dependency is made implicit for convenience. Also, $p^{\mathrm{A}}_{ju}(x_j, x_j)$ may be positive, i.e., an arrival may, with a positive probability, cause no change of state.

The dynamics of the network are described as follows. Class $u \in T_j$ entities from the outside, i.e., class u departures from node 0, arrive at the network according to a Poisson process with rate β_{0u}, and each is routed to node j as a class v arrival with probability $r_{0u,kv}$. A class u departure from node j, either triggered or non-triggered, joins node k as a class v arrival with probability $r_{ju,kv}$, $k = 0, 1, \ldots, N$, where

$$\sum_{k=0}^{N} \sum_{v \in T_k} r_{ju,kv} = 1, \qquad j = 0, 1, \ldots, N, \ u \in T_j.$$

Furthermore, whenever there is a class u arrival at node j, either from the outside or from other nodes, it causes the state of the node to change from x_j to x'_j with probability $p^{\mathrm{A}}_{ju}(x_j, x'_j)$, it also triggers a class v departure with probability $f_{ju,v}(x_j, x'_j)$, and it triggers no departure from node j with probability

$$1 - \sum_{v \in T_j} f_{ju,v}(x_j, x'_j), \qquad j = 0, 1, \ldots, N.$$

In this way, we associate the departures, both the regular departures and the triggered departures, from one node with the arrivals at another. We refer to this network as a *network with signals and instantaneous movements.*

A distinctive feature of this network is that there are simultaneous arrivals and departures at any number of nodes. For instance, if, for nodes $j_1, j_2, \ldots, j_k$ and classes $u_\ell, u'_\ell \in T_\ell$, $\ell = j_1, j_2, \ldots, j_k$,

$$\begin{aligned} &p^{\mathrm{A}}_{j_1u_1}(x_{j_1}, x'_{j_1}) f_{j_1u_1,u'_1}(x_{j_1}, x'_{j_1}) r_{j_1u'_1,j_2u_2} \times \cdots \times p^{\mathrm{A}}_{j_{k-1}u_{k-1}}(x_{j_{k-1}}, x'_{j_{k-1}}) \\ &\quad \times f_{j_{k-1}u_{k-1},u'_{k-1}}(x_{j_{k-1}}, x'_{j_{k-1}}) r_{j_{k-1}u'_{k-1},j_ku_k} p^{\mathrm{A}}_{j_ku_k}(x_{j_k}, x'_{j_k}) > 0, \end{aligned}$$

then a class u_1 arrival at node j_1 will simultaneously create arrivals at nodes j_2, j_3, $\ldots$, j_k, and change the states of these nodes to $x'_{j_1}, x'_{j_2}, \ldots, x'_{j_k}$ with a positive probability. Note that any given node may be visited several times on this route, and as a result the state of a node may change a number of times at one point in time.

We illustrate the notation using the bathroom problem discussed in Example 1.2.

Example 4.8 (Formulation of the bathroom problem) Consider the bathroom problem discussed in Example 1.2. Each bathroom has a capacity for one person, so anyone who arrives and finds the bathroom occupied has to wait in queue. Men arrive according to a Poisson process with rate λ_1, and women arrive according to a Poisson process with rate λ_2. Couples, consisting of a man and a woman, arrive according to a Poisson process with rate λ. Assume that a man uses the men's room for an exponentially distributed amount of time with rate μ_1, and a woman uses the ladies' room for an exponentially distributed amount of time with rate μ_2.

If $X_1(t)$ denotes the number of men and $X_2(t)$ the number of women in the system at time t, then the state of the system is $(X_1(t), X_2(t))$. We model this problem as a network with two nodes, one class of customers denoted by c, and one class of signals denoted by s^+. Customers arrive at the two nodes according to Poisson processes with rates λ_1 and λ_2, respectively. Signals arrive at node 1 according to a Poisson process with rate λ. The arrival of a positive signal at node 1 adds one customer at node 1, and then departs immediately for node 2. A positive signal that leaves node 1 joins node 2 as a regular customer. So $T_1 = \{c, s^+\}$ and $T_2 = \{c\}$. The two nodes are characterized as follows:

$$\begin{aligned}
p^{\text{A}}_{j,c}(n_j, n_j+1) &= 1, && j = 1, 2,\ n_j \ge 0, \\
p^{\text{A}}_{1,s^+}(n_1, n_1+1) &= 1, && n_1 \ge 0, \\
q^{\text{D}}_{j,c}(n_j, n_j-1) &= \mu_j, && j = 1, 2,\ n_j \ge 1, \\
f_{1s^+,s^+}(n_1, n_1+1) &= 1, && n_1 \ge 0.
\end{aligned}$$

All other rates are assumed to be zero. The nonzero routing probabilities are

$$r_{1c,0c} = 1, \quad r_{2c,0c} = 1, \quad r_{1s^+,2c} = 1. \qquad \square$$

Let

$$\boldsymbol{X}(t) = (X_1(t), X_2(t), \ldots, X_N(t))$$

denote the state of the network at time t, with $X_j(t)$ being the state of node j. Then $\boldsymbol{X}(t)$ is a Markov process on the state space $\mathcal{S}$, where

$$\mathcal{S} \equiv \prod_{j=1}^{N} \mathcal{S}_j.$$

To describe the stochastic process $\boldsymbol{X}(t)$ we extend the functions p^{A}_{ju}, $f_{ju.v}$, q^{D}_{ju}, and q^{I}_j to the network state space $\mathcal{S} \times \mathcal{S}$, and denote these new functions by $\boldsymbol{p}^{\text{A}}_{ju}$, $\boldsymbol{f}_{ju.v}$, $\boldsymbol{q}^{\text{D}}_{ju}$, and $\boldsymbol{q}^{\text{I}}_j$, respectively:

$$\begin{aligned}
\boldsymbol{p}^{\text{A}}_{ju}(\boldsymbol{x}, \boldsymbol{x}_j(x'_j)) &= p^{\text{A}}_{ju}(x_j, x'_j), && \boldsymbol{x} \in \mathcal{S}, \\
\boldsymbol{f}_{ju,v}(\boldsymbol{x}, \boldsymbol{x}_j(x'_j)) &= f_{ju,v}(x_j, x'_j), && \boldsymbol{x} \in \mathcal{S},
\end{aligned}$$

$$\boldsymbol{q}^{\mathrm{D}}_{ju}(\boldsymbol{x}, \boldsymbol{x}_j(x'_j)) = q^{\mathrm{D}}_{ju}(x_j, x'_j), \qquad \boldsymbol{x} \in \mathcal{S},$$

$$\boldsymbol{q}^{\mathrm{I}}_{j}(\boldsymbol{x}, \boldsymbol{x}_j(x'_j)) = q^{\mathrm{I}}_{j}(x_j, x'_j), \qquad \boldsymbol{x} \in \mathcal{S}.$$

Recall that $\boldsymbol{x}_j(x'_j)$ is the vector $\boldsymbol{x}$ with its jth element replaced by x'_j.

Let $j_1, j_2, \ldots$ be an arbitrary sequence of nodes to be visited and let $u_1(u'_1)$, $u_2(u'_2), \ldots$ be a sequence of arrival (departure) classes. Let $\boldsymbol{x}_k$ denote the state of the network after the kth visit when starting out from state $\boldsymbol{x}$. Let

$$\begin{aligned}
&\boldsymbol{p}_{\boldsymbol{x}}((j_1u_1, u'_1, \boldsymbol{x}_1), \ldots, (j_{\ell-1}u_{\ell-1}, u'_{\ell-1}, \boldsymbol{x}_{\ell-1}), (j_\ell u_\ell, \boldsymbol{x}_\ell)) \\
&\quad = \boldsymbol{p}^{\mathrm{A}}_{j_1,u_1}(\boldsymbol{x}, \boldsymbol{x}_1)\boldsymbol{f}_{j_1u_1,u'_1}(\boldsymbol{x}, \boldsymbol{x}_1) r_{j_1u'_1, j_2u_2} \times \cdots \\
&\qquad \times \boldsymbol{p}^{\mathrm{A}}_{j_{\ell-1}u_{\ell-1}}(\boldsymbol{x}_{\ell-2}, \boldsymbol{x}_{\ell-1})\boldsymbol{f}_{j_{\ell-1}u_{\ell-1},u'_{\ell-1}}(\boldsymbol{x}_{\ell-2}, \boldsymbol{x}_{\ell-1}) \\
&\qquad \times r_{j_{\ell-1}u'_{\ell-1}, j_\ell u_\ell}\boldsymbol{p}^{\mathrm{A}}_{j_\ell u_\ell}(\boldsymbol{x}_{\ell-1}, \boldsymbol{x}_\ell).
\end{aligned}$$

Clearly, this is the probability that, given that the network starts out from state $\boldsymbol{x}$, the arrival of a class u_1 entity at node j_1 causes at least ℓ transitions, in the form of class u'_m departures from node j_m $(m = 1, \ldots, \ell - 1)$ and class u_m arrivals at node j_m $(m = 1, \ldots, \ell)$. The sequence of states that the network simultaneously visits is $\boldsymbol{x}, \boldsymbol{x}_1, \ldots, \boldsymbol{x}_\ell$. Note that $\boldsymbol{p}_{\boldsymbol{x}}$ does not depend on u'_ℓ, the departure class at node j_ℓ.

The following technical assumption has to be made in order to avoid an infinite number of visits to a node at one time epoch. For any sequence of nodes $j_1, j_2, \ldots, j_\ell$, with arrival classes $u_1, \ldots, u_\ell$, and departure classes $u'_1, \ldots, u'_\ell$, and for any sequence of network states $\boldsymbol{x}, \boldsymbol{x}_1, \boldsymbol{x}_2, \ldots, \boldsymbol{x}_\ell$,

$$\lim_{\ell\to\infty} \boldsymbol{p}_{\boldsymbol{x}}((j_1u_1, u'_1, \boldsymbol{x}_1), \ldots, (j_\ell u_\ell, \boldsymbol{x}_\ell))\boldsymbol{f}_{j_\ell u_\ell, u'_\ell}(\boldsymbol{x}_{\ell-1}, \boldsymbol{x}_\ell) = 0. \tag{4.1}$$

In most applications this assumption is easily verified. For instance, if instantaneous movements always decrease the number of customers at the nodes and stop propagating when they arrive at empty nodes, then the network will be empty after a finite number of steps, so the sequence of nodes to be visited is of finite length. On the other hand, if each visit increases the number of customers, then, without the assumption above, the network will, with positive probability, explode at a single time epoch.

Let q denote the transition rate function of the Markov process $X(t)$. Even though this q is complicated, an expression can be found. For example, the transition rate from $\boldsymbol{x}$ to $\boldsymbol{x}'$ when there are ℓ simultaneous steps is

$$\sum_{v\in T_0}\sum_{j_1,\ldots,j_\ell}\sum_{u_1,\ldots,u_\ell}\sum_{u'_1,\ldots,u'_{\ell-1}}\sum_{\boldsymbol{x}_1,\ldots,\boldsymbol{x}_{\ell-1}} \beta_0(v) r_{0v, j_1u_1}$$

$$\times \boldsymbol{p_x}((j_1u_1, u_1', \boldsymbol{x}_1), \ldots, (j_\ell u_\ell, \boldsymbol{x}'))\Bigg(1 - \sum_{u_\ell'} \boldsymbol{f}_{j_\ell u_\ell, u_\ell'}(\boldsymbol{x}_{\ell-1}, \boldsymbol{x}')\Bigg)$$

$$+ \sum_{j_1,\ldots,j_{\ell+1}} \sum_{u_1,\ldots,u_{\ell+1}} \sum_{u_2',\ldots,u_\ell'} \sum_{\boldsymbol{x}_1,\ldots,\boldsymbol{x}_\ell} \boldsymbol{q}^{\mathrm{D}}_{j_1u_1}(\boldsymbol{x}, \boldsymbol{x}_1) r_{j_1u_1, j_2u_2}$$

$$\times \boldsymbol{p}_{\boldsymbol{x}_1}((j_2u_2, u_2', \boldsymbol{x}_2), \ldots, (j_{\ell+1}, u_{\ell+1}, \boldsymbol{x}'))\Bigg(1 - \sum_{u_{\ell+1}'} \boldsymbol{f}_{j_{\ell+1}u_{\ell+1}, u_{\ell+1}'}(\boldsymbol{x}_\ell, \boldsymbol{x}')\Bigg),$$

where $\boldsymbol{x}_0 = \boldsymbol{x}$ in the first summation when $\ell = 1$, and both summations are possible only for $\ell \geq 1$. Summing up the above rates for $\ell \geq 1$ gives the transition rate from $\boldsymbol{x}$ to $\boldsymbol{x}'$ when it involves interactions of two or more nodes. The transition rate from $\boldsymbol{x}$ to $\boldsymbol{x}'$ is easy when it involves a transition at only one node. Summing up all these transitions yields q. An explicit expression for q using matrix operators is given in Chapter 10.

As in Section 4.2, to compute the stationary distribution of the network, we need first to consider each individual node j with an auxiliary transition rate

$$q_j^{(\boldsymbol{\alpha}_j)}(x_j, x_j') = \sum_{u \in T_j} \Big(\alpha_{ju} p^{\mathrm{A}}_{ju}(x_j, x_j') + q^{\mathrm{D}}_{ju}(x_j, x_j')\Big) + q^{\mathrm{I}}_j(x_j, x_j'). \tag{4.2}$$

The $\boldsymbol{\alpha}_j = (\alpha_{ju}; u \in T_j)$ are considered as dummy parameters and their values are determined by the traffic equations. Assume $q_j^{(\boldsymbol{\alpha}_j)}$ has a stationary distribution $\pi_j^{(\boldsymbol{\alpha}_j)}$, $j = 1, \ldots, N$. We now require that $q_j^{(\boldsymbol{\alpha}_j)}$ be quasi-reversible. Note that $q^{\mathrm{A}}_{ju} \equiv \alpha_{ju} p^{\mathrm{A}}_{ju}$ satisfies condition (3.2) of Chapter 3. So the quasi-reversibility is equivalent to the existence of nonnegative numbers $\{\beta_{ju}; u \in T_j\}$ for all $j = 1, \ldots, N$ such that

$$\sum_{x_j' \in \mathcal{S}_j} \pi_j^{(\boldsymbol{\alpha}_j)}(x_j')\Bigg(q^{\mathrm{D}}_{ju}(x_j', x_j) + \sum_{v \in T_j} \alpha_{jv} p^{\mathrm{A}}_{jv}(x_j', x_j) f_{jv,u}(x_j', x_j)\Bigg) = \beta_{ju}\pi_j^{(\boldsymbol{\alpha}_j)}(x_j),$$
$$j = 1, \ldots, N,\ u \in T_j,\ x_j \in \mathcal{S}_j. \tag{4.3}$$

We call node j with signals to be quasi-reversible with $\boldsymbol{\alpha}_j$ if (4.3) is satisfied.

Since α_{ju} and β_{ju} are the arrival and departure rates of class u entities at node j, the traffic equations

$$\alpha_{ju} = \sum_{k=0}^{N} \sum_{v \in T_k} \beta_{kv} r_{kv, ju}, \qquad j = 0, 1, \ldots, N, \quad u \in T_j,$$

have to be satisfied. We need the following condition to ensure that the network process is regular:

$$\sum_{j=1}^{N} \sum_{x_j \in \mathcal{S}_j} \pi_j^{(\boldsymbol{\alpha}_j)}(x_j) \sum_{x_j' \in \mathcal{S}_j} q_j^{(\boldsymbol{\alpha}_j)}(x_j)(x_j, x_j') < \infty. \tag{4.4}$$

A simple sufficient condition for (4.4) is

$$\sum_{u\in T_j}(\alpha_{ju}+\beta_{ju})+\sum_{x_j'\in\mathcal{S}_j} q_j^{\mathrm{I}}(x_j)(x_j,x_j')<\infty, \qquad \text{for all } j=1,\ldots,N,$$

which is satisfied by all the examples in this book.

The following result for networks with signals and instantaneous movements is an extension of Theorem 4.3 for networks without signals.

Theorem 4.9 *If each node j with signals, $j=1,\ldots,N$, is quasi-reversible with $\boldsymbol{\alpha}_j$ that is the solution to the traffic equations (3.5), then the queueing network with signals has the product form stationary distribution*

$$\pi(\boldsymbol{x})=\prod_{j=1}^{N}\pi_j^{(\boldsymbol{\alpha}_j)}(x_j), \qquad \boldsymbol{x}\equiv(x_1,x_2,\ldots,x_N)\in\mathcal{S}. \tag{4.5}$$

where $\pi_j^{(\boldsymbol{\alpha}_j)}$ is the stationary distribution of $q_j^{(\boldsymbol{\alpha}_j)}$, $j=1,\ldots,N$.

PROOF. We use the detailed Kelly lemma. For convenience we drop the superscript $(\boldsymbol{\alpha}_j)$. Assume that the reversed process corresponds to a similar network that is characterized by

$$\tilde{q}_{ju}^{\mathrm{D}}(x_j',x_j)=\frac{\pi_j(x_j)\alpha_{ju}p_{ju}^{\mathrm{A}}(x_j,x_j')\left(1-\sum_v f_{ju,v}(x_j,x_j')\right)}{\pi_j(x_j')},$$

$$\tilde{q}_j^{\mathrm{I}}(x_j',x_j)=\frac{\pi_j(x_j)q_j^{\mathrm{I}}(x_j,x_j')}{\pi_j(x_j')},$$

$$\tilde{p}_{ju}^{\mathrm{A}}(x_j',x_j)=\frac{\pi_j(x_j)\left(q_{ju}^{\mathrm{D}}(x_j,x_j')+\sum_v\alpha_{jv}p_{ju}^{\mathrm{A}}(x_j,x_j')f_{jv,u}(x_j,x_j')\right)}{\pi_j(x_j')\beta_{ju}},$$

$$\tilde{f}_{ju,v}(x_j',x_j)=\frac{\alpha_{jv}p_{jv}^{\mathrm{A}}(x_j,x_j')f_{jv,u}(x_j,x_j')}{q_{ju}^{\mathrm{D}}(x_j,x_j')+\sum_w\alpha_{jw}p_{jw}^{\mathrm{A}}(x_j,x_j')f_{jw,u}(x_j,x_j')},$$

$$\tilde{r}_{ju,kv}=\frac{\beta_{kv}r_{kv,ju}}{\alpha_{ju}},$$

for $j,k=0,1,\ldots,N$. Because of the quasi-reversibility condition (4.3) and the traffic equations (3.5), $\tilde{p}_{ju}^{\mathrm{A}}(y_j,x_j)$ and $\tilde{r}_{ju,kv}$ are indeed probabilities. Thus they determine a queueing network with instantaneous movements.

To apply the detailed Kelly lemma, we need to verify two conditions. To check the first condition, i.e.,

$$\sum_{\boldsymbol{x}'}q(\boldsymbol{x},\boldsymbol{x}')=\sum_{\boldsymbol{x}'}\tilde{q}(\boldsymbol{x},\boldsymbol{x}'),$$

we first note that changes in the network state are initiated by either a departure or an internal transition, and terminated after a finite number of transitions. The latter is ensured by condition (4.1), which guarantees that the network process is well defined (see Chapter 10 for a more extensive discussion on this). Thus, similar to the proof of Theorem 4.3, we use the quasi-reversibility condition (4.3), the global balance of each node (3.2) and the total balance (3.7), to obtain

$$\begin{aligned}
\sum_{\boldsymbol{x}'} \tilde{q}(\boldsymbol{x}, \boldsymbol{x}') &= \sum_{j=0}^{N} \Bigg[\sum_{u \in T_j} \sum_{x_j'} \tilde{q}^{\mathrm{D}}_{ju}(x_j, x_j') + \sum_{x_j'} \tilde{q}^{\mathrm{I}}_j(x_j, x_j') \Bigg] \\
&= \sum_{j=0}^{N} \Bigg[\sum_{u \in T_j} \sum_{x_j'} \frac{\pi_j(x_j')\alpha_{ju} p^{\mathrm{A}}_{ju}(x_j', x_j)\Big(1 - \sum_{v \in T_j} f_{ju,v}(x_j', x_j)\Big)}{\pi_j(x_j)} \\
&\qquad + \sum_{x_j'} \frac{\pi_j(x_j') q^{\mathrm{I}}_j(x_j', x_j)}{\pi_j(x_j)} \Bigg] \\
&= \sum_{j=0}^{N} \Bigg[\Bigg(\sum_{u \in T_j} \sum_{x_j'} \pi_j(x_j')\alpha_{ju} p^{\mathrm{A}}_{ju}(x_j', x_j) + \sum_{u \in T_j} \sum_{x_j'} \pi_j(x_j') q^{\mathrm{D}}_{ju}(x_j', x_j) \\
&\qquad + \sum_{x_j'} \pi_j(x_j') q^{\mathrm{I}}_j(x_j', x_j) \Bigg) \Big/ \pi_j(x_j) - \sum_{u \in T_j} \beta_{ju} \Bigg] \\
&= \sum_{j=0}^{N} \Bigg[\sum_{u \in T_j} \sum_{x_j'} q^{\mathrm{D}}_{ju}(x_j, x_j') + \sum_{x_j'} q^{\mathrm{I}}_j(x_j, x_j') + \sum_{u \in T_j} (\alpha_{ju} - \beta_{ju}) \Bigg] \\
&= \sum_{j=0}^{N} \sum_{x_j'} \Bigg[\sum_{u \in T_j} q^{\mathrm{D}}_{ju}(x_j, x_j') + q^{\mathrm{I}}_j(x_j, x_j') \Bigg] \\
&= \sum_{\boldsymbol{x}'} q(\boldsymbol{x}, \boldsymbol{x}').
\end{aligned}$$

To check the second condition of the detailed Kelly lemma, we decompose the transition functions $q(\boldsymbol{x}, \boldsymbol{x}')$ and $\tilde{q}(\boldsymbol{x}, \boldsymbol{x}')$ according to sequences of simultaneous transitions. Clearly, a transition for this network is either an internal transition at a node, or a transition involving at least two nodes (possibly node 0, the outside). A transition involving two or more nodes takes the following form. The network starts out in state $\boldsymbol{x}$, a class u entity departs from node j ($j = 0, 1, \ldots, N$) and goes to node j_1 as an entity of class u_1, changes the state of the network to $\boldsymbol{x}_1$, and triggers at the same time a class u_1' departure from node j_1; this departing entity then goes to node j_2 as an entity of class u_2, changes the state of the network to $\boldsymbol{x}_2$, and triggers a class u_2' departure, etc. The string transition ends when a class v entity arrives at a node k and does not trigger an instantaneous departure; the state then changes to $\boldsymbol{x}'$. Let σ denote this sequence. See Figure 4.6(a). For the reversed process, $\tilde{\sigma}$ denotes the reversed sequence of σ which starts with a class v departure from node k when the network is in state $\boldsymbol{x}'$, triggers a string of transitions, and ends with a class u arrival at node j that does not trigger an instantaneous departure; the state of the network then changes to $\boldsymbol{x}$. See Figure 4.6(b).

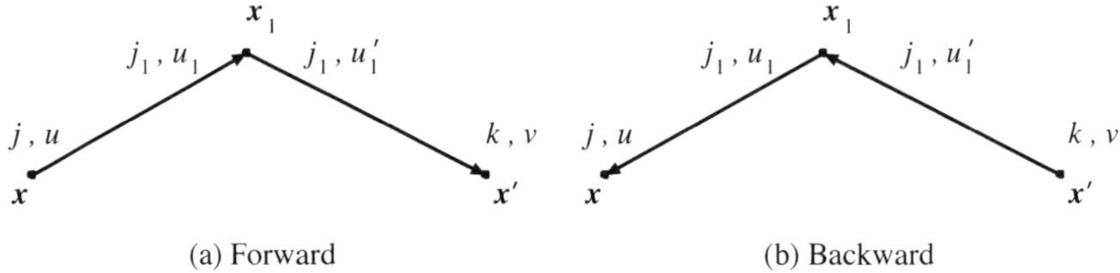

Figure 4.6 Transitions due to instantaneous triggering

Let q_σ be the transition function generated by the sequence σ, and $\tilde{q}_{\tilde{\sigma}}$ be the transition function of the time reversed process generated by the sequence $\tilde{\sigma}$. We now verify the second condition of the detailed Kelly lemma, i.e.,

$$\pi(\boldsymbol{x})q_\sigma(\boldsymbol{x}, \boldsymbol{x}') = \pi(\boldsymbol{x}')\tilde{q}_{\tilde{\sigma}}(\boldsymbol{x}', \boldsymbol{x}). \tag{4.6}$$

For an internal transition, (4.6) is easily seen to be satisfied by the definition of $\tilde{q}_j^{\mathrm{I}}$. For transitions that involve more than one node, we consider here only the case of a string transition that involves three nodes, as shown in Figure 4.6:

$$\begin{aligned}
&\pi(\boldsymbol{x})q_\sigma(\boldsymbol{x}, \boldsymbol{x}') \\
&= \pi_j(x_j)\pi_{j_1}(x_{j_1})\pi_k(x_k)q_{ju}^{\mathrm{D}}(x_j, x_j')r_{ju,j_1u_1}p_{j_1u_1}^{\mathrm{A}}(x_{j_1}, x_{j_1}')f_{j_1u_1,u_1'}(x_{j_1}, x_{j_1}') \\
&\qquad \times r_{j_1u_1',kv}p_{kv}^{\mathrm{A}}(x_k, x_k')\left(1 - \sum_{w\in T_k} f_{kv,w}(x_k, x_k')\right)\prod_{\ell\neq j,j_1,k}\pi_\ell(x_\ell).
\end{aligned}$$

Similarly, the right-hand side is

$$\begin{aligned}
&\pi(\boldsymbol{x}')\tilde{q}_{\tilde{\sigma}}(\boldsymbol{x}', \boldsymbol{x}) \\
&= \pi_k(x_k')\pi_{j_1}(x_{j_1}')\pi_j(x_j')\tilde{q}_{kv}^{\mathrm{D}}(x_k', x_k)\tilde{r}_{kv,j_1u_1'}\tilde{p}_{j_1u_1'}^{\mathrm{A}}(x_{j_1}', x_{j_1})\tilde{f}_{j_1u_1',u_1}(x_{j_1}', x_{j_1}) \\
&\qquad \times \tilde{r}_{j_1u_1,ju}\tilde{p}_{ju}^{\mathrm{A}}(x_j', x_j)\left(1 - \sum_{w\in T_j} \tilde{f}_{ju,w}(x_j', x_j)\right)\prod_{\ell\neq j,j_1,k}\pi_\ell(x_\ell).
\end{aligned}$$

It is straightforward to verify that these two terms are equal. In the case when the string contains more than 3 nodes, (4.6) can be verified in a similar way. This completes the proof of Theorem 4.9. □

Example 4.10 (A network with negative signals) Consider a queueing network with customers and negative signals, as described in Example 3.12. When a customer completes its service at node j, it goes to node k as a regular customer with probability $r_{jc,kc}$, and as a negative signal with probability $r_{jc,ks}$, $k = 0, 1, \ldots, N$. When a negative signal arrives at node j, it induces a customer, if there is one present, to depart. The customer then joins node k as a regular customer with probability $r_{js,kc}$, and as a negative signal with probability $r_{js,ks}$, $k = 0, 1, \ldots, N$. Let λ_j and λ_j^- be the exogenous arrival rates of customers and signals at node j. The state of the network is given by the vector $\boldsymbol{n} = (n_1, \ldots, n_N)$, where n_j is the number of customers at node j.

Node j, defined by (4.2), is the queue discussed in Example 3.12, with service rate μ_j, customer arrival rate α_j, and signal arrival rate α_j^-. It is quasi-reversible with departure rates of customers and signals given by

$$\beta_j = \frac{\alpha_j \mu_j}{\mu_j + \alpha_j^-}, \qquad \beta_j^- = \frac{\alpha_j \alpha_j^-}{\mu_j + \alpha_j^-}.$$

Thus the traffic equations are

$$\alpha_j = \lambda_j + \sum_{k=1}^{N} \frac{\alpha_k \mu_k}{\mu_k + \alpha_k^-} r_{kc,jc} + \sum_{k=1}^{N} \frac{\alpha_k \alpha_k^-}{\mu_k + \alpha_k^-} r_{ks,jc}, \tag{4.7}$$

$$\alpha_j^- = \lambda_j^- + \sum_{k=1}^{N} \frac{\alpha_k \mu_k}{\mu_k + \alpha_k^-} r_{kc,js} + \sum_{k=1}^{N} \frac{\alpha_k \alpha_k^-}{\mu_k + \alpha_k^-} r_{ks,js}, \tag{4.8}$$

$$j = 1, 2, \ldots, N.$$

Suppose these traffic equations have positive solutions α_j and α_j^- such that

$$\frac{\alpha_j}{\mu_j + \alpha_j^-} < 1, \qquad j = 1, 2, \ldots, N.$$

Since each node is uniformly quasi-reversible, applying Theorem 4.9 yields the stationary distribution

$$\pi(\boldsymbol{n}) = \prod_{j=1}^{N} \left(1 - \frac{\alpha_j}{\mu_j + \alpha_j^-}\right) \left(\frac{\alpha_j}{\mu_j + \alpha_j^-}\right)^{n_j}.$$

□

Example 4.11 (A finite buffer network with jump-over blocking) Consider a network of finite buffer queues discussed in Example 3.15. Assume that the network has one class of entities, i.e., customers, that arrive at node j from the outside according to a Poisson process with rate λ_j. Node j has a buffer of size K_j

and service rate $\mu_j(n_j)$ when n_j customers are present, where $\mu_j(0) = 0$. When a customer completes its service at node j, it joins node k with probability r_{jk}, $k = 0, 1, \ldots, N$, where $k = 0$ represents the outside. When a customer arrives at node j when there are fewer than K_j customers present, it joins the node. However, if a customer arrives at node j and finds K_j customers at the node, it jumps over the node and joins node k with probability r_{jk}, $k = 0, 1, \ldots, N$. This protocol is referred to as *jump-over blocking*.

Node j is characterized by

$$\begin{aligned} p^{\mathrm{A}}_{jc}(n_j, n_j+1) &= 1, & n_j &= 0, 1, \ldots, K_j - 1, \\ p^{\mathrm{A}}_{jc}(K_j, K_j) &= 1, & & \\ q^{\mathrm{D}}_{jc}(n_j, n_j-1) &= \mu_j(n_j), & n_j &= 1, 2, \ldots, K_j, \\ f_{jc,c}(K_j, K_j) &= 1. & & \end{aligned}$$

Node j defined by (4.2) is the queue with finite buffer and blocking discussed in Example 3.15, with Poisson arrivals of rate α_j. Its stationary distribution is

$$\pi_j(n_j) = \pi_j(0)\frac{\alpha_j^{n_j}}{\prod_{\ell=1}^{n_j}\mu_j(\ell)}, \qquad n_j = 0, 1, \ldots, K_j, \tag{4.9}$$

where

$$\pi_j(0) = \left(1 + \sum_{n=1}^{K_j}\frac{\alpha_j^n}{\prod_{\ell=1}^{n}\mu_j(\ell)}\right)^{-1}.$$

As shown in Example 3.15, this queue is quasi-reversible with departure rate $\beta_j = \alpha_j$. Hence if $\alpha_1, \ldots, \alpha_N$ are the solution to the traffic equations

$$\alpha_j = \lambda_j + \sum_{k=1}^{N}\alpha_k r_{kj}, \qquad j = 1, \ldots, N,$$

then the network has the product form stationary distribution

$$\pi(n_1, \ldots, n_N) = \prod_{j=1}^{N}\pi_j(n_j), \qquad 0 \le n_j \le K_j, \; j = 1, \ldots, N,$$

where π_j is given by (4.9). □

4.5 NETWORKS WITH POSITIVE AND NEGATIVE SIGNALS

In this section we apply Theorem 4.9 to a queueing network with two types of signals: positive signals and negative signals. The first subsection considers the case of a single class of positive signals and a single class of negative signals, and the second subsection considers multiple classes of positive and negative signals. As we will see in subsequent chapters, these models include most networks with batch movements as special cases.

4.5.1 Single Class of Positive and Negative Signals

Consider a network with N nodes and a single server at each node. There are three classes of entities: *customers, positive signals*, and *negative signals*, denoted by c, s^+ and s^-, respectively. Type c refers to the regular customers; their arrivals do not trigger any instantaneous movements. Positive and negative signals, however, represent signaling mechanisms that induce immediate transitions at the nodes where they arrive. Assume that the state of the network is $\boldsymbol{n} = (n_1, n_2, \ldots, n_N)$, where n_j is the number of customers at node j, $j = 1, 2, \ldots, N$. The arrival of a positive signal at a node increases the number of customers at that node by 1, and then leaves immediately for another node. The arrival of a negative signal at a node triggers a customer to depart, provided the node is not empty upon the signal's arrival. A negative signal disappears when it arrives at an empty node.

Assume that customers arrive from the outside at node j according to a Poisson process with rate λ_j, and positive and negative signals arrive from the outside at node j according to Poisson processes with rates λ_j^+ and λ_j^-. Node j, $j = 1, \ldots, N$, has exponential service times with rate μ_j. Upon a service completion at node j, a customer leaves for node k as a regular customer with probability $r_{jc,kc}$, as a positive signal with probability r_{jc,ks^+}, as a negative signal with probability r_{jc,ks^-}, and it leaves the network with probability $r_{jc,0}$, where

$$\sum_{k=1}^{N}(r_{jc,kc} + r_{jc,ks^+} + r_{jc,ks^-}) + r_{jc,0} = 1, \quad j = 1, \ldots, N.$$

When a positive signal arrives at node j, either from the outside or from another node, it adds one customer and then leaves immediately for node k as a regular customer with probability $r_{js^+,kc}$, as a positive signal with probability r_{js^+,ks^+}, as a negative signal with probability r_{js^+,ks^-}, and it leaves the network with probability $r_{js^+,0}$, where

$$\sum_{k=1}^{N}(r_{js^+,kc} + r_{js^+,ks^+} + r_{js^+,ks^-}) + r_{js^+,0} = 1, \quad j = 1, \ldots, N.$$

Finally, when a negative signal arrives at node j, either from the outside or from another node, it triggers a customer, if any, to depart. The departing customer goes to node k as a regular customer with probability $r_{js^-,kc}$, as a positive signal with probability r_{js^-,ks^+}, as a negative signal with probability r_{js^-,ks^-}, and it leaves the network with probability $r_{js^-,0}$, where again

$$\sum_{k=1}^{N}(r_{js^-,kc} + r_{js^-,ks^+} + r_{js^-,ks^-}) + r_{js^-,0} = 1, \quad j = 1, \ldots, N.$$

As indicated earlier, a negative signal that arrives at an empty node is assumed to be lost. We refer to this model as *a network with positive and negative signals*.

In this network there can be any number of customer additions or deletions at various nodes of the network at the same point in time. For instance, if there is a sequence $j_1, j_2, \ldots, j_k$ such that

$$r_{j_1s^+, j_2s^+} r_{j_2s^+, j_3s^+} \cdots r_{j_{k-1}s^+, j_ks^+} > 0,$$

then the arrival of a positive signal at node j_1 would, with positive probability, add one customer at each one of nodes $j_1, j_2, \ldots, j_k$. There can also be batch arrivals at node j if $r_{js^+,js^+} > 0$, and the arrival of a positive signal can add a batch of random size at a number of nodes. Unlike in networks with negative signals, in which a signal may be interrupted on its route once it hits an empty node, a positive signal is never interrupted. It disappears only when it is transformed into another class of entity or when it leaves the network.

The arrivals of couples in the bathroom problem are included in this framework by setting the network parameters equal to

$$\lambda_1^+ = \lambda, \qquad r_{1s^+,2c} = 1,$$
$$\lambda_1^- = \lambda_2^+ = \lambda_2^- = 0.$$

To ensure that the network is stable, i.e., that it will not be overloaded, we have to exclude the case when a positive signal from a node generates an infinite number of customers in the network at one point in time. Hence we make the following technical assumption in order to ensure that the stochastic process is regular: the Markov chain with state space $\{0, 1, \ldots, N\}$ and transition probabilities

$$p_{j,k} = r_{js^+,ks^+}, \qquad j, k = 1, \ldots, N,$$
$$p_{j,0} = 1 - \sum_{k=1}^{N} r_{js^+,ks^+} - \sum_{k=1}^{N} r_{js^+,kc}, \qquad j = 1, \ldots, N,$$
$$p_{0,0} = 1,$$

has only one recurrent state 0.

We are interested in the stationary probability of this network. However, it is known that such networks do not have closed form solutions. In the following theorem we modify the network process so as to obtain a product form solution for the network. This modification may appear artificial, and it is introduced purely to obtain the uniform quasi-reversibility of each node in order to apply Theorem 4.9. However, the product form solution serves as a stochastic upper bound for the original network.

Let $\{\alpha_j;\ j = 1, \ldots, N\}$, $\{\alpha_j^+;\ j = 1, \ldots, N\}$, and $\{\alpha_j^-;\ j = 1, \ldots, N\}$ be the solution to the traffic equations

$$\alpha_j = \lambda_j + \sum_{k=1}^N \rho_k \mu_k r_{kc,jc} + \sum_{k=1}^N \rho_k \alpha_k^- r_{ks^-,jc} + \sum_{k=1}^N \rho_k^{-1} \alpha_k^+ r_{ks^+,jc}, \tag{5.1}$$

$$\alpha_j^+ = \lambda_j^+ + \sum_{k=1}^N \rho_k \mu_k r_{kc,js^+} + \sum_{k=1}^N \rho_k \alpha_k^- r_{ks^-,js^+} + \sum_{k=1}^N \rho_k^{-1} \alpha_k^+ r_{ks^+,js^+}, \tag{5.2}$$

$$\alpha_j^- = \lambda_j^- + \sum_{k=1}^N \rho_k \mu_k r_{kc,js^-} + \sum_{k=1}^N \rho_k \alpha_k^- r_{ks^-,js^-} + \sum_{k=1}^N \rho_k^{-1} \alpha_k^+ r_{ks^+,js^-}, \tag{5.3}$$

for $j = 1, \ldots, N$, where

$$\rho_j = \frac{\alpha_j + \alpha_j^+}{\mu_j + \alpha_j^-}.$$

Now modify the network with positive and negative signals such that whenever node j is empty, a Poisson departure process of positive signals is activated with rate $\rho_j^{-1}\alpha_j^+$.

Theorem 4.12 *If the solution to the traffic equations satisfies* $\rho_j < 1$, $j = 1, \ldots, N$, *then the modified network described above has the product form stationary distribution*

$$\pi(n_1, \ldots, n_N) = \prod_{j=1}^N (1 - \rho_j)\rho_j^{n_j}. \tag{5.4}$$

PROOF. We use Theorem 4.9 to prove the result. It suffices to verify the quasi-reversibility of each node when it is in isolation and subject to Poisson arrivals of customers and signals. Let a service completion at node j be classified as a class c departure, and departures triggered by positive and negative signals as class s^+ and s^- departures, respectively. Then, node j is characterized by

$$p_{jc}^{\mathrm{A}}(n_j, n_j + 1) = 1, \qquad n_j \geq 0,$$

$$p^{\text{A}}_{js^+}(n_j, n_j+1) = 1, \qquad n_j \geq 0,$$
$$p^{\text{A}}_{js^-}(n_j, n_j-1) = 1, \qquad n_j \geq 1,$$
$$p^{\text{A}}_{js^-}(0, 0) = 1,$$
$$q^{\text{D}}_{jc}(n_j, n_j-1) = \mu_j, \qquad n_j \geq 1.$$

The triggering probabilities are

$$f_{jc,u}(n_j, n'_j) = 0, \qquad u = c, s^+, s^- \text{ and } n_j, n'_j \geq 0,$$
$$f_{js^+,s^+}(n_j, n_j+1) = 1, \qquad n_j \geq 0,$$
$$f_{js^-,s^-}(n_j+1, n_j) = 1, \qquad n_j \geq 0.$$

For convenience we let $\alpha_j = \alpha_{jc}$, $\alpha^+_j = \alpha_{js^+}$ and $\alpha^-_j = \alpha_{js^-}$. Node j, characterized by

$$q^{(\boldsymbol{\alpha}_j)}_j(n_j, n'_j) = \alpha_j p^{\text{A}}_{jc}(n_j, n'_j) + \alpha^+_j p^{\text{A}}_{js^+}(n_j, n'_j) + \alpha^-_j p^{\text{A}}_{js^-}(n_j, n'_j) + q^{\text{D}}_{jc}(n_j, n'_j),$$

is an $M/M/1$ queue with Poisson arrivals of three classes of entities, with respective rates α_j, α^+_j and α^-_j, and with service rate μ_j. As far as the stationary distribution of the node is concerned, this queue is the same as an $M/M/1$ queue with arrival rate $\alpha_j + \alpha^+_j$ and service rate $\mu_j + \alpha^-_j$. Hence its stationary distribution is

$$\pi_j(n_j) = (1 - \rho_j)\rho_j^{n_j}, \qquad n_j \geq 0,$$

provided the stability condition

$$\rho_j = (\alpha_j + \alpha^+_j)/(\mu_j + \alpha^-_j) < 1.$$

However, this node with positive and negative signals is not quasi-reversible, as we will see below. So we modify this queue by assuming that whenever it is empty, the node generates departures of class s^+ at a constant rate. That is, we modify the transition rate for node j such that $q^{\text{D}}_{js^+}(0, 0) > 0$. In what follows we show that this modified $M/M/1$ queue with positive and negative signals is quasi-reversible if and only if

$$q^{\text{D}}_{js^+}(0, 0) = \rho_j^{-1}\alpha^+_j. \tag{5.5}$$

It is easy to verify that the quasi-reversibility condition (4.3) is satisfied for $u = c, s^-$ with

$$\beta_j = \rho_j \mu_j, \qquad \beta^-_j (\equiv \beta_{js^-}) = \rho_j \alpha^-_j.$$

For $u = s^+$ and $n_j \geq 1$,

$$\sum_{n'_j} \pi_j(n'_j)\left(q^{\mathrm{D}}_{js^+}(n'_j, n_j) + \sum_{v=c,s^+,s^-} \alpha_{jv} p^{\mathrm{A}}_{jv}(n'_j, n_j) f_{jv,s^+}(n'_j, n_j)\right)$$
$$= \pi_j(n_j - 1)\alpha_j^+ p^{\mathrm{A}}_{js^+}(n_j - 1, n_j)$$
$$= \rho_j^{-1}\alpha_j^+ \pi_j(n_j).$$

And for $n_j = 0$,

$$\sum_{n'_j} \pi_j(n'_j)\left(q^{\mathrm{D}}_{js^+}(n'_j, 0) + \sum_{v=c,s^+,s^-} \alpha_{jv} p^{\mathrm{A}}_{jv}(n'_j, 0) f_{jv,s^+}(n'_j, 0)\right)$$
$$= q^{\mathrm{D}}_{js^+}(0, 0)\pi_j(0).$$

Thus for (4.3) to hold for $u = s^+$ and for all n_j, (5.5) is necessary and sufficient. Letting

$$\beta_j^+ (\equiv \beta_{js^+}) = \rho_j^{-1}\alpha_j^+,$$

we obtain that node j is uniformly quasi-reversible for all $\alpha_j, \alpha_j^+, \alpha_j^-$ and

$$\beta_j = \frac{\alpha_j + \alpha_j^+}{\mu_j + \alpha_j^-}\mu_j, \qquad j = 1, \ldots, N,$$
$$\beta_j^- = \frac{\alpha_j + \alpha_j^+}{\mu_j + \alpha_j^-}\alpha_j^-, \qquad j = 1, \ldots, N,$$
$$\beta_j^+ = \frac{\mu_j + \alpha_j^-}{\alpha_j + \alpha_j^+}\alpha_j^+, \qquad j = 1, \ldots, N,$$

provided $\rho_j < 1$ for all j. Hence, it follows from Theorem 4.9 that, if α_j, α_j^+ and α_j^- are the solution to the traffic equations (3.5), which are simplified to (5.1), (5.2) and (5.3), then the network has the geometric product form stationary distribution (5.4). This completes the proof of Theorem 4.12 □

Since the modified network has additional departures of positive signals, the network process stochastically dominates the corresponding process without the additional departures, provided a positive signal cannot be transformed into either a negative signal or a customer that can be transformed into a negative signal. This can be easily proved using a sample path stochastic comparison by constructing the two processes and coupling the number of customers in the two networks. Note that if the modified network has a stationary distribution, and if it stochastically dominates the original network, then the original network must also have a stationary distribution. Thus we obtain the following result.

Corollary 4.13 *Let π^0 be the stationary distribution of the network without the additional departures. If $r_{js^+,ks^-} = 0$ and $r_{js^+,kc}r_{kc,\ell s^-} = 0$ for all $j, k, \ell = 1, \ldots, N$, then π^0 is stochastically dominated by the product form geometric distribution obtained in Theorem 4.12, i.e.,*

$$\sum_{k_j \geq n_j, j=1,\ldots,N} \pi^0(k_1, \ldots, k_N) \leq \prod_{j=1}^{N} \left(\frac{\alpha_j + \alpha_j^+}{\mu_j + \alpha_j^-} \right)^{n_j}. \tag{5.6}$$

Remark 4.14 If a queueing network does not have positive signals, then it follows from Theorem 4.17 that the network has a product form stationary distribution without the additional departures.

Example 4.15 (Product form bound for the bathroom problem) Consider the bathroom problem described in Example 1.2. Let α_1, α_1^+ and α_2, α_2^+ denote the average arrival rates of customers and positive signals at the two nodes. From the traffic equations (5.1), (5.2), and (5.3), we obtain

$$\begin{aligned} \alpha_1 &= \lambda_1, \qquad \alpha_1^+ = \lambda, \\ \alpha_2 &= \lambda_2 + \rho_1^{-1}\alpha_1^+ r_{12} = \lambda_2 + \mu_1\lambda/(\lambda_1 + \lambda), \\ \alpha_2^+ &= 0. \end{aligned}$$

Hence it follows from $\alpha_1^- = \alpha_2^- = 0$ that the upper bound for the tail probability of $(X_1(t), X_2(t))$ is

$$\left(\frac{\lambda_1 + \lambda}{\mu_1} \right)^{n_1} \left(\frac{\lambda_2(\lambda_1 + \lambda) + \mu_1\lambda}{\mu_2(\lambda_1 + \lambda)} \right)^{n_2}.$$

Let $\pi_1^0(n_1|N_2 \geq n_2)$ (resp. $\pi_2^0(n_2|N_1 \geq n_1)$) be the conditional probability of n_1 (n_2) customers in bathroom 1 (2), given that there are at least n_2 (n_1) customers in bathroom 2 (1). The result above yields upper bounds for these conditional probabilities, and the bounds are exactly the marginal distributions of π:

$$\sum_{\ell=n_1}^{\infty} \pi_1^0(\ell|N_2 \geq n_2) \leq \left(\frac{\lambda_1 + \lambda}{\mu_1} \right)^{n_1},$$

$$\sum_{\ell=n_2}^{\infty} \pi_2^0(\ell|N_1 \geq n_1) \leq \left(\frac{\lambda_2(\lambda_1+\lambda)+\mu_1\lambda}{\mu_2(\lambda_1+\lambda)}\right)^{n_2}.$$

Note that these two bounds are independent of the conditional events.

Consider the special case $\lambda_1 = \lambda_2 = 0$, i.e., there is only one arrival process with each arrival bringing one customer to each bathroom. In this case,

$$\alpha_1 = 0, \quad \alpha_1^+ = \lambda, \quad \alpha_2 = \mu_1, \quad \alpha_2^+ = 0.$$

Hence under the additional assumption stated in Corollary 4.13, the stationary probability of the network is

$$\pi(n_1, n_2) = \left(1 - \frac{\lambda}{\mu_1}\right)\left(\frac{\lambda}{\mu_1}\right)^{n_1}\left(1 - \frac{\mu_1}{\mu_2}\right)\left(\frac{\mu_1}{\mu_2}\right)^{n_2},$$

provided $\lambda < \mu_1$ and $\mu_1 < \mu_2$.

What if $\mu_1 > \mu_2$? Note that the original bathroom problem is obtained by introducing positive signals that arrive at node 1 according to a Poisson process with rate λ, with each positive signal adding one customer at node 1 before leaving immediately for node 2 as a regular customer. It is clear that the same bathroom problem can be obtained by introducing a positive signal arrival process at node 2 with each positive signal adding one customer at node 2 and then leaving immediately for node 1 as a regular customer. The stationary probability of the modified bathroom problem is now

$$\pi(n_1, n_2) = \left(1 - \frac{\mu_2}{\mu_1}\right)\left(\frac{\mu_2}{\mu_1}\right)^{n_1}\left(1 - \frac{\lambda}{\mu_2}\right)\left(\frac{\lambda}{\mu_2}\right)^{n_2},$$

provided $\lambda < \mu_2$ and $\mu_2 < \mu_1$.

The case $\mu_1 = \mu_2$ cannot be solved within the framework of queueing networks with positive signals. □

This example suggests that there is a flexibility in constructing networks with positive and negative signals to model concurrent arrivals and concurrent departures. Different constructions result in different upper bounds.

4.5.2 Multiple Classes of Positive and Negative Signals

We extend the results of the previous subsection to networks with multiple classes of positive and negative signals. Suppose there is a single class of customers denoted by c, I^+ classes of positive signals denoted by $\{u^+; u = 1, 2, \ldots, I^+\}$, and I^- classes of negative signals denoted by $\{u^-; u = 1, 2, \ldots, I^-\}$, where I^+ and

I^- may be infinity. There is a single server at node j, and the service times at node j are exponentially distributed with rate μ_j. Customers arrive at node j from the outside according to a Poisson process with rate λ_j. Class u^+ positive signals, $u = 1, 2, \ldots, I^+$, arrive at node j from the outside according to a Poisson process with rate λ_{ju}^+, and class u^- negative signals, $u = 1, 2, \ldots, I^-$, arrive at node j from the outside according to a Poisson process with rate λ_{ju}^-.

The effects of positive and negative signals on a node are the same as in the previous subsection. That is, the arrival of a class u^+ positive signal at node j adds one customer to node j and then departs, whereas the arrival of a class u^- negative signal at node j triggers one customer, if at least one is present, to depart. If a negative signal arrives at an empty node, nothing happens and the signal disappears. Thus, node j is characterized by

$$\begin{aligned}
p_{jc}^{\mathrm{A}}(n_j, n_j+1) &= 1, && n_j \geq 0, \\
p_{ju^+}^{\mathrm{A}}(n_j, n_j+1) &= 1, && n_j \geq 0, u = 1, 2, \ldots, I^+, \\
p_{ju^-}^{\mathrm{A}}(n_j, n_j-1) &= 1, && n_j \geq 1, u = 1, 2, \ldots, I^-, \\
p_{ju^-}^{\mathrm{A}}(0, 0) &= 1, && u = 1, 2, \ldots, I^-, \\
q_{jc}^{\mathrm{D}}(n_j, n_j-1) &= \mu_j\,, && n_j \geq 1.
\end{aligned}$$

The triggering probabilities are

$$\begin{aligned}
f_{jc,w}(n_j, n_j') &= 0, && w = c, u^+, v^-, \text{ and } n_j, n_j' \geq 0, \\
f_{ju^+,u^+}(n_j, n_j+1) &= 1, && n_j \geq 0, u = 1, 2, \ldots, I^+, \\
f_{ju^-,u^-}(n_j, n_j-1) &= 1, && n_j > 0.
\end{aligned}$$

The routing probabilities are defined as follows. Upon a service completion at node j, a customer goes to node k as a regular customer with probability $r_{jc,kc}$, as a class v^+ positive signal with probability r_{jc,kv^+}, as a class v^- negative signal with probability r_{jc,kv^-}, and it leaves the network with probability $r_{jc,0}$. Clearly,

$$\sum_{k=1}^{N}\left(r_{jc,kc} + \sum_{v=1}^{I^+} r_{jc,kv^+} + \sum_{v=1}^{I^-} r_{jc,kv^-}\right) + r_{jc,0} = 1, \qquad j = 1, \ldots, N.$$

The arrival of a class u^+ positive signal at node j, either from the outside or from another node, adds one customer to node j, and the signal leaves immediately for node k as a customer with probability $r_{ju^+,kc}$, as a class v^+ positive signal with probability r_{ju^+,kv^+}, as a class v^- negative signal with probability r_{ju^+,kv^-}, and it leaves the network with probability $r_{ju^+,0}$. Again,

$$\sum_{k=1}^{N}\left(r_{ju^+,kc}+\sum_{v=1}^{I^+}r_{ju^+,kv^+}+\sum_{v=1}^{I^-}r_{ju^+,kv^-}\right)+r_{ju^+,0}=1,$$
$$j=1,\ldots,N,\ u=1,2,\ldots,I^+.$$

Finally, the arrival of a class u^- negative signal at node j, either from the outside or from another node, triggers one customer from the node to depart, provided the queue is not empty upon its arrival. The triggered customer then goes to node k as a customer with probability $r_{ju^-,kc}$, as a class v^+ positive signal with probability r_{ju^-,kv^+}, as a class v^- negative signal with probability r_{ju^-,kv^-}, and it leaves the network with probability $r_{ju^-,0}$. Again we must have

$$\sum_{k=1}^{N}\left(r_{ju^-,kc}+\sum_{v=1}^{I^+}r_{ju^-,kv^+}+\sum_{v=1}^{I^-}r_{ju^-,kv^-}\right)+r_{ju^-,0}=1,$$
$$j=1,\ldots,N,\ u=1,2,\ldots,I^-.$$

The arrival of a negative signal at an empty node does not have any effect and disappears.

Remark 4.16 The only additional feature of the network with multiple classes of positive and negative signals is the class-dependent routing. However, the class-dependent routing is very useful and it is general enough to include many queueing networks with batch arrivals and batch services as special cases.

Let α_{jc}, $\{\alpha^+_{ju};\ j=1,\ldots,N, u=1,\ldots,I^+\}$, and $\{\alpha^-_{ju};\ j=1,\ldots,N, u=1,\ldots,I^-\}$ denote the average arrival rates of customers, positive signals, and negative signals at node j. They are determined by the traffic equations

$$\alpha_j=\lambda_{jc}+\sum_{k=1}^{N}\rho_k\mu_k r_{kc,jc}+\sum_{k=1}^{N}\sum_{v=1}^{I^-}\rho_k\alpha^-_{kv}r_{kv^-,jc}+\sum_{k=1}^{N}\sum_{v=1}^{I^+}\rho_k^{-1}\alpha^+_{kv}r_{kv^-,ju},$$
$$j=1,\ldots,N; \tag{5.7}$$

$$\alpha^+_{ju}=\lambda^+_{ju}+\sum_{k=1}^{N}\rho_k\mu_k r_{kc,ju^+}+\sum_{k=1}^{N}\sum_{v=1}^{I^-}\rho_k\alpha^-_{kv}r_{kv^-,ju^+}+\sum_{k=1}^{N}\sum_{v=1}^{I^+}\rho_k^{-1}\alpha^+_{kv}r_{kv^+,ju^+},$$
$$j=1,\ldots,N, u=1,\ldots,I^+; \tag{5.8}$$

$$\alpha^-_{ju}=\lambda^-_{ju}+\sum_{k=1}^{N}\rho_k\mu_k r_{kc,ju^-}+\sum_{k=1}^{N}\sum_{v=1}^{I^-}\rho_k\alpha^-_{kv}r_{kv^-,ju^-}+\sum_{k=1}^{N}\sum_{v=1}^{I^+}\rho_k^{-1}\alpha^+_{kv}r_{kv^+,ju^-},$$
$$j=1,\ldots,N, u=1,2,\ldots,I^-, \tag{5.9}$$

where

$$\rho_j = \frac{\alpha_j + \alpha_j^+}{\mu_j + \alpha_j^-}, \qquad j = 1, \ldots, N,$$

and α_j^+ and α_j^- are defined as

$$\alpha_j^+ = \sum_{v=1}^{I^+} \alpha_{jv}^+, \qquad j = 1, \ldots, N,$$

$$\alpha_j^- = \sum_{v=1}^{I^-} \alpha_{jv}^-, \qquad j = 1, \ldots, N.$$

Clearly, α_j, α_j^+ and α_j^- are the average arrival rates of customers, positive signals and negative signals at node j. Also, the total average arrival rate of customers, including regular customers and those added by positive signals, is $\alpha_j + \alpha_j^+$. An argument similar to the proof of Theorem 4.12 leads to the following result.

Theorem 4.17 *Suppose the traffic equations (5.7), (5.8) and (5.9) have nonnegative solutions such that*

$$\rho_j \equiv \frac{\alpha_j + \alpha_j^+}{\mu_j + \alpha_j^-} < 1, \qquad \textit{for all } j = 1, \ldots, N.$$

If the network is modified so that whenever node j is empty there is an additional departure process of class u^+ positive signals with rate

$$\frac{\mu_j + \alpha_j^-}{\alpha_j + \alpha_j^+} \alpha_{ju}^+,$$

then the stationary probability of the network is

$$\pi(\boldsymbol{n}) = \prod_{j=1}^{N} \left(1 - \rho_j\right) \rho_j^{n_j}. \tag{5.10}$$

The following result gives conditions under which the product form distribution is a stochastic upper bound of the original network without additional departures.

Corollary 4.18 *Let π^0 be the stationary distribution of the network without the additional departures of positive signals. If $r_{ju^+,kv^-} = 0$ and $r_{ju^+,kc} r_{kc,\ell w^-} = 0$ for all j, k, ℓ u, v and w, then π^0 is stochastically dominated by the geometric product form π of (5.10).*

The following example illustrates how the multiple classes of signals can be used to model batch movements.

Example 4.19 (Networks with customer coalescence) Consider a network of N single-server nodes. Customers arrive at node j from the outside according to a Poisson process with rate λ_j, $j = 1, 2, \ldots, N$. The customers are served in batches of a fixed size K_j, and the service time of a batch is exponentially distributed with rate μ_j. Upon a service completion at node j, the K_j customers coalesce into a single customer, and this single customer goes to node k with probability r_{jk}, $k = 0, 1, \ldots, N$, where 0 is the outside world. If there are fewer than K_j customers at node j upon a service completion at the node, then these customers coalesce into a partial batch and are removed from the system. If a customer arrives at node j when the number of customers at that node is less than K_j, it joins the batch being served; otherwise it waits in the queue.

This model is a special case of the network with multiple classes of negative signals. To see this, consider a network with a single class of customers and $K_j - 1$ classes of negative signals at node j, denoted by u^- for $u = 1, 2, \ldots, K_j - 1$. Customers arrive at node j from the outside according to a Poisson process with rate λ_j. The customers are served one at a time and the service rate at node j is μ_j. The routing probabilities are

$$\begin{aligned} r_{jc,j(K_j-1)^-} &= 1, && j = 1, \ldots, N, \\ r_{ju^-,j(u-1)^-} &= 1, && u = 2, 3, \ldots, K_j - 1, \ j = 1, \ldots, N, \\ r_{j1^-,kc} &= r_{jk}, && k = 0, 1, \ldots, N, \ j = 1, \ldots, N. \end{aligned}$$

That is, upon a service completion at node j a customer goes back to node j as a class $(K_j - 1)^-$ negative signal with probability 1; the arrival of a class u^-, $u = 2, 3, \ldots, K_j - 1$, negative signal at node j removes one customer and then immediately goes to node j as a class $(u - 1)^-$ signal with probability 1; a class 1^- negative signal arrives at node j and reduces the number of customers by 1, then goes to node k as a regular customer with probability r_{jk}. This implies that a regular service completion at node j instantaneously removes K_j customers from node j, provided there are at least K_j customers present, and then goes to node k as a regular customer with probability r_{jk}. That a negative signal that arrives at an empty queue disappears translates into the fact that, when there are fewer than K_j customers at node j at a service completion, the entire batch is removed from the network.

By Theorem 4.17, this network has a geometric product form stationary distribution. Since there are no positive signals, no additional departure process is required.

This model is referred to as a *network with customer coalescence.* It will be discussed in more detail in Chapter 7. □

4.6 REFERENCE NOTES

There are several approaches for proving product form solutions for networks with quasi-reversible nodes. The delay line approach in Section 4.1 was first developed by Walrand (1983*b*) (see also Walrand (1988)) for conventional networks without signals and was extended by Chao and Pinedo (1993) for networks with signals and instantaneous movements. The formulation of the general queueing network in Section 4.2 is taken from Chao and Miyazawa (1996*a*). The reversed time approach employed in Section 4.3 is also taken from Chao and Miyazawa (1996*a*); it is motivated by Kelly (1979, 1982) and Henderson, Pearce, Pollett and Taylor (1992). The network with signals and instantaneous movements in Section 4.4 was first introduced and studied by Chao and Miyazawa (1996*b*), in which a different proof was given for Theorem 4.3. The reversed process proof in Section 4.4 is taken from Henderson and Taylor (1997). A more general approach was presented in Miyazawa (1998). Jump-over blocking has been analyzed by a number of researchers, see Van Dijk (1993) and Henderson and Taylor (1997), among others. Example 4.7 is taken from Gelenbe (1991), and Example 4.11 from Henderson and Taylor (1997).

The bathroom problem has been studied by a number of authors. The results in the literature are based on diffusion approximations under heavy traffic and asymptotic analysis using large deviations. See, for example, Flatto and Hahn (1984), Wright (1992), and Shwartz and Weiss (1993). The upper bound obtained in Example 4.15 is exactly the same asymptotic bound obtained by Shwartz and Weiss. The materials presented in Sections 4.4 and 4.5 are from Chao and Miyazawa (1996*b*), even though the notation is slightly different.

EXERCISES

4.1 Consider an $M/M/1$ queue with feedback. Exogenous customers arrive at the queue according to a Poisson process with rate λ, and their service times are i.i.d. and exponentially distributed with mean $1/\mu$. A customer that completes its service returns to the queue with probability p and leaves the queue with probability $q = 1 - p$. Let $X(t)$ be the number of customers in the system at time t.

(a) Compute the stationary distribution of $X(t)$.

(b) Argue that the overall arrival process at the queue is not Poisson.

4.2 Describe the Jackson network of Example 2.9, using the formulation in Section 4.2. Verify, using Theorem 4.3, the stationary distribution of the network obtained in Example 2.17.

4.3 Let queues 1 and 2 be the batch departure queues of Example 3.6 with arrival rates λ_1, λ_2, and service rates μ_1, μ_2. Characterize the two nodes using the notation p^{A}_{ju}, q^{D}_{ju} and q^{I}_{j}, and connect these two nodes into a tandem network. Assume that customers arrive at node 1 according to a Poisson process with rate λ, and full batch departures from node 1 join node 2 as single customers. Find the stationary distribution of the network.

4.4 Modify the tandem queue of Exercise 4.3 by adding a feedback loop from node 2 to node 1, such that a full batch departure from node 2 joins node 1 as a single customer with probability p, where $0 < p < 1$. Find the stationary distribution of the network.

4.5 Construct a time reversed network process for the network of Example 4.7 using the proof of Theorem 4.3.

4.6 Apply Theorem 4.17 to write down the traffic equations for the network with customer coalescence described in Example 4.19.

4.7 Consider a network of finite buffer queues and two classes of entities: customers and negative signals, denoted by c and s. Each node is of the class of queue discussed in Exercise 3.4. Node j has a single server with service rate μ_j, and a finite buffer of size K_j, $j = 1, 2, \ldots, N$. Customers and negative signals arrive at node j from the outside according to independent Poisson processes with rates λ_j and λ_j^-. When a customer completes its service at node j, it joins node k as a customer with probability $r_{jc,kc}$, and as a negative signal with probability $r_{jc,ks}$, $k = 0, 1, \ldots, N$. When a negative signal arrives at node j, it triggers a customer to depart for node k as a customer with probability $r_{js,kc}$ and as a negative signal with probability $r_{js,ks}$, $k = 0, 1, \ldots, N$. When a customer arrives at node j and node j has fewer than K_j customers present, the customer enters the queue; otherwise, it jumps over the queue and joins node k as a customer with probability

$$\frac{\mu_j}{\mu_j + \alpha_j^-} r_{jc,kc} + \frac{\alpha_j^-}{\mu_j + \alpha_j^-} r_{js,kc},$$

and as a negative signal with probability

$$\frac{\mu_j}{\mu_j + \alpha_j^-} r_{jc,ks} + \frac{\alpha_j^-}{\mu_j + \alpha_j^-} r_{js,ks},$$

where α_j and α_j^- are the solution to the traffic equations (4.7) and (4.8). Show that the stationary distribution of this network is

$$\pi(n_1, \ldots, n_N) = \prod_{j=1}^{N} \left(1 - \frac{\alpha_j}{\mu_j + \alpha_j^-}\right) \left(\frac{\alpha_j}{\mu_j + \alpha_j^-}\right)^{n_j},$$

$$n_j \leq K_j, \ j = 1, \ldots, N.$$

4.8 Consider a queueing network with N nodes and a single class of transitions. The state space of node j is $\mathcal{S}_j$, and node j is specified by $p_j^{\mathrm{A}}(x_j, x_j'), q_j^{\mathrm{D}}(x_j, x_j')$ and $q_j^{\mathrm{I}}(x_j, x_j')$, $x_j, x_j' \in \mathcal{S}_j$, $j = 0, 1, \ldots, N$, where node 0 represents the outside and assume that it is a Poisson source, i.e., $\mathcal{S}_0 = \{0\}$, $p_0^{\mathrm{A}}(0, 0) = 1$, and $q_0^{\mathrm{D}}(0, 0) = \beta_0$. A departure from node j joins node k as an arrival with probability r_{jk}, $j, k = 0, 1, \ldots, N$. Let $\boldsymbol{x} = (x_1, \ldots, x_N)$ be the state of the network. Let $\boldsymbol{x}_j(x_j')$ denote vector $\boldsymbol{x}$ with its j-th component replaced by x_j' and other components unchanged. Assume that the stationary distribution of the network exists and denote it by $\pi(\boldsymbol{x})$. Furthermore, assume that there exists a set of nonnegative numbers β_j, $j = 1, \ldots, N$ that satisfy

$$\sum_{x_j' \in \mathcal{S}_j} \pi(\boldsymbol{x}_j(x_j')) q_j^{\mathrm{D}}(x_j', x_j) = \beta_j \pi(\boldsymbol{x}), \qquad j = 1, \ldots, N.$$

Prove that the stationary distribution $\pi(\boldsymbol{x})$ takes the form $\pi(\boldsymbol{x}) = \prod_{j=1}^{N} \pi_j(x_j)$. (Note that the above condition is not quasi-reversibility!)

5

Networks with Exponential Service Times

This chapter focuses on a class of queueing networks with exponential service time distributions. Section 5.1 discusses multi-class Jackson networks and Kelly networks, and Section 5.2 covers networks with single-server nodes and negative signals. Section 5.3 considers networks with multi-server nodes.

5.1 JACKSON AND KELLY NETWORKS

Consider a network with N nodes and I classes of customers. Customers of class u arrive at node j from the outside according to a Poisson process with rate λ_{ju}, $u = 1, \ldots, I$, and $j = 1, \ldots, N$. Customers of each class have their own routing probabilities, and when a customer finishes service at one node, it may change its customer class. We call the amount of service a customer requires at a node its *service requirement*. Under the general service discipline considered in this chapter, the service requirement is different from the amount of time a customer spends in service, because the server may not be fully dedicated to a single customer at any point in time, e.g., processor sharing. The service requirement is, however, equal to its service time when the service discipline is First–Come First–Served (FCFS) and there is a single server at the node.

For each $j = 1, \ldots, N$, node j has an unlimited number of service positions which are numbered $1, 2, \ldots$. When there are n_j customers, positions $1, 2, \ldots, n_j$ are occupied. Assume that the following conditions are satisfied at node j when n_j customers are present.

(i) Class u customers have exponential service requirements with rate μ_{ju}.

(ii) The server provides service at rate $\phi_j(n_j)$, where $\phi_j(n_j) > 0$ if $n_j \geq 1$.

(iii) Upon arrival, a customer takes position ℓ with probability $\delta_j^{\mathrm{A}}(\ell, n_j + 1)$, $\ell = 1, 2, \ldots, n_j + 1$; customers previously in positions $\ell, \ell + 1, \ldots, n_j$ move into positions $\ell + 1, \ell + 2, \ldots, n_j + 1$, respectively. Clearly

$$\sum_{\ell=1}^{n_j+1} \delta_j^{\mathrm{A}}(\ell, n_j + 1) = 1.$$

(iv) A proportion $\delta_j^{\mathrm{D}}(\ell, n_j)$ of the total service rate is directed to the customer in position ℓ, $\ell = 1, \ldots, n_j$; when its service is completed, customers in positions $\ell+1, \ell+2, \ldots, n_j$ move into positions $\ell, \ell+1, \ldots, n_j - 1$, respectively. Again we must have

$$\sum_{\ell=1}^{n_j} \delta_j^{\mathrm{D}}(\ell, n_j) = 1.$$

Note that the superscripts "A" and "D" in the operating discipline stand for "arrival" and "departure". In condition (ii), if $\phi_j \equiv 1$, then node j is said to have a *single-server*. Thus, a single server node may serve more than one customer at once, but its total service effort is independent of the number of customers at the node.

Remark 5.1 To see the relationship between a customer's service time and its service requirement, assume that node j contains n_j customers, and consider the customer in position ℓ. The node is providing service at rate $\phi_j(n_j)$, and the service effort is allocated to position ℓ at a rate $\phi_j(n_j)\delta_j^{\mathrm{D}}(\ell, n_j)$ per unit time. When the amount of service the customer has received reaches its service requirement, the customer leaves the queue. The intensity at which a class u customer in position ℓ leaves is $\phi_j(n_j)\delta_j^{\mathrm{D}}(\ell, n_j)\mu_{ju}$. For example, when $\delta_j^{\mathrm{D}}(\ell, n_j) = 1/n_j$, the service effort is equally divided among all the n_j customers at node j. If, in addition, $\phi_j \equiv 1$, then node j is a single-server processor-sharing queue. In this case the departure rate of a class u customer at node j is, with n_j customers present, μ_{ju}/n_j.

By varying the functions δ_j^{D} and δ_j^{A}, we can construct various service disciplines. The following example illustrates how to construct a FCFS queue with multiple servers.

Example 5.2 (A FCFS multi-server queue) Assume node j is a multi-server queue with M_j servers. The service requirement of a customer is exponentially distributed with rate μ_j, independent of its class. The service discipline is FCFS. This can be achieved by setting

$$\phi_j(n) = \min\{n, M_j\}, \qquad n \geq 0,$$

$$\delta_j^{\mathrm{D}}(\ell, n) = \begin{cases} 1/n, & \ell = 1, 2, \ldots, n; n = 1, 2, \ldots, M_j, \\ 1/M_j, & \ell = 1, 2, \ldots, M_j; n \geq M_j + 1, \\ 0, & \text{otherwise}, \end{cases}$$

$$\delta_j^{\mathrm{A}}(\ell, n) = \begin{cases} 1, & \ell = n, \\ 0, & \text{otherwise}. \end{cases}$$

□

By changing ϕ_j we can allow the servers to work faster when the queue becomes longer, and by varying δ_j^{A} we can alter the queueing discipline to, for example, last in first served or service in random order. Note, however, that we cannot model priority disciplines, since both δ_j^{D} and δ_j^{A} are independent of the customer class.

Assume that when a class u customer completes its service at node j, it goes to node k as a class v customer with probability $r_{ju,kv}$, and leaves the system with probability $r_{ju,0}$, where

$$\sum_{k=1}^{N} \sum_{v=1}^{I} r_{ju,kv} + r_{ju,0} = 1, \qquad u = 1, \ldots, I, \;\; j = 1, \ldots, N.$$

This network is referred to as a *multi-class Jackson network with service positions* and exponential service times. In this chapter we focus on this network and its variants that include signals, and in the next chapter we consider the case of general service time distributions. Our main interest is to obtain the stationary distribution of the network process.

Example 5.3 (A fixed routing network) Consider the network with five nodes depicted in Figure 5.1. Customers enter the system from queues 1 and 2 according

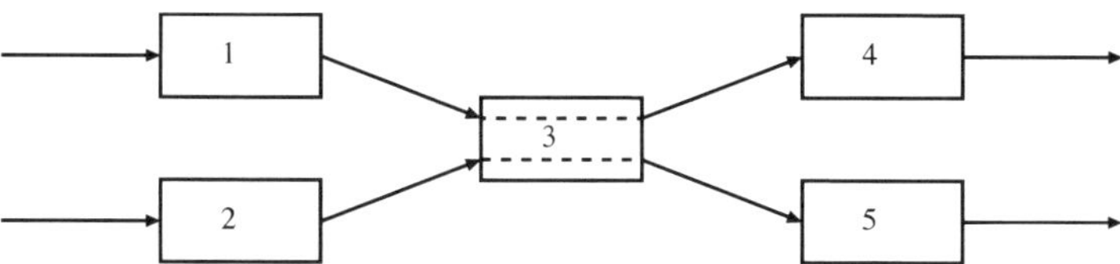

Figure 5.1 A fixed routing network

to independent Poisson processes, and they follow routes 1, 3, 4 and 2, 3, 5, respectively, before leaving the system. A customer that leaves node 3 does not select the next node randomly; it goes to node 4 if it has previously been at node 1, and it moves to node 5 if it has previously been at node 2.

To model this network, we need to define two classes of customers with class-dependent routing probabilities. Class 1 customers have the routing probabilities

$$r_{11,31} = 1, \qquad r_{31,41} = 1, \qquad r_{41,0} = 1,$$

and class 2 customers have the routing probabilities

$$r_{22,32} = 1, \qquad r_{32,52} = 1, \qquad r_{52,0} = 1.$$

Routing probabilities that are not defined are zero. □

Instead of a probabilistic routing, for each class of customers we may define a fixed route. A network with class-dependent fixed routings is referred to as a *Kelly network*. Here, assumptions (i)–(iv) are in effect. We could argue that a Kelly network is equivalent to the multi-class Jackson network just described. To this end, we introduce some notation for the Kelly network. Let class u customers arrive at node $u[1]$ according to a Poisson process with rate λ_u. They visit node $u[m]$ at stage m , $m = 2, \ldots, M(u)$, where $M(u)$ is the final stage on their route through the network. Let T be the set of customer classes. To describe this network as a multi-class Jackson network, we define a set of customer classes T^* by

$$T^* = \{(u, m); u \in T, m = 1, \ldots, M(u)\},$$

and the routing probabilities by

$$\begin{aligned} r_{u[m](u,m),u[m+1](u,m+1)} &= 1, \qquad m = 1, \ldots, M(u) - 1, \\ r_{u[M(u)](u,M(u)),0} &= 1. \end{aligned}$$

All other routing probabilities are zero. So, the Kelly network can be formulated as a multi-class Jackson network.

On the other hand, in the multi-class Jackson network, exogenous customers have at most countably many different routes through the system. If we define the customer class by its route through the network, then the network can be regarded as a Kelly network.

We now return to the multi-class Jackson network with service positions. Let $c_j(\ell)$ be the class of the customer in position ℓ at node j, and let

$$\boldsymbol{c}_j = (c_j(1), c_j(2), \ldots, c_j(n_j)),$$

where n_j is the number of customers at node j. The state of the network is represented by

$$\boldsymbol{c} = (\boldsymbol{c}_1, \ldots, \boldsymbol{c}_N).$$

Furthermore, let n_{ju} be the number of class u customers at node j, and let

$$\boldsymbol{n}_j = (n_{j1}, \ldots, n_{jI}),$$
$$\boldsymbol{n} = (\boldsymbol{n}_1, \ldots, \boldsymbol{n}_N).$$

Obviously

$$n_j = \sum_{u=1}^{I} n_{ju}.$$

This n_j is sometimes referred to as the population at node j.

Let α_{ju} be the overall arrival rate of class u customers at node j. Since any customer who enters a node will eventually leave, the overall departure rate from a node should remain the same, and, consequently, the traffic equations

$$\alpha_{ju} = \lambda_{ju} + \sum_{k=1}^{N}\sum_{v=1}^{I} \alpha_{kv} r_{kv,ju}, \qquad j = 1, 2, \ldots, N,\ u = 1, 2, \ldots, I, \quad (1.1)$$

have to be satisfied.

Note that traffic equations (1.1) uniquely determine the α_{ju}'s. Define the traffic intensity at node j by

$$\rho_j = \sum_{u=1}^{I} \frac{\alpha_{ju}}{\mu_{ju}}, \qquad j = 1, 2, \ldots, N.$$

The following is the main result of this section.

Theorem 5.4 *Consider a multi-class Jackson network with service positions and exponential service requirements. Assume that either one of the following two conditions holds:*

(a) μ_{ju} *is independent of the customer class* u.

(b) $\delta_j^{\mathrm{A}}(\ell, n_j) = \delta_j^{\mathrm{D}}(\ell, n_j)$ *for all* $\ell = 1, \ldots, n_j$. *This is called a symmetric service discipline.*

The stationary distribution of the network is then

$$\pi(\boldsymbol{c}) = \prod_{j=1}^{N} \pi_j(\boldsymbol{c}_j), \tag{1.2}$$

where $\pi_j(\boldsymbol{c}_j)$ *is the stationary distribution of node* j *assuming it is in isolation and subject to a Poisson arrival process of class* u *customers at rate* α_{ju}, *i.e.,*

$$\pi_j(\boldsymbol{c}_j) = b_j^{-1} \prod_{\ell=1}^{n_j} \frac{\alpha_{jc_j(\ell)}}{\phi_j(\ell)\mu_{jc_j(\ell)}}, \tag{1.3}$$

provided the normalization constant b_j *is finite, i.e.,*

$$b_j = \sum_{n=0}^{\infty} \frac{\rho_j^n}{\prod_{\ell=1}^{n} \phi_j(\ell)} < \infty.$$

PROOF. By Theorem 4.9 it suffices to verify the quasi-reversibility of each individual node. The operating disciplines are given by (i)–(iv) and either (a) or (b).

We first characterize node j using the notation of Chapter 4. For convenience, we define the operator $\ominus$ by

$$\boldsymbol{c}_j \ominus \mathbf{e}_\ell = (c_j(1), \ldots, c_j(\ell - 1), c_j(\ell + 1), \ldots, c_j(n_j)),$$

which is the state of the node after the customer in position ℓ leaves the node. The transition rate of such an event is

$$q^{\mathrm{D}}_{jc_j(\ell)}(\boldsymbol{c}_j, \boldsymbol{c}_j \ominus \mathbf{e}_\ell) = \mu_{jc_j(\ell)}\phi_j(n_j)\delta_j^{\mathrm{D}}(\ell, n_j). \tag{1.4}$$

Similarly, we define the operator $\oplus$ by

$$\boldsymbol{c}_j \oplus \mathbf{e}_m(u) = (c_j(1), \ldots, c_j(m - 1), u, c_j(m), \ldots, c_j(n_j)),$$

which represents the state of node j after a regular customer of class u arrives and takes position m, $m = 1, \ldots, n_j + 1$. The effect of a class u arrival is

$$p^{\mathrm{A}}_{ju}(\boldsymbol{c}_j, \boldsymbol{c}_j \oplus \mathbf{e}_m(u)) = \delta_j^{\mathrm{A}}(m, n_j + 1), \quad m = 1, \ldots, n_j + 1.$$

Node j, defined by

$$q_j(\boldsymbol{c}_j, \boldsymbol{c}_j') = \sum_{u=1}^{I} \alpha_{ju} p_{ju}^{\mathrm{A}}(\boldsymbol{c}_j, \boldsymbol{c}_j') + \sum_{\ell=1}^{n_j} q_{jc_j(\ell)}^{\mathrm{D}}(\boldsymbol{c}_j, \boldsymbol{c}_j'), \tag{1.5}$$

is subject to Poisson arrivals of class u customers with rate α_{ju}. The process obtained by reversing the time for queue q_j is expected to correspond to a similar queue but with δ_j^{A} and δ_j^{D} interchanged. Thus the departure rates of the reversed process of the queue are expected to be

$$\tilde{q}_{ju}^{\mathrm{D}}(\boldsymbol{c}_j \oplus \mathbf{e}_m(u), \boldsymbol{c}_j) = \mu_{ju}\phi_j(n_j+1)\delta_j^{\mathrm{A}}(m, n_j+1). \tag{1.6}$$

The arrival rate of class u customers in the reversed process remains α_{ju}, and the effect of arrivals is

$$\tilde{p}_{ju}^{\mathrm{A}}(\boldsymbol{c}_j, \boldsymbol{c}_j \oplus \mathbf{e}_\ell(u)) = \delta_j^{\mathrm{D}}(\ell, n_j+1). \tag{1.7}$$

Define the transition function $\tilde{q}_j$ by

$$\tilde{q}_j(\boldsymbol{c}_j, \boldsymbol{c}_j') = \sum_{u=1}^{I} \alpha_{ju} \tilde{p}_{ju}^{\mathrm{A}}(\boldsymbol{c}_j, \boldsymbol{c}_j') + \sum_{\ell=1}^{n_j} \tilde{q}_{jc_j(\ell)}^{\mathrm{D}}(\boldsymbol{c}_j, \boldsymbol{c}_j').$$

To show that π_j of (1.3) is indeed the stationary distribution of node j defined by q_j, it suffices to verify, by the detailed Kelly lemma (Lemma 2.23), the following conditions:

$$\pi_j(\boldsymbol{c}_j) q_{jc_j(\ell)}^{\mathrm{D}}(\boldsymbol{c}_j, \boldsymbol{c}_j \ominus \mathbf{e}_\ell) = \pi_j(\boldsymbol{c}_j \ominus \mathbf{e}_\ell)\alpha_{jc_j(\ell)} \tilde{p}_{jc_j(\ell)}^{\mathrm{A}}(\boldsymbol{c}_j \ominus \mathbf{e}_\ell, \boldsymbol{c}_j), \tag{1.8}$$

$$\pi_j(\boldsymbol{c}_j)\alpha_{ju} p_{ju}^{\mathrm{A}}(\boldsymbol{c}_j, \boldsymbol{c}_j \oplus \mathbf{e}_m(u)) = \pi_j(\boldsymbol{c}_j \oplus \mathbf{e}_m(u))\tilde{q}_{ju}^{\mathrm{D}}(\boldsymbol{c}_j \oplus \mathbf{e}_m(u), \boldsymbol{c}_j), \tag{1.9}$$

$$\sum_{\boldsymbol{c}_j'} q_j(\boldsymbol{c}_j, \boldsymbol{c}_j') = \sum_{\boldsymbol{c}_j'} \tilde{q}_j(\boldsymbol{c}_j, \boldsymbol{c}_j'), \tag{1.10}$$

$$\sum_{\boldsymbol{c}_j} \pi_j(\boldsymbol{c}_j) = 1. \tag{1.11}$$

Equality (1.11) follows from

$$1 + \sum_{n_j=1}^{\infty} \sum_{c_j(1)=1}^{I} \cdots \sum_{c_j(n_j)=1}^{I} \prod_{\ell=1}^{n_j} \frac{\alpha_{jc_j(\ell)}}{\phi_j(\ell)\mu_{jc_j(\ell)}} = \sum_{n_j=1}^{\infty} \frac{1}{\prod_{\ell=1}^{n_j} \phi_j(\ell)} \left(\sum_{u=1}^{I} \frac{\alpha_{ju}}{\mu_{ju}} \right)^{n_j}$$
$$= b_j\,.$$

Condition (1.10) follows from (a) or (b) since

$$\sum_{\boldsymbol{c}'_j} q_j(\boldsymbol{c}_j, \boldsymbol{c}'_j) = \sum_{u=1}^{I} \alpha_{ju} + \phi_j(n_j) \sum_{\ell=1}^{n_j} \mu_{jc_j(\ell)} \delta_j^{\mathrm{D}}(\ell, n_j), \tag{1.12}$$

$$\sum_{\boldsymbol{c}'_j} \tilde{q}_j(\boldsymbol{c}_j, \boldsymbol{c}'_j) = \sum_{u=1}^{I} \alpha_{ju} + \phi_j(n_j) \sum_{\ell=1}^{n_j} \mu_{jc_j(\ell)} \delta_j^{\mathrm{A}}(\ell, n_j). \tag{1.13}$$

Finally, conditions (1.8) and (1.9) can be verified using the transition rates (1.4) to (1.7). For instance, the left-hand side of (1.8) can be written as

$$\begin{aligned}
\pi_j(\boldsymbol{c}_j) q^{\mathrm{D}}_{jc_j(\ell)}(\boldsymbol{c}_j, \boldsymbol{c}_j \ominus \mathbf{e}_\ell) &= \left(\pi_j(\boldsymbol{c}_j \ominus \mathbf{e}_\ell) \frac{\alpha_{jc_j(\ell)}}{\phi_j(n_j)\mu_{jc_j(\ell)}} \right) \phi_j(n_j) \mu_{jc_j(\ell)} \delta_j^{\mathrm{D}}(\ell, n_j) \\
&= \pi_j(\boldsymbol{c}_j \ominus \mathbf{e}_\ell)\, \alpha_{jc_j(\ell)}\, \delta_j^{\mathrm{D}}(\ell, n_j) \\
&= \pi_j(\boldsymbol{c}_j \ominus \mathbf{e}_\ell) \alpha_{jc_j(\ell)} \tilde{p}^{\mathrm{A}}_{jc_j(\ell)}(\boldsymbol{c}_j \ominus \mathbf{e}_\ell, \boldsymbol{c}_j)\,.
\end{aligned}$$

This proves (1.8). Condition (1.9) can be shown in a similar way. Therefore, it follows from the detailed Kelly lemma that (1.3) is indeed the stationary distribution of the process q_j defined by (1.5).

Since

$$\begin{aligned}
\sum_{\boldsymbol{c}'_j} \pi_j(\boldsymbol{c}'_j) q^{\mathrm{D}}_{ju}(\boldsymbol{c}'_j, \boldsymbol{c}_j) &= \sum_{\ell=1}^{n_j+1} \pi_j(\boldsymbol{c}_j \oplus \mathbf{e}_\ell(u)) q^{\mathrm{D}}_{ju}(\boldsymbol{c}_j \oplus \mathbf{e}_\ell(u), \boldsymbol{c}_j) \\
&= \sum_{\ell=1}^{n_j+1} \pi_j(\boldsymbol{c}_j) \frac{\alpha_{ju}}{\phi_j(n_j+1)\mu_{ju}} \mu_{ju} \delta_j^{\mathrm{D}}(\ell, n_j+1) \phi_j(n_j+1) \\
&= \alpha_{ju} \pi_j(\boldsymbol{c}_j),
\end{aligned}$$

node j is uniformly quasi-reversible for all α_{ju} such that $b_j < \infty$, with departure rates

$$\beta_{ju} = \alpha_{ju}, \quad u = 1, \ldots, I.$$

Substituting this into (3.5) of Chapter 4 yields (1.1). Applying Theorem 4.9 we conclude that, if α_{ju}, $j = 1, \ldots, N$, $u = 1, \ldots, I$, is the solution to traffic equations (1.1), then the network has the product form stationary distribution (1.2) with π_j being the stationary distribution of node j defined by q_j. This completes the proof of Theorem 5.4. □

Remark 5.5 There is no condition other than (a) or (b) that yields the product form stationary distribution with the marginal distributions of the form (1.3). To see this, first note that if the network has a product form stationary distribution, then q_j has to be the transition rate of the marginal process at node j. This fact will be formally proved in Chapter 11 (see Theorem 11.2). If the marginal stationary distribution is given by (1.3), then $\tilde{q}_j$ has to be the time-reversed transition rate for q_j. So the rates (1.12) and (1.13) must be equal. Thus, for any choice of $\boldsymbol{c}_j$ and n_j, we have

$$\sum_{\ell=1}^{n_j} \mu_{jc_j(\ell)}(\delta_j^{\mathrm{D}}(\ell, n_j) - \delta_j^{\mathrm{A}}(\ell, n_j)) = 0. \tag{1.14}$$

If condition (a) is not satisfied, then there exist a μ_{ju} and a μ_{jv} such that $\mu_{ju} \neq \mu_{jv}$. Substituting $\boldsymbol{c}_j = (u, u, \ldots, u)$ and $\boldsymbol{c}'_j = (v, u, \ldots, u)$ into (1.14) and subtracting one from the other yields that when $\ell = 1$

$$\delta_j^{\mathrm{D}}(\ell, n_j) = \delta_j^{\mathrm{A}}(\ell, n_j). \tag{1.15}$$

A similar argument shows that (1.15) holds for any $\ell = 1, \ldots, n_j$, i.e., (b) is satisfied.

The symmetric service discipline plays a central role in the generalization of the service time distribution from the exponential to an arbitrary one. This will be discussed in Chapter 6.

The following corollary concerns the stationary distribution of the number of each class of customers at each node.

Corollary 5.6 *The stationary distribution $\pi(\boldsymbol{n})$ of the multi-class Jackson network with service positions is*

$$\pi(\boldsymbol{n}) = \prod_{j=1}^{N} \pi_j(\boldsymbol{n}_j),$$

where

$$\pi_j(\boldsymbol{n}_j) = b_j^{-1} \frac{n_j!}{n_{j1}!n_{j2}!\cdots n_{jI}!} \frac{\prod_{u=1}^{I}(\alpha_{ju}/\mu_{ju})^{n_{ju}}}{\prod_{\ell=1}^{n_j}\phi_j(\ell)}. \tag{1.16}$$

Moreover, the probability that node j contains n customers is

$$b_j^{-1}\frac{\rho_j^n}{\prod_{\ell=1}^{n}\phi_\ell(\ell)}.$$

Given that there is a customer in position ℓ of node j, it is of class u with probability

$$\frac{\alpha_{ju}/\mu_{ju}}{\sum_{v=1}^{I}\alpha_{jv}/\mu_{jv}}.$$

Theorem 5.4 and Corollary 5.6 claim that in steady state, each node behaves as if it were in isolation and subject to Poisson arrivals. However, it is known that arrival processes at any node are, in general, not Poisson (see Remark 4.1 of Chapter 4). Even though the arrival processes at node j are not Poisson, the Arrivals See Time Averages (ASTA) at each node, as shown in the next result.

Corollary 5.7 (ASTA) *Suppose the network is in steady state. Whenever a class u customer arrives at a node, either from another node or from outside the network, the probability that this customer sees the network in state $\boldsymbol{c}$ is $\pi(\boldsymbol{c})$.*

PROOF. If the customer arrives from the outside, then the result follows from the PASTA (Poisson Arrivals See Time Averages) property. Now assume that the customer arrives from within the network. Assume that a class u customer arrives at node j from node k. The probability that it observes the network in state $\boldsymbol{c}$ can be computed as follows:

$$
\begin{aligned}
&P(\text{the arrival sees state } \boldsymbol{c} \mid \text{a class } u \text{ customer arrives at node } j \text{ from node } k) \\
&= \frac{P(\text{a class } u \text{ customer arriving at node } j \text{ from node } k \text{ sees state } \boldsymbol{c})}{P(\text{a class } u \text{ customer arrives at node } j \text{ from node } k)} \\
&= \frac{\prod_{i\neq k}^{N} \pi_i(\boldsymbol{c}_i) \sum_{\ell=1}^{n_k+1} \sum_{v=1}^{I} \pi_k(\boldsymbol{c}_k \oplus \mathbf{e}_{\ell v})\mu_k\phi_k(n_k+1)\delta_k^{\mathrm{D}}(\ell, n_k+1) r_{kv,ju}}{\sum_{\boldsymbol{c}'} \prod_{i\neq k}^{N} \pi_i(\boldsymbol{c}'_i) \sum_{\ell=1}^{n'_k+1} \sum_{v=1}^{I} \pi_k(\boldsymbol{c}'_k \oplus \mathbf{e}_{\ell v})\mu_k\phi_k(n'_k+1)\delta_k^{\mathrm{D}}(\ell, n'_k+1) r_{kv,ju}} \\
&= \frac{\prod_{i=1}^{N} \pi_i(\boldsymbol{c}_i)\alpha_{ju}}{\sum_{\boldsymbol{c}'} \prod_{i=1}^{N} \pi_i(\boldsymbol{c}'_i)\alpha_{ju}} \\
&= \prod_{i=1}^{N} \pi_i(\boldsymbol{c}_i) \\
&= \pi(\boldsymbol{c}).
\end{aligned}
$$

This completes the proof. □

5.2 NETWORKS WITH NEGATIVE SIGNALS

In this section we incorporate negative signals into the queueing network models discussed in the previous section. In the remaining sections of this chapter we consider networks satisfying (a) or (b) of Theorem 5.4 under the assumption that the total service effort provided at each node is a constant, which, without loss of generality, is assumed to be one, i.e., $\phi_j(n_j) = 1$ for all n_j.

The following example illustrates the effect of a negative signal in a queueing network.

Example 5.8 (A quality control network) Consider the five-node network of Example 5.3. The two classes of customers (jobs) arrive at nodes 1 and 2, and pass through nodes 1, 3, 4 and 2, 3, 5, respectively, before leaving the system. Node 3 is a quality control station, which may detect some quality problems, and if one occurs, a message is sent back to either node 1 or node 2, depending upon where the customer came from. If a job came from node 1, it is defective with probability 1/8, and if the job came from node 2, it is defective with probability 1/4. Assume that when a job is found defective at node 3, a message is sent to its upstream node to trigger the customer being processed to leave the system. Furthermore, customers found defective are scrapped.

Clearly, the signals cannot be treated in the same way as customers, because they do not request services. In this example the signal is simply an instruction that the customer in service leaves the system. To model this problem as a queueing network with negative signals, we define two classes of customers and one class of signals. Customers of class 1 enter the system at node 1, and go through the network via nodes 1, 3, and 4; while customers of class 2 enter the system at node 2, and pass through the network via nodes 2, 3, and 5. When a class 1 customer completes service at node 3, it becomes a signal with probability 1/8 and returns to node 1, or it remains a customer and goes to node 4 with probability 7/8; when a class 2 customer completes its service at node 3, it becomes a signal and returns to node 2 with probability 1/4, or it remains a customer and goes to node 5 with probability 3/4. Customers completing service at nodes 4 and 5 leave the system. When a signal arrives at a non-empty node, it triggers the customer in service to leave the system; if a signal arrives at an empty node, nothing occurs. The network is depicted in Figure 5.2. □

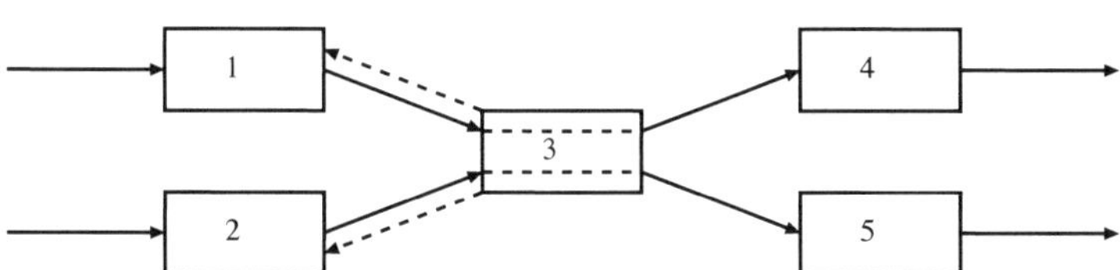

Figure 5.2 The quality control network

Consider a network with N nodes, I classes of customers, and one class of signals. Each node has $I + 1$ classes of arrivals and $2I$ classes of departures. A class u customer that departs due to a regular service completion is classified as a class u departure, and a class u customer that departs due to the triggering of a signal is classified as a class u^- departure. The I classes of customers are denoted by $\{u; u = 1, \ldots, I\}$, the single class of arrival signal is denoted by s^-, and the I

classes of departure triggered by signals are denoted by $\{u^-; u = 1, \ldots, I\}$. Thus the class of arrivals and departures is

$$T_j = \{u; u = 1, \ldots, I\} \cup \{s^-\} \cup \{u^-; u = 1, \ldots, I\}.$$

Customers of class u arrive at node j from outside the network according to a Poisson process with rate λ_{ju}, $u = 1, \ldots, I$, and $j = 1, \ldots, N$. Negative signals arrive at node j from the outside according to a Poisson process with rate λ_j^-. When a class u customer completes its service at node j, it goes to node k as a class v customer with probability $r_{ju,kv}$, as a signal with probability r_{ju,ks^-}, and it leaves the network with probability $r_{ju,0}$. These routing probabilities satisfy

$$\sum_{k=1}^{N}\sum_{v=1}^{I} r_{ju,kv} + \sum_{k=1}^{N} r_{ju,ks^-} + r_{ju,0} = 1, \quad j = 1, \ldots, N, \; u = 1, \ldots, I.$$

Whenever a signal arrives at node j, either from outside the network or from another node, it induces a service completion. In other words, when a signal arrives at node j, a customer departs immediately. If this customer is of class u, it departs as a class u^- signal, and goes to node k as a class v customer with probability $r_{ju^-,kv}$, as a signal with probability r_{ju^-,ks^-}, and leaves the network with probability $r_{ju^-,0}$. These probabilities satisfy

$$\sum_{k=1}^{N}\sum_{v=1}^{I} r_{ju^-,kv} + \sum_{k=1}^{N} r_{ju^-,ks^-} + r_{ju^-,0} = 1, \quad u = 1, \ldots, N, \; u = 1, \ldots, I.$$

When a signal arrives at an empty node, nothing happens.

Example 5.9 (Formulation of the quality control network) Consider the network of Example 5.8. There are five nodes and two classes of customers, so $N = 5$, and $I = 2$. The nonzero routing probabilities are

$$\begin{aligned} &r_{11,31} = 1, && r_{31,1s^-} = 1/8, && r_{31,41} = 7/8, && r_{41,0} = 1, \\ &r_{22,32} = 1, && r_{32,2s^-} = 1/4, && r_{32,52} = 3/4, && r_{52,0} = 1. \end{aligned}$$

□

The operational rules for the customers at each node are the same as those in Theorem 5.4 except that assumption (ii) is simplified to

(ii′) The total service effort at node j is provided at rate 1.

Assumptions (i), (iii) and (iv) remain unchanged, and either (a) or (b) of Theorem 5.4 is assumed. That is, the service requirement of a class u customer at node j has the exponential distribution with rate μ_{ju}. When a customer arrives at the node or completes its service, the state of the node changes according to the two sets of parameters $\delta_j^{\mathrm{A}}(\ell, n_j)$ and $\delta_j^{\mathrm{D}}(\ell, n_j)$, respectively, $\ell = 1, \ldots, n_j$. These two sets of parameters may be different when μ_{ju} is independent of u. Otherwise, they are identical. To specify the effect of the arrival of a signal, we need an additional set of parameters $\eta_j(\ell, n_j)$, $\ell = 1, \ldots, n_j$.

(v) When a signal arrives at node j, it triggers the customer in position ℓ to leave with probability $\eta_j(\ell, n_j)$, $\ell = 1, \ldots, n_j$; with its departure, customers in positions $\ell + 1, \ell + 2, \ldots, n_j$ move into positions $\ell, \ell + 1, \ldots, n_j - 1$, respectively. If the triggered customer is of class u, the departure is classified as a signal of class u^-.

Similar to the models in the previous section, this class of service disciplines, determined by δ_j^{D}, δ_j^{A}, and η_j, is very general. By properly selecting the parameters, we can obtain many different operational disciplines.

Example 5.10 (The effect of signals) Suppose node j is characterized by

$$\delta_j^{\mathrm{D}}(\ell, n) = 1/n, \qquad \ell = 1, \ldots, n,$$

$$\delta_j^{\mathrm{A}}(\ell, n) = \begin{cases} 1, & \ell = n, \\ 0, & \text{otherwise}, \end{cases}$$

and

$$\eta_j(\ell, n) = \begin{cases} 1, & \ell = [(n+1)/2], \\ 0, & \text{otherwise}, \end{cases}$$

where $[x]$ is the integer part of number x. Node j behaves like a single-server queue in which customers are served in random order, and an arriving signal triggers the customer in the middle of the queue to leave immediately. □

It should be noted that the effect of signals, which is determined by η_j, may be different from the service discipline, even though they both lead to a service completion.

If $c_j(\ell)$ denotes the class of the customer in position ℓ of node j, then the vector

$$\boldsymbol{c}_j = (c_j(1), c_j(2), \ldots, c_j(n_j))$$

describes the state of node j and the vector

$$\boldsymbol{c} = (\boldsymbol{c}_1, \boldsymbol{c}_2, \ldots, \boldsymbol{c}_N)$$

represents the state of the entire network.

Let α_{ju} denote the average arrival rate of class u customers at node j and α_j^- the average arrival rate of signals at node j. Since the signal arrival process at node j has a similar effect as the service process, the total potential customer departure rate from node j is $\alpha_j^- + \mu_{ju}$. Hence, the traffic intensity of class u customers at node j is

$$\rho_{ju} = \frac{\alpha_{ju}}{\alpha_j^- + \mu_{ju}},$$

and the overall traffic intensity of all customers at node j is

$$\rho_j = \sum_{u=1}^{I} \rho_{ju}\,.$$

Since μ_j is the maximum departure rate of customers due to service completions and α_j^- is the maximum departure rate of customers due to signal triggering, and the potential departures of classes u or u^- actually take place with probability ρ_{ju}, the departure rates of class u customers and class u^- signals at node j are expected to be $\rho_{ju}\mu_{ju}$ and $\rho_{ju}\alpha_j^-$, respectively. Thus, the arrival rates of customers and signals are determined by the traffic equations

$$\alpha_{ju} = \lambda_{ju} + \sum_{k=1}^{N}\sum_{v=1}^{I} \mu_{kv}\rho_{kv}r_{kv,ju} + \sum_{k=1}^{N}\sum_{v=1}^{I} \alpha_k^- \rho_{kv}r_{kv^-,ju}\,, \qquad j = 1, \ldots, N,\ u = 1, \ldots, I, \tag{2.1}$$

$$\alpha_j^- = \lambda_j^- + \sum_{k=1}^{N}\sum_{v=1}^{I} \mu_{kv}\rho_{kv}r_{kv,js^-} + \sum_{k=1}^{N}\sum_{v=1}^{I} \alpha_k^- \rho_{kv}r_{kv^-,js^-}, \qquad j = 1, \ldots, N. \tag{2.2}$$

Theorem 5.11 *Assuming either (a) or (b) of Theorem 5.4, i.e., either the service discipline of a node is symmetric or the service rate of a customer at the node does not depend on its class, the stationary distribution of the network described above is*

$$\pi(\boldsymbol{c}) = \prod_{j=1}^{N} \pi_j(\boldsymbol{c}_j), \tag{2.3}$$

where

$$\pi_j(\boldsymbol{c}_j) = (1 - \rho_j) \prod_{\ell=1}^{n_j} \rho_{jc_j(\ell)}\,. \tag{2.4}$$

Similar to Corollary 5.6, the stationary distribution of the number of customers at each node can be derived from Theorem 5.11.

Corollary 5.12 *The stationary probability for the network to be in state* $\boldsymbol{n}$ *is*

$$\pi(\boldsymbol{n}) = \prod_{j=1}^{N} \pi_j(\boldsymbol{n}_j),$$

where $\boldsymbol{n}_j = (n_{j1}, \ldots, n_{jI})$, n_{ju} *is the number of class* u *customers at node* j, *and*

$$\pi_j(\boldsymbol{n}_j) = (1 - \rho_j) \frac{n_j!}{n_{j1}! n_{j2}! \cdots n_{jI}!} \prod_{u=1}^{I} \rho_{ju}^{n_{ju}}.$$

Moreover, the stationary joint probability of the network having n_j *customers at node* j, $j = 1, \ldots, N$, *is*

$$\pi(n_1, \ldots, n_N) = \prod_{j=1}^{N} \left(1 - \rho_j\right) \rho_j^{n_j}.$$

The following corollary is left to the reader as an exercise. Note that if an entity is a signal, it may instantaneously pass through node j.

Corollary 5.13 *When an entity arrives at or passes through node* j, *either from the outside or from another node, the probability that it observes the network in state* $\boldsymbol{c}$ *just before it arrives or passes through node* j *is* $\pi(\boldsymbol{c})$.

Remark 5.14 If we view an $M/M/1$ queue with exponential service times as a system subject to a Poisson arrival process and a Poisson potential service process (with a dummy event taking place whenever a Poisson service epoch occurs and the system is empty), then Theorem 5.11 with assumption (a) can be stated as follows. When the system is in steady state, it behaves as if each node were in isolation subject to a Poisson arrival process of class u customers with rate α_{ju}, $u = 1, \ldots, I$, and a Poisson potential service process with rate $\mu_j + \alpha_j^-$. It can be argued that neither the arrival process nor the potential service process at a node are Poisson (see Exercise 5.2). Nevertheless, the system behaves as if they were! This is in contrast to Jackson networks. In a Jackson network each node behaves as if the arrivals are Poisson, even though they may not be Poisson. However, in a Jackson network the potential service process is always Poisson.

PROOF OF THEOREM 5.11 We prove Theorem 5.11 using Theorem 4.9. We need to show that each node, when in isolation, is a quasi-reversible queue with signals. There are $I+1$ classes of arrivals and $2I$ classes of departures, i.e., I classes of customers, denoted by $1, 2, \ldots, I$, one class of signal arrivals, denoted by s^-, and I classes of signal departures, denoted by $1^-, 2^-, \ldots, I^-$.

Node j, defined by

$$q_j(\boldsymbol{c}_j, \boldsymbol{c}_j') = \sum_{u=1}^{I} \alpha_{ju} p_{ju}^{\mathrm{A}}(\boldsymbol{c}_j, \boldsymbol{c}_j') + \alpha_j^- p_{js^-}^{\mathrm{A}}(\boldsymbol{c}_j, \boldsymbol{c}_j') + \sum_{\ell=1}^{n_j} q_{jc_j(\ell)}^{\mathrm{D}}(\boldsymbol{c}_j, \boldsymbol{c}_j'), \quad (2.5)$$

is subject to Poisson arrivals of class u customers with rate α_{ju} and Poisson arrivals of signals with rate α_j^-. The operational rules of this node are determined by δ_j^{A}, δ_j^{D}, and η_j. Let

$$\boldsymbol{c}_j = (c_j(1), c_j(2), \ldots, c_j(n_j))$$

be the state of the node when there are n_j customers present and the customer in position ℓ is of class $c_j(\ell)$. Recall that $\boldsymbol{c}_j \ominus \mathbf{e}_\ell$ denotes the state of the node after a customer of class $c_j(\ell)$ leaves position ℓ. The transition rate of this type of event, due to regular service completions, is

$$q_{jc_j(\ell)}^{\mathrm{D}}(\boldsymbol{c}_j, \boldsymbol{c}_j \ominus \mathbf{e}_\ell) = \mu_{jc_j(\ell)} \delta_j^{\mathrm{D}}(\ell, n_j). \quad (2.6)$$

The same state change may also be caused by the arrival of a signal, whose effect is described by

$$p_{js^-}^{\mathrm{A}}(\boldsymbol{c}_j, \boldsymbol{c}_j \ominus \mathbf{e}_\ell) = \eta_j(\ell, n_j). \quad (2.7)$$

Since $\boldsymbol{c}_j \oplus \mathbf{e}_m(u)$ denotes the state of the node after a class u customer arrives at position m, the probability function of the arrival effect is

$$p_{ju}^{\mathrm{A}}(\boldsymbol{c}_j, \boldsymbol{c}_j \oplus \mathbf{e}_m(u)) = \delta_j^{\mathrm{A}}(m, n_j+1), \qquad m = 1, \ldots, n_j+1.$$

Note, from (2.6) and (2.7), that the signal arrival process can actually be incorporated into the service process. Hence, queue j defined by q_j has transition rates:

$$q_j(\boldsymbol{c}_j, \boldsymbol{c}_j \ominus \mathbf{e}_\ell) = \mu_{jc_j(\ell)} \delta_j^{\mathrm{D}}(\ell, n_j) + \alpha_j^- \eta_j(\ell, n_j), \quad (2.8)$$

$$q_j(\boldsymbol{c}_j, \boldsymbol{c}_j \oplus \mathbf{e}_m(u)) = \alpha_{ju} \delta_j^{\mathrm{A}}(m, n_j+1). \quad (2.9)$$

Suppose the time reversed process is another queue with arrival rate α_{ju} and total departure rate $\alpha_j^- + \mu_{ju}$. Then, because arrivals and departures are interchanged under time reversal, the departure rate of a customer at position m is expected to be

$$\tilde{q}_j(\boldsymbol{c}_j \oplus \mathbf{e}_m(u), \boldsymbol{c}_j) = (\alpha_j^- + \mu_{ju})\delta_j^{\mathrm{A}}(m, n_j + 1), \qquad m = 1, \ldots, n_j + 1 . \tag{2.10}$$

Suppose the arrival of a class u customer in the reversed process chooses, when in state $\boldsymbol{c}_j$, position ℓ with probability

$$\frac{\alpha_j^- \eta_j(\ell, n_j) + \mu_{jc_j(\ell)}\delta_j^{\mathrm{D}}(\ell, n_j)}{\alpha_j^- + \mu_{jc_j(\ell)}},$$

which corresponds to the departure of a class u customer in the forward process. Thus the transition rate for the arrivals in the reversed process is

$$\tilde{q}_j(\boldsymbol{c}_j \ominus \mathbf{e}_\ell, \boldsymbol{c}_j) = \alpha_{jc_j(\ell)} \frac{\mu_{jc_j(\ell)}\delta_j^{\mathrm{D}}(\ell, n_j) + \alpha_j^- \eta_j(\ell, n_j)}{\alpha_j^- + \mu_{jc_j(\ell)}}, \tag{2.11}$$

or, equivalently,

$$\tilde{q}_j(\boldsymbol{c}_j, \boldsymbol{c}_j \oplus \mathbf{e}_\ell(u)) = \alpha_{ju} \frac{\mu_{ju}\delta_j^{\mathrm{D}}(\ell, n_j) + \alpha_j^- \eta_j(\ell, n_j)}{\alpha_j^- + \mu_{ju}} . \tag{2.12}$$

To show that (2.4) is indeed the stationary distribution of node j defined by (2.5), it suffices to verify the conditions of the Kelly lemma, i.e.,

$$\pi_j(\boldsymbol{c}_j)q_j(\boldsymbol{c}_j, \boldsymbol{c}_j \ominus \mathbf{e}_\ell) = \pi_j(\boldsymbol{c}_j \ominus \mathbf{e}_\ell)\tilde{q}_j(\boldsymbol{c}_j \ominus \mathbf{e}_\ell, \boldsymbol{c}_j), \tag{2.13}$$

$$\pi_j(\boldsymbol{c}_j)q_j(\boldsymbol{c}_j, \boldsymbol{c}_j \oplus \mathbf{e}_m(u)) = \pi_j(\boldsymbol{c}_j \oplus \mathbf{e}_m(u))\tilde{q}_j(\boldsymbol{c}_j \oplus \mathbf{e}_m(u), \boldsymbol{c}_j), \tag{2.14}$$

$$\sum_{\boldsymbol{c}_j'} q_j(\boldsymbol{c}_j, \boldsymbol{c}_j') = \sum_{\boldsymbol{c}_j'} \tilde{q}_j(\boldsymbol{c}_j, \boldsymbol{c}_j'). \tag{2.15}$$

The verification of these equations is straightforward. Since

$$\pi_j(\boldsymbol{c}_j \ominus \mathbf{e}_\ell)\frac{\alpha_{jc_j(\ell)}}{\alpha_j^- + \mu_{jc_j(\ell)}} = \pi_j(\boldsymbol{c}_j),$$

(2.13) follows from (2.8) and (2.11). Similarly, (2.14) can be verified, for each u and m, using (2.9) and (2.10). Finally, it follows from the definitions of q_j and $\tilde{q}_j$ together with (2.12) that

$$\sum_{\boldsymbol{c}_j'} q_j(\boldsymbol{c}_j, \boldsymbol{c}_j') = \sum_{u=1}^{I} \alpha_{ju} + \sum_{\ell=1}^{n_j} \mu_{jc_j(\ell)}\delta_j^{\mathrm{D}}(\ell, n_j) + \alpha_j^- 1[n_j > 0],$$

$$\sum_{\boldsymbol{c}_j'} \tilde{q}_j(\boldsymbol{c}_j, \boldsymbol{c}_j') = \sum_{u=1}^{I} \alpha_{ju} + \sum_{\ell=1}^{n_j} \mu_{jc_j(\ell)}\delta_j^{\mathrm{A}}(\ell, n_j) + \alpha_j^- 1[n_j > 0].$$

Hence either assumption (a) or (b) implies (2.15).

We now show that the queue is quasi-reversible. The triggering probabilities of the negative signals are

$$f_{js^-,u^-}(\mathbf{c}_j \oplus \mathbf{e}_{j\ell}(u), \mathbf{c}_j) = 1, \qquad u = 1, \dots, I.$$

It follows from (2.6) and (2.7) that

$$\begin{aligned}
\sum_{\mathbf{c}'_j} \pi_j(\mathbf{c}'_j) q^{\mathrm{D}}_{ju}(\mathbf{c}'_j, \mathbf{c}_j) &= \sum_{\ell=1}^{n_j+1} \frac{\alpha_{ju}}{\mu_{ju} + \alpha_j^-} \pi_j(\mathbf{c}_j) \mu_{ju} \delta_j^{\mathrm{D}}(\ell, n_j + 1) \\
&= \alpha_{ju} \frac{\mu_{ju}}{\mu_{ju} + \alpha_j^-} \pi_j(\mathbf{c}_j), \\
\sum_{\mathbf{c}'_j} \pi_j(\mathbf{c}'_j) q^{\mathrm{A}}_{js^-}(\mathbf{c}'_j, \mathbf{c}_j) f_{js^-,u^-}(\mathbf{c}'_j, \mathbf{c}_j) &= \sum_{\ell=1}^{n_j+1} \frac{\alpha_{jc_j(\ell)}}{\mu_{jc_j(\ell)} + \alpha_j^-} \pi_j(\mathbf{c}_j) \alpha_j^- \eta_j(\ell, n_j + 1) \\
&= \alpha_{jc_j(\ell)} \frac{\alpha_j^-}{\mu_{jc_j(\ell)} + \alpha_j^-} \pi_j(\mathbf{c}_j).
\end{aligned}$$

Hence the node is uniformly quasi-reversible with departure rates

$$\begin{aligned}
\beta_{ju} &= \mu_{ju} \frac{\alpha_{ju}}{\mu_{ju} + \alpha_j^-} = \mu_{ju} \rho_{ju}, \quad u = 1, \dots, I, \ j = 1, \dots, N, \\
\beta_{ju^-} &= \alpha_j^- \frac{\alpha_{ju}}{\mu_{ju} + \alpha_j^-} = \alpha_j^- \rho_{ju}, \quad u = 1, \dots, I, \ j = 1, \dots, N.
\end{aligned}$$

Applying Theorem 4.9, the stationary distribution of the network is given by (2.3) with α_{ju}, $u = 1, \dots, I$, and α_j^- being the solution to the traffic equations (2.2). □

The following two examples illustrate the results of this section.

Example 5.15 (A tandem queue with signals) Consider the tandem queue with two nodes in series discussed in Example 2.16. Customers arrive at node 1 according to a Poisson process with rate λ, and signals arrive at node 1 according to a Poisson process with rate λ^-. The service time distribution at node j is exponential with rate μ_j, $j = 1, 2$. When a customer completes its service at node 1, it goes to node 2 as a signal with probability p, and it goes to node 2 as a customer with probability $1 - p$. When a signal arrives at node 1, a customer, if at least one is present, is induced to leave the system immediately. A customer that either completes its service at node 2 or is induced to move by a signal at node 2 leaves the system. See Figure 5.3.

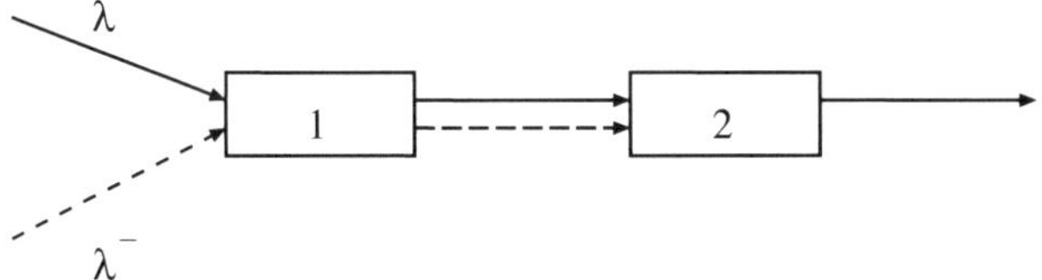

Figure 5.3 A tandem queue with signals

Denote the customers by c and signals by s^-. The routing probabilities of this network are

$$\begin{aligned} &r_{1c,2c} = 1 - p, \quad r_{1c,2s^-} = p, \quad r_{1s^-,0} = 1, \\ &r_{2c,0} = 1, \qquad\qquad r_{2s^-,0} = 1. \end{aligned}$$

All other routing probabilities are zero. The solution to the traffic equations is

$$\begin{aligned} \alpha_{1c} &= \lambda\,, \\ \alpha_1^- &= \lambda^-\,, \\ \alpha_{2c} &= (1-p)\alpha_{1c}\frac{\mu_1}{\alpha_1^- + \mu_1} = \frac{(1-p)\lambda\mu_1}{\lambda^- + \mu_1}\,, \\ \alpha_2^- &= p\alpha_{1c}\frac{\mu_1}{\alpha_1^- + \mu_1} = \frac{p\lambda\mu_1}{\lambda^- + \mu_1}\,. \end{aligned}$$

By Theorem 5.11, the stationary distribution for this tandem network is

$$\begin{aligned} \pi(n_1, n_2) &= \left(1 - \frac{\alpha_{1c}}{\alpha_1^- + \mu_1}\right)\left(\frac{\alpha_{1c}}{\alpha_1^- + \mu_1}\right)^{n_1}\left(1 - \frac{\alpha_{2c}}{\alpha_2^- + \mu_2}\right)\left(\frac{\alpha_{2c}}{\alpha_2^- + \mu_2}\right)^{n_2} \\ &= \left(1 - \frac{\lambda}{\lambda^- + \mu_1}\right)\left(\frac{\lambda}{\lambda^- + \mu_1}\right)^{n_1} \\ &\quad \times \left(1 - \frac{(1-p)\lambda\mu_1}{\lambda^-\mu_2 + \mu_1\mu_2 + p\lambda\mu_1}\right)\left(\frac{(1-p)\lambda\mu_1}{\lambda^-\mu_2 + \mu_1\mu_2 + p\lambda\mu_1}\right)^{n_2}. \end{aligned}$$

The stationary distribution π enables us to compute various performance measures of interest. For example, the average number of customers in the system is

$$\frac{\lambda}{\mu_1 + \lambda^- - \lambda} + \frac{(1-p)\lambda\mu_1}{\lambda^-\mu_2 + \mu_1\mu_2 + p\lambda\mu_1 - (1-p)\lambda\mu_1}\,.$$

We now modify the tandem queue model so that when a customer at node 2

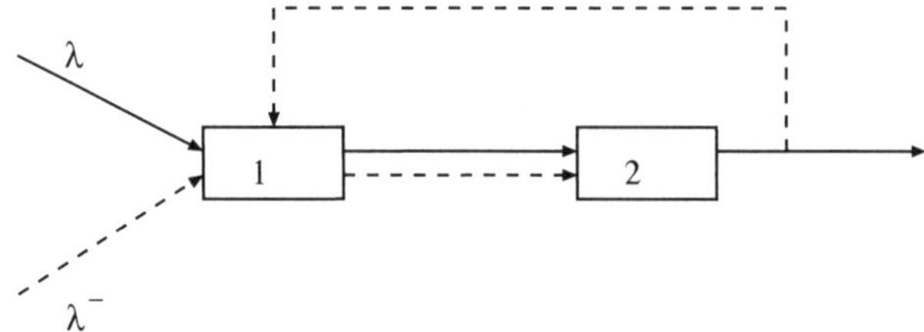

Figure 5.4 The modified tandem queue with signals

is induced by a signal to leave, it returns to node 1 as a signal. Hence the routing probabilities are the same as before except that now $r_{2s^-,1s^-} = 1$. See Figure 5.4.

For this model, the traffic equations are

$$\begin{aligned}
\alpha_{1c} &= \lambda, \\
\alpha_1^- &= \lambda^- + \alpha_{2c}\frac{\alpha_2^-}{\alpha_2^- + \mu_2}, \\
\alpha_{2c} &= \alpha_{1c}\frac{\mu_1}{\alpha_1^- + \mu_1}(1-p), \\
\alpha_2^- &= \alpha_{1c}\frac{\mu_1}{\alpha_1^- + \mu_1}p.
\end{aligned}$$

The stationary distribution of the network is

$$\pi(n_1, n_2) = \left(1 - \frac{\alpha_{1c}}{\alpha_1^- + \mu_1}\right)\left(\frac{\alpha_{1c}}{\alpha_1^- + \mu_1}\right)^{n_1}\left(1 - \frac{\alpha_{2c}}{\alpha_2^- + \mu_2}\right)\left(\frac{\alpha_{2c}}{\alpha_2^- + \mu_2}\right)^{n_2}. \tag{2.16}$$

We examine the potential customer departure epochs from node 1. When a customer leaves node 1 and goes to node 2, this customer becomes a signal with probability p and induces a customer at node 2, if at least one is present, to leave for node 1 as a signal. This process results in another potential customer departure epoch at node 1. This implies that, when the system is in steady state, given that there is a departure from node 1 at time t, there is a positive probability that there will be another potential departure from this node at the same point in time. This shows that the potential customer departure epochs from node 1 are not Poisson. Actually, both nodes 1 and 2 have batch departure processes. Similarly, both nodes 1 and 2 have batch arrival processes. Nevertheless, by (2.16), the system behaves as if node j were subject to a Poisson arrival process with rate α_{jc} and a Poisson potential service process with rate $\alpha_j^- + \mu_j$, $j = 1, 2$. □

Example 5.16 (The quality control network continued) Consider the quality control network discussed in Example 5.8. There are five nodes, two classes of jobs, and a single class of signals. The routing probabilities are given in Example 5.9.

The service times at the five nodes are exponentially distributed with rates μ_1 to μ_5, which are class independent. That is, assumption (a) is in force. The two classes of jobs arrive from the outside according to Poisson processes with rates λ_1 and λ_2, i.e., $\lambda_{11} = \lambda_1$ and $\lambda_{22} = \lambda_2$. The traffic equations are

$$\begin{aligned} \alpha_{11} &= \lambda_1, & \alpha_{31} &= \mu_1\rho_1, \\ \alpha_{22} &= \lambda_2, & \alpha_{32} &= \mu_2\rho_2, \\ \alpha_{41} &= \frac{7}{8}\alpha_{31}, & \alpha_{52} &= \frac{3}{4}\alpha_{32}, \\ \alpha_1^- &= \frac{1}{8}\alpha_{31}, & \alpha_2^- &= \frac{1}{4}\alpha_{32}, \end{aligned}$$

where

$$\begin{aligned} \rho_1 &= \frac{\alpha_{11}}{\mu_1 + \alpha_1^-}, & \rho_2 &= \frac{\alpha_{22}}{\mu_2 + \alpha_2^-}, & & \\ \rho_{31} &= \frac{\alpha_{31}}{\mu_3}, & \rho_{32} &= \frac{\alpha_{32}}{\mu_3}, & \rho_3 &= \rho_{31} + \rho_{32}, \\ \rho_4 &= \frac{\alpha_{41}}{\mu_4}, & \rho_5 &= \frac{\alpha_{52}}{\mu_5}. & & \end{aligned}$$

Solving these equations yields

$$\begin{aligned} \rho_1 &= 4\left(\sqrt{1 + \frac{\lambda_1}{2\mu_1}} - 1\right), \qquad \rho_2 = \sqrt{1 + \frac{\lambda_2}{\mu_2}} - 1, \\ \rho_3 &= \frac{\mu_1}{\mu_3}\rho_1 + \frac{\mu_2}{\mu_3}\rho_2, \quad \rho_4 = \frac{7\mu_1}{8\mu_4}\rho_1, \quad \rho_5 = \frac{3\mu_2}{4\mu_5}\rho_2\,. \end{aligned}$$

By Theorem 5.11, the stationary probability that there are n_j customers at node j, $j = 1, \ldots, 5$, is

$$\pi(\boldsymbol{n}) = \prod_{j=1}^{5} (1 - \rho_j)\rho_j^{n_j}.$$

The average number of jobs in the system is

$$\frac{\rho_1}{\rho_1 - 1} + \frac{\rho_2}{\rho_2 - 1} + \frac{\mu_1\rho_1 + \mu_2\rho_2}{\mu_3 - \mu_1\rho_1 - \mu_2\rho_2} + \frac{7\mu_1\rho_1}{8\mu_4 - 7\mu_1\rho_1} + \frac{3\mu_2\rho_2}{4\mu_5 - 3\mu_2\rho_2}\,.$$

Since the arrival of a signal at node 1 removes one job with probability ρ_1 (the probability that node 1 is not empty), the rate at which jobs are scrapped at node 1 is

$$\alpha_1^- \rho_1 = 2\mu_1 \left(\sqrt{1 + \frac{\lambda_1}{2\mu_1}} - 1 \right)^2 .$$

Similarly, the rate at which jobs are scrapped at node 2 is

$$\alpha_2^- \rho_2 = \mu_2 \left(\sqrt{1 + \frac{\lambda_2}{\mu_2}} - 1 \right)^2 .$$

□

5.3 NETWORKS WITH MULTI-SERVER NODES AND SIGNALS

In the previous section we assumed that each node has a single server and that service is provided at a constant rate. We now relax this assumption and assume that each node provides service at a state-dependent rate. As a result, we need to strengthen the conditions on the routing probabilities in order to obtain a stationary distribution that is product form.

The operational rules are the same as those of the network in Section 5.2 except that (ii′) is now changed back to (ii), i.e.,

(ii) The total service effort at node j, when there are n_j customers present, is provided at rate $\phi_j(n_j)$.

As in Section 5.2, there are I classes of customer arrivals and one class of signal arrivals. Class u customers arrive at node j from the outside according to a Poisson process with rate λ_{ju}, and signals arrive at node j from the outside according to a Poisson process with rate λ_j^-. Assume that the customer routing probabilities are independent of the cause of the departure. That is, when a class u customer departs from node j, either due to a service completion or due to a signal triggering, it goes to node k as a class v customer with probability $r_{ju,kv}$, as a signal with probability r_{ju,ks^-}, and it leaves the network with probability $r_{ju,0}$. So any class u customer leaving node j follows exactly the same pattern of movement. Note that this is equivalent to assuming that the departure of a class u customer due to a signal triggering is classified as the same class of departure as a regular service completion of a class u customer.

Again, let α_{ju} be the average arrival rate of class u customers at node j and λ_j^- the average arrival rate of signals at node j. Since any customer entering the node will eventually leave the node, either after its service is completed or when it is induced to move by a signal, the rate α_{ju} should also be the total departure rate of class u customers from node j. Since we assume that all customers departing from node j have the same routing probabilities $r_{ju,kv}$, r_{ju,ks^-}, and $r_{ju,0}$, we have the linear traffic equations

$$\alpha_{ju} = \lambda_{ju} + \sum_{k=1}^{N} \sum_{v=1}^{I} \alpha_{kv} r_{kv,ju}, \quad j = 1, \ldots, N, \ u = 1, \ldots, I, \tag{3.1}$$

$$\alpha_j^- = \lambda_j^- + \sum_{k=1}^{N} \sum_{v=1}^{I} \alpha_{kv} r_{kv,js^-}, \quad j = 1, \ldots, N. \tag{3.2}$$

Note that these equations can be obtained from the traffic equations (2.1) and (2.2) by assuming $r_{ju,kv} = r_{ju^-,kv}$ and $r_{ju,kv^-} = r_{ju^-,kv^-}$ for all j, k, u and v.

Theorem 5.17 *The stationary distribution of the network described above is*

$$\pi(\boldsymbol{c}) = \prod_{j=1}^{N} \pi_j(\boldsymbol{c}_j), \tag{3.3}$$

where

$$\pi_j(\boldsymbol{c}_j) = b_j^{-1} \prod_{\ell=1}^{n_j} \frac{\alpha_{jc_j(\ell)}}{\alpha_j^- + \mu_j \phi_j(\ell)}, \tag{3.4}$$

and

$$b_j = \sum_{n=0}^{\infty} \frac{(\sum_{u=1}^{I} \alpha_{ju})^n}{\prod_{\ell=1}^{n} (\alpha_j^- + \mu_j \phi_j(\ell))}.$$

Let $\boldsymbol{n}_j = (n_{j1}, \ldots, n_{jI})$ *for* $j = 1, \ldots, N$ *and* $\boldsymbol{n} = (\boldsymbol{n}_1, \ldots, \boldsymbol{n}_N)$. *The stationary probability for the network to be in state* $\boldsymbol{n}$ *is*

$$\pi(\boldsymbol{n}) = \prod_{j=1}^{N} \pi_j(\boldsymbol{n}_j),$$

where

$$\pi_j(\boldsymbol{n}_j) = b_j^{-1} \frac{n_j!}{n_{j1}! \cdots n_{jI}!} \frac{\prod_{u=1}^{I} \alpha_{ju}^{n_{ju}}}{\prod_{\ell=1}^{n_j} (\alpha_j^- + \mu_j \phi_j(\ell))}.$$

PROOF. The proof of this result is similar to that of Theorem 5.11. The main difference lies in the classification of departure classes: now there are only I classes of departures since service completions and signal-triggered departures are classified as being of the same class (see Example 3.13). Let a class u customer that completes its service or is triggered to leave be classified as a class u departure. The classes of arrivals and departures are

$$T_j = \{1, 2, \ldots, I\} \cup \{s^-\}.$$

Node j is characterized by the transition rates and arrival effect functions

$$\begin{aligned} q^{\mathrm{D}}_{ju}(\boldsymbol{c}_j, \boldsymbol{c}_j \ominus \mathbf{e}_\ell) &= \mu_j \phi_j(n_j) \delta^{\mathrm{D}}_j(\ell, n_j), \\ p^{\mathrm{A}}_{ju}(\boldsymbol{c}_j, \boldsymbol{c}_j \oplus \mathbf{e}_m(u)) &= \delta^{\mathrm{A}}_j(m, n_j + 1), \\ p^{\mathrm{A}}_{js^-}(\boldsymbol{c}_j, \boldsymbol{c}_j \ominus \mathbf{e}_\ell) &= \eta_j(\ell, n_j). \end{aligned}$$

Since departures triggered by signals are classified in the same way as regular service completions, the triggering probabilities are

$$f_{js^-,u}(\boldsymbol{c}_j, \boldsymbol{c}_j \ominus \mathbf{e}_\ell) = 1,$$

where we have assumed that $c_j(\ell) = u$. Node j, defined by

$$q_j(\boldsymbol{c}_j, \boldsymbol{c}'_j) = \sum_{u=1}^{I} \alpha_{ju} p^{\mathrm{A}}_{ju}(\boldsymbol{c}_j, \boldsymbol{c}'_j) + \alpha^-_j p^{\mathrm{A}}_{js^-}(\boldsymbol{c}_j, \boldsymbol{c}'_j) + \sum_{u=1}^{I} q^{\mathrm{D}}_{ju}(\boldsymbol{c}_j, \boldsymbol{c}'_j),$$

is an $M/M(n)/1$ queue with signals. Class u customers arrive according to a Poisson process with rate α_{ju} and signals arrive according to a Poisson process with rate α^-_j. It is left to the reader to show that node j has the stationary distribution π_j given by (3.4). Since

$$\begin{aligned} &\sum_{\boldsymbol{c}'_j} \pi_j(\boldsymbol{c}'_j)\Big(q^{\mathrm{D}}_{ju}(\boldsymbol{c}'_j, \boldsymbol{c}_j) + \alpha^-_j p^{\mathrm{A}}_{js^-}(\boldsymbol{c}'_j, \boldsymbol{c}_j) f_{js^-,u}(\boldsymbol{c}'_j, \boldsymbol{c}_j)\Big) \\ &\quad = \pi_j(\boldsymbol{c}_j) \sum_{\ell=1}^{n_j} \frac{\alpha_{ju}}{\alpha^-_j + \phi_j(n_j)\mu_j} \Big(\alpha^-_j \eta_j(\ell, n_j) + \mu_j \phi_j(n_j) \delta^{\mathrm{D}}_j(\ell, n_j)\Big) \\ &\quad = \pi_j(\boldsymbol{c}_j) \frac{\alpha_{ju}}{\alpha^-_j + \phi_j(n_j)\mu_j} \Big(\alpha^-_j + \mu_j \phi_j((n_j)\Big) \\ &\quad = \alpha_{ju} \pi_j(\boldsymbol{c}_j), \end{aligned}$$

node j is uniformly quasi-reversible with departure rate $\beta_{ju} = \alpha_{ju}$. Substituting this into the traffic equations (3.5) of Chapter 4, we obtain, by applying Theorem 4.9, that the network has the product form stationary distribution π of (3.3) with traffic equations given by (3.1) and (3.2). □

Example 5.18 (A multi-server tandem queue with signals) Consider the same tandem system as in Example 5.15 but now assume that there are C_1 servers at node 1 and C_2 servers at node 2. When a customer at node 1 completes its service or is triggered to leave by a signal when the node is not empty, it joins node 2 as a customer with probability $1 - p$, and as a signal with probability p. A customer that completes its service at node 2 or that is triggered to move from node 2 leaves the network.

The routing probabilities for this model are

$$r_{1c,2c} = 1 - p, \qquad r_{1c,2s^-} = p, \qquad r_{2c,0} = 1.$$

The service rates at the two nodes are

$$\phi_1(n_1) = \min\{C_1, n_1\}, \qquad n_1 \geq 0,$$
$$\phi_2(n_2) = \min\{C_2, n_2\}, \qquad n_2 \geq 0.$$

Solving the traffic equations we obtain

$$\alpha_{1c} = \lambda,$$
$$\alpha_{2c} = (1 - p)\alpha_{1c} = (1 - p)\lambda,$$
$$\alpha_1^- = \lambda^-,$$
$$\alpha_2^- = p\alpha_{1c} = p\lambda.$$

It follows from Theorem 5.17 that the stationary distribution of this tandem network is

$$\pi(n_1, n_2) = \pi_1(n_1)\pi_2(n_2), \qquad n_1 \geq 0, n_2 \geq 0,$$

where

$$\pi_1(n_1) = b_1^{-1} \frac{\lambda^{n_1}}{\prod_{\ell=1}^{n_1}(\lambda^- + \min\{\ell, C_1\}\mu_1)}, \qquad n_1 \geq 0,$$
$$\pi_2(n_2) = b_2^{-1} \frac{((1 - p)\lambda)^{n_2}}{\prod_{\ell=1}^{n_2}(p\lambda + \min\{\ell, C_2\}\mu_2)}, \qquad n_2 \geq 0,$$

and b_1 and b_2 are the normalization constants. □

The assumption that the routing probabilities are independent of the source of departure can be relaxed, provided we impose conditions on the effect of a signal (see Exercise 5.3.)

5.4 REFERENCE NOTES

The pioneering work on product form results for networks of queues was reported in Jackson (1957, 1963), and his model has ever since been referred to as the Jackson network. Gordon and Newell (1967*a*, 1967*b*) extend Jackson's result to closed queueing networks. The network with fixed routes was first introduced in Kelly (1975), and it can also be found in Kelly (1979). The networks with negative signals described in Sections 5.2 and 5.3 were introduced in Chao and Pinedo (1993). The proofs of these results, however, have been modified to fit within the framework of Chapters 3 and 4.

EXERCISES

5.1 Prove Corollary 5.13. That is, a stronger version of the *Arrival Theorem* holds for the networks with signals discussed in Section 5.2: When an entity (a customer, or a signal) arrives at a node, it observes the network in state $\boldsymbol{c}$ with probability $\pi(\boldsymbol{c})$. Note that this entity may be an entity that arrives at a node and leaves immediately.

5.2 Consider an $M/M/1$ queue with feedback. Upon a service completion, a customer is fed back to the end of the queue as another customer with probability p, or it is fed back to the end of the queue as a signal with probability p^-, or it leaves the system with probability $1 - p - p^-$.

(a) Compute the stationary distribution of this system.

(b) If only the departure epochs of the customers are of interest, what class of departure process do you observe? Argue that all the potential customer departure epochs do not constitute a Poisson process.

(c) Relate this model to Section 2.6, and derive the stationary distribution from Lemma 2.29.

5.3 Consider the model of Section 5.3, and assume that $\phi_j(n_j) \le B_j$ for all n_j. Assume that when a class u customer completes its service at node j, it goes to node k as a class v customer with probability $r_{ju,kv}$, as a signal with probability r_{ju,ks^-}, and it leaves the network with probability $r_{ju,0}$. However, when a class u customer is induced to move by a signal, it goes to node k as a class v customer with probability $r_{ju^-,kv}$, as a signal with probability r_{ju^-,ks^-}, and it leaves the network with probability $r_{ju^-,0}$, where

$$\sum_{k=1}^{N}\left(\sum_{v=1}^{I} r_{ju^-,kv} + r_{ju^-,ks^-}\right) + r_{ju^-,0} = 1, \qquad j = 1, \ldots, N, \ u = 1, \ldots, I.$$

Furthermore, assume that given there are n_j customers at node j and a signal arrives at position ℓ of node j, it triggers the customer to move with probability $\phi_j(n_j)/B_j$, and there is no effect with probability $1 - \phi_j(n_j)/B_j$. Show that this network has a product form stationary distribution $\prod_{j=1}^{N} \pi_j(n_j)$ and compute $\pi_j(n_j)$.

5.4 Show that the result of Theorem 5.11 still holds when the functions $\delta_j^{\mathrm{A}}(\ell, n_j)$, $\delta_j^{\mathrm{D}}(\ell, n_j)$, and $\eta_j(\ell, n_j)$ are relaxed to functions of $\boldsymbol{c}_j$, i.e., $\delta_j^{\mathrm{A}}(\ell, \boldsymbol{c}_j)$, $\delta_j^{\mathrm{D}}(\ell, \boldsymbol{c}_j)$, and $\eta_j(\ell, \boldsymbol{c}_j)$, as long as they remain unchanged under permutations of $\boldsymbol{c}_j$, and

$$\sum_{\ell=1}^{n_j} \delta_j^{\mathrm{A}}(\ell, \boldsymbol{c}_j) = \sum_{\ell=1}^{n_j} \delta_j^{\mathrm{D}}(\ell, \boldsymbol{c}_j) = \sum_{\ell=1}^{n_j} \eta_j(\ell, \boldsymbol{c}_j) = 1, \qquad n_j = 1, 2, \ldots.$$

5.5 Corollary 5.13 has a corresponding result for closed queueing networks with instantaneous movements. The arrival of an entity, however, observes the network in steady state but with itself excluded from the network. Consider, for instance, a closed Jackson network with N nodes and M customers circulating in the network. Each station has a finite capacity of C_j, and a customer that arrives at node j and finds C_j customers immediate leaves (considered as a signal). This customer then goes to node k with probability r_{jk}, and again, if node k has C_k customers present it leaves immediately, and so on. Show that, given a customer arrives at node j, it observes the network in state $\boldsymbol{n}$ ($\sum_{j=1}^{N} n_j = M-1$) with probability $\pi_{M-1}(\boldsymbol{n})$, where $\pi_{M-1}(\boldsymbol{n})$ is the stationary distribution of the same network but with a population size $M - 1$.

6

Multiple Customer Classes and Arbitrary Service Times

This chapter extends the product form results of the previous chapter to networks with multiple customer classes and arbitrary service times. Of central importance is the concept of symmetric service disciplines. Not only can the service times under these disciplines be distributed arbitrarily, they may also depend on the customer class. The first two sections focus on networks without signals. Section 6.1 describes the so-called BCMP networks and networks with symmetric service disciplines, and Section 6.2 establishes the product form results for these models. Section 6.3 considers networks with signals and single-server nodes, and their product form results are proved in Section 6.4. The final section discusses possible extensions to networks with signals and multi-server nodes.

6.1 BCMP AND NETWORKS WITH SYMMETRIC SERVICES

The first set of queueing networks that include non-exponential service times was introduced by Baskett, Chandy, Muntz, and Palacios. In the literature these networks are referred to as BCMP networks. BCMP networks may consist of a number of different types of nodes including those discussed in Chapter 5.

The network consists of N nodes and I classes of customers. Class u customers arrive at node j from the outside according to a Poisson process with rate λ_{ju}, $j = 1, \ldots, N$, and $u = 1, \ldots, I$. As in Chapter 5, a customer may change its class as it moves from one node to another. More precisely, a class u customer, upon completing its service at node j, goes to node k as a class v customer with probability $r_{ju,kv}$, and it leaves the network with probability $r_{ju,0}$. These routing probabilities satisfy

$$\sum_{k=1}^{N}\sum_{v=1}^{I} r_{ju,kv} + r_{ju,0} = 1, \qquad j = 1, \ldots, N,\ u = 1, \ldots, I.$$

Let n_{ju} be the number of class u customers at node j, and

$$\boldsymbol{n}_j = (n_{j1}, \ldots, n_{jI}).$$

Furthermore, we use n_j to denote the total number of customers at node j, i.e.,

$$n_j = \sum_{u=1}^{I} n_{ju}.$$

In BCMP networks, the type of nodes with exponential service requirements discussed in Chapter 5 (see (a) of Theorem 5.4) are called type 0 nodes. In addition to type 0 nodes, a BCMP network may consist of four other types of nodes.

Type 1 node. This type of node consists of a single processor-sharing server. The service requirements of each class of customer are arbitrarily distributed, and may depend on the class of the customer. The distribution of the service requirements of class u customers at a type 1 node j is F_{ju}. When there are n_j customers present at node j, the service effort is divided equally among all the customers. For instance, when there are n_{ju} class u customers at node j and the service requirement of a class u customer is exponential with rate μ_{ju}, then the service completion rate of class v customers is

$$\frac{n_{ju}}{n_j}\mu_{ju}, \qquad u = 1, \ldots, I.$$

Type 2 node. A single server processes customers according to the Last–Come First–Served Preemptive resume (LCFS-P) rule. The service requirements of each class of customer are arbitrarily distributed and may depend on the class of the customer. The distribution of the service requirements of class u customers at a type 2 node j is F_{ju}, $u = 1, \ldots, I$.

Type 3 node. This type of node has an infinite number of servers. Each customer, upon arrival, starts its service immediately. The service requirements of the customers are arbitrarily distributed and may depend on the class of the customer. The distribution of service requirements of class u customers at a type 3 node j is F_{ju}, $u = 1, \ldots, I$.

Type 4 node. This type of node has a finite number of servers, say K_j, but no waiting room for additional customers. When a customer arrives and finds no server available, it leaves immediately as if its service had been completed. The service requirement of a customer is arbitrarily distributed and may depend on the class of the customer. The distribution of service requirements of class u customers at a type 4 node j is F_{ju}, $u = 1, \ldots, I$.

Let α_{ju} denote the average arrival rate of class u customers at node j. Since any customer who enters a node will eventually leave the node, α_{ju} should also be the average departure rate of class u customers from node j. Thus the traffic equation

$$\alpha_{ju} = \lambda_{ju} + \sum_{k=1}^{N} \sum_{v=1}^{I} \alpha_{kv} r_{kv,ju} \tag{1.1}$$

must be satisfied.

Let S_{ju} be a generic service time of a class u customer at node j. If node j is of type 0, i.e., the type of nodes with exponential service times satisfying (i′) of Chapter 5, then S_{ju} has an exponential distribution with mean $E(S_{ju}) = 1/\mu_j$ that is independent of the class u. If node j is one of the four types described above, then it has an arbitrary distribution with mean $E(S_{ju})$.

For $j = 1, 2, \ldots, N$, denote traffic intensities at node j by

$$\rho_{ju} = \alpha_{ju} E(S_{ju}), \qquad u = 1, 2, \ldots, I,$$

$$\rho_j = \sum_{u=1}^{I} \rho_{ju} .$$

The following result is known as the BCMP theorem.

Theorem 6.1 *Assume that the network consists of nodes of types 0, 1, 2, 3, and 4, and α_{ju}, $u = 1, \ldots, I$, and $j = 1, \ldots, N$, are the solution to the traffic equations (1.1) with $\rho_j < 1$ when node j is of type 0, 1, or 2. Then the stationary distribution of the network is*

$$\pi(\boldsymbol{n}) = \prod_{j=1}^{N} \pi_j(\boldsymbol{n}_j),$$

where $\pi_j(\boldsymbol{n}_j)$ depends on the type of node j. If node j is of type 0, then π_j is given by (1.16) of Corollary 5.6 with $\mu_{ju} = \mu_j$; if node j is of type 1 or type 2, then

$$\pi_j(\boldsymbol{n}_j) = (1 - \rho_j) \frac{n_j!}{n_{j1}! n_{j2}! \cdots n_{jI}!} \prod_{u=1}^{I} \rho_{ju}^{n_{ju}};$$

if node j is of type 3, then

$$\pi_j(\boldsymbol{n}_j) = e^{-\rho_j} \prod_{u=1}^{I} \frac{\rho_{ju}^{n_{ju}}}{n_{ju}!};$$

and if node j is of type 4, then

$$\pi_j(\boldsymbol{n}_j) = \frac{\prod_{u=1}^{I} \rho_{ju}^{n_{ju}}/n_{ju}!}{\sum_{k=0}^{K_j} \rho_j^k/k!}, \qquad \sum_{u=1}^{I} n_{ju} \le K_j.$$

We now take a closer look at the four types of nodes in the BCMP networks. The service requirements in these nodes may be arbitrarily distributed, but the service disciplines at these nodes are very restrictive. For instance, a type 1 node has a processor-sharing discipline, and its $\delta_j^{\mathrm{D}}(\ell, n_j)$, using the notation introduced in Section 5.1, is given by

$$\delta_j^{\mathrm{D}}(\ell, n_j) = \frac{1}{n_j}, \qquad n_j = 1, 2, \ldots, \quad \ell = 1, 2, \ldots, n_j,$$

since the service effort is shared equally among all customers in the queue. Moreover, since the server does not make any distinction between the positions in the queue, the arriving customers may be viewed as joining the queue in a random fashion, i.e.,

$$\delta_j^{\mathrm{A}}(\ell, n_j) = \frac{1}{n_j}, \qquad n_j = 1, 2, \ldots, \quad \ell = 1, 2, \ldots, n_j.$$

An important observation here is that

$$\delta_j^{\mathrm{A}}(\ell, n_j) = \delta_j^{\mathrm{D}}(\ell, n_j), \qquad n_j = 1, 2, \ldots, \quad \ell = 1, 2, \ldots, n_j. \tag{1.2}$$

Similarly, type 2 nodes can be characterized by

$$\phi_j(n_j) = 1, \qquad n_j = 1, 2, \ldots,$$
$$\delta_j^{\mathrm{A}}(n_j, n_j) = \delta_j^{\mathrm{D}}(n_j, n_j) = 1, \qquad n_j = 1, 2, \ldots,$$

and type 3 nodes can be characterized by

$$\phi_j(n_j) = n_j, \qquad n_j = 1, 2, \ldots,$$
$$\delta_j^{\mathrm{A}}(\ell, n_j) = \delta_j^{\mathrm{D}}(\ell, n_j) = \frac{1}{n_j}, \qquad n_j = 1, 2, \ldots, \quad \ell = 1, 2, \ldots, n_j.$$

A node of type 4, which is also known as Erlang's loss model, is quite similar to a node of type 3. An immediate departure when there are K_j customers present can be viewed as the result of an infinite service capacity, i.e.,

$$\phi_j(n_j) = n_j, \qquad n_j = 1, 2, \ldots, K_j,$$

$$\phi_j(n_j) = \infty, \qquad n_j > K_j.$$

Thus the four types of nodes all satisfy (1.2), i.e., their service disciplines are symmetric. This naturally leads to the question: Can the BCMP theorem be extended to any node with a symmetric service discipline ? The answer is positive.

Consider a queue with service positions as defined in Chapter 5. That is, customers are arranged in positions, and when there are n customers at the node, the first n positions are occupied. The services are performed according to conditions (ii), (iii) and (iv) of Section 5.1, but condition (i) is relaxed so that service time distributions are arbitrary. Assume that such a node has a symmetric service discipline, i.e., (1.2) holds. In this chapter, we denote δ_j^{A} and δ_j^{D}, which are identical, by δ_j. Clearly, $\delta_j(\ell, n_j)$ is a nonnegative number and it satisfies

$$\sum_{\ell=1}^{n_j} \delta_j(\ell, n_j) = 1.$$

In summary, node j is said to operate according to a symmetric service discipline if the following conditions are satisfied when there are n_j customers present.

(i) The service requirement S_{ju} of a class u customer at node j has an arbitrary distribution function F_{ju} with finite mean $E(S_{ju})$.

(ii) The total service effort is provided at rate $\phi_j(n_j)$.

(iii) A proportion $\delta_j(\ell, n_j)$ of the total service effort is directed to the customer in position ℓ, $\ell = 1, \ldots, n_j$. When its service is completed, customers in positions $\ell+1, \ell+2, \ldots, n_j$ move to positions $\ell, \ell+1, \ldots, n_j - 1$, respectively.

(iv) When a customer arrives at node j, it moves into position ℓ, $\ell = 1, 2, \ldots, n_j + 1$ with probability $\delta_j(\ell, n_j + 1)$. Customers previously in positions $\ell, \ell + 1, \ldots, n_j$ move to positions $\ell + 1, \ell + 2, \ldots, n_j + 1$, respectively.

Symmetric service excludes many service disciplines, e.g., First–Come First–Served. However, it does include many interesting special cases. For instance, as we have already seen, it includes all the BCMP nodes with arbitrarily distributed service time requirements, e.g., nodes of types 1, 2, 3, and 4.

Suppose the network has N nodes and I classes of customers. The nodes in the network are classified as either symmetric or non-symmetric. In the latter case the node satisfies condition (i′) of Section 5.2, i.e., the service requirements of all customers at the node are exponentially distributed with the same mean. The nodes are connected via Markovian routing, as in BCMP networks. When a class u customer completes its service at node j, it goes to node k as a class v customer with probability $r_{ju,kv}$ and it leaves the network with probability $r_{ju,0}$, where

$$\sum_{k=1}^{N}\sum_{v=1}^{I} r_{ju,kv} + r_{ju,0} = 1, \qquad j = 1, \dots, N,\ u = 1, \dots, I.$$

As in Section 5.1, we call this network a *multiclass Jackson network with service positions and arbitrary service requirements.*

Let $c_j(\ell)$ be the class of customer in position ℓ of node j when there are n_j customers present, $\ell = 1, \dots, n_j$. The state of node j is

$$\boldsymbol{c}_j = (c_j(1), \dots, c_j(n_j)),$$

and the state of the network is

$$\boldsymbol{c} = (\boldsymbol{c}_1, \dots, \boldsymbol{c}_N).$$

Note that the process with state $\boldsymbol{c}$ is not Markovian. To make the process Markovian, we need a supplementary variable for the customer with non-exponential service requirements. This variable can be either the amount of service already received, or the amount of work still to be done.

Theorem 6.2 *Consider a queueing network with service positions. Each node is either a non-symmetric node with class-independent exponentially distributed service requirements or a symmetric node with class-dependent arbitrarily distributed service requirements. Let α_{ju} be the mean arrival rate of class u customers at node j, which is determined by the linear traffic equations (1.1). The stationary distribution of the network is*

$$\pi(\boldsymbol{c}) = \prod_{j=1}^{N} \pi_j(\boldsymbol{c}_j),$$

where, using the notation $\rho_{ju} = \alpha_{ju}E(S_{ju})$ and $\rho_j = \sum_{u=1}^{I}\rho_{ju}$, the marginal distribution $\pi_j(\boldsymbol{c}_j)$ is given by

$$\pi_j(\boldsymbol{c}_j) = b_j^{-1}\prod_{\ell=1}^{n_j}\frac{\rho_{jc_j(\ell)}}{\phi_j(\ell)}, \tag{1.3}$$

provided

$$b_j = \sum_{n=0}^{\infty}\frac{\rho_j^n}{\prod_{\ell=1}^{n}\phi_j(\ell)} < \infty.$$

The proofs of Theorems 6.1 and 6.2 are given in the next section.

6.2 QUASI-REVERSIBILITY OF SYMMETRIC QUEUES

In this section we prove the product form results for queueing networks with service positions and arbitrary service requirements. Since Theorem 6.1 is a special case of Theorem 6.2, we only prove Theorem 6.2.

By Theorem 4.3 it suffices to show that node j, defined by (4.2) of Section 4.4, is quasi-reversible with $\beta_{ju} = \alpha_{ju}$, $u = 1, \ldots, I$.

The case that node j is a non-symmetric queue has been shown in Theorem 5.4. Assume node j is a symmetric queue. The proof follows three steps.

Step 1. The result holds when the service times have Erlang distributions.

Step 2. The result holds when the service times are mixtures of Erlang distributions.

Step 3. The result holds when the service times have arbitrary distributions.

For convenience, in what follows we drop the index j for the node under consideration. First we assume that the service times S_u of class u customers, $u = 1, 2, \ldots, I$, have an Erlang distribution with k_u phases and each phase is exponentially distributed with parameter ζ_u, i.e., E_{k_u,ζ_u} (see (2.3) of Appendix B for its formal definition). The node defined by (4.2) of Chapter 4 is subject to Poisson arrivals of class u customers with rate α_u. Let $c(\ell)$ be the class of the customer in position ℓ and $w(\ell)$ be the number of remaining phases to be completed, where $1 \leq w(\ell) \leq k_{c(\ell)}$. Let $\boldsymbol{x}(\ell) = (c(\ell), w(\ell))$, and

$$\boldsymbol{x} = (\boldsymbol{x}(1), \boldsymbol{x}(2), \ldots, \boldsymbol{x}(n)),$$

where n is the number of customers at the node. Clearly, $\boldsymbol{x}$ represents the state of the node and it is a Markov chain. Let

$$a = \sum_{u=1}^{I} \alpha_u \frac{k_u}{\zeta_u} = \sum_{u=1}^{I} \alpha_u E(S_u),$$

which represents the average amount of service requirements arriving at the node per unit time.

Lemma 6.3 *If*

$$b = \sum_{n=0}^{\infty} \frac{a^n}{\prod_{\ell=1}^{n} \phi(\ell)} < \infty,$$

then the stationary distribution of the node is

$$\pi(\boldsymbol{x}) = b^{-1} \prod_{\ell=1}^{n} \frac{\alpha_{c(\ell)}/\zeta_{c(\ell)}}{\phi(\ell)}, \qquad 1 \leq w(\ell) \leq k_{c(\ell)}, \tag{2.1}$$

and the node is quasi-reversible with $\alpha_u = \beta_u$, i.e., the departure rate of class u customers is equal to the arrival rate.

PROOF. We use the detailed Kelly lemma, i.e., Lemma 2.23, to prove this result. The arrival of a class u customer brings k_u service phases. Each phase is processed at rate ζ_u, and the completion of the last phase results in a class u departure. Hence, for $\boldsymbol{x} = (\boldsymbol{x}(1), \boldsymbol{x}(2), \ldots, \boldsymbol{x}(n))$, we can decompose the transition rate q according to arrival, departure and intermediate service phase changes, for each class of customers, as follows:

$$\begin{aligned} q_u^{\mathrm{A}}(\boldsymbol{x}, \boldsymbol{x} \oplus \mathbf{e}_\ell(u, k_u)) &= \alpha_u \delta(\ell, n+1), \\ q_u^{\mathrm{D}}(\boldsymbol{x} \oplus \mathbf{e}_\ell(u, 1), \boldsymbol{x}) &= \phi(n+1)\delta(\ell, n+1)\zeta_u, \\ q_u^{\mathrm{I}}(\boldsymbol{x}, \boldsymbol{x} - \mathbf{e}_\ell) &= \phi(n)\delta(\ell, n)\zeta_u, \qquad 2 \leq w(\ell) \leq k_u, \end{aligned}$$

where $\boldsymbol{x} \oplus \mathbf{e}_\ell(u, w)$ is the state obtained from $\boldsymbol{x}$ after a customer of class u with w service phases is inserted into position ℓ and customers in positions $\ell, \ell+1, \ldots, n$ move to positions $\ell+1, \ell+2, \ldots, n+1$; and $\boldsymbol{x} - \mathbf{e}_\ell$ is the state obtained from $\boldsymbol{x}$ after an intermediate phase of a class u customer in position ℓ has been completed. Clearly,

$$q(\boldsymbol{x}, \boldsymbol{x}') = \sum_{u=1}^{I} \left(q_u^{\mathrm{A}}(\boldsymbol{x}, \boldsymbol{x}') + q_u^{\mathrm{D}}(\boldsymbol{x}, \boldsymbol{x}') + q_u^{\mathrm{I}}(\boldsymbol{x}, \boldsymbol{x}') \right).$$

The reversed process is supposed to be a queue that is subject to Poisson arrivals of class u customers with rate α_u and it operates in exactly the same manner as the original queue except that now $w(\ell)$ represents the service phase the customer in position ℓ has reached. The transition rate $\tilde{q}$ can be decomposed into the components

$$\begin{aligned} \tilde{q}_u^{\mathrm{A}}(\boldsymbol{x}, \boldsymbol{x} \oplus \mathbf{e}_\ell(u, 1)) &= \alpha_u \delta(\ell, n+1), \\ \tilde{q}_u^{\mathrm{D}}(\boldsymbol{x} \oplus \mathbf{e}_\ell(u, k_u), \boldsymbol{x}) &= \phi(n+1)\delta(\ell, n+1)\zeta_u, \\ \tilde{q}_u^{\mathrm{I}}(\boldsymbol{x} - \mathbf{e}_\ell, \boldsymbol{x}) &= \phi(n)\delta(\ell, n)\zeta_u, \qquad 2 \leq w(\ell) \leq k_u. \end{aligned}$$

The first condition of the detailed Kelly lemma follows from

$$\sum_{x'} q(x, x') = \sum_{u=1}^{I} \alpha_u + \sum_{\ell=1}^{n} \phi(n)\zeta_{c(\ell)}\delta(\ell, n)$$
$$= \sum_{x'} \tilde{q}(x, x').$$

The second condition is also immediate, since the expression for π in (2.1) implies

$$\pi(x)q_u^{A}(x, x \oplus \mathbf{e}_\ell(u, k_u)) = \pi(x \oplus \mathbf{e}_\ell(u, k_u))\tilde{q}_u^{D}(x \oplus \mathbf{e}_\ell(u, k_u), x),$$
$$\pi(x \oplus \mathbf{e}_\ell(u, 1))q_u^{D}(x \oplus \mathbf{e}_\ell(u, 1), x) = \pi(x)\tilde{q}_u^{A}(x, x \oplus \mathbf{e}_\ell(u, 1)),$$
$$\pi(x)q_u^{I}(x, x - \mathbf{e}_\ell) = \pi(x - \mathbf{e}_\ell)\tilde{q}_u^{I}(x - \mathbf{e}_\ell, x).$$

Thus it follows from the Kelly lemma that π of (2.1) is the stationary distribution of the queue. That the node is quasi-reversible follows from the structure of the reversed process. □

Since each class u customer passes through k_u phases to complete service and it is in any phase with equal probability, the following result follows immediately from Lemma 6.3.

Corollary 6.4 *The probability that there are n_u customers of class u, $u = 1, \ldots, I$, in the queue is*

$$\pi(n_1, n_2, \ldots, n_I) = b^{-1} \frac{(n_1 + n_2 + \cdots + n_I)!}{n_1! n_2! \cdots n_I!} \frac{\prod_{u=1}^{I} (\alpha_u E(S_u))^{n_u}}{\prod_{\ell=1}^{\sum_{u=1}^{I} n_u} \phi(\ell)}. \quad (2.2)$$

The probability that there are n customers at the node is

$$\pi(n) = b^{-1} \frac{a^n}{\prod_{\ell=1}^{n} \phi(\ell)}. \quad (2.3)$$

Given that there are n customers at the node, the numbers of customers belonging to the same classes are independent, and each customer in a given position is of class u with probability

$$\frac{\alpha_u E(S_u)}{a}. \quad (2.4)$$

The probabilities (2.2), (2.3) and (2.4) depend on the values of $E(S_u)$, the mean service time of class u customers.

We now consider the case when the service requirement of each class of customers has a mixed Erlang distribution. The service requirement S_u of a class u customer is Erlang $E_{k_{u,z},\zeta_{u,z}}$ with probability $p_{u,z}$, where z belongs to a countable set, and $\sum_z p_{u,z} = 1$. The arrival process of class u customers is Poisson with rate α_u. If we establish for customers of class u a finer classification scheme (u, z), with a customer belonging to class (u, z) with probability $p_{u,z}$, then class (u, z) customers arrive at the system according to a Poisson process with rate $\alpha_u p_{u,z}$, and each class (u, z) customer has an Erlang service requirement $E_{k_{u,z},\zeta_{u,z}}$. Let $S_{u,z}$ be a generic service requirement of a class (u, z) customer. The queueing system with mixed Erlang service requirements is, therefore, equivalent to another queueing system with Erlang service requirements. Note that the average service requirement of a class u customer is

$$E(S_u) = \sum_z p_{u,z} k_{u,z}/\zeta_{u,z} = \sum_z p_{u,z} E(S_{u,z}).$$

Applying Corollary 6.4 with regard to the more refined customer classification scheme, we find that the probability of the node containing n customers is

$$b^{-1} \frac{a^n}{\prod_{\ell=1}^n \phi(\ell)},$$

where

$$\begin{aligned} a &= \sum_u \sum_z \alpha_u p_{u,z} E(S_{u,z}) \\ &= \sum_u \alpha_u E(S_u). \end{aligned}$$

Furthermore, given that there are n customers in the queue, the classes of the n customers are independent and a customer in a given position is of class u with probability

$$\frac{\sum_z \alpha_u p_{u,z} E(S_{u,z})}{a} = \frac{\alpha_u E(S_u)}{a}.$$

The queue is quasi-reversible with respect to the classification u as well as the more refined classification (u, z). Thus the result also holds when the service requirements have mixed Erlang distributions.

Finally, consider the case where the service requirements of class u customers are arbitrarily distributed positive random variables. To avoid mathematical technicalities we shall be content with an intuitive argument. It is known that mixed Erlang distributions are dense in the class of probability distribution functions. That is, for any distribution function F, we can choose a sequence of mixed Erlang distribution functions $F_n(x)$, $n = 1, 2, \ldots$, such that

$$\lim_{n\to\infty} F_n(x) = F(x)$$

for all x where F is continuous. Thus we can show by a limiting argument that the quasi-reversibility property as well as the properties of Corollary 6.4 hold even when the service requirements are arbitrarily distributed.

To summarize, a symmetric queue with arbitrarily distributed service requirements is quasi-reversible with $\beta_{ju} = \alpha_{ju}$. Therefore, the application of Theorem 4.3 shows that the network of quasi-reversible nodes has a product form solution. This completes the proof of Theorem 6.2.

6.3 NETWORKS WITH NEGATIVE SIGNALS AND SINGLE-SERVER NODES

We extend the results of Section 6.1 to include negative signals. We first consider the case where each node has a single server. The multi-server case is studied in Section 6.5.

Consider a queueing network with N nodes, I classes of customers, denoted by $u, u = 1, \ldots, I$, and a single class of negative signals, denoted by s^-. The signals carry instructions to the nodes and induce customers to move instantaneously. Class u customers arrive at node j from the outside according to a Poisson process with rate λ_{ju}, and negative signals arrive at node j from the outside according to a Poisson process with rate λ_j^-, $j = 1, \ldots, N$. Upon a service completion at node j, a class u customer goes to node k as a class v customer with probability $r_{ju,kv}$, as a negative signal with probability r_{ju,ks^-}, and it leaves the system with probability $r_{ju,0}$, where

$$\sum_{k=1}^{N}\sum_{v=1}^{I} r_{ju,kv} + \sum_{k=1}^{N} r_{ju,ks^-} + r_{ju,0} = 1, \qquad j = 1, \ldots, N, \ u = 1, \ldots, I.$$

Whenever a signal arrives at node j, either from the outside or from another node, it triggers a service completion. In other words, when a signal arrives at node j, a customer departs immediately without finishing its service. If this customer is of class u, it goes to node k as a class v customer with probability $r_{ju^-,kv}$, as a signal with probability r_{ju^-,ks^-}, and it leaves the network with probability $r_{ju^-,0}$. Again we have

$$\sum_{k=1}^{N}\sum_{v=1}^{I} r_{ju^-,kv} + \sum_{k=1}^{N} r_{ju^-,ks^-} + r_{ju^-,0} = 1, \qquad j = 1, \ldots, N, \ u = 1, \ldots, I.$$

When a signal arrives at an empty node, nothing happens.

Recall that when there are no signals, the operational rules of a node are specified by two sets of parameters, i.e., $\delta_j^{\mathrm{A}}(\ell, n_j)$ and $\delta_j^{\mathrm{D}}(\ell, n_j)$. With signals, we have to specify the effect of an arriving signal as well. Under the symmetric service discipline of the previous section, the proportion of service effort allocated to a customer in a certain position is equal to the probability of a customer arriving at that same position (or the probability of that position being created). Since the effect of an arriving signal is the same as that of a service completion, we might expect that, for a node to remain "symmetric", an arriving signal should join a position with the same probability as the proportion of service effort allocated to that position. However, this intuition turns out to be incorrect. As we will show in this section and the next, if the total service effort at a node is a constant, the probability that a signal arrives in a position may be arbitrary for the product form result to hold; if the total service effort is state-dependent, then the probability that an arriving signal is effective must depend on the total service effort, although the probability that it joins a position may be arbitrary. In this section, we consider the first case, while the second case will be considered in Section 6.5.

Thus we define node j with signals as operating according to a symmetric discipline if conditions (i), (iii) and (iv) of Section 6.1 are satisfied, condition (ii) is replaced by

(ii′) The total service effort is supplied at rate 1, i.e., $\phi_j(n_j) = 1[n_j \geq 1]$,

and the following condition with regard to the signals is added:

(v) When a signal arrives at node j, it triggers the customer in position ℓ to leave with probability $\eta_j(\ell, n_j)$. With its departure, customers in positions $\ell+1, \ell+2, \ldots, n_j$ move to positions $\ell, \ell+1, \ldots, n_j - 1$, respectively, where

$$\sum_{\ell=1}^{n_j} \eta_j(\ell, n_j) = 1, \qquad n_j = 1, 2, \ldots.$$

Note that $\eta_j(\ell, n_j)$ is defined arbitrarily. Condition (i) is needed because in this section we focus on single-server nodes. This condition will be removed later on in Section 6.5.

Similar to the notation in Section 6.1, let

$$\boldsymbol{c} = (\boldsymbol{c}_1, \ldots, \boldsymbol{c}_N)$$

represent the state of the network, where

$$\boldsymbol{c}_j = (c_j(1), \ldots, c_j(n_j)),$$

and $c_j(\ell)$ is the class of the customer in position ℓ of node j, $\ell = 1, \ldots, n_j$, and n_j is the number of customers at node j, $j = 1, \ldots, N$. Let

$$\boldsymbol{n} = (\boldsymbol{n}_1, \ldots, \boldsymbol{n}_j),$$

where

$$\boldsymbol{n}_j = (n_{j1}, \ldots, n_{jI}),$$

and n_{ju} is the number of class u customers at node j. Clearly,

$$n_j = \sum_{u=1}^{I} n_{ju}\,.$$

Let $\hat{F}_{ju}$ be the Laplace transform of the distribution function F_{ju}, i.e.,

$$\hat{F}_{ju}(s) = \int_0^\infty e^{-st} dF_{ju}(t).$$

The following is the main result of this section.

Theorem 6.5 *Assume*

$$\sum_{u=1}^{I} \alpha_{ju} E(\min\{S_j^-, S_{ju}\}) < 1, \qquad j = 1, \ldots, N, \tag{3.1}$$

where S_j^- is an exponential random variable with parameter α_j^- that is independent of S_{ju}, a generic service requirement of a class u customer at node j, and α_{ju} and α_j^- are the solutions to the traffic equations

$$\alpha_{ju} = \lambda_{ju} + \sum_{k=1}^{N}\sum_{v=1}^{I} \alpha_{kv}\hat{F}_{kv}(\alpha_k^-) r_{kv,ju} + \sum_{k=1}^{N}\sum_{v=1}^{I} \alpha_{kv}(1 - \hat{F}_{kv}(\alpha_k^-)) r_{kv^-,ju}\,, \quad j = 1, \ldots, N,\ u = 1, \ldots, I, \tag{3.2}$$

$$\alpha_j^- = \lambda_j^- + \sum_{k=1}^{N}\sum_{v=1}^{I} \alpha_{kv}\hat{F}_{kv}(\alpha_k^-) r_{kv,js^-} + \sum_{k=1}^{N}\sum_{v=1}^{I} \alpha_{kv}(1 - \hat{F}_{kv}(\alpha_k^-)) r_{kv^-,js^-}, \quad j = 1, \ldots, N. \tag{3.3}$$

The stationary distributions of the network are

$$\pi(\boldsymbol{c}) = \prod_{j=1}^{N} \pi_j(\boldsymbol{c}_j), \tag{3.4}$$

$$\pi(\boldsymbol{n}) = \prod_{j=1}^{N} \pi_j(\boldsymbol{n}_j), \tag{3.5}$$

where $\pi_j(\boldsymbol{c}_j)$ and $\pi_j(\boldsymbol{n}_j)$ are the probabilities for node j to be in states $\boldsymbol{c}_j$ and $\boldsymbol{n}_j$

when it is in isolation and subject to Poisson arrivals of class u customers with rate α_{ju} and Poisson arrivals of signals with rate α_j^-, i.e.,

$$\pi_j(\boldsymbol{c}_j) = b_j^{-1} \prod_{\ell=1}^{n_j} \alpha_{jc_j(\ell)} E(\min\{S_j^-, S_{jc_j(\ell)}\}),$$

$$\pi_j(\boldsymbol{n}_j) = b_j^{-1} \frac{n_j!}{n_{j1}! \cdots n_{jI}!} \prod_{u=1}^{I} \Big(\alpha_{ju} E(\min\{S_j^-, S_{ju}\})\Big)^{n_{ju}},$$

where

$$b_j = \frac{1}{1 - \sum_{u=1}^{I} \alpha_{ju} E(\min\{S_j^-, S_{ju}\})}.$$

Remark 6.6 It is worth noting that the stationary distribution of the network process is the same as that of a similar network without signals, but with the service requirement of a class u customer at node j, originally S_{ju}, replaced by $\min\{S_j^-, S_{ju}\}$. The expectation of this random variable is calculated in terms of the Laplace transform of its distribution as

$$\begin{aligned} E(\min\{S_j^-, S_{jc_j(\ell)}\}) &= \int_0^\infty P(S_j > x, S_{ju} > x)dx \\ &= \int_0^\infty e^{-\alpha_j^- x} P(S_{ju} > x)dx \\ &= E\left(\int_0^\infty e^{-\alpha_j^- x} 1(S_{ju} > x)dx\right) \\ &= E\left(\int_0^{S_{ju}} e^{-\alpha_j^- x} dx\right) \\ &= \frac{1 - \hat{F}_{ju}(\alpha_j^-)}{\alpha_j^-}. \end{aligned} \tag{3.6}$$

A formal proof of Theorem 6.5 will be given in the next section. Here we provide an intuitive argument why the result has to be expected for the case $\eta_j = \delta_j$.

By Theorem 4.9 we need to show that each node in isolation is quasi-reversible. Consider node j in isolation that is subject to Poisson arrivals of class u customers with rate α_{ju} and Poisson arrivals of signals with rate α_j^-. If

$$\eta_j(\ell, n_j) = \delta_j(\ell, n_j), \qquad \ell = 1, 2, \ldots, n_j,$$

then the probability that a signal arrives at position ℓ is the same as the proportion of service effort allocated to position ℓ. Since the customer in position ℓ leaves the node either when its service is completed, or when a signal arrives at that position, the time until its departure is the minimum of these two random variables. Therefore, by incorporating the signaling effect into the service process and taking into account that the time until the arrival of the next signal is exponentially distributed, we can consider $\min\{S_j^-, S_{ju}\}$ as the modified service requirement of a class u customer at node j, where S_j^- is the time until the arrival of the next signal, which is exponentially distributed with rate α_j^-. Furthermore, from the analysis in Section 6.2 it follows that this node with modified service requirements is quasi-reversible with departure rate α_{ju}. Clearly, a class u departure from node j can be either due to a regular service completion or due to the arrival of a signal. It is due to a regular service completion with probability

$$P(S_{ju} < S_j^-) = \hat{F}_{ju}(\alpha_j^-),$$

and it is due to the arrival of a signal with probability

$$P(S_j^- < S_{ju}) = 1 - \hat{F}_{ju}(\alpha_j^-).$$

This implies that node j in isolation is quasi-reversible with the departure rate of class u customers due to service completions being $\alpha_{ju}\hat{F}_{ju}(\alpha_j^-)$ and the departure rate of class u customers due to signal triggers being $\alpha_{ju}(1 - \hat{F}_{ju}(\alpha_j^-))$. Applying Theorem 4.9 yields that if α_{ju} and α_j^- are the solutions to the traffic equations (3.2) and (3.3), then the network has the product form stationary distribution given in Theorem 6.5.

Theorem 6.5 enables us to compute various performance measures of interest. For instance, the probability that there are n customers at node j is

$$\left(1 - \sum_{u=1}^{I} \alpha_{ju} E(\min\{S_j^-, S_{ju}\})\right)\left(\sum_{u=1}^{I} \alpha_{ju} E(\min\{S_j^-, S_{ju}\})\right)^n,$$

and the average number of customers in the entire system is

$$\sum_{j=1}^{N} \frac{\sum_{u=1}^{I} \alpha_{ju} E(\min\{S_j^-, S_{ju}\})}{1 - \sum_{u=1}^{I} \alpha_{ju} E(\min\{S_j^-, S_{ju}\})}.$$

We can also calculate the average amount of time a customer spends in the network until it disappears, i.e., when it leaves the system, or when it is removed by a signal, or when it is transformed into a signal. Furthermore, given that there are

n customers at node j, the classes of these customers are independent, and each customer is of class u with probability

$$\frac{\alpha_{ju} E(\min\{S_j^-, S_{ju}\})}{\sum_{u=1}^{I} \alpha_{ju} E(\min\{S_j^-, S_{ju}\})}.$$

6.4 QUASI-REVERSIBILITY OF SYMMETRIC QUEUES WITH SIGNALS

This section proves the product form results of the previous section. The proof is, again, based on Theorem 4.9, which implies that it suffices to prove the quasi-reversibility of the nodes when they are in isolation. A phase-type argument is used in the proof. First, the result is established for the class of mixed Erlang distributions, and then it is shown that the result can be extended to arbitrary distributions. For simplicity the node index j is dropped in the following argument.

Consider a single node queueing system with I classes of customers and a single class of negative signals. Class v customers arrive according to a Poisson process with rate α_v, and negative signals arrive according to a Poisson process with rate α^-. We start with the assumption that class u customers have the service requirement distributions

$$F_u(x) = \sum_{k=1}^{\infty} E_{k,\zeta}(x) f_u(k), \tag{4.1}$$

where f_u is a discrete probability mass function, and $E_{k,\zeta}$ is an Erlang distribution with k phases and phase transition rate ζ. This means that a class u arrival requires an Erlang service time $E_{k,\zeta}$ with probability $f_u(k)$, $k = 1, 2, \ldots$.

Assume that this queue operates according to the same rules as the node in Theorem 6.5. We actually prove a more general result than is needed for Theorem 6.5. That is, we allow a state-dependent service effort, provided the total signal arrival rate is proportional to the service effort. Specifically, instead of (ii′) and (v) of Section 6.3, we make the following two assumptions when there are n customers in the queue.

(ii) The total service effort is supplied at rate $\phi(n)$.

(v′) The probability that an arriving signal joins position ℓ is $\eta(\ell, n)$, and the total signal arrival rate is $\alpha^-\phi(n)$ for some $\alpha^- \geq 0$.

If $\phi(n) \leq B$ for all $n \geq 1$ and for some $B > 0$, then (v′) is equivalent to that a signal arrives at rate $\alpha^- B$ and removes the customer in position ℓ with probability $\eta(\ell, n)\phi(n)/B$.

To obtain a Markov chain we supplement the state of the queue by the remaining service phases for each customer. Let $(c(\ell), w(\ell))$ represent a class $c(\ell)$ customer in position ℓ with $w(\ell)$ service phases to complete, and $\boldsymbol{x}(\ell) = (c(\ell), w(\ell))$ for $\ell = 1, 2, \ldots, n$, where $w(\ell) \geq 1$ for all $\ell = 1, 2, \ldots, n$. The state of the queue is

$$\boldsymbol{x} = (\boldsymbol{x}(1), \boldsymbol{x}(2), \ldots, \boldsymbol{x}(n)).$$

As in the proof of Lemma 6.3, we use the following vector operators:

$$\begin{aligned}
\boldsymbol{x} \ominus \mathbf{e}_\ell &= (\boldsymbol{x}(1), \ldots, \boldsymbol{x}(\ell-1), \boldsymbol{x}(\ell+1), \ldots, \boldsymbol{x}(n)),\\
\boldsymbol{x} \oplus \mathbf{e}_\ell(u,k) &= (\boldsymbol{x}(1), \ldots, \boldsymbol{x}(\ell-1), (u,k), \boldsymbol{x}(\ell), \boldsymbol{x}(\ell+1), \ldots, \boldsymbol{x}(n)),\\
\boldsymbol{x} + \mathbf{e}_\ell &= (\boldsymbol{x}(1), \ldots, \boldsymbol{x}(\ell-1), (c(\ell), w(\ell)+1), \boldsymbol{x}(\ell+1), \ldots, \boldsymbol{x}(n)).
\end{aligned}$$

The first operation represents the state after the customer in the ℓth position leaves. Note that after this customer leaves, customer $(c(\ell+1), w(\ell+1))$ moves into position ℓ. The second operation is the state after a class u customer arrives at position ℓ and it has k phases. The last operator has the same number of customers but the customer in the ℓth position has one more remaining phase to complete.

Suppose that the queue has a stationary distribution $\pi(\boldsymbol{x})$. To derive the global balance equation, we calculate the rates going into and out of each state. The transition rate going out of $\boldsymbol{x}$ is

$$\left(\sum_{u=1}^{I} \alpha_u + \sum_{\ell=1}^{n} \left(\delta(\ell, n)\zeta + \eta(\ell, n)\alpha^-\right)\phi(n)\right)\pi(\boldsymbol{x}), \tag{4.2}$$

and the rate going into $\boldsymbol{x}$ is

$$\begin{aligned}
&\sum_{\ell=1}^{n} \alpha_{c(\ell)}\delta(\ell, n)\pi(\boldsymbol{x} \ominus \mathbf{e}_\ell) f_{c(\ell)}(k_\ell) + \sum_{\ell=1}^{n} \delta(\ell, n)\phi(n)\zeta\pi(\boldsymbol{x} + \mathbf{e}_\ell)\\
&\quad + \sum_{\ell=1}^{n+1}\sum_{u=1}^{I} \delta(\ell, n+1)\phi(n+1)\zeta\pi(\boldsymbol{x} \oplus \mathbf{e}_\ell(u, 1))\\
&\quad + \sum_{\ell=1}^{n+1}\sum_{u=1}^{I}\sum_{k=1}^{\infty} \eta(\ell, n+1)\alpha^-\phi(n+1)\pi(\boldsymbol{x} \oplus \mathbf{e}_\ell(u, k)),
\end{aligned} \tag{4.3}$$

where the first term is due to the arrival of a customer, the second due to the completion of a phase, the third due to the departure of a customer, and the last due to the arrival of a signal.

We now assume that $\pi(\boldsymbol{x})$ is decomposable, i.e., it can be expressed as the product form

$$\pi(\boldsymbol{x}) = p(\mathbf{c}) \prod_{\ell=1}^{n} h_{c(\ell)}(w(\ell)), \tag{4.4}$$

where $\boldsymbol{c} = (c(1), c(2), \ldots, c(n))$ and $p(\cdot)$ is a probability distribution of the form

$$p(\boldsymbol{c}) = p(\mathbf{0}) \prod_{\ell=1}^{n} \frac{\alpha_{c(\ell)}}{\phi(\ell)\bar{\mu}_{c(\ell)}} \tag{4.5}$$

for some positive number $\bar{\mu}_u$, and positive function h_u, which is a probability distribution of the remaining service phases of class u, $u = 1, \ldots, I$. The $\bar{\mu}_u$ and h_u are unknown at the moment, and will be determined later.

After substituting (4.4) and (4.5) into (4.3) and dividing by $\pi(\boldsymbol{x})$, we obtain the rate going into $\boldsymbol{x}$:

$$\begin{aligned}
&\sum_{\ell=1}^{n} \alpha_{c(\ell)}\delta(\ell, n) \frac{\phi(n)\bar{\mu}_{c(\ell)}}{\alpha_{c(\ell)}} \frac{f_{c(\ell)}(w(\ell))}{h_{c(\ell)}(w(\ell))} + \sum_{\ell=1}^{n} \delta(\ell, n)\phi(n) \frac{h_{c(\ell)}(w(\ell)+1)}{h_{c(\ell)}(w(\ell))} \\
&\quad + \sum_{\ell=1}^{n+1} \sum_{u=1}^{I} \delta(\ell, n+1)\phi(n+1)\zeta \frac{\alpha_u}{\phi(n+1)\bar{\mu}_u} h_u(1) \\
&\quad + \sum_{\ell=1}^{n+1} \sum_{u=1}^{I} \sum_{k=1}^{\infty} \eta(\ell, n+1)\phi(n+1)\alpha^- \frac{\alpha_u}{\phi(n+1)\bar{\mu}_u} h_u(k) \\
&= \sum_{\ell=1}^{n} \delta(\ell, n)\phi(n)\bar{\mu}_{c(\ell)} \frac{f_{c(\ell)}(w(\ell))}{h_{c(\ell)}(w(\ell))} + \sum_{\ell=1}^{n} \delta(\ell, n)\phi(n)\zeta \frac{h_{c(\ell)}(w(\ell)+1)}{h_{c(\ell)}(w(\ell))} \\
&\quad + \sum_{u=1}^{I} \zeta \frac{\alpha_u}{\bar{\mu}_u} h_u(1) + \sum_{u=1}^{I} \frac{\alpha^- \alpha_u}{\bar{\mu}_u}.
\end{aligned} \tag{4.6}$$

The rate going out of state $\boldsymbol{x}$ can be written, after dividing (4.2) by $\pi(\boldsymbol{x})$, as

$$\sum_{u=1}^{I} \alpha_u + \sum_{\ell=1}^{n} \left(\delta(\ell, n)\zeta + \eta(\ell, n)\alpha^-\right) = \sum_{u=1}^{I} \alpha_u + \sum_{\ell=1}^{n} \delta(\ell, n) \left(\zeta + \alpha^-\right). \tag{4.7}$$

Note how η is changed to δ, from which it is seen that a "symmetry" assumption with regard to the effect of a signal is not necessary. Comparing (4.6) with (4.7) and matching the coefficients of $\delta(\ell, n)$ as well as the remaining terms we obtain

$$\begin{aligned}
\sum_{u=1}^{I} \alpha_u &= \sum_{u=1}^{I} \frac{\alpha_u}{\bar{\mu}_u} \left(\zeta h_u(1) + \alpha^-\right), \\
\delta(\ell, n)(\zeta + \alpha^-) &= \delta(\ell, n) \left(\bar{\mu}_{c(\ell)} \frac{f_{c(\ell)}(w(\ell))}{h_{c(\ell)}(w(\ell))} + \zeta \frac{h_{c(\ell)}(w(\ell)+1)}{h_{c(\ell)}(w(\ell))}\right).
\end{aligned}$$

These equalities hold if

$$\bar{\mu}_u = \zeta h_u(1) + \alpha^-, \tag{4.8}$$

$$(\zeta + \alpha^-)h_u(k) = \bar{\mu}_u f_u(k) + \zeta h_u(k+1), \tag{4.9}$$

for all $u = 1, \ldots, I$ and $k \geq 1$. Since

$$\sum_{k=1}^{\infty} h_u(k) = \sum_{k=1}^{\infty} f_u(k) = 1,$$

equation (4.9) leads to (4.8). So (4.9) is a sufficient condition for the global balance equation.

Let $\tilde{h}_u$ and $\tilde{f}_u$ be the generating functions of h_u and f_u, respectively, i.e.,

$$\tilde{h}_u(z) = \sum_{k=1}^{\infty} h_u(k) z^k,$$
$$\tilde{f}_u(z) = \sum_{k=1}^{\infty} f_u(k) z^k.$$

Multiplying both sides of (4.9) by z^{k+1} and summing over all $k \geq 1$ yields

$$\begin{aligned} z(\zeta + \alpha^-)\tilde{h}_u(z) &= z\bar{\mu}_u \tilde{f}_u(z) + \zeta(\tilde{h}_u(z) - z h_u(1)) \\ &= z\bar{\mu}_u \tilde{f}_u(z) + (\zeta \tilde{h}_u(z) - z(\bar{\mu}_u - \alpha^-)). \end{aligned}$$

Thus we obtain

$$\tilde{h}_u(z) = \frac{z(\alpha^- - \bar{\mu}_u(1 - \tilde{f}_u(z)))}{z(\zeta + \alpha^-) - \zeta}. \tag{4.10}$$

Let H_u denote the remaining service time distribution of a class u customer, and let $\hat{F}_u$ and $\hat{H}_u$ denote the Laplace transform of F_u and H_u. It is easily checked that

$$\hat{F}_u(s) = \tilde{f}_u\left(\frac{\zeta}{\zeta + s}\right),$$
$$\hat{H}_u(s) = \tilde{h}_u\left(\frac{\zeta}{\zeta + s}\right).$$

Substituting $z = \zeta/(\zeta + s)$ in (4.10) yields

$$\hat{H}_u(s) = \frac{\alpha^- - \bar{\mu}_u(1 - \hat{F}_u(s))}{\alpha^- - s}. \tag{4.11}$$

Since $\hat{H}_u(s)$ is finite for $s > 0$, the denominator of the right-hand side of (4.11) has to be zero at $s = \alpha^-$. Hence, from (3.6), we obtain

$$\bar{\mu}_u = \frac{\alpha^-}{1 - \hat{F}_u(\alpha^-)}. \tag{4.12}$$

This and (4.8) imply that

$$h_u(1) = \frac{\overline{\mu}_u - \alpha^-}{\zeta} = \frac{\alpha^- \hat{F}(\alpha^-)}{\zeta(1 - \hat{F}(\alpha^-))}. \tag{4.13}$$

Substituting (4.12) into (4.11) gives

$$\hat{H}_u(s) = \frac{\bar{\mu}_u\left(\hat{F}_u(\alpha^-) - \hat{F}_u(s)\right)}{s - \alpha^-}. \tag{4.14}$$

We now prove that the node is a quasi-reversible queue with signals. There are $I + 1$ classes of arrivals, denoted by $\{1, \ldots, I\} \cup \{s^-\}$, where s^- denotes the negative signals, and $2I$ classes of departures, denoted by $\{1, 2, \ldots, I\}$ and $\{1^-, 2^-, \ldots, I^-\}$, where we define a class u service completion as a class u departure, and a class u customer departure due to the trigger of a signal as a class u^- departure. Using the notation of Chapter 3, the nonzero, non-internal transition rates of the node are

$$\begin{aligned}
q_u^{\mathrm{D}}(\boldsymbol{x} \oplus \mathbf{e}_\ell(u, 1), \boldsymbol{x}) &= \delta(\ell, n)\phi(n+1)\zeta, \\
q_u^{\mathrm{A}}(\boldsymbol{x}, \boldsymbol{x} \oplus \mathbf{e}_\ell(u, k)) &= \alpha_u \delta(\ell, n+1) f_u(k), \\
q_{s^-}^{\mathrm{A}}(\boldsymbol{x} \oplus \mathbf{e}_\ell(u, k), \boldsymbol{x}) &= \delta(\ell, n+1)\phi(n+1)\alpha^-, \\
f_{s^-,s^-}(\boldsymbol{x} \oplus \mathbf{e}_\ell(u, k), \boldsymbol{x}) &= 1.
\end{aligned}$$

Hence, the departure rates are

$$\begin{aligned}
\sum_{\boldsymbol{x}'} \pi(\boldsymbol{x}') q_u^{\mathrm{D}}(\boldsymbol{x}', \boldsymbol{x}) &= \sum_{\ell=1}^{n+1} \pi(\boldsymbol{x} \oplus \mathbf{e}_\ell(u, 1)) q_u^{\mathrm{D}}(\boldsymbol{x} \oplus \mathbf{e}_\ell(u, 1)) \\
&= \frac{\alpha_u}{\bar{\mu}_u} \zeta h_u(1) \pi(\boldsymbol{x}) \\
&= \alpha_u \hat{F}_u(\alpha^-) \pi(\boldsymbol{x}),
\end{aligned}$$

and

$$\begin{aligned}
&\sum_{\ell=1}^{n+1} \pi(\boldsymbol{x} \oplus \mathbf{e}_\ell(u, 1)) q_{s^-}^{\mathrm{A}}(\boldsymbol{x} \oplus \mathbf{e}_\ell(u, k), \boldsymbol{x}) f_{s^-,s^-}(\boldsymbol{x} \oplus \mathbf{e}_\ell(u, k), \boldsymbol{x}) \\
&\quad = \sum_{\ell=1}^{n+1} \pi(\boldsymbol{x} \oplus \mathbf{e}_\ell(u, 1)) \alpha^- \delta(\ell, n+1) \\
&\quad = \frac{\alpha_u}{\bar{\mu}_u} \alpha^- \pi(\boldsymbol{x}) \\
&\quad = \alpha_u (1 - \hat{F}_u(\alpha^-)) \pi(\boldsymbol{x}),
\end{aligned}$$

where (4.13) is used to obtain the last equalities. Thus the node is quasi-reversible with

$$\beta_u = \alpha_u \hat{F}_u(\alpha^-), \qquad u = 1, \ldots, I,$$
$$\beta_{u^-} = \alpha_u(1 - \hat{F}_u(\alpha^-)), \qquad u = 1, \ldots, I.$$

By Theorem 4.9, the network has a product form stationary distribution. Substituting (after restoring the node index)

$$\beta_{ju} = \alpha_{ju} \hat{F}_{ju}(\alpha_j^-), \qquad j = 1, \ldots, N, \ u = 1, \ldots, I,$$
$$\beta_{ju^-} = \alpha_{ju}(1 - \hat{F}_{ju}(\alpha_j^-)), \qquad j = 1, \ldots, N, \ u = 1, \ldots, I,$$

into the traffic equations we obtain (3.2) and (3.3). This completes the proof of Theorem 6.5 when the service requirements have phase-type distributions. The case of arbitrary distributions follows from a limiting argument, since phase-type distributions are dense in the class of distribution functions.

6.5 NETWORKS WITH SIGNALS AND MULTI-SERVER NODES

In this section we assume that the service supplied at node j is state-dependent with rate $\phi_j(n_j)$ when there are n_j customers present. If all other assumptions remain as in Section 6.3, then it can be shown that the network does not have a product form stationary distribution. Thus additional assumptions have to be imposed to obtain a product form solution. In this section we modify the effect of the arrival of a signal in such a way that it depends on the service effort allocated at the time of its arrival. In other words, the probability that a customer in position ℓ is induced to leave by a signal depends not only on $\delta_j(\ell, n_j)$, but also on $\phi_j(n_j)$.

Example 6.7 (Tandem queue with signals and arbitrary service times) Consider a tandem queue with two nodes in series. There are two classes of customers that arrive at node 1 according to Poisson processes with rates λ_1 and λ_2. Signals arrive at node 1 according to a Poisson process with rate λ^-. The arrival of a signal triggers a customer at the node where it arrives, provided the node is not empty, to leave the system. The arrival of a signal at an empty node have no effect. The nonzero routing probabilities are

$$r_{1u,2u} = 1, \quad r_{2u,0} = 1, \qquad u = 1, 2.$$

Let

$$\phi_1(n_1) = \min\{n_1, 2\}, \qquad \phi_2(n_2) = 1[n_2 \geq 1].$$

This implies that there are two servers at node 1 and a single server at node 2. Assume that the service disciplines at both nodes are processor-sharing, and the service times are all arbitrarily distributed. It can be shown that this network does not have a product form solution (see Exercise 11.1). □

Assume that class u customers arrive at node j according to a Poisson process with rate λ_{ju}, $u = 1, \ldots, I$, and signals arrive at node j according to a Poisson process with rate λ_j^-, $j = 1, \ldots, N$. The routing probabilities are defined in the same way as in Section 6.3. Furthermore, when there are n_j customers at node j, node j operates in the same way as in Section 6.3 except that assumptions (ii′) and (v) are now modified as follows.

(ii″) The total service effort is supplied at rate $\phi_j(n_j)$, and $\phi_j(n_j)$ is bounded by B_j.

(v″) When a signal arrives at node j, it induces a customer to leave with probability $\phi_j(n_j)/B_j$. An effective signal is directed to position ℓ with probability $\eta_j(\ell, n_j)$, $\ell = 1, \ldots, n_j$. When this customer leaves, the customers in positions $\ell+1, \ell+2, \ldots, n_j$ move to positions $\ell, \ell+1, \ldots, n_j - 1$, respectively.

Note that in the multi-server case an arriving signal can still join a position with an arbitrary probability $\eta_j(\ell, n_j)$. The only requirement is that the total signaling intensity has to be proportional to the total service effort and it has to be bounded. The intuition here is that the effect of the signal increases with an increase in the service effort at that node. Under these assumptions the following can be concluded from the proof of Section 6.4.

Theorem 6.8 *The queueing network with the signal effects satisfying (ii″) and (v″) has the product form stationary distributions (3.4) and (3.5) with*

$$\pi_j(\boldsymbol{c}_j) = b_j^{-1} \prod_{\ell=1}^{n_j} \frac{\alpha_{jc_j(\ell)} E(\min\{S_j^-, S_{jc_j(\ell)}\})}{\phi_j(\ell)}, \tag{5.1}$$

$$\pi_j(\boldsymbol{n}_j) = b_j^{-1} \frac{n_j!}{n_{j1}! \cdots n_{jI}!} \frac{\prod_{u=1}^{I} (\alpha_{ju} E(\min\{S_j^-, S_{ju}\}))^{n_{ju}}}{\prod_{\ell=1}^{n_j} \phi_j(\ell)}, \tag{5.2}$$

where S_j^- is an exponential random variable with mean B_j/α_j^- which is independent of S_{ju}, b_j is the normalization constant which is assumed to be finite, i.e.,

$$b_j = \sum_{n=0}^{\infty} \frac{\left(\sum_{u=1}^{I} \alpha_{ju} E(\min\{S_j^-, S_{ju}\})\right)^n}{\prod_{\ell=1}^{n} \phi_j(\ell)} < \infty,$$

and α_{ju}, $j = 1, \ldots, N$, $u = 1, \ldots, I$, and α_j^-, $j = 1, \ldots, N$, are the solutions to the traffic equations (3.2) and (3.3).

Example 6.9 (Product form of the tandem queue with signals) Consider the tandem queue of Example 6.7. Assume that the service discipline is processor-sharing, and an arriving customer takes any position with equal probability, i.e.,

$$\delta_j(\ell, n_j) = \frac{1}{n_j}, \qquad j = 1, 2, \ \ell = 1, \ldots, n_j.$$

The solutions to the traffic equations are

$$\begin{aligned} \alpha_{1u} &= \lambda_u, && u = 1, 2, \\ \alpha_1^- &= \lambda^-, && \\ \alpha_{2u} &= \lambda_u \hat{F}_{1u}(\lambda^-), && u = 1, 2, \end{aligned}$$

where F_{1u} is the service time distributions of class u customers at node 1. We set $B_j = 2$ for the condition (v$''$), so the arrival of a signal at any position in node 1 triggers the customer to leave with probability

$$\frac{\min\{n_1, 2\}}{2} = \begin{cases} 1/2, & \text{if } n_1 = 1; \\ 1, & \text{if } n_1 > 1. \end{cases}$$

The stationary distribution of the network is

$$\pi(\boldsymbol{n}_1, \boldsymbol{n}_2) = \pi_1(\boldsymbol{n}_1)\pi_2(\boldsymbol{n}_2),$$

where

$$\pi_1(\boldsymbol{n}_1) = \begin{cases} \pi_1(\boldsymbol{0}) \prod_{u=1}^{2} (\lambda_u E(\min\{S_1^-, S_{1u}\}))^{n_{1u}}, & \text{if } n_{11} + n_{12} \le 1; \\ \pi_1(\boldsymbol{0}) \dfrac{n_1!}{n_{11}! n_{12}!} \dfrac{\prod_{u=1}^{2} (\lambda_u E(\min\{S_1^-, S_{1u}\}))^{n_{1u}}}{2^{n_1 - 1}}, & \text{if } n_{11} + n_{12} \ge 2, \end{cases}$$

$$\pi_2(\boldsymbol{n}_2) = \left(1 - \sum_{u=1}^{2} \alpha_{2u} E(S_{2u})\right) \prod_{u=1}^{2} (\alpha_{2u} E(S_{2u}))^{n_{2u}},$$

where $\mathbf{0} = (0, 0)$, S_1^- is an exponential random variable with mean $2/\lambda^-$, and

$$\pi_1(\mathbf{0}) = \left(1 + \sum_{u=1}^{2} \lambda_u E(\min\{S_1^-, S_{1u}\}) + \frac{\left(\sum_{u=1}^{2} \lambda_u E(\min\{S_1^-, S_{1u}\})\right)^2}{2\left(1 - \sum_{u=1}^{2} \lambda_u E(\min\{S_1^-, S_{1u}\})\right)}\right)^{-1}. \qquad \square$$

In the above analysis the signaling mechanisms are independent of the class of customer. When the service times at a node are exponentially distributed, we can also have class-dependent signaling effects. In the remaining part of this section we assume

(i′) The service requirements of the customers are exponentially distributed, and the service rate of class u customers at node j is μ_{ju}.

The effect of signals is now modified as follows:

(v‴) When a signal arrives at node j, it triggers the customer of class u in position ℓ to leave with a probability that is proportional to $\mu_{ju}\delta_j(\ell, n_j)$.

If, for example, $\mu_{\max}$ is the maximum service rate of all classes of customers at node j, then the probability that a signal triggers a class u customer in position ℓ to leave can be defined as $\delta_j(\ell, n_j)\mu_{ju}/\mu_{\max}$. That is, the effect of the signal on a customer is larger when the mean service time is smaller. For simplicity, let this triggering probability (or rate) be written as $\mu_{ju}\delta_j(\ell, n_j)$. We leave it to the reader to show that under assumptions (i′) and (v‴), the stationary distribution of the network has a product form with

$$\pi_j(\boldsymbol{c}_j) = b_j^{-1} \prod_{\ell=1}^{n_j} \frac{\rho_{jc_j(\ell)}}{\alpha_j^- + \phi_j(\ell)},$$

$$\pi_j(\boldsymbol{n}_j) = b_j^{-1} \frac{n_j!}{n_{j1}! \cdots n_{jI}!} \frac{\prod_{u=1}^{I} \rho_{ju}^{n_{ju}}}{\prod_{\ell=1}^{n_j} (\alpha_j^- + \phi_j(\ell))},$$

where

$$\rho_{ju} = \frac{\alpha_{ju}}{\mu_{ju}}, \qquad u = 1, \ldots, I,$$

and b_j is the normalization constant which is assumed to be finite and is given by

$$b_j = \sum_{n=0}^{\infty} \frac{\left(\sum_{u=1}^{I} \rho_{ju}\right)^n}{\prod_{\ell=1}^{n} (\alpha_j^- + \phi_j(\ell))} < \infty,$$

and α_{ju}, $u = 1, \ldots, I$, α_j^-, $j = 1, \ldots, N$, are the solutions to the traffic equations (3.2) and (3.3).

6.6 REFERENCE NOTES

The pioneering work extending Jackson networks to include non-exponential service times is BCMP networks, reported in Baskett, Chandy, Muntz and Palacios (1975), though earlier discussions also appeared in the literature, e.g., Baskett and Palacios (1972), Chandy and Palacios (1972), and Muntz (1972). Considerable extensions were also made by Kelly (1975, 1976), and a thorough discussion is contained in Kelly (1979). Queueing networks were also considered from the viewpoint that the network distributions depend on the service time distributions only through their means, which is called *insensitivity*. This direction was first studied by Matthes (1962), and followed by a group of the east German school (see König, Matthes and Nawrotzki (1967), Jansen, König and Nawrotzki (1979) and Franken et al. (1982)). Schassberger (1977) studied this problem using the so-called Generalized Semi-Markov processes (GSMP), see also Schassberger (1986). The results were extended to GSMP with interruptions by Henderson and Taylor (1989), and further to the so-called reallocatable GSMP by Miyazawa (1993). These two papers also consider networks with negative customers and arbitrary service times. More general signals for such networks are discussed in Chao (1995), in which it is shown that the product form result holds under the assumption $\delta^{\mathrm{A}} = \delta^{\mathrm{D}} = \eta$. The proof of Theorem 6.5 presented in this chapter is new. Some other related papers are Chao and Pinedo (1995*b*), and Miyazawa and Wolff (1996). Exercise 6.5 is taken from Chao (1994).

EXERCISES

6.1 Assume the service requirements are exponentially distributed with class-dependent rates. Prove that Theorem 6.5 still holds under the assumption

$$\delta_j^{\mathrm{A}}(\ell, n_j) = \frac{\alpha_j^-}{\mu_j + \alpha_j^-}\eta_j(\ell, n_j) + \frac{\mu_j}{\mu_j + \alpha_j^-}\delta_j^{\mathrm{D}}(\ell, n_j)$$

for $\ell = 1, \ldots, n_j$ and $n_j = 1, 2, \ldots$.

6.2 Prove that the results of Sections 6.2 and 6.3 still hold when $\delta_j(\ell, n_j)$ is replaced by $\delta_j(\ell, \boldsymbol{x}_j)$, as long as $\delta_j(\ell, \boldsymbol{x}_j)$ is invariant under permutations of $\boldsymbol{x}_j$ and

$$\sum_{\ell=1}^{n_j} \delta_j(\ell, \boldsymbol{x}_j) = 1.$$

6.3 Prove Theorem 6.8 by showing that each node as defined is quasi-reversible.

6.4 Argue that in a queueing network, the service requirements of a customer at the symmetric queues along its route through the network may be dependent. (Hint:

construct the dependency by using different classes of customers corresponding to each pattern of dependence).

6.5 **(Networks with random signal triggering times)** Consider a network with signals and *random triggering times*. That is, the amount of time to trigger a customer is an arbitrarily distributed random variable that depends on the source of the signal. The network consists of N nodes, I different classes of customers and a single class of signals. Customers of class u arrive at node j according to a Poisson process with rate λ_{ju}. Upon a service completion at node j, a customer of class u goes to node k as a class v customer with probability $r_{ju,kv}$, as a signal with probability r_{ju,ks^-}, and it leaves the system with probability $r_{ju,0}$. Signals arrive at node j from the outside according to a Poisson process with rate λ_j^-. Whenever a signal arrives at node j, either from outside the system or from another node, it triggers a customer to leave. The triggering times are arbitrarily distributed random variables with distributions depending on the source of the signal. Specifically, when a class u customer at node j is triggered by a signal from the outside (node i), the customer leaves after a random amount of time $T_{ju}(0)$ ($T_{ju}(i)$). This customer then goes to node k as a class v customer with probability $r_{ju^-,kv}$, as a signal with probability r_{ju^-,ks^-}, and it leaves the system with probability $r_{ju^-,0}$. Assume that a triggered customer stops receiving service. Therefore, the service effort is only allocated to the non-triggered customers. Assume that there is a single server at each node. Also, assume that if the service requirements at node j are exponential, then all the customer classes have the same service rate μ_j. However, when the service requirements at node j are arbitrarily distributed, the service discipline is symmetric among the non-triggered customers. In this case the service requirement S_{ju} of a class u customer at node j has distribution F_{ju}. Define the state of the network as follows. Each node consists of two types of customers: regular customers and triggered customers. Let $\boldsymbol{n}_j = (\boldsymbol{n}_j^+, \boldsymbol{n}_j^-)$, where $\boldsymbol{n}_j^+ = (n_{j1}, \ldots, n_{jI})$ represents the number of each class of regular customers, and $\boldsymbol{n}_j^- = (n_{j1}^-, \ldots, n_{jI}^-)$ represents the number of each class of triggered customers at node j. The state of the network is $\boldsymbol{n} = (\boldsymbol{n}_1, \ldots, \boldsymbol{n}_N)$. Prove that if the stability condition

$$\sum_{u=1}^{I} \alpha_{ju} E(\min\{S_j^-, S_{ju}\}) < 1, \qquad j = 1, \ldots, N,$$

is satisfied, where S_j^- is an exponential random variable with rate α_j^- that is independent of S_{ju}, and α_{ju} and α_j^- are the solutions to the traffic equations

$$\alpha_{ju} = \lambda_{ju} + \sum_{k=1}^{I}\sum_{v=1}^{N} \hat{F}(\alpha_k^-)\alpha_{kv} r_{kv,ju} + \sum_{k=1}^{I}\sum_{v=1}^{N} (1 - \hat{F}(\alpha_k^-))\alpha_{kv} r_{kv^-,ju},$$

$$\alpha_j^- = \lambda_j^- + \sum_{k=1}^{I}\sum_{v=0}^{N} \hat{F}(\alpha_k^-)\alpha_{kv} r_{kv,jc^-}^- + \sum_{k=1}^{I}\sum_{v=1}^{N} (1 - \hat{F}(\alpha_k^-))\alpha_{kv} r_{kv^-,jc^-},$$

then the stationary distribution of the network is

$$\pi(\boldsymbol{n}) = \prod_{j=1}^{N} \pi_j(\boldsymbol{n}_j, \boldsymbol{n}_j^-),$$

where

$$\pi_j(\boldsymbol{n}_j, \boldsymbol{n}_j^-) = b_j(n_{j1} + \cdots + n_{jI})! \times \prod_{u=1}^{I} \left(\frac{(\alpha_{ju} E(\min\{S_j^-, S_{ju}\}))^{n_{ju}}}{n_{ju}!} \frac{(\alpha_{ju} \hat{F}_{ju}(\alpha_j^-) E(T_{ju}))^{n_{ju}^-}}{n_{ju}^-!} \right),$$

and

$$b_j = \left(1 - \sum_{u=1}^{I} \alpha_{ju} E\big(\min\{S_j^-, S_{ju}\}\big)\right) \exp\left(-\sum_{u=1}^{I} \alpha_{ju} \hat{F}_{ju}(\alpha_j^-) E(T_{ju})\right),$$

$E(T_{ju})$ is given by

$$E(T_{ju}) = \sum_{i=0}^{N} \frac{\alpha_j^-(i)}{\alpha_j^-} E(T_{ju}(i)),$$

and $\alpha_j^-(i)$ is defined as

$$\alpha_j^-(i) = \sum_{u=1}^{I} \Big(\alpha_{iu} r_{iu,jc^-} \hat{F}_{iu}(\alpha_i^-) + \alpha_{iu} r_{iu^-,jc^-}(1 - \hat{F}_{iu}(\alpha_i^-)) \Big),$$

for $i = 1, \ldots, N$ and $\alpha_j^-(0) = \lambda_j^-$, $j = 1, \ldots, N$. Extend this result to allow the routing probabilities $r_{ju^-,kv}$ and $r_{ju^-,0}$ of a triggered customer to depend on the source of the triggering signal and on the amount of triggering time (i.e., i and $T_{ju}(i)$).

7

Networks with Batch Services and Negative Signals

This chapter considers a class of queueing networks with batch services. The major feature of these networks is that there may be an arbitrary number of batch departures occurring throughout the network simultaneously. However, these departing batches do not arrive at other nodes as batches; they arrive either as a single customer or as a signal. The case of batch arrivals as well as batch departures will be discussed in the next chapter.

We start in Section 7.1 with a simple network without signals. In this model, the services at each node are in batches and all the customers in a batch, upon a service completion, coalesce into a single customer. Section 7.2 extends this model to include negative signals. The arrival of a negative signal at a node triggers a batch of customers to coalesce into a single customer, and the customer then either leaves the network or joins another node. When it joins another node, it may go either as a customer or as a signal that triggers another batch of customers. This cascading effect results in a sequence of batch departures at a random number of nodes throughout the network. Several examples are presented in Section 7.3, including networks with string transitions and networks with catastrophes. Three of the problems discussed in Chapter 1, i.e., the computer virus model, the production distribution system, and the quality control problem, are also solved in Section 7.3.

7.1 BATCH SERVICES AND CUSTOMER COALESCENCE

Consider a network with N single server nodes. Customers arrive at node j from the outside according to a Poisson process with rate λ_j, $j = 1, 2, \ldots, N$. The service process at each node occurs in a batch mode. For simplicity we first assume that the batch sizes at node j are fixed, say K_j. The service time of each batch of customers at node j is exponentially distributed with rate μ_j. If a customer arrives at node j when the number of customers is less than K_j, then it joins the batch currently in service; otherwise it waits in the queue. Upon a service completion, the entire batch coalesces into a single customer, provided there are at least K_j customers at the node. For convenience these K_j customers are called a full batch. A full batch, upon service completion, goes to node k as *a single customer* with probability r_{jk}, $k = 1, \ldots, N$, and it leaves the network with probability r_{j0}, where

$$\sum_{k=0}^{N} r_{jk} = 1, \qquad j = 1, \ldots, N.$$

If, however, the number of customers present at the service completion is strictly less than K_j, all the customers present at node j coalesce into an incomplete unit, called a partial batch. A partial batch leaves the network with probability 1.

Let n_j be the number of customers at node j, and let

$$\boldsymbol{n} = (n_1, \ldots, n_N)$$

denote the state of the network. Recall that $\mathbf{e}_j$ is the N-dimensional unit vector with a 1 in the jth element and a 0 everywhere else.

Theorem 7.1 *If the traffic equations*

$$\alpha_j = \lambda_j + \sum_{k=1}^{N} \rho_k^{K_k} \mu_k r_{kj}, \qquad j = 1, 2, \ldots, N, \tag{1.1}$$

have solutions $(\alpha_1, \ldots, \alpha_N)$ *and* $(\rho_1, \ldots, \rho_N)$ *such that*

$$\alpha_j = \mu_j \left(\rho_j + \rho_j^2 + \cdots + \rho_j^{K_j}\right), \qquad j = 1, \ldots, N, \tag{1.2}$$

and

$$\rho_j < 1, \qquad j = 1, \ldots, N, \tag{1.3}$$

then the stationary distribution of the network is

$$\pi(\boldsymbol{n}) = \prod_{j=1}^{N} (1 - \rho_j)\rho_j^{n_j}. \tag{1.4}$$

Furthermore, π satisfies the following balance equations, which are referred to as the cross local balance equations:

$$\sum_{\ell=0}^{K_j-1} \pi(\boldsymbol{n} + \ell \mathbf{e}_j)\mu_j = \pi(\boldsymbol{n} - \mathbf{e}_j)\lambda_j + \sum_{k=1}^{N} \pi(\boldsymbol{n} + K_k \mathbf{e}_k - \mathbf{e}_j)\mu_k r_{kj}\,, \qquad (1.5)$$

$$j = 1, 2, \ldots, N.$$

Remark 7.2 Balance equations (1.5) state that for any state $\boldsymbol{n}$, the rate going into state $\boldsymbol{n}$ due to an arrival at node j is equal to the rate going across state $\boldsymbol{n}$ from above due to a service completion at node j. This is not the local balance of Definition 2.12; it represents a balance of rates crossing a given state. For this reason we call this type of local balance the *cross local balance*. Cross local balance equations will be studied in Chapter 10; they play an important role in extending product form results to networks with state-dependent transitions.

Remark 7.3 The inequalities $\rho_j < 1,\ j = 1, \ldots, N$, are equivalent to the stability condition

$$\alpha_j < K_j \mu_j\,, \qquad j = 1, \ldots, N.$$

The stability condition basically requires, therefore, that the total arrival rate is less than the maximum departure rate, which is achieved when all the servers are busy continuously. Stability issues will be discussed in Chapter 11.

PROOF OF THEOREM 7.1 We use Theorem 4.3 to prove this result. Since a partial batch leaves the network and does not participate in the routing, it is classified as an internal transition. Thus node j is characterized by

$$p_j^{\mathrm{A}}(n_j, n_j + 1) = 1, \qquad n_j \geq 0,$$

$$q_j^{\mathrm{D}}(n_j + K_j, n_j) = \mu_j, \qquad n_j \geq 0,$$

$$q_j^{\mathrm{I}}(n_j, 0) = \mu_j, \qquad n_j = 1, 2, \ldots, K_j - 1.$$

Node j, defined by

$$q_j(n_j, n_j') = \alpha_j p_j^{\mathrm{A}}(n_j, n_j') + q_j^{\mathrm{D}}(n_j, n_j') + q_j^{\mathrm{I}}(n_j, n_j'),$$

is subject to a Poisson arrival process with rate α_j. By Lemma 2.32 it has the stationary distribution

$$\pi_j(n_j) = (1 - \rho_j)\rho_j^{n_j}, \tag{1.6}$$

where ρ_j is the solution of equation (1.2). Since

$$\begin{aligned}\sum_{n'_j} \pi_j(n'_j) q_j^{\mathrm{D}}(n'_j, n_j) &= \pi_j(n_j + K_j) q_j^{\mathrm{D}}(n_j + K_j, n_j) \\ &= \rho_j^{K_j} \mu_j \pi_j(n_j), \qquad n_j \geq 0,\end{aligned}$$

node j is quasi-reversible with

$$\beta_j = \rho_j^{K_j} \mu_j, \qquad j = 1, \ldots, N. \tag{1.7}$$

Substituting (1.7) into the traffic equations (3.5) of Chapter 4 yields (1.1). Hence applying Theorem 4.3 we conclude that if $\alpha_1, \ldots, \alpha_N$ are the solutions to the traffic equations (1.1) with ρ_j determined by (1.2) and condition (1.3) satisfied, then the stationary distribution of the network is given by (1.4).

It is easily seen that (1.5) can be written as

$$\mu_j \sum_{\ell=1}^{K_j} \rho_j^{\ell} = \lambda_j + \sum_{k=1}^{N} \mu_k \rho_k^{K_k} r_{kj},$$

which follows immediately from (1.1) and (1.2). □

Example 7.4 (A tandem queue with batch departures) Consider two queues in tandem with batch processing at both nodes. The batch sizes at the two nodes are both equal to two, i.e.,

$$K_1 = K_2 = 2,$$

and the service time distributions are exponential with rates μ_1 and μ_2, respectively. The arrivals at the first node follow a Poisson process with rate λ.

Equations (1.1) and (1.2) are

$$\begin{aligned}\alpha_1 &= \lambda, & \alpha_2 &= \rho_1^2 \mu_1, \\ \alpha_1 &= \mu_1(\rho_1 + \rho_1^2), & \alpha_2 &= \mu_2(\rho_2 + \rho_2^2).\end{aligned}$$

Solving for ρ_1 and ρ_2 yields

$$\rho_1 = \frac{1}{2}\sqrt{1 + 4\frac{\lambda}{\mu_1}} - \frac{1}{2},$$

$$\rho_2 = \frac{1}{2}\sqrt{1 + 4\frac{\lambda}{\mu_2} + 2\frac{\mu_1}{\mu_2} - 2\frac{\mu_1}{\mu_2}\sqrt{1 + 4\frac{\lambda}{\mu_1}}} - \frac{1}{2}.$$

Thus, if $\rho_1 < 1$ and $\rho_2 < 1$, then the stationary distribution of the tandem network is

$$\pi(n_1, n_2) = (1 - \rho_1)\rho_1^{n_1}(1 - \rho_2)\rho_2^{n_2}.$$

Using the stationary distribution we can compute various quantities of interest. For instance, the rate at which partial batches are produced and discarded at node 1 is

$$\mu_1\pi_1(1) = \mu_1(1 - \rho_1)\rho_1,$$

and the proportion of customers at node 1 that have to leave the system is

$$\frac{\mu_1(1 - \rho_1)\rho_1}{\lambda}.$$

Similarly, the proportion of partial batches that have to be discarded at node 2 is

$$\frac{\mu_2(1 - \rho_2)\rho_2}{\rho_1^2\mu_1}. \qquad \square$$

We relax the fixed batch size to an arbitrarily distributed integer valued random variable, which is denoted by B_j. Assume that the batch size B_j at node j has the probability mass function

$$P(B_j = n) = b_j(n), \qquad n = 1, 2, \ldots, \tag{1.8}$$

i.e., the batch size is n with probability $b_j(n)$. If $B_j = n$ and the number of customers at node j upon the service completion is greater than or equal to the required size n, then n customers coalesce into a single customer and this customer proceeds according to the probabilities r_{jk}, $k = 0, 1, \ldots, N$. Otherwise, if the number of customers present is less than n, node j is emptied out and all the customers at the node leave the system. The remaining assumptions are the same as in the fixed batch size case.

Define

$$\bar{b}_j(n) = \sum_{\ell=n}^{\infty} b_j(\ell),$$

$$\tilde{B}_j(z) = \sum_{n=1}^{\infty} b_j(n) z^n.$$

The next result follows from the same line of proof as Theorem 7.1, and is a special case of Theorem 7.6.

Theorem 7.5 *If the traffic equations*

$$\alpha_j = \lambda_j + \sum_{k=1}^{N} \tilde{B}_k(\rho_k)\mu_k r_{kj}, \qquad j = 1, \ldots, N,$$

have solutions $(\alpha_1, \ldots, \alpha_N)$ *and* $(\rho_1, \ldots, \rho_N)$ *satisfying*

$$\alpha_j(1 - \rho_j) = \mu_j \rho_j (1 - \tilde{B}_j(\rho_j)), \quad j = 1, \ldots, N,$$

and

$$\rho_j < 1, \qquad j = 1, \ldots, N,$$

then the stationary distribution π *of the network is given by the product form (1.4) and it satisfies the cross local balance equations*

$$\mu_j \sum_{\ell=0}^{\infty} \pi(\boldsymbol{n} + \ell \mathbf{e}_j)\bar{b}_j(\ell + 1) = \lambda_j \pi(\boldsymbol{n} - \mathbf{e}_j) + \sum_{k=1}^{N}\sum_{\ell=1}^{\infty} \mu_k \pi(\boldsymbol{n} + \ell \mathbf{e}_k - \mathbf{e}_j) b_k(\ell) r_{kj}, \quad j = 1, \ldots, N. \quad (1.9)$$

Clearly, equation (1.9) has the same interpretation as equation (1.5) (see Remark 7.2).

7.2 BATCH SERVICES AND NEGATIVE SIGNALS

This section extends the results of the previous section to include negative signals. This enables us to model simultaneous batch departures throughout the network.

Assume that customers arrive at node j from the outside according to a Poisson process with rate λ_j, and negative signals arrive at node j from the outside according to a Poisson process with rate λ_j^-, $j = 1, 2, \ldots, N$. As in the previous section, the services at each node occur in batches, and the batch sizes are arbitrarily distributed. The batch size probability mass function at node j is given by (1.8). If the number of customers present at node j at the completion of a service is greater than or equal to the batch size requirement B_j, then B_j customers form a full batch and coalesce into a single entity; otherwise all customers at node j form a partial batch and coalesce into an incomplete entity and this entity leaves the network. Service times at node j are exponential with rate μ_j. When a full batch finishes service at node j, it joins node k as a single customer with probability $r_{jc,kc}$, as a negative signal with probability r_{jc,ks^-}, and it leaves the network with probability $r_{jc,0}$. Clearly,

$$\sum_{k=1}^{N} r_{jc,kc} + \sum_{k=1}^{N} r_{jc,ks^-} + r_{jc,0} = 1, \qquad j = 1, \ldots, N.$$

The effect of a negative signal can be described as follows. When a negative signal arrives at node j, either from the outside or from another node, it intends to trigger a batch of A_j^- customers to move. The sizes of the triggered batches are arbitrarily distributed with probability mass function

$$P(A_j^- = \ell) = a_j^-(\ell), \qquad \ell = 1, 2, \ldots. \tag{2.1}$$

If a negative signal arrives at node j when the number of customers present is greater than or equal to A_j^-, then A_j^- customers depart immediately as a full batch. This full batch joins node k as a single customer with probability $r_{js^-,kc}$, as a negative signal with probability r_{js^-,ks^-}, and it leaves the network with probability $r_{js^-,0}$. Again,

$$\sum_{k=1}^{N} r_{js^-,kc} + \sum_{k=1}^{N} r_{js^-,ks^-} + r_{js^-,0} = 1, \qquad j = 1, \ldots, N.$$

If, however, the number of customers present at node j upon the signal's arrival is less than A_j^-, then all the customers form a partial batch and they leave the network.

This model includes simultaneous batch departures from any number of nodes in the network. For instance, if

$$a_{j_1}(n_1) r_{j_1c,i_2s^-} a_{j_2}^-(n_2) r_{j_2s^-,j_3s^-} a_{j_3}^-(n_3) \cdots r_{j_{m-1}s^-,j_ms^-} a_{j_m}^-(n_m) > 0,$$

then a service completion at node j_1 will, with a positive probability, remove n_ℓ customers from node j_ℓ, $\ell = 1, 2, \ldots, m$, provided there is a sufficient number of customers at all these nodes.

Define, for $j = 1, 2, \ldots, N$,

$$\bar{b}_j(n) = \sum_{\ell=n}^{\infty} b_j(\ell), \qquad n = 1, 2, \ldots,$$

$$\bar{a}_j^-(n) = \sum_{\ell=n}^{\infty} a_j^-(\ell), \qquad n = 1, 2, \ldots,$$

$$\tilde{B}_j(z) = \sum_{\ell=1}^{\infty} b_j(\ell) z^{\ell},$$

$$\tilde{A}_j^-(z) = \sum_{\ell=1}^{\infty} a_j^-(\ell) z^{\ell}.$$

The following theorem contains the main result of this section.

Theorem 7.6 *Assume that the traffic equations*

$$\alpha_j = \lambda_j + \sum_{k=1}^{N} \mu_k \tilde{B}_k(\rho_k) r_{kc,jc} + \sum_{k=1}^{N} \alpha_k^- \tilde{A}_k^-(\rho_k) r_{ks^-,jc}, \qquad j = 1, \ldots, N, \quad (2.2)$$

$$\alpha_j^- = \lambda_j^- + \sum_{k=1}^{N} \mu_k \tilde{B}_k(\rho_k) r_{kc,js^-} + \sum_{k=1}^{N} \alpha_k^- \tilde{A}_k^-(\rho_k) r_{ks^-,js^-}, \quad j = 1, \ldots, N, \quad (2.3)$$

have solutions $(\alpha_1, \ldots, \alpha_N)$, $(\alpha_1^-, \ldots, \alpha_N^-)$ *and* $(\rho_1, \ldots, \rho_N)$ *such that*

$$\alpha_j(1 - \rho_j) = \mu_j \rho_j (1 - \tilde{B}_j(\rho_j)) + \alpha_j^- \rho_j (1 - \tilde{A}_j^-(\rho_j)), \quad j = 1, \ldots, N. \quad (2.4)$$

If

$$\rho_j < 1, \qquad j = 1, \ldots, N,$$

then the stationary distribution of the network is

$$\pi(\boldsymbol{n}) = \prod_{j=1}^{N} (1 - \rho_j) \rho_j^{n_j}. \quad (2.5)$$

PROOF. Since signals trigger instantaneous movements in the network, we use Theorem 4.9 to prove this result. We only need to focus on one node of the network in isolation, say node j. Let the customers be classified as class c, and signals that intend to trigger u customers to depart as class u^-, $u = 1, 2, \ldots$. Clearly, the arrival of a negative signal at node j is classified as class u^- with probability

$a_j^-(u)$, $u = 1, 2, \ldots$. Also, let the regular service completions of full batches be classified as class c and the departures of full batches that are triggered by signals as class s^-. The class of entities is

$$T_j = \{c, s^-\} \cup \{u^-; u = 1, 2, \ldots\}.$$

Using the notation of Chapter 3, node j is characterized by

$$\begin{aligned}
p_{jc}^{\mathrm{A}}(n_j, n_j + 1) &= 1, & n_j &\geq 0, \\
p_{ju^-}^{\mathrm{A}}(n_j + u, n_j) &= 1, & n_j &\geq 0,\ u = 1, 2, \ldots, \\
p_{ju^-}^{\mathrm{A}}(n_j, 0) &= 1, & u &> n_j = 1, 2, \ldots, \\
q_{jc}^{\mathrm{D}}(n_j + u, n_j) &= \mu_j b_j(u), & n_j &\geq 0, u = 1, 2, \ldots, \\
q_j^{\mathrm{I}}(n_j, 0) &= \mu_j \bar{b}_j(n_j + 1), & n_j &= 1, 2, \ldots, \\
f_{ju^-, s^-}(n_j + u, n_j) &= 1, & n_j &\geq 0,\ u \geq 1.
\end{aligned}$$

Node j, defined by

$$q_j(n_j, n_j') = \alpha_j p_{jc}^{\mathrm{A}}(n_j, n_j') + \sum_{u=1}^{\infty} \alpha_j^- a_j^-(u) p_{ju^-}^{\mathrm{A}}(n_j, n_j') + q_{jc}^{\mathrm{D}}(n_j, n_j') + q_j^{\mathrm{I}}(n_j, n_j'),$$

is subject to Poisson arrivals of customers with rate α_j, and Poisson arrivals of class u^- negative signals with rate $\alpha_j^- a_j^-(u)$. It is not hard to see that, as far as the stationary distribution of the node is concerned, q_j is equivalent to another queue without signals that has exponentially distributed service times with rate $\mu_j + \alpha_j^-$, and batch services of size Z_j, where

$$P(Z_j = n) = c_j(n) = \frac{\mu_j}{\mu_j + \alpha_j^-} b_j(n) + \frac{\alpha_j^-}{\mu_j + \alpha_j^-} a_j^-(n), \quad n = 1, 2, \ldots. \tag{2.6}$$

Hence, by Lemma 2.32, the stationary distribution of the node is

$$\pi_j(n_j) = (1 - \rho_j)\rho_j^{n_j},$$

where ρ_j is the solution of

$$\begin{aligned}
\alpha_j &= (\mu_j + \alpha_j^-) \sum_{n=1}^{\infty} \sum_{\ell=n}^{\infty} c_j(\ell) \rho_j^n \\
&= \mu_j \sum_{n=1}^{\infty} \bar{b}_j(n) \rho_j^n + \alpha_j^- \sum_{n=1}^{\infty} \bar{a}_j^-(n) \rho_j^n,
\end{aligned}$$

which is the same as (2.4). Since

$$\begin{aligned}\sum_{n'_j} \pi_j(n'_j) q^{\mathrm{D}}_{jc}(n'_j, n_j) &= \sum_{u=1}^{\infty} \pi_j(n_j+u) q^{\mathrm{D}}_{jc}(n_j+u, n_j)\\ &= \pi_j(n_j)\mu_j \sum_{u=1}^{\infty} \rho_j^u b_j(u)\\ &= \pi_j(n_j)\mu_j \tilde{B}_j(\rho_j),\end{aligned}$$

and

$$\begin{aligned}\sum_{n'_j} \pi_j(n'_j) &\left(q^{\mathrm{D}}_{js^-}(n'_j, n_j) + \sum_{u=1}^{\infty} \alpha_j^- a_j^-(u) f_{ju^-,s^-}(n'_j, n_j) \right)\\ &= \sum_{u=1}^{\infty} \pi_j(n_j+u)\alpha_j^- a_j^-(u) f_{ju^-,s^-}(n_j+u, n_j)\\ &= \pi_j(n_j)\alpha_j^- \sum_{u=1}^{\infty} \rho_j^u a_j^-(u)\\ &= \pi_j(n_j)\alpha_j^- \tilde{A}_j(\rho_j),\end{aligned}$$

the node is a quasi-reversible queue with signals and departure rates

$$\beta_{jc} = \mu_j \tilde{B}_j(\rho_j), \qquad \beta_{js^-} = \alpha_j^- \tilde{A}_j(\rho_j). \tag{2.7}$$

Hence, by Theorem 4.9, the stationary distribution of the network is given by (2.5), where α_j and α_j^- are the solutions of the traffic equations (3.5) of Chapter 4, i.e.,

$$\begin{aligned}\alpha_j &= \lambda_j + \sum_{k=1}^{N} \beta_{kc} r_{kc,jc} + \sum_{k=1}^{N} \beta_{ks^-} r_{ks^-,jc},\\ \alpha_j^- &= \lambda_j^- + \sum_{k=1}^{N} \beta_{kc} r_{kc,js^-} + \sum_{k=1}^{N} \beta_{ks^-} r_{ks^-,js^-}.\end{aligned}$$

Substituting the departure rates (2.7) into these equations yields (2.2) and (2.3). This completes the proof of the theorem. □

Remark 7.7 The inequality $\rho_j < 1$ is equivalent to

$$\alpha_j < \mu_j E(B_j) + \alpha_j^- E(A_j^-). \tag{2.8}$$

Its proof is similar to that of Lemma 2.32. The stability condition (2.8) is intuitively appealing: for this network to be stable, the average arrival rate of customers must be less than the average departure rate due to service completions and signal triggerings, assuming the servers are always busy.

We now extend Theorem 7.6 so that the routing probabilities of a batch depend on the batch size. Customers and signals arrive at node j from the outside according to Poisson processes with rates λ_j and λ_j^-. When a signal arrives at node j from the outside, it triggers a batch of customers to move as a single unit. The probability mass function of the batch sizes is $\{a_j^-(n); n \geq 1\}$. The services at node j occur in batches and the batch size distribution is $\{b_j(n); n \geq 1\}$. Upon a service completion, a full batch of size u ($u = 1, 2, \ldots$) coalesces into a single entity that joins node k as a customer with probability $r_{ju,kc}$, as a signal that triggers a batch of v customers with probability r_{ju,kv^-}, and it leaves the network with probability $r_{ju,0}$. These routing probabilities satisfy

$$\sum_{k=1}^{N} r_{ju,kc} + \sum_{k=1}^{N}\sum_{v=1}^{\infty} r_{ju,kv^-} + r_{ju,0} = 1, \qquad j = 1, \ldots, N, \; u \geq 1.$$

When a signal arrives at node k to trigger u customers, u customers coalesce into a single entity, provided there are u or more customers present. This single entity then joins node k as a customer with probability $r_{ju^-,kc}$, as a signal that triggers a batch of v ($v = 1, 2, \ldots$) customers with probability r_{ju^-,kv^-}, and it leaves the network with probability $r_{ju^-,0}$. Again, we must have

$$\sum_{k=1}^{N} r_{ju^-,kc} + \sum_{k=1}^{N}\sum_{v=1}^{\infty} r_{ju^-,kv^-} + r_{ju^-,0} = 1, \qquad j = 1, \ldots, N, \; u \geq 1.$$

If a signal arrives at node j to trigger u customers and there are fewer than u customers present, then all the customers at node j leave the system.

Theorem 7.8 *Assume that the traffic equations*

$$\alpha_j = \lambda_j + \sum_{k=1}^{N}\sum_{v=1}^{\infty} \mu_k \rho_k^v b_k(v) r_{kv,jc} + \sum_{k=1}^{N}\sum_{v=1}^{\infty} \alpha_{kv}^- \rho_k^v r_{kv^-,jc}, \qquad j = 1, \ldots, N,$$

$$\alpha_{ju}^- = \lambda_j^- a_j^-(u) + \sum_{k=1}^{N}\sum_{v=1}^{\infty} \mu_k \rho_k^v b_k(v) r_{kv,ju^-} + \sum_{k=1}^{N}\sum_{v=1}^{\infty} \alpha_{kv}^- \rho_k^v r_{kv^-,ju^-}, \qquad j = 1, \ldots, N, \; u \geq 1,$$

have solutions $\{\alpha_j; 1 \le j \le N\}$, $\{\alpha_{ju}^-; 1 \le j \le N, u \ge 1\}$ *and* $\{\rho_j; 1 \le j \le N\}$ *such that*

$$\alpha_j = \mu_j \sum_{\ell=1}^{\infty} \sum_{u=\ell}^{\infty} b_j(u) \rho_j^{\ell} + \sum_{\ell=1}^{\infty} \sum_{u=\ell}^{\infty} \alpha_{ju}^- \rho_j^{\ell} .$$

If $\rho_j < 1$ *for* $j = 1, 2, \ldots, N$, *then the stationary distribution of the network has the geometric product form*

$$\pi(\boldsymbol{n}) = \prod_{j=1}^{N} (1 - \rho_j) \rho_j^{n_j} .$$

We leave the proof of this theorem to the reader as an exercise.

7.3 SPECIAL CASES AND EXAMPLES

This section presents several special cases and examples of the results in the previous section. Among these are three of the examples discussed in Chapter 1, i.e., the computer virus problem, the production and distribution problem, and the quality control problem.

Example 7.9 (A computer virus model) Consider a network with a single server at each node. Customers (jobs) arrive at node j according to a Poisson process with rate λ_j, and the processing times at node j are exponentially distributed with rate μ_j. Upon a service completion at node j, the customer goes to node k as a customer (a job without a virus) with probability r_{jk}, as a signal (a job with a virus) with probability r_{jk}^-, and it leaves the network with probability r_{j0}. Signals arrive at node j from the outside according to a Poisson process with rate λ_j^-, which represents newly created jobs with viruses. When a signal arrives at node j, either from another node or from the outside, a batch of jobs is infected. We assume that the infected jobs are discarded. In the next section we consider the case where infected jobs can be repaired or partially repaired. The size of the infected batch is random with probability mass function $\{a_j^-(n); n = 1, 2, \ldots\}$, where

$$\sum_{n=1}^{\infty} a_j^-(n) = 1, \qquad j = 1, \ldots, N. \qquad \square$$

The following is a corollary of Theorem 7.6. Its verification is left to the reader as an exercise.

Proposition 7.10 *The stationary distribution of the queueing network with computer viruses described above is*

$$\pi(\boldsymbol{n}) = \prod_{j=1}^{N} (1-\rho_j)\rho_j^{n_j},$$

where ρ_j is determined by

$$\rho_j = \frac{\alpha_j}{\mu_j + \alpha_j^-(1-\tilde{A}_j^-(\rho_j))/(1-\rho_j)}, \tag{3.1}$$

and the traffic equations

$$\alpha_j = \lambda_j + \sum_{k=1}^{N} \rho_k \mu_k r_{kj}, \qquad j = 1, \ldots, N,$$

$$\alpha_j^- = \lambda_j^- + \sum_{k=1}^{N} \rho_k \mu_k r_{kj}^-, \qquad j = 1, \ldots, N.$$

Example 7.11 (A two-node tandem queue with signals) Consider a tandem queue with two nodes in series. Customers arrive at node 1 according to a Poisson process with rate λ, and the service times at the two nodes are exponentially distributed with rates μ_1 and μ_2, respectively. Upon a service completion at node 1, the customer goes to node 2 as a customer with probability p_1, as a signal with probability p_2, and it returns to node 1 as a signal with probability p_3, where

$$p_1 + p_2 + p_3 = 1.$$

Whenever a signal arrives at a node, it induces two customers to leave the network, provided there are two or more customers present; otherwise the node is emptied out.

Since $\tilde{B}_j(\rho_j) = \rho_j$ and $\tilde{A}_j^-(\rho_j) = \rho_j^2$, the stationary distribution of the network is, by Proposition 7.10, of a geometric product form. The traffic equations are

$$\alpha_1 = \lambda,$$

$$\alpha_1^- = \rho_1 \mu_1 p_3,$$

$$\alpha_2 = \rho_1 \mu_1 p_1,$$

$$\alpha_2^- = \rho_1 \mu_1 p_2,$$

and the ρ_j's are determined by

$$\rho_1 = \frac{\alpha_1}{\mu_1 + \alpha_1^-(1-\rho_1^2)/(1-\rho_1)} = \frac{\lambda}{\mu_1 + \mu_1\rho_1 p_3 + \rho_1^2\mu_1 p_3},$$

$$\rho_2 = \frac{\alpha_2}{\mu_2 + \alpha_2^-(1-\rho_2^2)/(1-\rho_2)} = \frac{\rho_1\mu_1 p_1}{\mu_2 + \rho_1\mu_1 p_2 + \rho_1\rho_2\mu_1 p_2}. \qquad \square$$

Example 7.12 (A production–distribution system) Consider the production and distribution system discussed in Example 1.3. Parts (raw materials) arrive at the manufacturing facility, denoted by node 1, according to a Poisson process with rate λ, and the processing times at the facility are exponentially distributed with rate μ_1. There is a single machine at the facility. A job that has completed its processing at node 1 is stored in a warehouse, denoted by node 2. Demand arrives at an order processing station, denoted by node 3, according to a Poisson process with rate λ^-, and the order processing times are exponentially distributed with rate μ_3. The demand size is random with probability mass function $\{a^-(\ell);\ \ell \geq 1\}$. The orders are processed one at a time. After an order of size n has been processed, the queue size at the order processing station (number of orders waiting to be processed), i.e., node 3, is reduced by 1, and the queue size at the warehouse, i.e., node 2, is reduced by ℓ, provided at least ℓ finished jobs are available; otherwise node 2 is emptied out. This is equivalent to a partial fulfillment.

Let n_1 be the number of parts at station 1, n_2 the number of units stored at station 2, and n_3 the number of orders at the order processing station 3. Clearly, this problem is a special case of Example 7.9 with three nodes, and the service rate at node 2 is zero. The jobs are regular customers and the demands are the negative signals. One can easily see that the average arrival rate of jobs and the average arrival rate of demands at node 2 are λ and λ^-, respectively. Thus the stationary distribution of the network is

$$\pi(n_1, n_2, n_3) = \prod_{j=1}^{3}(1-\rho_j)\rho_j^{n_j},$$

where

$$\rho_1 = \frac{\lambda}{\mu_1}, \qquad \rho_2 = \frac{\lambda(1-\rho_2)}{\lambda^-(1-\tilde{A}^-(\rho_2))}, \qquad \rho_3 = \frac{\lambda^-}{\mu_3}.$$

Using the stationary distribution of the system we can compute various performance measures of interest. For instance, the probability that an order is completely satisfied (not just partially) is, by conditioning on the size of the demand,

$$\sum_{\ell=1}^{\infty} a^-(\ell)\rho_2^\ell.$$

We can extend the model by assuming that an order, after being processed, can wait for jobs to be finished. Assume that R is the maximum number of processed orders that are allowed to wait. The queue size at node 2 is then extended to

$$\{-R, -R+1, \ldots, -2, -1, 0, 1, 2, \ldots\}.$$

The stationary distribution for this case is

$$\pi(n_1, n_2, n_3) = \rho_2^R \prod_{j=1}^{3} (1-\rho_j)\rho_j^{n_j}.$$

The probability that an order is completely satisfied is

$$\sum_{n=R}^{\infty} a^-(n) P(\text{the number of units at node } 2 \geq n - R) = \sum_{n=1}^{\infty} a^-(n)\rho_2^n.$$

The probability that an order has to wait before it is completely satisfied is

$$\sum_{n=1}^{\infty} a^-(n) \sum_{\ell=n-R}^{n-1} (1-\rho_2)\rho_2^{\ell+R},$$

and the probability that an order is completely satisfied and does not have to wait is

$$\sum_{n=1}^{\infty} a^-(n)\rho_2^{n+R}.$$

Given that an order is completely satisfied, the probability that it has to wait is

$$\frac{\sum_{n=1}^{\infty} a^-(n) \sum_{\ell=n-R}^{n-1} (1-\rho_2)\rho_2^{\ell+R}}{\sum_{n=1}^{\infty} a^-(n)\rho_2^n} = 1 - \rho_2^R,$$

and the probability that it does not have to wait is ρ_2^R. □

A case even more special than Example 7.9 is the network with so-called catastrophes; the arrival of a catastrophe destroys all the customers at a node.

Example 7.13 (A network subject to catastrophes) Assume that customers arrive at node j according to a Poisson process with rate λ_j, and the service time distribution at node j is exponential with rate μ_j, $j = 1, \ldots, N$. After a customer finishes its service at node j, it either goes to another node for another service or leaves the system. However, when this customer goes to another node, with a certain probability it causes a catastrophe. Specifically, we assume that upon a service completion at node j, the customer goes to node k as a regular customer with probability r_{jk}, and it goes there to cause a catastrophe with probability r_{jk}^-. The customer leaves the system with probability r_{j0}. The routing probabilities satisfy

$$\sum_{k=1}^{N} r_{jk} + \sum_{k=1}^{N} r_{jk}^- + r_{j0} = 1, \qquad j = 1, \ldots, N.$$

In addition, catastrophes may also occur at node j because of Poisson arrivals from the outside with rate λ_j^-. Whenever a catastrophe occurs at node j, either due to an arrival from the outside or due to an arrival from another node, all the customers at the node are destroyed and lost, and the server is sent for repair. We assume that the repair time is negligible. □

The following proposition is a limiting case of Proposition 7.10.

Proposition 7.14 *For the network with catastrophes, suppose the traffic equations*

$$\alpha_j = \lambda_j + \sum_{k=1}^{N} \rho_k \mu_k r_{kj}, \qquad j = 1, \ldots, N,$$

$$\alpha_j^- = \lambda_j^- + \sum_{k=1}^{N} \rho_k \mu_k r_{kj}^-, \qquad j = 1, \ldots, N,$$

have solutions α_j and α_j^-, which are the average arrival rates of customers and catastrophes at node j, where ρ_j is given by

$$\rho_j = \frac{\alpha_j + \alpha_j^- + \mu_j - \sqrt{(\alpha_j + \alpha_j^- + \mu_j)^2 - 4\alpha_j \mu_j}}{2\mu_j}. \tag{3.2}$$

If the solutions of the traffic equations satisfy either

(i) $\alpha_j^- > 0$, *or*

(ii) $\alpha_j^- = 0$, *and* $\alpha_j < \mu_j$,

then the stationary distribution of the network has the product form

$$\pi(\boldsymbol{n}) = \prod_{j=1}^{N} (1 - \rho_j) \rho_j^{n_j}.$$

PROOF. We apply an argument similar to the proof of Theorem 7.6. The only difference is that the departure batch size due to a signal is infinity. Consider node j in isolation with customers and signals arriving according to Poisson processes with rates α_j^+ and α_j^-, respectively. Using the batch size distribution calculated by (2.6), from Lemma 2.32 we obtain that ρ_j is the solution of the equation

$$\begin{aligned}\alpha_j &= \frac{\mu_j}{\mu_j + \alpha_j^-}\rho_j + \frac{\alpha_j^-}{\mu_j + \alpha_j^-}\sum_{n=1}^{\infty}\rho_j^n \\ &= \frac{\rho_j(\mu_j(1-\rho_j) + \alpha_j^-)}{(1-\rho_j)(\mu_j + \alpha_j^-)}.\end{aligned}$$

This can be written as

$$\mu_j\rho_j^2 - (\alpha_j + \alpha_j^- + \mu_j)\rho_j + \alpha_j = 0. \tag{3.3}$$

Let

$$f(\rho_j) = \mu_j\rho_j^2 - (\alpha_j + \alpha_j^- + \mu_j)\rho_j + \alpha_j .$$

Since $f(0) = \alpha_j > 0$ and $f(1) = -\alpha_j^- \leq 0$, equation (3.3) has a solution ρ_j such that $0 \leq \rho_j < 1$ if and only if (i) or (ii) holds. Thus, under either one of these conditions, ρ_j is the smaller root of (3.3), which is given by (3.2), and the marginal stationary distribution of node j is geometric with decay rate ρ_j. Since each node is quasi-reversible with departure rate $\rho_j\mu_j$, it follows from Theorem 4.3 that the network has a product form stationary distribution. Note that here we do not need Theorem 4.9, since in this network there are no departures triggered by signals. □

Remark 7.15 Conditions (i) and (ii) are quite intuitive. If (i) is satisfied, then no matter how large the average arrival rate at node j is, the node will be stable since the next catastrophe will destroy all the accumulated customers. If $\alpha_j = 0$, then condition $\alpha_j < \mu_j$ is just the usual "average arrival rate is less than service rate" condition. In this case, $\rho_j = \alpha_j/\mu_j$.

Example 7.16 (A quality control problem) Consider the quality control problem discussed in Example 1.5. It is a serial system with $N + 1$ stations. Raw materials arrive at the first station according to a Poisson process with rate λ. Each of the first N stations performs an operation on the job on its way to becoming a finished product. The final station is a quality control station. The processing times at station j are exponentially distributed with rate μ_j, and there is a single server at each station.

Assume that each job is inspected at the quality control station. With probability p_0 a job passes the quality control station and with probability p_j^- it is a defect and the source of the defect is identified as station j. Call such a defect a defect of type j, $j = 0, 1, 2, \ldots, N$. A defect of type 0 is defined as a defect in the raw materials. Clearly

$$p_0 + \sum_{j=0}^{N} p_j^- = 1.$$

First we assume that a consequence of a type j defect is the elimination of all jobs at station j, $j = 0, 1, \ldots, N$. To model this, we introduce a negative signal that is a catastrophe. Upon a processing completion at station $N+1$, the job leaves the system with probability p_0 and it goes to station j as a catastrophe with probability p_j^-, $j = 0, 1, \ldots, N$. Let n_j be the number of jobs at station j, $j = 1, 2, \ldots, N+1$. By Proposition 7.14 the stationary distribution of the network is given by

$$\pi(\boldsymbol{n}) = \prod_{j=1}^{N+1} (1 - \rho_j)\rho_j^{n_j}, \tag{3.4}$$

where

$$\begin{aligned}
\alpha_1 &= \lambda, \\
\alpha_{j+1} &= \mu_j \rho_j, \qquad & j = 1, 2, \ldots, N, \\
\alpha_j^- &= \mu_{N+1} p_j^-, \qquad & j = 1, 2, \ldots, N,
\end{aligned}$$

and ρ_j is determined by (3.2) for $j = 1, \ldots, N+1$.

We next consider a situation where a consequence of a type j defect is the elimination of a random number of jobs at stations $j, j+1, \ldots, N+1$, due to the fact that any job that has passed station j may have a similar type of defect. To model this situation we introduce $N+1$ classes of signals. A job completion at node $N+1$ leaves the system as a quality product with probability p_0, and it leaves as a class j signal that goes to station j with probability p_j^-, $j = 0, 1, \ldots, N$. A class j signal implies that the source of the defect is station j, and a class 0 signal indicates that the raw materials may be defective. The effect of an arrival of a class j signal at station j is the removal of one job, provided the station is not empty; the job then joins station $j+1$ as a class j signal and removes one job from station $j+1$, provided that station is not empty, and so on. It continues until either it reaches an empty station or it reaches station $N+1$, where it removes one job from station $N+1$, and then returns to station j, again, as a class j signal. This process continues, and as a result it removes a random number of jobs from stations

$j, j+1, \ldots, N+1$. It stops only when one of the stations $j, j+1, \ldots, N+1$ empties out.

Let α_{ju}^- be the average arrival rate of class u negative signals at station j ($j > u$). Then

$$\begin{aligned}
\alpha_1 &= \lambda, \\
\alpha_{j+1} &= \mu_j \rho_j, && j = 1, 2, \ldots, N, \\
\alpha_{jj}^- &= \mu_{N+1} p_j^- + \alpha_{(N+1)j}^- \rho_{N+1}, && j = 1, 2, \ldots, N, \\
\alpha_{(j+1)u}^- &= \alpha_{ju}^- \rho_j, && j = 1, 2, \ldots, N, \quad 1 \le u \le j-1,
\end{aligned}$$

where

$$\rho_j = \frac{\alpha_j}{\mu_j + \sum_{u=1}^{j} \alpha_{ju}^-}, \qquad j = 1, 2, \ldots, N+1.$$

If the solution of these traffic equations satisfies $\rho_j < 1$ for all j, then the stationary distribution of the network is (3.4). □

Example 7.17 (A network with string transitions) Consider a network with N nodes and state space $\mathcal{S} = \mathbb{Z}_+^N$, where $\mathbb{Z}_+ = \{0, 1, \ldots\}$. Instead of assuming that each node has a server with a given service rate as well as with given routing probabilities, we now assume that the network transitions are defined by strings. Specifically, there is a set of strings $\mathcal{A}$ that contains M elements, where M maybe infinity. Each string in $\mathcal{A}$ takes the form

$$\boldsymbol{s} = (s_1, s_2, \ldots, s_{L(\boldsymbol{s})}),$$

where $L(\boldsymbol{s})$ is the length of the string $\boldsymbol{s}$. $L(\boldsymbol{s}) = 1$ is considered an exogenous arrival. Associated with a string $\boldsymbol{s} \in \mathcal{A}$ is a transition rate $\lambda_{\boldsymbol{s}}$ and a state transition

$$\boldsymbol{n} \to \boldsymbol{n} - \mathbf{e}_{s_1} - \mathbf{e}_{s_2} - \cdots - \mathbf{e}_{s_{L(\boldsymbol{s})-1}} + \mathbf{e}_{s_{L(\boldsymbol{s})}},$$

provided the sequence of subtractions does not jump out of the state space $\mathcal{S}$. Otherwise, i.e., if

$$\boldsymbol{n} - \mathbf{e}_{s_1} - \cdots - \mathbf{e}_{s_h} \in \mathcal{S}$$

for some $h \leq L(s) - 2$, and

$$\boldsymbol{n} - \mathbf{e}_{s_1} - \cdots - \mathbf{e}_{s_{h+1}} \notin \mathcal{S},$$

the transition stops at the last state before it jumps out of the state space $\mathcal{S}$, i.e., state

$$\boldsymbol{n} \to \boldsymbol{n} - \mathbf{e}_{s_1} - \cdots - \mathbf{e}_{s_h}.$$

So each string transition attempts to remove a sequence of customers from the network, and then adds one customer to a node (if the final node is node 0, then the customer leaves the network). Since the s_ℓ's may repeat, such a string transition may remove any number of customers from a node. □

Proposition 7.18 *Let ρ_j be the solution to the traffic equations*

$$\rho_j \sum_{\ell \geq 0} \sum_{(s_1,\ldots,s_\ell,j,s') \in \mathcal{A}} \rho_{s_1} \cdots \rho_{s_\ell} \lambda_{(s_1,\ldots,s_\ell,j,s')} = \sum_{\ell \geq 0} \sum_{(s_1,\ldots,s_\ell,j) \in \mathcal{A}} \rho_{s_1} \cdots \rho_{s_\ell} \lambda_{(s_1,\ldots,s_\ell,j)}, \quad j = 1, \ldots, N, \qquad (3.5)$$

where $\boldsymbol{s}' = (h_1, \ldots, h_m)$ for some $m \geq 1$. If $0 \leq \rho_j < 1$ for $j = 1, \ldots, N$, then the stationary distribution of the string transition network is given by the geometric product form (2.5).

PROOF. We use Theorem 4.17 to prove this result. Define a class $(\boldsymbol{s}, \ell)$ $(1 \leq \ell \leq L(\boldsymbol{s}))$ entity as the ℓth departure (or arrival if $\ell = L(\boldsymbol{s})$) of string $\boldsymbol{s}$. Clearly, a class $(\boldsymbol{s}, \ell)$ entity is a negative signal if $\ell < L(\boldsymbol{s})$ and a regular customer if $\ell = L(\boldsymbol{s})$. Thus

$$T_j = \{c\} \cup \{(\boldsymbol{s}, \ell); \boldsymbol{s} \in \mathcal{A}; \ell = 1, \ldots, L(\boldsymbol{s})\}.$$

The nonzero routing probabilities under such a classification are

$$r_{j(\boldsymbol{s},\ell),k(\boldsymbol{s},\ell+1)} = 1[j = s_\ell, k = s_{\ell+1}].$$

This network is a special case of the network of Theorem 4.17 without positive signals. So it has a product form stationary distribution (2.5), and no extra arrival conditions are needed. Because only the final event in each transition is an arrival, the traffic equations are

$$\alpha_j = \sum_{\ell \geq 0, s_1, \ldots, s_\ell} \rho_{s_1} \cdots \rho_{s_\ell} \lambda_{(s_1, \ldots, s_\ell, j)}, \tag{3.6}$$

$$\alpha^-_{j(s,\ell)} = \rho_{j_1} \cdots \rho_{s_\ell} \lambda_{(s_1, \ldots, s_\ell, j, s')}, \qquad \ell = 1, \ldots, L(s), \tag{3.7}$$

where

$$s = (s_1, \ldots, s_\ell, j, s') \in \mathcal{A},$$

and

$$\rho_j = \frac{\alpha_j}{\sum_{\ell \geq 0} \sum_{s \in \mathcal{A}} \alpha^-_{j(s,\ell)}}.$$

Substituting α_j and $\alpha^-_{j(s,\ell)}$ into this equation yields the traffic equations (3.5). On the other hand, if equations (3.5) have the solution ρ_j such that $0 \leq \rho_j < 1$, then we define α_j and $\alpha^-_{j(s,\ell)}$ by (3.6) and (3.7). Thus the proposition follows from Theorem 4.17. □

7.4 REFERENCE NOTES

The results of Section 7.1 are taken from Chao, Pinedo and Shaw (1996). Section 7.2 is based on Chao and Pinedo (1995*a*). The result of Section 7.2 has been extended to arbitrary service time distributions by Miyazawa and Wolff (1996) under the assumption of symmetric service disciplines as defined in Chapter 6. The queueing networks subject to catastrophes of Section 7.3 were first studied in Chao (1995). Example 7.17 on networks with string transitions is taken from Serfozo and Yang (1998), in which the global balance is directly verified. A special case of this type of signals is also studied in Gelenbe (1993). It is worth noting that all these results are proved here in a unified framework using quasi-reversibility.

EXERCISES

7.1 Consider the network of Theorem 7.1 with constant service batch sizes.

(a) Show that the global balance equation is equivalent to

$$\sum_{j=1}^{N}(\lambda_j + \mu_j 1[n_j \geq 1])\pi(\boldsymbol{n}) + \sum_{j=1}^{N}\sum_{\ell=1}^{K_j-1} \mu_j \pi(\boldsymbol{n} + \ell \mathbf{e}_j) 1[n_j \geq 1]$$
$$= \sum_{j=1}^{N}\left(\lambda_j \pi(\boldsymbol{n} - \mathbf{e}_j) + \sum_{k=1}^{N} \mu_k \pi(\boldsymbol{n} + K_k \mathbf{e}_k - \mathbf{e}_j) r_{kj}\right) 1[n_j \geq 1]$$
$$+ \sum_{j=1}^{N}\sum_{\ell=1}^{K_j-1} \mu_j \pi(\boldsymbol{n} + \ell \mathbf{e}_j).$$

(b) Show that the stationary distribution of (1.4) satisfies

$$\sum_{j=1}^{N} \lambda_j \pi(\boldsymbol{n}) = \sum_{j=1}^{N}\sum_{\ell=1}^{K_j-1} \mu_j \pi(\boldsymbol{n} + \ell \mathbf{e}_j).$$

(c) Prove that the stationary distribution of (1.4) satisfies the global balance obtained in (a).

7.2 Show that the cross local balance (1.5) and the equation given in (b) of Exercise 7.1 are also necessary if the stationary distribution of the network in Theorem 7.1 is of a geometric product form, i.e., of the form (1.4).

7.3 Use Theorem 4.9 to prove Theorem 7.5.

7.4 Prove Theorem 7.8.

7.5 Prove Proposition 7.10.

8

Batch Arrivals, Batch Services and Concurrent Movements

This chapter focuses on networks with batch arrivals, batch services and concurrent batch movements. In Section 8.1 we illustrate through a simple example why additional conditions are required for the network to have a geometric product form solution. Section 8.2 considers a network with batch arrivals and batch services, which is referred to as an assembly-transfer network. In assembly-transfer networks, a batch of customers that completes service at one node may go to another node in the form of a batch of a different size. Section 8.3 extends these results by introducing both positive and negative signals, resulting in a network with arbitrary concurrent batch movements. Two of the examples presented in Chapter 1, i.e., the production and inventory control problem and the computer virus network, are analyzed in Section 8.4.

8.1 INTRODUCTION

The two features discussed in the previous chapter are customer coalescence and concurrent batch departures. The first feature implies that after a batch of customers has been served, the entire batch coalesces into a single customer. The second feature is that both a service completion and the arrival of a signal may trigger batch departures at a random number of nodes throughout the network. The major restrictions in these networks are that

(1) customers can only arrive at a node one at a time, i.e., there are no batch arrivals, and

(2) customers cannot arrive concurrently at different nodes, i.e., there are no concurrent arrivals.

A natural question: Do the product form results still hold for networks with batch arrivals, batch services, and concurrent movements ? Before answering this question, we consider the following example.

Example 8.1 (A batch arrival queue) Consider a single node queue where customers arrive according to a Poisson process with rate λ. However, each arrival brings in two customers. Customers are served one at a time and the service times are exponentially distributed with rate μ. Thus the transition rates for this system are

$$q^{\mathrm{A}}(n, n+2) = \lambda, \qquad n = 0, 1, 2, \ldots,$$
$$q^{\mathrm{D}}(n, n-1) = \mu, \qquad n = 1, 2, \ldots.$$

According to the discussion in Chapter 4 concerning a product form network, we need to show that each node in isolation is quasi-reversible. For this particular example, the queue is quasi-reversible if and only if its stationary distribution is of a geometric form. To see this, suppose the queue has the stationary distribution π. Then the queue is quasi-reversible if and only if there exists a nonnegative constant α such that

$$\pi(n+1)\mu = \alpha\pi(n), \qquad n = 0, 1, \ldots.$$

This implies that

$$\pi(n) = (1-\rho)\rho^n, \qquad n = 0, 1, \ldots, \tag{1.1}$$

where $\rho = \alpha/\mu$. Hence, the stationary distribution has to be geometric for the queue to be quasi-reversible (see Figure 8.1).

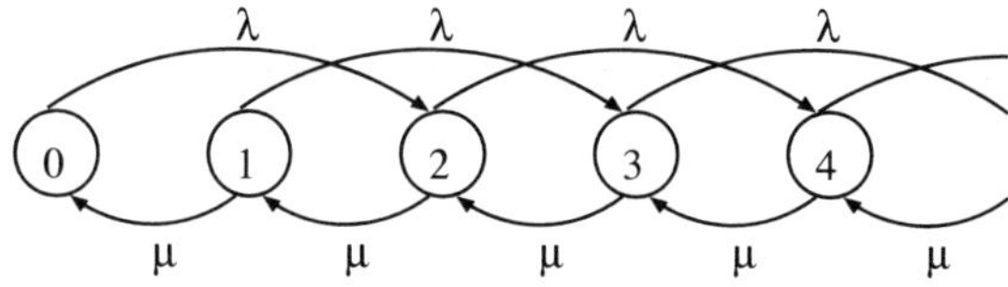

Figure 8.1 The transition diagram for the batch arrival queue

From the cross balance equations (1.10) in Chapter 2, we have that for each n, the rate going from state n or below to a state above n must be equal to the rate going from a state above n to state n or below. Thus for $n \geq 1$ we have

$$\pi(n-1)\lambda + \pi(n)\lambda = \pi(n+1)\mu, \tag{1.2}$$

and for $n = 0$, we have

$$\pi(0)\lambda = \pi(1)\mu. \tag{1.3}$$

Substituting (1.1) into these two equations yields a contradiction. So the queue cannot have a geometric stationary distribution and cannot be quasi-reversible.

Actually, as shown in Theorem 2.33 of Chapter 2, no queue with batch arrivals has a geometric stationary distribution, but a queue does have such a distribution if additional batch arrivals are introduced when the queue is empty. Let us consider why such additional batch arrivals are needed.

Compare (1.2) with (1.3). To make these two equations consistent, we add an additional term on the left-hand side of (1.3). This term corresponds to arrivals when the state is 0. Let its rate be λ^*, i.e.,

$$q^{\mathrm{A}}(0, 1) = \lambda^*.$$

With this modification, the balance equation (1.3) becomes

$$\pi(0)\lambda^* + \pi(0)\lambda = \pi(1)\mu \tag{1.4}$$

and (1.2) remains unchanged (see Figure 8.2). Substituting (1.1) into (1.2) and (1.4) we conclude that the only choice for λ^* is

$$\lambda^* = \lambda\rho^{-1},$$

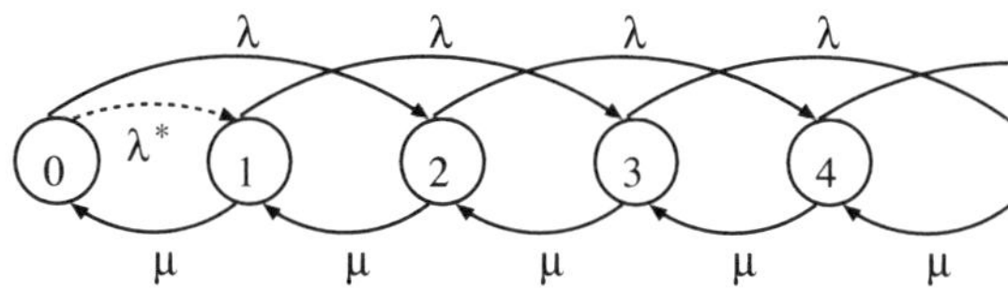

Figure 8.2 The transition diagram for the modified queue

where ρ is the solution of

$$\lambda\rho^{-1} + \lambda = \mu\rho.$$

Thus

$$\rho = \frac{\lambda + \sqrt{\lambda^2 + 4\lambda\mu}}{2\mu}. \tag{1.5}$$

That is, if there is an additional arrival process of rate λ^* at the node whenever it is empty, then the queue has the geometric stationary distribution (1.1) with ρ given by (1.5).

Since the assumption of additional arrivals can only make the node more congested, the geometric stationary distribution (1.1) is a stochastic upper bound for the original queue. That is, if π^0 is the stationary distribution of the original queue without the additional arrivals, then

$$\sum_{n' \geq n} \pi^0(n') \leq \rho^n, \qquad n = 0, 1, 2, \ldots. \qquad \square$$

All the results obtained for this simple example can be extended to general networks. In the following sections we show that networks with batch arrivals, batch services, and concurrent batch movements have a product form geometric stationary distribution, provided we introduce an additional arrival process at each node whenever it is empty.

8.2 ASSEMBLY-TRANSFER NETWORKS

Consider a network with N nodes. Customers arrive at node j from the outside according to a batch Poisson process with rate λ_j. The batch sizes are i.i.d. with probability mass function

$$P(A_j = n) = a_j(n), \qquad n = 1, 2, \ldots.$$

Customers are served at each node in batches, and the service times of the different batches at node j are independent and exponentially distributed with rate μ_j, regardless of the number of customers in the batch. The service batch sizes at node j are i.i.d. random variables with probability mass function

$$P(B_j = n) = b_j(n), \qquad n = 1, 2, \ldots.$$

If there are at least B_j customers at node j at the time of a service completion, then these customers depart simultaneously from that node, and the queue length decreases by B_j. These customers constitute a full batch, and the routing probabilities of this batch depend on the batch size. Specifically, when a full batch of u customers leaves node j, it joins node k as a batch of v customers with probability $r_{ju,kv}$, and it leaves the network with probability $r_{ju,0}$. Clearly

$$\sum_{k=1}^{N}\sum_{v=1}^{\infty} r_{ju,kv} + r_{ju,0} = 1, \qquad j = 1, \ldots, N, u = 1, 2, \ldots.$$

If, however, there are fewer than B_j customers at node j upon the service completion, the node is emptied out, and all customers at the node, referred to as a partial batch, leave the network. This network is called an *assembly-transfer network.*

As we showed in the previous section, queues with batch arrivals do not have a geometric stationary distribution, not even in the single node case. To make the network process analytically tractable, we introduce an additional batch Poisson arrival process with rate λ_j^* at node j. However, this additional arrival process is activated only when node j is empty. The batch sizes of this additional arrival process are i.i.d. random variables with probability mass function

$$P(A_j^* = n) = a_j^*(n), \qquad n = 1, 2, \ldots.$$

Let n_j denote the number of customers at node j and let

$$\boldsymbol{n} = (n_1, \ldots, n_N)$$

denote the state of the network. If $r_{jj} = 0$ for $j = 1, 2, \ldots, N$, then the transition rate function q for the network process is given by

$$\begin{aligned} q(\boldsymbol{n}, \boldsymbol{n} - u\mathbf{e}_j + v\mathbf{e}_k) &= \mu_j b_j(u) r_{ju,kv}, \\ q(\boldsymbol{n}, \boldsymbol{n} - u\mathbf{e}_j) &= \mu_j b_j(u) r_{ju,0} + \mu_j \bar{b}_j(u) 1[u = n_j], \\ q(\boldsymbol{n}, \boldsymbol{n} + u\mathbf{e}_j) &= \lambda_j a_j(u) + \lambda_j^* a_j^*(u) 1[n_j = 0], \end{aligned}$$

where $\bar{b}_j(u) = \sum_{\ell=u}^{\infty} b_j(\ell)$. However, for the following results, r_{jj} may be positive. This network is referred to as an *assembly-transfer network with additional arrivals.*

Theorem 8.2 *Assume that the equations*

$$\lambda_j \sum_{\ell=1}^{\infty} a_j(\ell)\rho_j^{-\ell} + \sum_{k=1}^{N}\sum_{m=1}^{\infty}\sum_{\ell=1}^{\infty} \mu_k b_k(m) \rho_k^m r_{km,j\ell}(\rho_j^{-\ell} - 1) + \mu_j \sum_{\ell=1}^{\infty} b_j(\ell)\rho_j^{\ell}$$
$$= \lambda_j + \mu_j, \qquad j = 1, 2, \ldots, N, \quad (2.1)$$

have a solution with $\rho_j < 1$ for $j = 1, \ldots, N$. Then, under the condition that whenever node j is empty an additional batch arrival process is activated with rate

$$\lambda_j^* = \sum_{n=1}^{\infty} \sum_{\ell=n+1}^{\infty} \left(\lambda_j a_j(\ell) + \sum_{k=1}^{N} \sum_{m=1}^{\infty} \mu_k b_k(m) \rho_k^m r_{km,j\ell} \right) \rho_j^{-(\ell-n)} \tag{2.2}$$

and batch size distribution

$$a_j^*(n) = \frac{\sum_{\ell=n+1}^{\infty} \left(\lambda_j a_j(\ell) + \sum_{k=1}^{N} \sum_{m=1}^{\infty} \mu_k b_k(m) \rho_k^m r_{km,j\ell} \right) \rho_j^{-(\ell-n)}}{\lambda_j^*}, \tag{2.3}$$

the stationary distribution of the assembly-transfer network with the additional arrivals has the geometric product form

$$\pi(\boldsymbol{n}) = \prod_{j=1}^{N} (1 - \rho_j) \rho_j^{n_j}. \tag{2.4}$$

PROOF. Use Theorem 4.3 and consider node j in isolation. To analyze the quasi-reversibility of node j, we first introduce classes for arrivals and departures. Call a batch arrival or batch departure of size u an arrival or departure of class u. Node j is then described by

$$\begin{aligned} p_{ju}^{\mathrm{A}}(n_j, n_j + u) &= 1, && n_j \geq 0, u \geq 1, \\ q_{ju}^{\mathrm{D}}(n_j + u, n_j) &= \mu_j b_j(u), && n_j \geq 0, u \geq 1, \\ q_j^{\mathrm{I}}(0, n_j) &= \lambda_j^* a_j^*(n_j), && n_j \geq 0, \\ q_j^{\mathrm{I}}(n_j, 0) &= \mu_j \sum_{\ell=n_j+1}^{\infty} b_j(\ell), && n_j \geq 1. \end{aligned}$$

Note that departures of partial batches are classified as internal transitions since they are not routed to other nodes. Similarly, the extra arrivals are also classified as internal transitions. Node j, defined by

$$q_j(n_j, n_j') = \sum_{u=1}^{\infty} \alpha_{ju} p_{ju}^{\mathrm{A}}(n_j, n_j') + \sum_{u=1}^{\infty} q_{ju}^{\mathrm{D}}(n_j, n_j') + q_j^{\mathrm{I}}(n_j, n_j'),$$

is subject to Poisson arrivals of size u batches with rate α_{ju}, and an additional batch arrival process is activated whenever it is empty. This is equivalent to the single node queue, discussed in Section 2.7, that is subject to batch Poisson arrivals with rate

$$\alpha_j = \sum_{u=1}^{\infty} \alpha_{ju}$$

and batch size distribution

$$a_j(u) = \frac{\alpha_{ju}}{\alpha_j}, \qquad u = 1, 2, \ldots.$$

It follows from Theorem 2.33 that the stationary distribution of this node is

$$\pi_j(n_j) = (1 - \rho_j)\rho_j^{n_j},$$

where ρ_j is the solution of

$$\sum_{\ell=1}^{\infty} \alpha_{j\ell}\rho_j^{-\ell} + \mu_j \sum_{\ell=1}^{\infty} b_j(\ell)\rho_j^{\ell} = \sum_{\ell=1}^{\infty} \alpha_{j\ell} + \mu_j, \tag{2.5}$$

provided that, whenever the node is empty, there are additional batch Poisson arrivals with rate

$$\lambda_j^* = \sum_{n=1}^{\infty} \left(\sum_{\ell=n+1}^{\infty} \alpha_{j\ell}\rho_j^{-(\ell-n)} \right) \tag{2.6}$$

and batch size distribution

$$a_j^*(n) = \frac{\sum_{\ell=n+1}^{\infty} \alpha_{j\ell}\rho_j^{-(\ell-n)}}{\lambda_j^*}. \tag{2.7}$$

Since

$$\begin{aligned}\sum_{n_j'} \pi_j(n_j') q_{ju}^{\mathrm{D}}(n_j', n_j) &= \pi_j(n_j + u) q_{ju}^{\mathrm{D}}(n_j + u, n_j) \\ &= \pi_j(n_j)\mu_j\rho_j^u b_j(u), \qquad u = 1, 2, \ldots,\end{aligned}$$

node j is quasi-reversible with class u departure rate

$$\beta_{ju} = \mu_j \rho_j^u b_j(u), \qquad u = 1, 2, \ldots.$$

Thus it follows from Theorem 4.3 that, if the α_{ju}'s are the solutions to the traffic equations

$$\alpha_{ju} = \lambda_j a_j(u) + \sum_{i=1}^{N} \sum_{v=1}^{\infty} \mu_i \rho_i^v b_i(v) r_{iv,ju}, \tag{2.8}$$

then the stationary distribution of the network is given by (2.4) with ρ_j being the solution of (2.5), provided the extra arrival conditions (2.6) and (2.7) are in effect. Substituting (2.8) into (2.5) yields equation (2.1). Note that (2.1) is equivalent to (2.5) and (2.8). If ρ_j solves (2.1), then $\alpha_j(u)$, defined by (2.8), satisfies (2.5). This completes the proof of Theorem 8.2. □

The additional arrival conditions (2.6) and (2.7) are given in terms of ρ_j, which is determined by (2.1). This implies that the additional arrival process at each node actually depends on the dynamics of the entire network. This arrival process is defined artificially and it is introduced only for the network to be analytically tractable. As we discussed earlier, without these additional arrivals, a product form solution is not possible.

Since the additional arrivals can only make the network more congested, the original network has to be stochastically dominated by the network subject to these additional arrivals. A sample path can be constructed on the same probability space in such a way that the number of customers at the N nodes in the original network is always less than or equal to the number of customers at the N nodes in the network with the additional arrivals. Thus the same argument as in Theorem 2.34 of Chapter 2 yields the following result.

Theorem 8.3 *The stationary distribution π^0 for the queueing network without the additional arrivals is stochastically less than the stationary distribution of the network with the additional arrivals, i.e.,*

$$\sum_{\boldsymbol{n}' \geq \boldsymbol{n}} \pi^0(\boldsymbol{n}') \leq \prod_{j=1}^{N} \rho_j^{n_j}, \qquad \textit{for all } \boldsymbol{n}.$$

We now consider the structure of the reversed process of the network. Because the batch arrivals and batch departures in the original process correspond to batch departures and batch arrivals in the reversed process, we expect that the reversed process is also a queueing network with batch arrivals and batch departures. We may actually expect that the arrival process at each node is Poisson. However, a little reflection shows that, when a node is empty, the arrivals in the reversed process are somewhat different. To see this, note that when the node is empty, the arrivals in the reversed process correspond to the departures in the original process. However, there are two types of departures that leave the node in an empty state: departures of full batches and departures of partial batches. The partial batch departures, according to our assumption, leave the network. This implies that in the reversed process, when a node is empty, there is an additional arrival process to the node from the outside. This is exactly the same structure as the original process. The following theorem provides the network parameters of the reversed process.

Theorem 8.4 *The reversed process of the assembly-transfer network with additional arrivals is another assembly-transfer network with additional arrivals. The arrival rate $\tilde{\lambda}_j$, the arrival batch size distribution $\{\tilde{a}_j(\ell); \ell \geq 1\}$, the service rate $\tilde{\mu}_j$, the service batch size distribution $\{\tilde{b}_j(\ell); \ell \geq 1\}$, the routing probabilities*

$\tilde{r}_{jk,im}$, *the rate* $\tilde{\lambda}_j^*$ *of the additional arrivals, and the distribution* $\{\tilde{a}_j^*(n); n \geq 1\}$ *of the batch sizes of the additional arrivals are, for* $j, k = 1, 2, \ldots, N$ *and for* $\ell, m = 1, 2, \ldots,$

$$\tilde{\lambda}_j = \mu_j \sum_{\ell'=1}^{\infty} \rho_j^{\ell'} b_j(\ell') r_{j\ell',0}, \qquad \tilde{a}_j(\ell) = \frac{\mu_j \rho_j^{\ell} b_j(\ell) r_{j\ell,0}}{\tilde{\lambda}_j},$$

$$\tilde{\mu}_j = \sum_{k'=1}^{N} \sum_{m'=1}^{\infty} \sum_{\ell'=1}^{\infty} \mu_{k'} \rho_{k'}^{m'} b_j(m') \rho_j^{-\ell'} r_{k'm',j\ell'} + \lambda_j \sum_{\ell'=1}^{\infty} \rho_j^{-\ell'} a_j(\ell'),$$

$$\tilde{b}_j(\ell) = \frac{\sum_{k'=1}^{N} \sum_{m'=1}^{\infty} \mu_{k'} \rho_{k'}^{m'} b_j(m') \rho_j^{-\ell} r_{k'm',j\ell} + \lambda_j \rho_j^{-\ell} a_j(\ell)}{\tilde{\mu}_j},$$

$$\tilde{r}_{j\ell,km} = \frac{\mu_k \rho_k^m b_j(m) \rho_j^{-\ell} r_{km,j\ell}}{\tilde{\mu}_j \tilde{b}_j(\ell)}, \qquad \tilde{r}_{j\ell,0} = \frac{\lambda_j \rho_j^{-\ell} a_j(\ell)}{\tilde{\mu}_j \tilde{b}_j(\ell)},$$

$$\tilde{\lambda}_j^* = \mu_j \sum_{\ell'=1}^{\infty} \rho_j^{\ell'} \bar{b}_j(\ell'+1), \qquad \tilde{a}_j^*(\ell) = \frac{\mu_j \rho_j^{\ell} \bar{b}_j(\ell+1)}{\tilde{\lambda}_j^*}.$$

The proof of Theorem 8.4 follows directly from the given transition rates and the Kelly lemma, and is left as an exercise. From the structure of the reversed process, the following result is immediate.

Corollary 8.5 *The departure process from node* j *to the outside in the assembly-transfer network with additional arrivals is Poisson with rate* $\mu_j \sum_{\ell=1}^{\infty} b_j(\ell) \rho_j^{\ell} r_{j\ell,0}$. *Furthermore, these departure processes are independent of one another.*

8.3 CONCURRENT BATCH MOVEMENTS

This section extends the assembly-transfer networks to include multiple classes of positive and negative signals. The signals trigger instantaneous movements and yield a network with concurrent batch departures and concurrent batch arrivals. This class of network models is referred to as *networks with concurrent batch movements*.

Consider a queueing network with N nodes. There are a single class of customers and multiple classes of positive and negative signals. Customers arrive in batches at node j from the outside according to a Poisson process with rate λ_j, and each batch is of size u with probability $a_j(u)$, where

$$\sum_{u=1}^{\infty} a_j(u) = 1, \qquad j = 1, \ldots, N.$$

Positive signals of class u, denoted by u^+, arrive at node j from the outside according to a Poisson process with rate $\lambda_j^+ a_j^+(u)$, where

$$\sum_{u=1}^{\infty} a_j^+(u) = 1, \qquad j = 1, \ldots, N.$$

This implies that positive signals arrive at node j from the outside according to a Poisson process with rate λ_j^+, and each is of class u with probability $a_j^+(u)$, $u = 1, 2, \ldots$. Negative signals of class u, denoted by u^-, arrive at node j according to a Poisson process with rate $\lambda_j^- a_j^-(u)$, where

$$\sum_{u=1}^{\infty} a_j^-(u) = 1, \qquad j = 1, \ldots, N.$$

Again, this means that negative signals arrive at node j from the outside according to a Poisson process with rate λ_j^-, and each is of class u with probability $a_j^-(u)$, $u = 1, 2, \ldots$.

There is a single server at node j, and the services occur in batches. The service batch sizes at node j have probability mass function $\{b_j(u);\ u \geq 1\}$, and the service times of the batches are exponentially distributed with rate μ_j. If, upon a service completion, the required batch size is u and there are u or more customers present, then u customers depart as a full batch. A full batch joins node k as a batch of v customers with probability $r_{ju,kv}$, as a class v positive signal with probability r_{ju,kv^+}, as a class v negative signal with probability r_{ju,kv^-}, and the full batch leaves the system with probability $r_{ju,0}$. Clearly,

$$\sum_{k=1}^{N} \sum_{v=1}^{\infty} (r_{ju,kv} + r_{ju,kv^+} + r_{ju,kv^-}) + r_{ju,0} = 1, \qquad j = 1, \ldots, N, u \geq 1.$$

If, however, fewer than u customers are present at node j, the node is emptied out and its customers form a partial batch and leave the system.

The effects of the positive and negative signals can be described as follows. When a positive signal of class u arrives at node j, either from the outside or from another node, it adds u customers to the node. It then leaves immediately for node k as a batch of v customers with probability $r_{ju^+,kv}$, as a positive signal of class v with probability r_{ju^+,kv^+}, as a negative signal of class v with probability r_{ju^+,kv^-}, and leaves the system with probability $r_{ju^+,0}$, where

$$\sum_{k=1}^{N} \sum_{v=1}^{\infty} (r_{ju^+,kv} + r_{ju^+,kv^+} + r_{ju^+,kv^-}) + r_{ju^+,0} = 1, \quad j = 1, \ldots, N, u \geq 1.$$

When a negative signal of class u arrives at node j, either from the outside or from another node, it intends to delete u customers from the node. If there are u or more customers present at node j upon the signal's arrival, the queue length is reduced by u and the signal leaves immediately. It then goes to node k as a batch of v customers with probability $r_{ju^-,kv}$, as a positive signal of class v with probability r_{ju^-,kv^+}, as a negative signal of class v with probability r_{ju^-,kv^-}, and it leaves the system with probability $r_{ju^-,0}$, where

$$\sum_{k=1}^{N}\sum_{v=1}^{\infty}(r_{ju^-,kv}+r_{ju^-,kv^+}+r_{ju^-,kv^-})+r_{ju^-,0}=1, \quad j=1,2,\ldots,N, u\geq 1.$$

If, however, fewer than u customers are present at node j when a negative signal of class u arrives, the node is emptied out and the signal disappears from the system.

In this model there may be simultaneous batch arrivals at any number of nodes throughout the network. For instance, if

$$\prod_{\ell=1}^{m-1} r_{j_\ell u_\ell^+, j_{\ell+1}u_{\ell+1}^+} > 0$$

for some positive integers $u_1, u_2, \ldots, u_m$, then the arrival of a positive signal of class u_1 at node j_1 adds u_ℓ customers at node j_ℓ, $\ell = 1, 2, \ldots, m$. Similarly, if

$$\prod_{\ell=1}^{m-1} r_{j_\ell u_\ell^-, j_{\ell+1}u_{\ell+1}^-} > 0,$$

then the arrival of a negative signal of class u_1 at node j_1 deletes u_ℓ customers at node j_ℓ, $\ell = 1, \ldots, m$, provided there is a sufficient number of customers at all these nodes. Clearly, we can also have mixed concurrent arrivals and concurrent departures. By appropriately selecting the routing probabilities we can construct networks with arbitrary concurrent batch arrivals and batch departures along any fixed or random route in the network.

For the network to be stable, we have to assume that the Markov chain with state space

$$\{(j,u),\, j=1,\ldots,N, u\geq 1\}\cup\{0\}$$

and transition probabilities

$$\begin{aligned}
p_{ju,kv} &= r_{ju^+,kv^+}, && j,k=1,\ldots,N,\ u,v\geq 1,\\
p_{ju,0} &= 1-\sum_{v=1}^{\infty}\sum_{k=1}^{N} r_{ju^+,kv^+}, && j=1,\ldots,N,\ u\geq 1,\\
p_{0,0} &= 1,
\end{aligned}$$

has only one recurrent state 0. Intuitively, this implies that any customer or positive signal that enters the network will either ultimately leave the network, or transform itself into a negative signal.

Let α_{ju}, $j = 1, \ldots, N, u \geq 1$, α^+_{ju}, $j = 1, \ldots, N, u \geq 1$, and α^-_{ju}, $j = 1, \ldots, N, u \geq 1$, be the solutions to the traffic equations

$$\alpha_{ju} = \lambda_j a_j(u) + \sum_{k=1}^{N} \sum_{v=1}^{\infty} \mu_k b_k(v) \rho_k^v r_{kv,ju} + \sum_{k=1}^{N} \sum_{v=1}^{\infty} \alpha^+_{kv} \rho_k^{-v} r_{kv^+,ju} + \sum_{k=1}^{N} \sum_{v=1}^{\infty} \alpha^-_{kv} \rho_k^v r_{kv^-,ju}, \qquad j = 1, \ldots, N, u \geq 1, \tag{3.1}$$

$$\alpha^+_{ju} = \lambda^+_j a^+_j(u) + \sum_{k=1}^{N} \sum_{v=1}^{\infty} \mu_k b_k(v) \rho_k^v r_{kv,ju^+} + \sum_{k=1}^{N} \sum_{v=1}^{\infty} \alpha^+_{kv} \rho_k^{-v} r_{kv^+,ju^+} + \sum_{k=1}^{N} \sum_{v=1}^{\infty} \alpha^-_{kv} \rho_k^v r_{kv^-,ju^+}, \qquad j = 1, \ldots, N, u \geq 1, \tag{3.2}$$

$$\alpha^-_{ju} = \lambda^-_j a^-_j(u) + \sum_{k=1}^{N} \sum_{v=1}^{\infty} \mu_k b_k(v) \rho_k^v r_{kv,ju^-} + \sum_{k=1}^{N} \sum_{v=1}^{\infty} \alpha^+_{kv} \rho_k^{-v} r_{kv^+,ju^-} + \sum_{k=1}^{N} \sum_{v=1}^{\infty} \alpha^-_{kv} \rho_k^v r_{kv^-,ju^-}, \qquad j = 1, \ldots, N, u \geq 1, \tag{3.3}$$

where ρ_j is the solution to the equation

$$\sum_{u=1}^{\infty} \alpha_{ju} \rho_j^{-u} + \sum_{u=1}^{\infty} \alpha^+_{ju} \rho_j^{-u} + \mu_j \sum_{u=1}^{\infty} b_j(u) \rho_j^u + \sum_{u=1}^{\infty} \alpha^-_{ju} \rho_j^u = \sum_{u=1}^{\infty} \alpha_{ju} + \sum_{u=1}^{\infty} \alpha^+_{ju} + \sum_{u=1}^{\infty} \alpha^-_{ju} + \mu_j \, . \tag{3.4}$$

Theorem 8.6 *Assume that, whenever node j is empty, an additional batch arrival process is activated with rate*

$$\lambda_j^* = \sum_{n=1}^{\infty} \sum_{\ell=n+1}^{\infty} (\alpha_{j\ell} + \alpha^+_{j\ell}) \rho_j^{-(\ell-n)} \tag{3.5}$$

and batch size distribution

$$a_j^*(n) = \frac{\sum_{\ell=n+1}^{\infty} (\alpha_{j\ell} + \alpha^+_{j\ell}) \rho_j^{-(\ell-n)}}{\lambda_j^*}, \qquad n = 1, 2, \ldots, \tag{3.6}$$

and that whenever there are n_j customers present at node j, an additional departure process of class u positive signals with rate $\alpha^+_{ju}\rho_j^{-u}$, $u \geq n_j + 1$, is activated and that these departures do not change the state of node j, $j = 1, \ldots, N$. Then the modified network process has the stationary distribution of the geometric product form

$$\pi(n_1, \ldots, n_N) = \prod_{j=1}^{N} (1 - \rho_j)\rho_j^{n_j}, \tag{3.7}$$

where ρ_j, $j = 1, \ldots, N$, is the solution to the traffic equations (3.1) to (3.4).

PROOF. By Theorem 4.9 it suffices to verify the quasi-reversibility of each node in the network. Call the arrival of a batch of u customers and the departure of a full batch of u customers due to a service completion a class u arrival and a class u departure, respectively. Then the class of entities is

$$T_j = \{u; u \geq 1\} \cup \{u^+; u \geq 1\} \cup \{u^-; u \geq 1\}.$$

Node j, with the additional arrivals and departures as described in the theorem, is characterized by

$$\begin{aligned}
p^{\text{A}}_{ju}(n_j, n_j + u) &= 1, && n_j \geq 0, u \geq 1,\\
p^{\text{A}}_{ju^+}(n_j, n_j + u) &= 1, && n_j \geq 0, u \geq 1,\\
p^{\text{A}}_{ju^-}(n_j, n_j - u) &= 1, && n_j \geq u \geq 1,\\
p^{\text{A}}_{ju^-}(n_j, 0) &= 1, && 0 \leq n_j < u,\\
f_{ju^+,u^+}(n_j, n_j + u) &= 1, && n_j \geq 0, u \geq 1,\\
f_{ju^-,u^-}(n_j, n_j - u) &= 1, && n_j \geq u \geq 1,\\
q^{\text{D}}_{ju}(n_j, n_j - u) &= \mu_j b_j(u), && n_j \geq u \geq 1,\\
q^{\text{D}}_{ju^+}(n_j, n_j) &= \alpha_j^+(u)\rho_j^{-u}, && u \geq n_j + 1,\\
q^{\text{I}}_j(0, n_j) &= \lambda_j^* a_j^*(n_j), && n_j \geq 1,\\
q^{\text{I}}_j(n_j, 0) &= \mu_j \textstyle\sum_{\ell=n_j+1}^{\infty} b_j(\ell), && n_j \geq 1,
\end{aligned}$$

where ρ_j is the solution to equation (3.4). We assume that the α_{ju} are chosen so that $\rho_j < 1$.

For a fixed $\boldsymbol{\alpha}_j = (\alpha_{ju}, \alpha^+_{ju}, \alpha^-_{ju}; u \geq 1)$, node j, defined by

$$\begin{aligned}
q_j^{(\boldsymbol{\alpha}_j)}(n_j, n_j') = \sum_{u=1}^{\infty} \left(\alpha_{ju} p^{\text{A}}_{ju}(n_j, n_j') + \alpha^+_{ju} p^{\text{A}}_{ju^+}(n_j, n_j') + \alpha^-_{ju} p^{\text{A}}_{ju^-}(n_j, n_j')\right)) \\
+ \sum_{u=1}^{\infty} \left(q^{\text{D}}_{ju}(n_j, n_j') + q^{\text{D}}_{ju^+}(n_j, n_j')\right) + q^{\text{I}}_j(n_j, n_j'),
\end{aligned} \tag{3.8}$$

is an $M^X/M^Y/1$ queue with signals. Batches of customers of size u arrive according to a Poisson process with rate α_{ju}. The service rate is μ_j, and the sizes of the service batches have probability mass function $\{b_j(\ell); \ell \geq 1\}$. Moreover, positive and negative signals of class u arrive according to Poisson processes with rates α^+_{ju} and α^-_{ju}, $u = 1, 2, \ldots$.

It is not hard to see that, as far as the stationary distribution of (3.8) is concerned, this node is equivalent to another $M^X/M^Y/1$ model without signals, but with different arrival and service processes. Batches of customers of size u arrive according to a Poisson process with rate

$$\alpha_{ju} + \alpha^+_{ju},$$

and the departure rate of batches of size u is

$$\mu_j b_j(u) + \alpha^-_{ju}.$$

It follows from Theorem 2.33 that this node has the geometric stationary distribution

$$\pi_j(n_j) = (1 - \rho_j)\rho_j^{n_j}, \qquad n_j \geq 0,$$

provided that, whenever it is empty, there is an additional arrival process of size u with rate $\lambda^*_j a^*_j(u)$, $u = 1, 2, \ldots$, which is specified by (3.5) and (3.6). Note that since the additional departures due to $q^{\mathrm{D}}_{ju^+}$ do not change the state of the queue, they do not directly affect π_j.

To check the quasi-reversibility, we perform the following verifications:

$$\begin{aligned}\sum_{n'_j} \pi_j(n'_j) q^{\mathrm{D}}_{ju}(n'_j, n_j) &= \pi_j(n_j + u)\mu_j b_j(u) \\ &= \mu_j \rho_j^u b_j(u)\pi_j(n_j), \qquad n_j \geq 0,\end{aligned}$$

$$\begin{aligned}\sum_{n'_j} \pi_j(n'_j)\alpha^-_{ju} f_{ju^-,u^-}(n'_j, n_j) &= \pi_j(n_j + u)\alpha^-_{ju} f_{ju^-,u^-}(n_j + u, n_j) \\ &= \rho_j^u \alpha^-_{ju}\pi_j(n_j), \qquad n_j \geq 0.\end{aligned}$$

So the departure rates for classes u and u^- are

$$\begin{aligned}\beta_{ju} &= \rho_j^u \mu_j b_j(u), \qquad & u \geq 1, \\ \beta_{ju^-} &= \rho_j^u \alpha^-_{ju}, \qquad & u \geq 1.\end{aligned}$$

In the case of positive signals, if $u \leq n_j$, then

$$\sum_{n'_j} \pi_j(n'_j)\big((q^{\mathrm{D}}_{ju^+}(n'_j, n_j) + \alpha^+_{ju} f_{ju^+,u^+}(n'_j, n_j)\big)$$
$$= \pi_j(n_j - u)\alpha^+_{ju} f_{ju^+,u^+}(n_j - u, n_j)$$
$$= \rho_j^{-u}\alpha^+_{ju}\pi_j(n_j).$$

By the definition of $q^{\mathrm{D}}_{ju^+}(n_j, n_j)$, if $u \geq n_j + 1$, there are additional departures of class u^+ positive signals with rate $\rho_j^{-u}\alpha^+_{ju}$ which do not change the state of the node. Thus we have

$$\sum_{n'_j} \pi_j(n'_j)\big(q^{\mathrm{D}}_{ju^+}(n'_j, n_j) + \alpha^+_{ju^+} f_{ju^+,u^+}(n'_j, n_j)\big)$$
$$= \pi_j(n_j)q^{\mathrm{D}}_{ju^+}(n_j, n_j)$$
$$= \rho_j^{-u}\alpha^+_{ju}\pi_j(n_j), \qquad u > n_j \geq 0.$$

Hence the quasi-reversibility condition is also satisfied for class u^+ with departure rate

$$\beta_{ju^+} = \rho_j^{-u}\alpha^+_{ju}, \qquad u \geq 1.$$

To summarize, we have shown that node j is a quasi-reversible queue with signals for any $\boldsymbol{\alpha}_j$ such that $\rho_j < 1$. Substituting β_{ju}, β_{ju^+} and β_{ju^-} into the traffic equations (3.5) of Chapter 4 we obtain the traffic equations (3.1), (3.2) and (3.3). If we connect the N nodes as described at the beginning of this section, the resulting network has, by Theorem 4.9, the geometric product form stationary distribution (3.7). This completes the proof of Theorem 8.6. □

Similar to Theorem 8.3, the geometric product form solution (3.7) may often be used as an upper bound for the original network without the modification.

Corollary 8.7 *If* $r_{ju^+,kv^-} = 0$ *and* $r_{jc,kv^-} = 0$ *for* j, k, u *and* v, *then the product form probability distribution (3.7) is a stochastic upper bound for the original network without the modification stated in Theorem 8.6. That is, if* $\pi^0(\boldsymbol{n})$ *is the stationary distribution of the original network, then*

$$\sum_{\boldsymbol{n}' \geq \boldsymbol{n}} \pi^0(\boldsymbol{n}') \leq \prod_{j=1}^{N} \rho_j^{n_j}.$$

Remark 8.8 We indicated at the beginning of this chapter that the extra transitions at a node occur only when it is empty. However, there is an extra transition of positive signals in Theorem 8.6 when there are n_j customers at node j for any n_j. This appears contradictory. However, the extra rate can also be viewed as being added when node j is empty by introducing class (n_j, u) positive signals, $n_j \leq u$, and whenever node j is empty, there are extra departures of class (n_j, u) positive signals with rate $\alpha^+_{ju}\rho_j^{-(u-n_j)}$. A departure of a class (n_j, u) positive signal returns to node j and adds n_j customers there, then leaves node j as a regular class u positive signal. It is left as an exercise for the reader to show that such a modification is equivalent to the one introduced in Theorem 8.6. The structure of these modifications will be explored further in Chapter 10.

8.4 EXAMPLES

We apply the results of this chapter to two of the examples discussed in Chapter 1, namely the production and inventory control problem and the computer virus model. For each problem we assume additional arrivals and obtain a product form stationary distribution. This distribution serves as a stochastic upper bound for the original network. We first consider the production and inventory control problem of Example 1.4.

Example 8.9 (Production and inventory control model continued) Demands of class k, $k = 1, 2, \ldots, I$, arrive according to a Poisson process with rate λ_k, and each class k arrival adds $u_{k\ell}$ customers to node $j_{k\ell}$, $\ell = 1, 2, \ldots, m_k$. For convenience, we assume

$$j_{k1} \leq j_{k2} \leq \cdots \leq j_{km_k}.$$

The processing times at node j are exponentially distributed with rate μ_j, $j = 1, 2, \ldots, N$. This problem can be modeled as a queueing network with positive signals. Introduce I classes of positive signals. Class k signals arrive at node j_{k1} according to a Poisson process with rate λ_k, add u_{k1} customers to node j_{k1} and then leave for node j_{k2}, add u_{k2} to node j_{k2} and then leave for node j_{k3}, and so forth. This process continues until the signals arrive at node j_{km_k}, where they add u_{km_k} customers and then leave the system.

Let α^+_{jk} be the overall arrival rate of class k positive signals at node j. They are determined by the traffic equations

$$\alpha^+_{jk} = \lambda_k 1[j = j_{k1}] + \sum_{\ell=1}^{m_k-1} \lambda_{j_{k\ell}} \rho_{j_{k\ell}}^{-u_{k\ell}} 1[j = j_{k(\ell+1)}],$$
$$j = 1, 2, \ldots, N, \; k = 1, \ldots, I,$$

and

$$\sum_{k=1}^{I} \alpha_{jk}^{+} \rho_j^{-u_{kj}} + \mu_j \rho_j = \sum_{k=1}^{I} \alpha_{jk}^{+} + \mu_j \,, \qquad j = 1, 2, \ldots, N,$$

where u_{kj} is zero whenever it is not defined. If n_j denotes the number of customers at node j, then the stationary distribution of the network is, after introducing the additional arrivals at the node when it is empty, given by

$$\pi(n_1, \ldots, n_N) = \prod_{j=1}^{N} (1 - \rho_j) \rho_j^{n_j} .$$

This is a stochastic upper bound for the original model. □

The following example illustrates the production and inventory control model.

Example 8.10 (A three-node network with concurrent arrivals) Consider a network with three nodes. The service rates at the three nodes are μ_1, μ_2, and μ_3, respectively. There is a concurrent Poisson arrival process at nodes 1 and 2 with rate λ_{12}, and each adds two customers to node 1, and one customer to node 2. There is also a concurrent Poisson arrival process at nodes 2 and 3 with rate λ_{23}, and at each arrival epoch it adds three customers to node 2, and either one customer (with probability 1/3) or two customers (with probability 2/3) to node 3.

This is a production and inventory control model with $N = 3$, $I = 3$, and arrival rates

$$\lambda_1 = \lambda_{12}, \qquad \lambda_2 = \frac{2}{3}\lambda_{23}, \qquad \lambda_3 = \frac{1}{3}\lambda_{23} .$$

The batch sizes are

$$\begin{aligned} a_{11} &= 2, & a_{12} &= 1, \\ a_{22} &= 3, & a_{23} &= 1, \\ a_{32} &= 3, & a_{33} &= 2. \end{aligned}$$

To obtain a stochastic upper bound for the network, we introduce positive signals of class 1 that arrive at node 1 according to a Poisson process with rate λ_{12}, add two customers to node 1, then leave for node 2 and add one customer to node 2 before exiting the system. We also introduce positive signals of class 2 that arrive at node 2 according to a Poisson process with rate $2\lambda_{23}/3$, add three customers to node 2, then join node 3 to add one customer before leaving the system. Finally, positive signals of class 3 arrive at node 2 according to a Poisson process with rate $\lambda_{23}/3$, add three customers to node 2, then leave for node 3, and add two customers to node 3 before leaving the system. There is no regular customer arrival process.

If α^+_{ju} denotes the average arrival rate of class u positive signals at node j, then

$$\begin{aligned} \alpha^+_{11} &= \lambda_{12}, & \alpha^+_{21} &= \alpha^+_{11}\rho_1^{-2}, \\ \alpha^+_{22} &= \frac{2}{3}\lambda_{23}, & \alpha^+_{32} &= \alpha^+_{22}\rho_2^{-3}, \\ \alpha^+_{23} &= \frac{1}{3}\lambda_{23}, & \alpha^+_{33} &= \alpha^+_{23}\rho_2^{-3}, \end{aligned}$$

and

$$\begin{aligned} \alpha^+_{11}\rho_1^{-2} + \mu_1\rho_1 &= \alpha^+_{11} + \mu_1, \\ \alpha^+_{21}\rho_2^{-1} + \alpha^+_{22}\rho_2^{-3} + \alpha^+_{23}\rho_2^{-3} + \mu_2\rho_2 &= \alpha^+_{21} + \alpha^+_{22} + \alpha^+_{23} + \mu_2, \\ \alpha^+_{32}\rho_3^{-1} + \alpha^+_{33}\rho_3^{-2} + \mu_3\rho_3 &= \alpha^+_{32} + \alpha^+_{33} + \mu_3. \end{aligned}$$

The stochastic upper bound for the original network process is

$$\pi(n_1, n_2, n_3) = \prod_{j=1}^{3}(1-\rho_j)\rho_j^{n_j}. \qquad \square$$

Example 8.11 (Computer virus problem continued) We consider again the computer virus problem discussed in Chapter 1. In Example 7.9 we studied this problem under the assumption that any job that is infected at a station is removed from the system. We now assume that a job that is infected at station j is sent for repair. For simplicity we assume that there is a single repair facility at node j. Only one job can be repaired at any point in time, and the repair times at node j are exponentially distributed with rate ν_j.

Node j may be regarded as consisting of two substations, one being the original processing facility, and the other the repair facility. When a batch of jobs is infected at the processing facility, it is transferred to the repair facility. At the end of a repair, a job is recovered with probability ϑ_j or is found to be nonrecoverable with probability $1-\vartheta_j$. This is modeled through a routing probability that any job that finishes repair at the repair facility of node j leaves the system with probability $1-\vartheta_j$ and joins the processing facility at node j with probability ϑ_j.

As a result, the computer virus network with repair is modeled as a queueing network with concurrent movements. If n_{jc} denotes the number of jobs at the processing facility of node j, and n_{jr} the number of infected jobs at the repair facility of node j, then $\boldsymbol{n}_j = (n_{jc}, n_{jr})$ is the state of node j and

$$\boldsymbol{n} = (\boldsymbol{n}_1, \boldsymbol{n}_2, \ldots, \boldsymbol{n}_N)$$

is the state of the network. $\square$

Let α_j and α_j^- denote the average arrival rates of jobs, without and with virus infections, respectively, at the processing facility of node j, and let α_{ju}^+ denote the average arrival rate of infected batches of size u at the repair facility of node j. Recall that $\{a_j^-(u); u \geq 1\}$ is the infected batch size probability. If additional arrivals are introduced whenever nodes are empty, they satisfy the traffic equations

$$\alpha_j = \lambda_j + \sum_{k=1}^{N} \mu_k \rho_{kc} r_{kj} + \nu_j \rho_{jr} \vartheta_j,$$

$$\alpha_j^- = \lambda_j^- + \sum_{k=1}^{N} \mu_k \rho_{kc} r_{kj}^-,$$

$$\alpha_{ju}^+ = \alpha_j^- a_j^-(u) \rho_{jc}^u,$$

where ρ_{jc} and ρ_{jr} are determined by

$$\alpha_j \rho_{jc}^{-1} + \alpha_j^- \sum_{u=1}^{\infty} a_j^-(u) \rho_{jc}^u + \mu_j \rho_{jc} = \alpha_j + \alpha_j^- + \mu_j,$$

$$\sum_{u=1}^{\infty} \alpha_{ju}^+ \rho_{jr}^{-u} + \nu_j \rho_{jr} = \sum_{u=1}^{\infty} \alpha_{ju}^+ + \nu_j, \qquad j = 1, 2, \ldots, N.$$

The proof of the following result is left to the reader as an exercise.

Proposition 8.12 *Under the assumption that whenever the repair station at node j is empty there is an additional arrival process of batches of size n with rate*

$$\alpha_j^- \sum_{\ell=n+1}^{\infty} a_j^-(\ell) \rho_{jc}^{\ell} \rho_{jr}^{-(\ell-n)}, \qquad j = 1, 2, \ldots, N, n \geq 1,$$

the stationary distribution of the network is

$$\pi(\boldsymbol{n}) = \prod_{j=1}^{N} (1 - \rho_{jc})(1 - \rho_{jr}) \rho_{jc}^{n_{jc}} \rho_{jr}^{n_{jr}}. \tag{4.1}$$

The stationary distribution (4.1) is a stochastic upper bound for the original network without the additional arrivals.

We may have the situation where the repairs of infected jobs at each node occur in batches, and a batch contains the jobs that are infected by the same virus. To deal with such a situation, we assume that an infected batch of size u is transferred to the repair facility as a single job of class u. The jobs at the repair facility of node j are repaired at rate ν_j. A repaired job of class u joins the processing facility as a batch of u jobs with probability ϑ_j, and leaves the network with probability $1 - \vartheta_j$. Clearly, this model has batch arrivals, batch departures, and negative signals.

Let α_{ju}, α_j^- and α_{ju}^+ be the solutions to the traffic equations

$$\alpha_{ju} = \left(\lambda_j + \sum_{k=1}^{N} \mu_k \rho_{kc} r_{kj}\right) 1[u = 1] + \nu_j \rho_{ju}^+ \vartheta_j,$$

$$\alpha_j^- = \lambda_j^- + \sum_{k=1}^{N} \mu_k \rho_{kc} r_{kj}^-,$$

$$\alpha_{ju}^+ = \alpha_j^- a_j^-(u) \rho_{jc}^u,$$

where ρ_{jc} and ρ_{ju}^+ are determined by

$$\sum_{u=1}^{\infty} \alpha_{ju} \rho_{jc}^{-u} + \alpha_j^- \sum_{u=1}^{\infty} a_j^-(u) \rho_{jc}^u + \mu_j \rho_{jc} = \sum_{u=1}^{\infty} \alpha_{ju} + \alpha_j^- + \mu_j, \quad j = 1, 2, \ldots, N,$$

$$\rho_{ju}^+ = \frac{\alpha_{ju}^+}{\nu_j} = \frac{\alpha_j^-}{\nu_j} a_j^-(u) \rho_{jc}^u, \qquad j = 1, \ldots, N, u \geq 1.$$

Define ρ_{jr} as

$$\rho_{jr} = \sum_{u=1}^{\infty} \rho_{ju}^+, \qquad j = 1, \ldots, N.$$

It is left to the reader to show the following proposition.

Proposition 8.13 *Assume that, whenever the processing station of node j is empty, i.e., when there are no noninfected jobs at node j, there is an additional arrival process of batches of size n with rate*

$$\alpha_j^- \vartheta_j \rho_{jc}^n \sum_{u=n+1}^{\infty} a_j^-(u), \qquad j = 1, 2, \ldots, N, u \geq 1.$$

If the solutions to the traffic equations satisfy $\rho_{jc} < 1$ and $\rho_{jr} < 1$ for $j = 1, \ldots, N$, then the stationary distribution of the network is given by (4.1), and it is a stochastic upper bound for the original network without the additional arrivals.

8.5 REFERENCE NOTES

The assembly-transfer network is first studied by Miyazawa and Taylor (1997). The material of Section 8.3 is taken from Chao and Miyazawa (1996*b*). The production and inventory control model is studied by Song, Xu and Liu (1999), where a matrix geometric approach is used in analyzing its performance.

EXERCISES

8.1 Consider the assembly-transfer network of Section 8.2.

(a) Classify batches of size u as class u for both arrivals and departures, and formulate this network as a network with multiple classes of customers.

(b) Under the formulation of (a), find conditions on the additional batch arrivals so that each node is quasi-reversible and the network has a product form stationary distribution.

(c) Change the service mechanism so that each batch is served as a single customer but departs as a batch of the same size. Find the stationary distribution when all the batches at each node have the same exponential service time distribution, and are served according to the First–Come First–Served discipline.

8.2 In Section 8.3 we assumed that the triggering movement does not depend on the source of the signal. For instance, when a class u negative signal arrives at node j and triggers a batch of u customers to leave, the routing probability depends only on u and j. In reality, the routing probability may depend on the source of the signal, i.e., where the signal comes from. Modify the results of Section 8.3 so that the routing probabilities of the triggered movements depend on the source of signals.

8.3 Prove Propositions 8.12 and 8.13.

9

State-Dependent and History-Dependent Transitions

In this chapter we extend the results of the preceding chapters to networks with state-dependent transition rates. The approach, which is based on local balance equations, is illustrated in Section 9.1 through a Jackson network without signals. Jackson networks with state-dependent transition rates belong to a larger class of networks called linear traffic networks, which are discussed in Section 9.2. Networks with signals, however, generally belong to the class of nonlinear models. Section 9.3 illustrates how a new type of local balance can be used to study nonlinear networks. The full generality of this local balance will be explored in the next chapter. In Sections 9.4 and 9.5, state-dependent networks with negative and positive signals are discussed, respectively. Finally, Sections 9.6 and 9.7 describe two networks with history-dependent routing probabilities. Most results in this chapter are presented, for convenience, in terms of stationary measures (see Chapter 2).

9.1 STATE-DEPENDENT JACKSON NETWORKS

In this section we show how to construct tractable networks with state-dependent transition rates from networks with state-independent transition rates.

Consider a Jackson network with N nodes. The state of the network is represented by $\boldsymbol{n} = (n_1, \ldots, n_N)$, where n_j is the number of customers at node j, $j = 1, \ldots, N$. Customers arrive at node j from the outside according to a Poisson process with rate λ_j. There is a single server at node j with service rate μ_j. The routing probabilities are r_{jk}, $k = 0, 1, \ldots, N$, i.e., a customer finishing service at node j goes to node k with probability r_{jk}, where $k = 0$ represents the outside. The stationary distribution of this network is

$$\pi(\boldsymbol{n}) = \prod_{j=1}^{N} \left(1 - \frac{\alpha_j}{\mu_j}\right) \left(\frac{\alpha_j}{\mu_j}\right)^{n_j},$$

where α_j is the solution to the traffic equations

$$\alpha_j = \lambda_j + \sum_{k=1}^{N} \alpha_k r_{kj}, \qquad j = 1, \ldots, N. \tag{1.1}$$

Furthermore, it follows from (3.14) and (3.13) of Chapter 2 that the stationary distribution $\pi(\boldsymbol{n})$ satisfies the local balance equations

$$\pi(\boldsymbol{n} + \mathbf{e}_j)\mu_j = \pi(\boldsymbol{n})\lambda_j + \sum_{k=1}^{N} \pi(\boldsymbol{n} + \mathbf{e}_k)\mu_k r_{kj}, \qquad j = 1, \ldots, N, \tag{1.2}$$

$$\pi(\boldsymbol{n}) \sum_{j=1}^{N} \lambda_j = \sum_{k=1}^{N} \pi(\boldsymbol{n} + \mathbf{e}_k)\mu_k r_{k0}. \tag{1.3}$$

A close look at (1.2) reveals that if the service rate μ_j when the state of the network is $\boldsymbol{n}$ is replaced by

$$\mu_j(\boldsymbol{n}) = \mu_j \frac{\Psi(\boldsymbol{n} - \mathbf{e}_j)}{\Phi(\boldsymbol{n})}, \qquad j = 1, \ldots, N, \tag{1.4}$$

and the arrival rate λ_j when the state of the network is $\boldsymbol{n}$ is replaced by

$$\lambda_j(\boldsymbol{n}) = \lambda_j \frac{\Psi(\boldsymbol{n})}{\Phi(\boldsymbol{n})}, \qquad j = 1, \ldots, N, \tag{1.5}$$

where Ψ and Φ are arbitrary nonnegative and positive functions, respectively, then equations (1.2) imply that

$$\hat{\pi}(\boldsymbol{n})\mu_j(\boldsymbol{n} + \mathbf{e}_j) = \hat{\pi}(\boldsymbol{n})\lambda_j(\boldsymbol{n}) + \sum_{k=1}^{N} \hat{\pi}(\boldsymbol{n} + \mathbf{e}_k)\mu_k(\boldsymbol{n} + \mathbf{e}_k) r_{kj}, \tag{1.6}$$

$$j = 1, \ldots, N,$$

$$\hat{\pi}(\boldsymbol{n}) \sum_{j=1}^{N} \lambda_j(\boldsymbol{n}) = \sum_{k=1}^{N} \hat{\pi}(\boldsymbol{n} + \mathbf{e}_k)\mu_k(\boldsymbol{n} + \mathbf{e}_k) r_{k0}, \tag{1.7}$$

where

$$\hat{\pi}(\boldsymbol{n}) = \Phi(\boldsymbol{n})\pi(\boldsymbol{n}).$$

This is true because the denominator of each term in the new transition rate cancels, and the extra part in the numerators on both sides, i.e., $\Psi(\boldsymbol{n})$, is common to every term. Since summing (1.6) over all j and (1.7) yields the global balance equation, $\hat{\pi}$ is indeed the stationary measure for the network with the modified transition rates (1.4) and (1.5).

Summarizing, we have the following result.

Theorem 9.1 *Assume that when the state of the network is* $\boldsymbol{n}$, *the arrival process at node* j *is state-dependent with rate (1.5) and the service rate at node* j *is (1.4). Then, the stationary measure of the network is*

$$\pi(\boldsymbol{n}) = \Phi(\boldsymbol{n}) \prod_{j=1}^{N} \left(\frac{\alpha_j}{\mu_j}\right)^{n_j}, \tag{1.8}$$

where α_j *is the solution to the traffic equations (1.1).*

Example 9.2 (Jackson network with state-dependent arrival rates) Consider the special case where customers arrive at the network from the outside, when there is a total of n customers in the entire network, according to a Poisson process with rate $\lambda(n)$, and each customer joins node j with probability r_{0j}, $j = 1, \ldots, N$. The service rate at node j is, when there are n_j customers at node j, $\mu_j(n_j)$. This network fits in the framework above by setting $\lambda_j = r_{0j}$ and $\mu_j = 1$, and

$$\Phi(\boldsymbol{n}) = \frac{\prod_{\ell=0}^{|\boldsymbol{n}|-1} \lambda(\ell)}{\prod_{j=1}^{N} \prod_{\ell=1}^{n_j} \mu_j(\ell)},$$

$$\Psi(\boldsymbol{n}) = \frac{\prod_{\ell=0}^{|\boldsymbol{n}|} \lambda(\ell)}{\prod_{j=1}^{N} \prod_{\ell=1}^{n_j} \mu_j(\ell)},$$

where $|\boldsymbol{n}|$ is the total number of customers in the network, i.e.,

$$|\boldsymbol{n}| = \sum_{j=1}^{N} n_j .$$

We leave it to the reader to check that $\lambda_j(\boldsymbol{n})$ and $\mu_j(\boldsymbol{n})$ of (1.4) and (1.5) are equal to $\lambda(n) r_{0j}$ and $\mu_j(n_j)$, respectively. By Theorem 9.1, the stationary distribution of this network is

$$\pi(\boldsymbol{n}) = b^{-1} \prod_{\ell=1}^{|\boldsymbol{n}|} \lambda(\ell - 1) \prod_{j=1}^{N} \frac{\alpha_j^{n_j}}{\prod_{\ell=1}^{n_j} \mu_j(\ell)},$$

where α_j is the solution to the traffic equations

$$\alpha_j = r_{0j} + \sum_{k=1}^{N} \alpha_k r_{kj}, \qquad j = 1, \ldots, N,$$

and b is the normalization constant defined by

$$b = 1 + \sum_{n=1}^{\infty} \prod_{\ell=1}^{n} \lambda(\ell - 1) \sum_{\sum_{j=1}^{I} n_j = n} \frac{\alpha_j^{n_j}}{\prod_{\ell=1}^{n_j} \mu_j(\ell)}.$$

Furthermore, in the case when

$$\lambda(n) = \lambda, \qquad n = 0, 1, 2, \ldots,$$

the stationary distribution reduces to

$$\pi(\boldsymbol{n}) = \prod_{j=1}^{N} \pi_j(n_j),$$

where

$$\pi_j(n_j) = \pi_j(0) \frac{\alpha_j^{n_j}}{\prod_{\ell=1}^{n_j} \mu_j(\ell)},$$

and

$$\pi_j(0) = \left(1 + \sum_{n=1}^{\infty} \frac{\alpha_j^n}{\prod_{\ell=1}^{n} \mu_j(\ell)}\right)^{-1}.$$

This model is a special case of the multi-class Jackson network studied in Section 5.1, and its stationary distribution has also been obtained in Theorem 5.4. □

Example 9.3 (A finite-source queueing network) Following the notation of the previous example, consider now the special case where

$$\lambda(n) = 0, \qquad \text{for } n \geq M,$$

and $\lambda(n)$ is positive and finite for $n \leq M - 1$, where M is a given positive integer. In this case there is a maximum number of customers that the system can hold, which can be considered to be the capacity of the system. Customers who find the system full upon arrival are lost. In this case the stationary distribution is

$$\pi(\boldsymbol{n}) = b^{-1} \prod_{\ell=1}^{|\boldsymbol{n}|} \lambda(\ell - 1) \prod_{j=1}^{N} \frac{\alpha_j^{n_j}}{\prod_{\ell=1}^{n_j} \mu_j(\ell)}, \qquad \text{for } \sum_{j=1}^{N} n_j \leq M,$$

where b is the normalization constant defined by

$$b = \sum_{n_1 + \cdots + n_N \leq M} \prod_{\ell=1}^{|\boldsymbol{n}|} \lambda(\ell - 1) \prod_{j=1}^{N} \frac{\alpha_j^{n_j}}{\prod_{\ell=1}^{n_j} \mu_j(\ell)}.$$

This special case is often referred to as a *finite-source queueing network*, where M is the capacity of the exogenous source.

If we consider the outside also as a regular node with service rate

$$\mu_0(n_0) = \lambda(M - n_0),$$

where $n_0 = M - \sum_{j=1}^{N} n_j$ is the number of customers at node 0 (the outside), then the network becomes a *closed queueing network* with a total of M customers. Using the notation

$$\bar{\boldsymbol{n}} = (n_0, n_1, \ldots, n_N),$$

it follows from the above result that the stationary distribution of the closed network is

$$\begin{aligned}
\pi(\bar{\boldsymbol{n}}) &= \left(b^{-1} \prod_{\ell=1}^{M} \lambda(\ell - 1) \right) \frac{1}{\prod_{\ell=1}^{M-|\boldsymbol{n}|} \lambda(M - \ell)} \prod_{j=1}^{N} \frac{\alpha_j^{n_j}}{\prod_{\ell=1}^{n_j} \mu_j(\ell)} \\
&= \left(b^{-1} \prod_{\ell=1}^{M} \lambda(\ell - 1) \right) \frac{1}{\prod_{\ell=1}^{n_0} \mu_0(\ell)} \prod_{j=1}^{N} \frac{\alpha_j^{n_j}}{\prod_{\ell=1}^{n_j} \mu_j(\ell)} \\
&= \tilde{b}^{-1} \prod_{j=0}^{N} \pi_j(n_j), \qquad \text{for } \sum_{j=0}^{N} n_j = M,
\end{aligned}$$

where

$$\pi_j(n_j) = \frac{\alpha_j^{n_j}}{\prod_{\ell=1}^{n_j} \mu_j(\ell)}, \qquad j = 0, 1, \ldots, N,$$

with $\alpha_0 = 1$ and the normalization constant is now given by

$$\tilde{b} = \sum_{n_0 + \cdots + n_N = M} \prod_{j=0}^{N} \pi_j(n_j).$$

Note that all nodes including node 0 are quasi-reversible in this network. Hence the product form result can also be obtained by applying the proof of Theorem 4.3 (see Remark 4.4). □

Theorem 9.1 states that the forms of the service rates (1.4) and arrival rates (1.5), for some arbitrarily defined functions $\Phi(\boldsymbol{n})$ and $\Psi(\boldsymbol{n})$, are sufficient for the stationary measure (1.8) to satisfy the local balance equations (1.6) and (1.7). The following result states that these forms are also necessary. Assume that, when the state of the network is $\boldsymbol{n} = (n_1, \ldots, n_N)$, the departure rate from node j is $\mu_j(\boldsymbol{n})$, which depends on the state of the network, $j = 0, 1, \ldots, N$. Node 0 is the outside and $\mu_0(\boldsymbol{n})$ denotes the arrival rate at the network when the state is $\boldsymbol{n}$. Assume that a departure from node j joins node k with probability r_{jk} for $j, k = 0, 1, 2, \ldots, N$. Under this notation, the local balance equations (1.6) and (1.7) can be written as

$$\pi(\boldsymbol{n} + \mathbf{e}_j)\mu_j(\boldsymbol{n} + \mathbf{e}_j) = \sum_{k=0}^{N} \pi(\boldsymbol{n} + \mathbf{e}_k)\mu_k(\boldsymbol{n} + \mathbf{e}_k)r_{kj}, \quad j = 1, 2, \ldots, N. \tag{1.9}$$

Theorem 9.4 *Assume that the routing matrix $\{r_{jk}\}$ is irreducible and α_j is the solution to the traffic equations*

$$\alpha_j = \mu_0 r_{0j} + \sum_{k=1}^{N} \alpha_k r_{kj}, \qquad j = 1, 2, \ldots, N, \tag{1.10}$$

where μ_0 is a positive constant. A necessary and sufficient condition for the stationary measure of the network to satisfy the local balance equations (1.9) is that the arrival and service rates $\mu_j(\boldsymbol{n})$ are of the form

$$\mu_j(\boldsymbol{n}) = \frac{\Psi(\boldsymbol{n} - \mathbf{e}_j)}{\Phi(\boldsymbol{n})}\mu_j, \qquad j = 0, 1, \ldots, N. \tag{1.11}$$

Under this condition the stationary measure π of the network is given by (1.8).

PROOF. Sufficiency has been proved in Theorem 9.1. To prove necessity, divide both sides of (1.9) by $\pi(\boldsymbol{n})\mu_0(\boldsymbol{n})$:

$$\frac{\pi(\boldsymbol{n} + \mathbf{e}_j)\mu_j(\boldsymbol{n} + \mathbf{e}_j)}{\pi(\boldsymbol{n})\mu_0(\boldsymbol{n})} = r_{0j} + \sum_{k=1}^{N} \frac{\pi(\boldsymbol{n} + \mathbf{e}_k)\mu_k(\boldsymbol{n} + \mathbf{e}_k)}{\pi(\boldsymbol{n})\mu_0(\boldsymbol{n})} r_{kj},$$
$$j = 0, 1, \ldots, N.$$

Since the routing matrix is irreducible, the traffic equations (1.10) have a unique solution. This implies

$$\frac{\pi(\boldsymbol{n} + \mathbf{e}_j)\mu_j(\boldsymbol{n} + \mathbf{e}_j)}{\pi(\boldsymbol{n})\mu_0(\boldsymbol{n})} = \frac{\alpha_j}{\mu_0}, \qquad j = 1, \ldots, N.$$

Define the function Ψ as

$$\Psi(\boldsymbol{n}) = \pi(\boldsymbol{n})\frac{\mu_0(\boldsymbol{n})}{\mu_0}\prod_{j=1}^{N}\left(\frac{\mu_j}{\alpha_j}\right)^{n_j},$$

then

$$\mu_j(\boldsymbol{n}+\mathbf{e}_j) = \frac{\Psi(\boldsymbol{n})}{\pi(\boldsymbol{n}+\mathbf{e}_j)}\alpha_j\prod_{j=1}^{N}\left(\frac{\alpha_j}{\mu_j}\right)^{n_j}.$$

Thus if we define the function Φ by

$$\Phi(\boldsymbol{n}) = \pi(\boldsymbol{n})\prod_{j=1}^{N}\left(\frac{\mu_j}{\alpha_j}\right)^{n_j},$$

then the arrival and departure rates take the form (1.11), and the stationary measure is given by (1.8). □

Remark 9.5 To present Theorem 9.4 in the form of Theorem 9.1, we only need to set $\lambda_j = \mu_0 r_{0j}$.

The same argument can be applied to networks with multiple classes of customers. The state for the multiple-class case is $\boldsymbol{n} = (\boldsymbol{n}_1, \ldots, \boldsymbol{n}_N)$, where $\boldsymbol{n}_j = (n_{j1}, \ldots, n_{jI})$ and n_{ju} is the number of class u customers at node j. When the state of the network is $\boldsymbol{n}$, class u customers arrive at node j at rate

$$\lambda_{ju}(\boldsymbol{n}) = \lambda_{ju}\frac{\Psi(\boldsymbol{n})}{\Phi(\boldsymbol{n})}, \tag{1.12}$$

and the service rate of class u customers at node j is

$$\mu_{ju}(\boldsymbol{n}) = \mu_{ju}\frac{\Psi(\boldsymbol{n}-\mathbf{e}_{ju})}{\Phi(\boldsymbol{n})}, \tag{1.13}$$

where $\mathbf{e}_{ju}$ is an $(I \times N)$-dimensional vector with a 1 in position ju, referring to a class u customer at node j, and a 0 everywhere else. The customer routing probabilities are defined as follows. When a class u customer completes its service at node j, it joins node k as a class v customer with probability $r_{ju,kv}$, and it leaves the network with probability $r_{ju,0}$, where

$$\sum_{k=1}^{N}\sum_{v=1}^{I} r_{ju,kv} + r_{ju,0} = 1, \qquad j = 1, \ldots, N,\ u = 1, \ldots, I.$$

Theorem 9.6 *If α_{ju} is the solution to the traffic equations*

$$\alpha_{ju} = \lambda_{ju} + \sum_{k=1}^{N} \sum_{v=1}^{I} \alpha_{kv} r_{kv,ju}, \qquad j = 1, \ldots, N,\ u = 1, \ldots, I,$$

then the stationary measure of the network is

$$\pi(\boldsymbol{n}) = \Phi(\boldsymbol{n}) \prod_{j=1}^{N} \prod_{u=1}^{I} \left(\frac{\alpha_{ju}}{\mu_{ju}} \right)^{n_{ju}}.$$

A result similar to Theorem 9.4 can also be obtained for networks with multiple classes of customers. However, in that case additional information is needed with regard to how a customer circulates and how it changes its class. This will be discussed in Chapter 12 within a more general framework.

9.2 LINEAR NETWORKS WITH BATCH MOVEMENTS

The result of the previous section easily extends to networks with batch movements. Let $\mathcal{A}$ be a set of $(N + 1)$-dimensional vectors, which denotes the set of batches that may move together. In the single movement case discussed in the previous section, only one customer can depart from any node, including node 0 (the outside). Hence in the single movement case $\mathcal{A} = \{\bar{\mathbf{e}}_0, \bar{\mathbf{e}}_1, \ldots, \bar{\mathbf{e}}_N\}$, where $\bar{\mathbf{e}}_j$ is the $(N+1)$-dimensional vector with a 1 in position j and a 0 everywhere else, including node 0. Here we use $\bar{\mathbf{e}}_j$ instead of $\mathbf{e}_j$ since $\mathbf{e}_j$ has been used throughout the book as the unit vector excluding node 0, whereas $\bar{\mathbf{e}}_j$ includes node 0.

Let n_j be the number of customers at node j, $j = 1, \ldots, N$, and let $\boldsymbol{n} = (n_1, \ldots, n_N)$ be the state of the network with state space $\mathcal{S}$. The following is the system dynamics when the state of the network is $\boldsymbol{n}$.

(i) The departure rate of a vector $\bar{\boldsymbol{a}} \in \mathcal{A}$ is $\mu(\boldsymbol{n}, \bar{\boldsymbol{a}})$, where

$$\bar{\boldsymbol{a}} = (a_0, a_1, \ldots, a_N),$$

and a_j is the number of departures from node j, $j = 0, 1, \ldots, N$. The first component, a_0, is the number of departures from node 0 (the number of arrivals at the network from the outside).

(ii) The departure vector $\bar{\boldsymbol{a}}$ is transformed into another $(N + 1)$-dimensional arrival vector $\bar{\boldsymbol{a}}' = (a'_0, a'_1, \ldots, a'_N)$ with probability $r(\bar{\boldsymbol{a}}, \bar{\boldsymbol{a}}')$, where a'_j is the number of arrivals at node j, $j = 0, 1, \ldots, N$, and a'_0, which is the number of arrivals at node 0, is the number of customers leaving the network. In this section we assume

$$\sum_{\bar{\boldsymbol{a}}' \in \mathcal{A}} r(\bar{\boldsymbol{a}}, \bar{\boldsymbol{a}}') = 1, \qquad \bar{\boldsymbol{a}} \in \mathcal{A}, \tag{2.1}$$

$$\sum_{j=0}^{N} a_j = \sum_{j=0}^{N} a'_j. \tag{2.2}$$

For a batch $\bar{\boldsymbol{a}} = (a_0, a_1, \ldots, a_N) \in \mathcal{A}$, we use $\boldsymbol{a}$ to denote the vector $\bar{\boldsymbol{a}}$ with the first element removed, i.e.,

$$\boldsymbol{a} = (a_1, \ldots, a_N).$$

Define a Markov chain $(\boldsymbol{n}, \bar{\boldsymbol{a}})$ with transition probabilities

$$p((\boldsymbol{n}, \bar{\boldsymbol{a}}), (\boldsymbol{n}', \bar{\boldsymbol{a}}')) = r(\bar{\boldsymbol{a}}, \bar{\boldsymbol{a}}') 1[\boldsymbol{n}' = \boldsymbol{n} - \boldsymbol{a} + \boldsymbol{a}'].$$

It is clear that state $(\boldsymbol{n}, \bar{\boldsymbol{a}})$ communicates with state $(\boldsymbol{n}, \bar{\boldsymbol{a}}')$ only if

$$\boldsymbol{n} + \boldsymbol{a} = \boldsymbol{n}' + \boldsymbol{a}'$$

and (2.2) is satisfied. For simplicity, assume the class of states communicating with $(\boldsymbol{n}, \bar{\boldsymbol{a}})$ to be irreducible, and let $\langle \boldsymbol{n}, \bar{\boldsymbol{a}} \rangle$ denote the class of states that communicate with $(\boldsymbol{n}, \bar{\boldsymbol{a}})$. Let $\nu(\bar{\boldsymbol{a}})$ be the solution to the traffic equations

$$\nu(\bar{\boldsymbol{a}}) = \sum_{\bar{\boldsymbol{a}}'} \nu(\bar{\boldsymbol{a}}') r(\bar{\boldsymbol{a}}', \bar{\boldsymbol{a}}), \qquad \bar{\boldsymbol{a}} \in \mathcal{A}. \tag{2.3}$$

The following result is an extension of Theorem 9.4.

Theorem 9.7 *The stationary measure of the network π satisfies the local balance equations*

$$\pi(\boldsymbol{n}) \mu(\boldsymbol{n}, \bar{\boldsymbol{a}}) = \sum_{\boldsymbol{n}', \bar{\boldsymbol{a}}'} \pi(\boldsymbol{n}') \mu(\boldsymbol{n}', \bar{\boldsymbol{a}}') r(\bar{\boldsymbol{a}}', \bar{\boldsymbol{a}}), \tag{2.4}$$

if and only if there exist a nonnegative function Ψ and a positive measure Φ such that the departure functions take the form

$$\mu(\boldsymbol{n}, \bar{\boldsymbol{a}}) = \frac{\Psi(\langle \boldsymbol{n}, \bar{\boldsymbol{a}} \rangle)}{\Phi(\boldsymbol{n})} \nu(\bar{\boldsymbol{a}}), \qquad \bar{\boldsymbol{a}} \in \mathcal{A}, \boldsymbol{n} \in \mathcal{S}, \tag{2.5}$$

where ν is the stationary measure of the traffic equations (2.3). When this is true, the stationary measure of the network is

$$\pi(\boldsymbol{n}) = \Phi(\boldsymbol{n}).$$

PROOF. Since the routing probabilities $r(\bar{\boldsymbol{a}}, \bar{\boldsymbol{a}}')$ correspond to an irreducible Markov chain, it has a unique stationary measure up to a multiplication factor. If the local balance is satisfied, then

$$\frac{\pi(\boldsymbol{n})\mu(\boldsymbol{n}, \bar{\boldsymbol{a}})}{\nu(\bar{\boldsymbol{a}})} = \frac{\pi(\boldsymbol{n}')\mu(\boldsymbol{n}', \bar{\boldsymbol{a}}')}{\nu(\bar{\boldsymbol{a}}')} \tag{2.6}$$

for any $(\boldsymbol{n}', \bar{\boldsymbol{a}}') \in \langle \boldsymbol{n}, \bar{\boldsymbol{a}} \rangle$. We denote this quantity by $\Psi(\langle \boldsymbol{n}, \bar{\boldsymbol{a}} \rangle)$ and obtain (2.5) by letting $\Phi = \pi$.

On the other hand, if $\mu(\boldsymbol{n}, \bar{\boldsymbol{a}})$ takes the form (2.5), then $\pi(\boldsymbol{n})\mu(\boldsymbol{n}, \bar{\boldsymbol{a}})$ clearly satisfies the local balance equations (2.4) provided $\pi(\boldsymbol{n})$ is defined as $\Phi(\boldsymbol{n})$. Summing (2.4) over all $\bar{\boldsymbol{a}}$ yields the global balance. This shows that π is indeed a stationary measure for the network and it satisfies the local balance equations. □

In the the local balance (2.4), the rate $\pi(\boldsymbol{n})\mu(\boldsymbol{n}, \bar{\boldsymbol{a}})$ can be interpreted as the arrival as well as departure rates of batches $\bar{\boldsymbol{a}}$ when the state of the network is $\boldsymbol{n}$, and the local balance equations can be interpreted as linear traffic equations. Because of this, a batch movement network is said to be *linear* or to have linear traffic equations if it satisfies the local balance (2.4). A complete characterization of linear networks with batch movements will be discussed for discrete time models in Section 12.8.

Since any $(\boldsymbol{n}', \bar{\boldsymbol{a}}') \in \langle \boldsymbol{n}, \bar{\boldsymbol{a}} \rangle$ satisfies $\boldsymbol{n} - \boldsymbol{a} = \boldsymbol{n}' - \boldsymbol{a}'$ and $|\bar{\boldsymbol{a}}| = |\bar{\boldsymbol{a}}'|$, where $|\bar{\boldsymbol{a}}| = \sum_{i=0}^{N} a_i$, an example of Ψ is

$$\Psi(\langle \boldsymbol{n}, \bar{\boldsymbol{a}} \rangle) = f(\boldsymbol{n} - \boldsymbol{a}, |\bar{\boldsymbol{a}}|),$$

where f is a nonnegative function. An even simpler case of Ψ is

$$\Psi(\langle \boldsymbol{n}, \bar{\boldsymbol{a}} \rangle) = g(\boldsymbol{n} - \boldsymbol{a})$$

for some nonnegative function g.

An important special case is the one in which the solution ν of the traffic equations takes the product form

$$\nu(\bar{\boldsymbol{a}}) = \prod_{j=0}^{N} \nu_j(a_j) \tag{2.7}$$

for some positively valued functions ν_j. This is the case in most of the linear networks discussed earlier in this book, e.g., in Jackson networks $\nu(\bar{\boldsymbol{a}}) = \prod_{j=0}^{N} \rho_j^{a_j}$ and $\rho_0 = 1$. In this special case

$$a_0 = |\bar{\boldsymbol{a}}| - |\boldsymbol{a}|$$

and

$$\nu(\bar{\boldsymbol{a}}) = \frac{\nu(\boldsymbol{n})}{\nu(\boldsymbol{n} - \boldsymbol{a})} \nu_0(a_0).$$

Since $\nu(\boldsymbol{n})$ can be included in $\Phi(\boldsymbol{n})$ and $\nu(\boldsymbol{n} - \boldsymbol{a})$ can be included in $\Psi(\langle \boldsymbol{n}, \bar{\boldsymbol{a}} \rangle)$, Theorem 9.7 states that if the departure rates take the form

$$\mu(\boldsymbol{n}, \bar{\boldsymbol{a}}) = \frac{\Psi(\langle \boldsymbol{n}, \bar{\boldsymbol{a}} \rangle)}{\Phi(\boldsymbol{n})} \nu_0(a_0),$$

then the stationary measure of the network is

$$\pi(\boldsymbol{n}) = \Phi(\boldsymbol{n}) \nu(\boldsymbol{n}).$$

In all the applications of linear networks discussed in this book $\nu_0(a_0) \equiv 1$ so the theorem is even further simplified: if the departure rates of the network take the form

$$\mu(\boldsymbol{n}, \bar{\boldsymbol{a}}) = \frac{\Psi(\langle \boldsymbol{n}, \bar{\boldsymbol{a}} \rangle)}{\Phi(\boldsymbol{n})}, \tag{2.8}$$

then the stationary measure is

$$\pi(\boldsymbol{n}) = \Phi(\boldsymbol{n}) \prod_{j=1}^{N} \nu_j(n_j). \tag{2.9}$$

Thus, we have the following result.

Corollary 9.8 *If the solution ν of traffic equations (2.3) has the form (2.7) with $\nu_0(a_0) \equiv 1$, then the network with the departure function (2.8) has the stationary measure π given by (2.9).*

The following two examples illustrate the use of Theorem 9.7.

Example 9.9 (Clustering process) Suppose there are N possible cluster types, labeled $i = 1, 2, \ldots, N$. Two clusters may merge and form a single cluster or a cluster may break up into two separate clusters. Let n_i be the number of type i clusters, and $\boldsymbol{n} = (n_1, n_2, \ldots, n_N)$. The transition rates of this process are

$$\begin{aligned} q(\boldsymbol{n}, \boldsymbol{n} - \mathbf{e}_i - \mathbf{e}_j + \mathbf{e}_k) &= \lambda_{ij,k} \frac{\Psi(\boldsymbol{n} - \mathbf{e}_i - \mathbf{e}_j)}{\Phi(\boldsymbol{n})}, \\ q(\boldsymbol{n}, \boldsymbol{n} - \mathbf{e}_k + \mathbf{e}_i + \mathbf{e}_j) &= \lambda_{k,ij} \frac{\Psi(\boldsymbol{n} - \mathbf{e}_k)}{\Phi(\boldsymbol{n})}, \\ q(\boldsymbol{n}, \boldsymbol{n} - \mathbf{e}_i + \mathbf{e}_j) &= \lambda_{i,j} \frac{\Psi(\boldsymbol{n} - \mathbf{e}_i)}{\Phi(\boldsymbol{n})}, \end{aligned} \tag{2.10}$$

where the first transition represents the merging of two clusters, and the second represents the breakup of one cluster into two, and the last transition corresponds to the transformation of one cluster into another cluster. The parameter $\lambda_{ij,k}$ may be regarded as the rate at which a type i cluster attaches itself to a type j cluster to become a type k cluster, and $\lambda_{ii,k}$ may be regarded as the rate at which two type i clusters react to produce a type k cluster. Similarly, $\lambda_{k,ij}$ can be regarded as the rate at which a type k cluster breaks up into a type i cluster and a type j cluster, and $\lambda_{i,j}$ is the rate at which a type i cluster transforms itself into a type j cluster. We call this process a *clustering process*.

For instance, when Ψ and Φ take the form

$$\Phi(\boldsymbol{n}) = \Psi(\boldsymbol{n}) = \frac{1}{\prod_{i=1}^{N} n_i!},$$

the transition rates are

$$\begin{aligned} q(\boldsymbol{n}, \boldsymbol{n} - \mathbf{e}_i - \mathbf{e}_j + \mathbf{e}_k) &= \lambda_{ij,k} n_i n_j, \\ q(\boldsymbol{n}, \boldsymbol{n} - \mathbf{e}_k + \mathbf{e}_i + \mathbf{e}_j) &= \lambda_{k,ij} n_k, \\ q(\boldsymbol{n}, \boldsymbol{n} - \mathbf{e}_i + \mathbf{e}_j) &= \lambda_{i,j} n_i. \end{aligned} \tag{2.11}$$

To describe the clustering process in our formulation, we normalize $\lambda_{i,j}$, $\lambda_{ij,k}$ and $\lambda_{k,ij}$. Let the departure rates be defined as

$$\begin{aligned} \overline{\lambda}_i &= \sum_{j=1}^{N} \lambda_{i,j} + \sum_{j=1}^{N} \sum_{k=1}^{N} \lambda_{i,jk}, \\ \overline{\lambda}_{ij} &= \sum_{k=1}^{N} \lambda_{ij,k}, \end{aligned}$$

and the routing probabilities as

$$r_{i,j} = \lambda_{i,j}/\overline{\lambda}_i, \qquad r_{i,jk} = \lambda_{i,jk}/\overline{\lambda}_i, \qquad r_{ij,k} = \lambda_{ij,k}/\overline{\lambda}_{ij}.$$

Now, if we let $\alpha_i = \nu(\mathbf{e}_i)$ and $\alpha_{ij} = \nu(\mathbf{e}_i + \mathbf{e}_j)$, then the traffic equations (2.3) take the form

$$\alpha_{ij} = \sum_{k=1}^{N} \alpha_k r_{k,ij}, \qquad i, j = 1, \ldots, N, \tag{2.12}$$

$$\alpha_k = \sum_{i=1}^{N}\sum_{j=1}^{N} \alpha_{ij} r_{ij,k} + \sum_{j=1}^{N} \alpha_j r_{j,k}, \qquad k = 1, \ldots, N. \tag{2.13}$$

These equations have a solution $\{\alpha_i, \alpha_{ij};\, i, j = 1, 2, \ldots, N\}$ that is unique up to a multiplication factor, provided the routing transitions are irreducible.

Note that the communication class $\langle \boldsymbol{n} + \mathbf{e}_i, \mathbf{e}_i \rangle$ is given by

$$\langle \boldsymbol{n} + \mathbf{e}_i, \mathbf{e}_i \rangle = \{(\boldsymbol{n} + \mathbf{e}_j, \mathbf{e}_j), (\boldsymbol{n} + \mathbf{e}_j + \mathbf{e}_k, \mathbf{e}_j + \mathbf{e}_k);\quad j, k = 1, 2, \ldots, N\}.$$

Thus, each $\boldsymbol{n}$ represents one communication class, so Ψ is a function of $\boldsymbol{n}$. This means that the setting of (2.10) has the form of the departure function (2.5), provided the following conditions are satisfied for some functions f and g of $\boldsymbol{n}$:

$$\begin{aligned} \overline{\lambda}_{ij} &= \frac{f(\boldsymbol{n} - \mathbf{e}_i - \mathbf{e}_j)}{g(\boldsymbol{n})} \alpha_{ij}, \\ \overline{\lambda}_i &= \frac{f(\boldsymbol{n} - \mathbf{e}_i)}{g(\boldsymbol{n})} \alpha_i . \end{aligned} \tag{2.14}$$

From these equations, we have

$$\begin{aligned} f(\boldsymbol{n} - \mathbf{e}_i) &= \frac{\overline{\lambda}_i \alpha_{ij}}{\alpha_i \overline{\lambda}_{ij}} f(\boldsymbol{n} - \mathbf{e}_i - \mathbf{e}_j), \\ g(\boldsymbol{n} + \mathbf{e}_i + \mathbf{e}_j) &= \frac{\overline{\lambda}_i \alpha_{ij}}{\alpha_i \overline{\lambda}_{ij}} g(\boldsymbol{n} + \mathbf{e}_i). \end{aligned}$$

For these to be true for all $\boldsymbol{n}$, f and g must take the form

$$\frac{f(\boldsymbol{n})}{f(\mathbf{0})} = \frac{g(\boldsymbol{n})}{g(\mathbf{0})} = \prod_{j=1}^{N} \beta_j^{n_j},$$

where $\beta_j = \overline{\lambda}_i \alpha_{ij} / (\alpha_i \overline{\lambda}_{ij})$. Substituting these f and g into (2.14) shows that the departure function has the form (2.5) if and only if there exist β_j's satisfying

$$\begin{aligned} \overline{\lambda}_i &= c\alpha_i / \beta_i, & i &= 1, 2, \ldots, N, \\ \overline{\lambda}_{ij} &= c\alpha_{ij} / (\beta_i \beta_j), & i, j &= 1, 2, \ldots, N, \end{aligned}$$

where $c = f(\mathbf{0})/g(\mathbf{0})$. From the first equation, it follows that $\beta_i = c\alpha_i/\overline{\lambda}_i$. Hence these conditions can be reduced to

$$\overline{\lambda}_{ij} = \frac{\alpha_{ij}}{c\alpha_i\alpha_j}\overline{\lambda}_i\overline{\lambda}_j\,, \qquad i, j = 1, 2, \ldots, N. \tag{2.15}$$

This implies that we cannot freely choose $\lambda_{ij,k}$ in this model to apply Theorem 9.7. The departure rate functions (2.5) are given by

$$\begin{aligned}\mu(\boldsymbol{n}, \mathbf{e}_i + \mathbf{e}_j) &= \frac{\Psi(\boldsymbol{n} - \mathbf{e}_i - \mathbf{e}_j) f(\boldsymbol{n} - \mathbf{e}_i - \mathbf{e}_j)}{g(\boldsymbol{n})\Phi(\boldsymbol{n})}\alpha_{ij}\,, \\ \mu(\boldsymbol{n}, \mathbf{e}_k) &= \frac{\Psi(\boldsymbol{n} - \mathbf{e}_k) f(\boldsymbol{n} - \mathbf{e}_k)}{g(\boldsymbol{n})\Phi(\boldsymbol{n})}\alpha_k\,.\end{aligned}$$

Thus, it follows from Theorem 9.7 that the clustering process satisfies the local balance equations (2.4) if and only if condition (2.15) is satisfied for some positive constant c. In this case, the stationary measure π is given by

$$\pi(\boldsymbol{n}) = \Phi(\boldsymbol{n}) \prod_{j=1}^{N} \left(\frac{c\alpha_j}{\overline{\lambda}_j}\right)^{n_j}. \tag{2.16}$$

Note that if we let $\hat{\alpha}_j = c\alpha_j/\overline{\lambda}_j$, then the traffic equations (2.12) and (2.13) can be written as

$$\hat{\alpha}_i\hat{\alpha}_j \sum_{k=1}^{N} \lambda_{ij,k} = \sum_{k=1}^{N} \hat{\alpha}_k\lambda_{k,ij}, \quad i, j = 1, \ldots, N, \tag{2.17}$$

$$\hat{\alpha}_k \left(\sum_{i=1}^{N}\sum_{j=1}^{N} \lambda_{k,ij} + \sum_{j=1}^{N} \lambda_{k,j}\right) = \sum_{i=1}^{N}\sum_{j=1}^{N} \hat{\alpha}_i\hat{\alpha}_j\lambda_{ij,k} + \sum_{j=1}^{N} \hat{\alpha}_j\lambda_{j,k}, \quad k = 1, \ldots, N. \tag{2.18}$$

Thus, condition (2.15) is equivalent to the traffic equations (2.17) and (2.18) having a solution. Obviously, the former condition is more informative than the latter, since α_i and α_{ij} are uniquely determined by the linear equations (2.12) and (2.13) up to a multiplication factor. In other words, condition (2.15) specifies the precise constraints that have to be satisfied with regard to the network parameters in order for the results to hold. □

Example 9.10 (A circuit switching network) Suppose there are N nodes, labeled $1, 2, \ldots, N$, in a telecommunication network. Node j has a limited number of resources, say C_j. They are, for example, links of telecommunication lines or switches. Calls are distinguished by classes, labeled by $u = 1, 2, \ldots, I$. When a class u call arrives at the network, it simultaneously requests $a_j(u)$ resources from node j, $j = 1, 2, \ldots, N$. We use the vector

$$\boldsymbol{a}(u) = (a_1(u), a_2(u), \ldots, a_N(u))$$

to represent the resource requirement of a class u call. If there is a sufficiently large number of resources available in the network, the call is accepted and the number of available resources at node j is reduced by $a_j(u)$, $j = 1, \ldots, N$. Otherwise, the call is rejected. After a call completes its service, it leaves the system and returns the occupied resources to the network.

Let n_u be the number of class u calls in the network, and

$$\boldsymbol{n} = \{n_u;\, u = 1, 2, \ldots, I\}.$$

Let a_j be the available resources at node j, and

$$\boldsymbol{a} = (a_1, a_2, \ldots, a_N).$$

We define $\bar{\boldsymbol{n}} = (\boldsymbol{n}, \boldsymbol{a})$ as the state of the network. Clearly the vector $\boldsymbol{a}$ is uniquely determined by $\boldsymbol{n}$ through the equations

$$a_j = C_j - \sum_{u=1}^{I} n_u a_j(u), \qquad j = 1, \ldots, N.$$

Assume class u calls arrive at the network according to a state-dependent process with rate $\lambda(n) p_u$, where $\{p_u\}_{u=1}^{I}$ is a probability mass function, and n is the total number of calls in the network, i.e.,

$$n = \sum_{u=1}^{I} n_u.$$

This implies that the total arrival rate of calls is $\lambda(n)$, and each call is of class u with probability p_u. Furthermore, assume that the class u calls are processed at rate $\mu_u(n_u)$ when there are n_u class u calls present, where μ_u is an arbitrary function such that $\mu_u(0) = 0$ and $\mu_u(\ell) > 0$ for $\ell \geq 1$.

Because of the capacity constraints on the resources, the network state space $\mathcal{S}$ is

$$\mathcal{S} = \left\{ \bar{\boldsymbol{n}} = (\boldsymbol{n}, \boldsymbol{a}) \geq \boldsymbol{0};\;\; a_j = C_j - \sum_{u=1}^{I} n_u a_j(u) \geq 0,\, j = 1, 2, \ldots, N \right\}.$$

This network is referred to as a *circuit switching network with state-dependent arrivals*.

We formulate this network as a linear network. Since there are only two types of batch movements, arrivals from the outside and departures to the outside, the routing function r is defined as

$$r(\bar{\boldsymbol{a}}(u), \bar{\mathbf{e}}_u) = r(\bar{\mathbf{e}}_u, \bar{\boldsymbol{a}}(u)) = 1, \qquad u = 1, 2, \ldots, N,$$

where $\bar{\mathbf{e}}_u = (\mathbf{e}_u, \mathbf{0}_N)$, $\bar{\boldsymbol{a}}(u) = (\mathbf{0}_I, \boldsymbol{a}(u))$, and $\mathbf{0}_\ell$ is the ℓ-dimensional zero vector. It is easy to verify that $\nu(\bar{\mathbf{e}}_u) = \nu(\bar{\boldsymbol{a}}(u)) = 1$, $u = 1, \ldots, I$, is the stationary measure of traffic equations (2.3), and

$$\{(\bar{\boldsymbol{n}} + \bar{\mathbf{e}}_u, \bar{\mathbf{e}}_u), (\bar{\boldsymbol{n}} + \bar{\boldsymbol{a}}(u), \bar{\boldsymbol{a}}(u))\}$$

is a closed communication class for each $\boldsymbol{n} \in \mathcal{S}$ and each $u = 1, 2, \ldots, I$. Hence, the communication classes are specified by $\boldsymbol{n}$ and u. From Theorem 9.6, the network is linear if the departure functions take the forms

$$\mu(\bar{\boldsymbol{n}}, \bar{\mathbf{e}}_u) = \frac{\Psi(\bar{\boldsymbol{n}} - \bar{\mathbf{e}}_u, u)}{\Phi(\bar{\boldsymbol{n}})}, \qquad u = 1, 2, \ldots, N, \ \bar{\boldsymbol{n}} \in \mathcal{S}, \tag{2.19}$$

$$\mu(\bar{\boldsymbol{n}}, \bar{\boldsymbol{a}}(u)) = \frac{\Psi(\bar{\boldsymbol{n}} - \bar{\boldsymbol{a}}(u), u)}{\Phi(\bar{\boldsymbol{n}})}, \qquad u = 1, 2, \ldots, N, \ \bar{\boldsymbol{n}} \in \mathcal{S}, \tag{2.20}$$

where Ψ and Φ are arbitrary nonnegative and positive functions, respectively. Note that the rates (2.19) and (2.20) represent the departure rate and arrival rate of class u calls when the network is in state $\bar{\boldsymbol{n}}$.

Define Ψ and Φ as

$$\Psi((\boldsymbol{n}, \boldsymbol{a}), u) = 1[\boldsymbol{a} \geq \mathbf{0}] p_u \prod_{\ell=1}^{n+1} \lambda(\ell) \prod_{v=1}^{I} \prod_{m=1}^{n_v} \frac{p_v}{\mu_v(m)},$$

$$\Phi((\boldsymbol{n}, \boldsymbol{a})) = \prod_{\ell=1}^{n} \lambda(\ell) \prod_{v=1}^{I} \prod_{m=1}^{n_v} \frac{p_v}{\mu_v(m)}.$$

Substituting these functions into (2.19) and (2.20) yields

$$\mu(\bar{\boldsymbol{n}}, \bar{\mathbf{e}}_u) = \mu_u(n_u),$$

$$\mu(\bar{\boldsymbol{n}}, \bar{\boldsymbol{a}}(u)) = \lambda(n) 1[\boldsymbol{a} - \boldsymbol{a}(u) \geq \mathbf{0}].$$

This is precisely the circuit switching network we have introduced. Thus it follows from Theorem 9.6 that the stationary measure of the network is

$$\pi((\boldsymbol{n}, \boldsymbol{a})) = \Phi((\boldsymbol{n}, \boldsymbol{a})) = \prod_{\ell=1}^{n} \lambda(\ell) \prod_{v=1}^{I} \prod_{m=1}^{n_v} \frac{p_v}{\mu_v(m)}.$$

A typical choice of μ_u is

$$\mu_u(\ell) = \ell\mu_u 1[\ell \geq 1], \qquad \ell = 0, 1, \ldots.$$

In this case, each class u call is processed independently at rate μ_u, i.e., each node is an infinite-server queue. Since an infinite-server queue has a symmetric service discipline, the service time distribution for each class of calls can be arbitrary as long as it has a finite mean.

Clearly, a more general state-dependent circuit switching network can be obtained by varying the functions Ψ and Φ in (2.19) and (2.20). □

9.3 NONLINEAR NETWORKS WITH STATE-DEPENDENT TRANSITIONS

The approach used in Section 9.1 for networks with state-dependent transitions can be extended to networks with signals and instantaneous movements, i.e., nonlinear networks. In contrast to linear networks where we used local balance equations, we have to use another type of local balance for nonlinear networks, called the *cross local balance*, which will be discussed in a unified framework in Chapter 10. In this and the two subsequent sections we present the state-dependent versions of the network models discussed in earlier chapters.

In this section we illustrate the use of the cross local balance through the simple nonlinear queueing network described in Example 4.7, i.e., the network with negative customers. The network has N nodes, and customers and negative customers arrive at node j from the outside according to Poisson processes with rates λ_j and λ_j^-, respectively. Denote a regular customer by c and a negative customer by c^-. Since arrivals from outside can be considered as departures from node 0, the total departure rate from node 0 is

$$q_{0c}^{\mathrm{D}}(0, 0) = \sum_{j=1}^{N}(\lambda_j + \lambda_j^-) \equiv \mu_0,$$

and each joins node j as a regular customer with probability $r_{0,jc}$ and as a negative customer with probability r_{0,jc^-}, where

$$r_{0,jc} = \frac{\lambda_j}{\mu_0}, \qquad j = 1, 2, \ldots, N,$$

$$r_{0,jc^-} = \frac{\lambda_j^-}{\mu_0}, \qquad j = 1, 2, \ldots, N.$$

The service rate at node j is μ_j, i.e.,

$$q^{\mathrm{D}}_{jc}(n_j, n_j - 1) = \mu_j 1[n_j > 0].$$

When a customer completes its service at node j, it joins node k as a regular customer with probability $r_{jc,kc}$, as a negative customer with probability r_{jc,kc^-}, and it leaves the network with probability $r_{jc,0}$. The arrival of a negative customer at node j removes one customer, provided one or more customers are present. Thus

$$p^{\mathrm{A}}_{jc}(n_j, n_j + 1) = 1,$$
$$p^{\mathrm{A}}_{jc^-}(n_j, n_j - 1) = 1[n_j > 0].$$

It is shown in Example 4.7 that the stationary distribution of this network is

$$\pi(\boldsymbol{n}) = \prod_{j=1}^{N}(1 - \rho_j)\rho_j^{n_j}, \tag{3.1}$$

where

$$\rho_j = \frac{\alpha_j}{\alpha_j^- + \mu_j}$$

and α_j and α_j^- are the solutions to the traffic equations

$$\alpha_j = \sum_{k=0}^{N} \mu_k \rho_k r_{kc,jc}, \qquad j = 1, \ldots, N, \tag{3.2}$$

$$\alpha_j^- = \sum_{k=0}^{N} \mu_k \rho_k r_{kc,jc^-}, \qquad j = 1, \ldots, N, \tag{3.3}$$

where we used the convention that $r_{0c,jc} = r_{0,jc}$ and $r_{0c,jc^-} = r_{0,jc^-}$. Substituting α_j and α_j^- into the relationship

$$\alpha_j^- + \mu_j = \rho_j^{-1}\alpha_j$$

yields

$$\mu_j + \sum_{k=0}^{N} \mu_k \rho_k r_{kc,jc^-} = \sum_{k=0}^{N} \mu_k \rho_k \rho_j^{-1} r_{kc,jc}.$$

Multiplying both sides of this equation by $\pi(\boldsymbol{n})$ of (3.1), we obtain the local balance equations

$$\begin{aligned}\pi(\boldsymbol{n})q^{\mathrm{D}}_{jc}(n_j, n_j-1)&\\ +\sum_{k=0}^{N}\pi(\boldsymbol{n}+\mathbf{e}_k)q^{\mathrm{D}}_{kc}(n_k+1[k\neq 0], n_k)r_{kc,jc^-}p^{\mathrm{A}}_{jc^-}(n_j, n_j-1)&\\ =\sum_{k=0}^{N}\pi(\boldsymbol{n}+\mathbf{e}_k-\mathbf{e}_j)q^{\mathrm{D}}_{kc}(n_k+1[k\neq 0], n_k)r_{kc,jc}p^{\mathrm{A}}_{jc}(n_j-1, n_j),&\quad (3.4)\end{aligned}$$

where $\mathbf{e}_0$ is the zero vector, i.e., $\mathbf{e}_0 = \mathbf{0}$. We refer to these equations as the cross local balance equations. These equations are different from the local balance equations defined in Section 2.3. They may be considered as a special case of the cross balance (1.10) of Chapter 2. Furthermore, summing equations (3.4) over $j = 1, 2, \ldots, N$ does not directly lead to a global balance. Hence, equations (3.4) may not be a full set of balance equations as described in Section 2.3. This type of local balance will be discussed in more detail, within the so-called *biased local balance*, in the next section.

We modify this network in such a way that it has state-dependent transitions in the following way:

$$\hat{q}^{\mathrm{D}}_{jc}(\boldsymbol{n}, \boldsymbol{n}-\mathbf{e}_j) = \frac{\Psi(\boldsymbol{n}-\mathbf{e}_j)}{\Phi(\boldsymbol{n})}\mu_j, \qquad j = 0, 1, \ldots, N, \tag{3.5}$$

where we have included $1[n_j > 0]$ in $\Psi(\boldsymbol{n})$ for $j \neq 0$, i.e., $\Psi(\boldsymbol{n}) = 0$ if $\boldsymbol{n} \not> \mathbf{0}$. Modify the probability that the arrival of a negative customer at node j removes one customer to

$$\hat{p}^{\mathrm{A}}_{jc^-}(\boldsymbol{n}, \boldsymbol{n}-\mathbf{e}_j) = \frac{\Psi(\boldsymbol{n}-\mathbf{e}_j)}{\Psi(\boldsymbol{n})}, \tag{3.6}$$

while the arrival effect probabilities of customers are unchanged, i.e.,

$$\hat{p}^{\mathrm{A}}_{jc}(\boldsymbol{n}, \boldsymbol{n}+\mathbf{e}_j) = 1. \tag{3.7}$$

Note that for $\Psi(\boldsymbol{n}-\mathbf{e}_j)/\Psi(\boldsymbol{n})$ to qualify as a probability, we have to assume that

(i) $\Psi(\boldsymbol{n})$ is nondecreasing.

This is a condition we impose on all state-dependent nonlinear traffic models with negative signals in this and the following chapters. The routing probabilities for the modified network remain the same.

If $j = 0$, the transition rates (3.5) imply that, when the network is in state $\boldsymbol{n}$, customers arrive at node j according to a state-dependent process with rate

$$\lambda_j \frac{\Psi(\boldsymbol{n})}{\Phi(\boldsymbol{n})},$$

and negative customers arrive at node j at rate

$$\lambda_j^- \frac{\Psi(\boldsymbol{n})}{\Phi(\boldsymbol{n})}.$$

Since the arrival of a negative customer at node j removes, by (3.6), one customer with probability $\Psi(\boldsymbol{n} - \mathbf{e}_j)/\Psi(\boldsymbol{n})$, the rate at which negative customers arrive at node j and change the state of the network from $\boldsymbol{n}$ to $\boldsymbol{n} - \mathbf{e}_j$ is

$$\lambda_j^- \frac{\Psi(\boldsymbol{n} - \mathbf{e}_j)}{\Phi(\boldsymbol{n})}. \tag{3.8}$$

Theorem 9.11 *The network with negative customers and state-dependent transition rates (3.5) to (3.7) has the stationary measure*

$$\hat{\pi}(\boldsymbol{n}) = \Phi(\boldsymbol{n})\pi(\boldsymbol{n}) = \Phi(\boldsymbol{n}) \prod_{j=1}^{N} (1 - \rho_j)\rho_j^{n_j}, \tag{3.9}$$

where $\rho_1, \ldots, \rho_N$ are the solutions to the traffic equations (3.2) and (3.3).

PROOF. We use two approaches to prove this result: one is to directly check the global balance equations, while the other is to use the cross local balance equations. The first approach, even for this simple model, is already cumbersome. In contrast with the global balance approach, the cross local balance approach is extendable to more general models. The first approach is helpful in the understanding of the model structure; the second approach explains how the cross local balance structure can be utilized to establish the global balance.

We first use the global balance approach. The rate going out of state $\boldsymbol{n}$ includes the transition rates due to the arrivals of effective negative customers from the outside, arrivals of regular customers from the outside, as well as service completions; while the rate going into state $\boldsymbol{n}$ includes the components due to exogenous arrivals of customers and effective negative customers, and regular service completions that join other nodes as regular customers, as effective negative customers, or as ineffective negative customers. Thus we obtain the following global balance equations:

$$\hat{\pi}(\boldsymbol{n}) \left(\sum_{j=1}^{N} \lambda_j^- \frac{\Psi(\boldsymbol{n} - \mathbf{e}_j)}{\Phi(\boldsymbol{n})} + \sum_{j=1}^{N} \lambda_j \frac{\Psi(\boldsymbol{n})}{\Phi(\boldsymbol{n})} + \sum_{j=1}^{N} \mu_j \frac{\Psi(\boldsymbol{n} - \mathbf{e}_j)}{\Phi(\boldsymbol{n})} \right)$$

$$
\begin{aligned}
&= \sum_{j=1}^{N} \hat{\pi}(\boldsymbol{n}-\mathbf{e}_j)\lambda_j \frac{\Psi(\boldsymbol{n}-\mathbf{e}_j)}{\Phi(\boldsymbol{n}-\mathbf{e}_j)} + \sum_{j=1}^{N} \hat{\pi}(\boldsymbol{n}+\mathbf{e}_j)\lambda_j^- \frac{\Psi(\boldsymbol{n})}{\Phi(\boldsymbol{n}+\mathbf{e}_j)} \\
&+ \sum_{k=1}^{N}\sum_{j=0}^{N} \hat{\pi}(\boldsymbol{n}+\mathbf{e}_k-\mathbf{e}_j)\mu_k \frac{\Psi(\boldsymbol{n}-\mathbf{e}_j)}{\Phi(\boldsymbol{n}+\mathbf{e}_k-\mathbf{e}_j)} r_{kc,jc} \\
&+ \sum_{k=1}^{N}\sum_{j=1}^{N} \hat{\pi}(\boldsymbol{n}+\mathbf{e}_k+\mathbf{e}_j)\mu_k \frac{\Psi(\boldsymbol{n}+\mathbf{e}_j)}{\Psi(\boldsymbol{n}+\mathbf{e}_k+\mathbf{e}_j)} r_{kc,jc^-} \frac{\Psi(\boldsymbol{n})}{\Psi(\boldsymbol{n}+\mathbf{e}_j)} \\
&+ \sum_{k=1}^{N}\sum_{j=1}^{N} \hat{\pi}(\boldsymbol{n}+\mathbf{e}_k)\mu_k \frac{\Psi(\boldsymbol{n})}{\Phi(\boldsymbol{n}+\mathbf{e}_k)} r_{kc,jc^-} \frac{\Psi(\boldsymbol{n})-\Psi(\boldsymbol{n}-\mathbf{e}_j)}{\Psi(\boldsymbol{n})}. \qquad (3.10)
\end{aligned}
$$

We show that the $\hat{\pi}$ of (3.9) satisfies these global balance equations. For convenience we drop the common term $\prod_{j=1}^{N} \rho_j^{n_j}$ from both the rates in and out of state $\boldsymbol{n}$. Substituting (3.9) into the left-hand side of (3.10) gives

$$
\sum_{j=1}^{N} [\lambda_j^- + \mu_j]\Psi(\boldsymbol{n}-\mathbf{e}_j) + \left[\sum_{j=1}^{N} \lambda_j\right]\Phi(\boldsymbol{n}), \qquad (3.11)
$$

and substituting $\hat{\pi}$ into the right-hand side of (3.10) yields

$$
\begin{aligned}
&\sum_{j=1}^{N} \left[\frac{1}{\rho_j}\left(\lambda_j + \sum_{k=1}^{K} \mu_k \rho_k r_{kc,jc}\right) - \sum_{k=1}^{N} \mu_k \rho_k r_{kc,jc^-}\right] \Psi(\boldsymbol{n}-\mathbf{e}_j) \\
&\quad + \left[\sum_{j=1}^{N} \lambda_j^- \rho_j + \sum_{k=1}^{N} \mu_k \rho_k r_{kc,0} \right. \\
&\quad \left. + \sum_{k=1}^{N}\sum_{j=1}^{N} \mu_k \rho_k \rho_j r_{kc,jc^-} + \sum_{k=1}^{N}\sum_{j=1}^{N} \mu_k \rho_k r_{kc,jc^-}\right] \Phi(\boldsymbol{n}). \qquad (3.12)
\end{aligned}
$$

That the terms in the first square brackets of expressions (3.11) and (3.12) are equal follows immediately from the traffic equations and the fact that $\rho_j = \alpha_j/(\alpha_j^- + \mu_j)$. Thus it suffices to prove that the terms in the second square brackets of the two expressions are also equal, i.e.,

$$
\begin{aligned}
\sum_{j=1}^{N} \lambda_j &= \sum_{j=1}^{N} \lambda_j^- \rho_j + \sum_{k=1}^{N} \mu_k \rho_k r_{kc,0} \\
&\quad + \sum_{k=1}^{N}\sum_{j=1}^{N} \mu_k \rho_k \rho_j r_{kc,jc^-} + \sum_{k=1}^{N}\sum_{j=1}^{N} \mu_k \rho_k r_{kc,jc^-}.
\end{aligned}
$$

This can be verified as follows:

$$\begin{aligned}
&\sum_{j=1}^{N} \lambda_j^- \rho_j + \sum_{k=1}^{N} \mu_k \rho_k r_{kc,0} + \sum_{k=1}^{N}\sum_{j=1}^{N} \mu_k \rho_k \rho_j r_{kc,jc^-} + \sum_{k=1}^{N}\sum_{j=1}^{N} \mu_k \rho_k r_{kc,jc^-} \\
&= \sum_{j=1}^{N} \lambda_j^- \rho_j + \sum_{k=1}^{N} \mu_k \rho_k \left(r_{kc,0} + \sum_{j=1}^{N} r_{kc,jc^-} \right) + \sum_{k=1}^{N}\sum_{j=1}^{N} \mu_k \rho_k \rho_j r_{kc,jc^-} \\
&= \sum_{j=1}^{N} \lambda_j^- \rho_j + \sum_{k=1}^{N} \mu_k \rho_k \left(1 - \sum_{j=1}^{N} r_{kc,jc} \right) + \sum_{k=1}^{N}\sum_{j=1}^{N} \mu_k \rho_k \rho_j r_{kc,jc^-} \\
&= \sum_{j=1}^{N} \rho_j \left(\lambda_j^- + \sum_{k=1}^{N} \mu_k \rho_j r_{kc,jc^-} + \mu_j \right) - \sum_{j=1}^{N}\sum_{k=1}^{N} \mu_k \rho_k r_{kc,jc^-} \\
&= \sum_{j=1}^{N} \alpha_j - \sum_{j=1}^{N}\sum_{k=1}^{N} \mu_k \rho_k r_{kc,jc^-} \\
&= \sum_{j=1}^{N} \lambda_j.
\end{aligned}$$

Thus (3.9) is indeed a stationary measure of the network. This completes the first proof of Theorem 9.11.

We now use the cross local balance equations to give a second proof for Theorem 9.11. It is easily checked that the local balance equations (3.4) also hold for the modified network with state-dependent transition rates if $\hat{\pi}$ of (3.9), $\hat{q}^{\mathrm{D}}_{ju}$ and $\hat{p}^{\mathrm{A}}_{ju}$ are used, i.e.,

$$\begin{aligned}
&\hat{\pi}(\boldsymbol{n})\hat{q}^{\mathrm{D}}_{jc}(\boldsymbol{n}, \boldsymbol{n} - \mathbf{e}_j) + \sum_{k=0}^{N} \hat{\pi}(\boldsymbol{n} + \mathbf{e}_k)\hat{q}^{\mathrm{D}}_{kc}(\boldsymbol{n} + \mathbf{e}_k, \boldsymbol{n}) r_{kc,jc^-} \hat{p}^{\mathrm{A}}_{jc^-}(\boldsymbol{n}, \boldsymbol{n} - \mathbf{e}_j) \\
&\quad = \sum_{k=0}^{N} \hat{\pi}(\boldsymbol{n} + \mathbf{e}_k - \mathbf{e}_j)\hat{q}^{\mathrm{D}}_{kc}(\boldsymbol{n} + \mathbf{e}_k - \mathbf{e}_j, \boldsymbol{n} - \mathbf{e}_j) r_{kc,jc} \hat{p}^{\mathrm{A}}_{jc}(\boldsymbol{n} - \mathbf{e}_j, \boldsymbol{n}). \qquad (3.13)
\end{aligned}$$

To see this, substitute $\hat{\pi}$ into the balance equations and cancel common terms; (3.13) is then reduced to (3.4) after dividing both sides by $\Psi(\boldsymbol{n} - \mathbf{e}_j)$.

Similar to (3.4), the local balance (3.14) itself does not, in general, imply the global balance equation. Thus we need another balance equation. Because each node in the original network is quasi-reversible, we have

$$\pi_j(n_j + 1) q^{\mathrm{D}}_{jc}(n_j + 1, n_j) = \mu_j \rho_j \pi_j(n_j) = \beta_j \pi_j(n_j),$$

where $\beta_j = \mu_j \rho_j$. Furthermore, the rate at which customers are removed from node j by negative customers is $\alpha_j - \beta_j$, i.e.,

$$\alpha_j^- \pi_j(n_j + 1) p^{\mathrm{A}}_{jc^-}(n_j + 1, n_j) = (\alpha_j - \beta_j)\pi_j(n_j).$$

Translating these into the modified network with state-dependent transition rates yields

$$\hat{\pi}(\boldsymbol{n}+\mathbf{e}_j)\hat{q}^{\mathrm{D}}_{jc}(\boldsymbol{n}+\mathbf{e}_j,\boldsymbol{n}) = \beta_j \frac{\Psi(\boldsymbol{n})}{\Phi(\boldsymbol{n})}\hat{\pi}(\boldsymbol{n}), \tag{3.14}$$

$$\alpha_j^{-}\hat{\pi}(\boldsymbol{n}+\mathbf{e}_j)\hat{p}^{\mathrm{A}}_{jc^-}(\boldsymbol{n}+\mathbf{e}_j,\boldsymbol{n}) = (\alpha_j-\beta_j)\frac{\Psi(\boldsymbol{n})}{\Phi(\boldsymbol{n})}\hat{\pi}(\boldsymbol{n}). \tag{3.15}$$

Adding (3.15) to (3.13) and using (3.14) and the traffic equations we obtain

$$\begin{aligned}
&\hat{\pi}(\boldsymbol{n})\hat{q}^{\mathrm{D}}_j(\boldsymbol{n},\boldsymbol{n}-\mathbf{e}_j) + \alpha_j^{-}\frac{\Psi(\boldsymbol{n})}{\Phi(\boldsymbol{n})}\hat{\pi}(\boldsymbol{n})\hat{p}^{\mathrm{A}}_{jc^-}(\boldsymbol{n},\boldsymbol{n}-\mathbf{e}_j) + (\alpha_j-\beta_j)\frac{\Psi(\boldsymbol{n})}{\Phi(\boldsymbol{n})}\hat{\pi}(\boldsymbol{n})\\
&\quad= \sum_{k=0}^{N}\hat{\pi}(\boldsymbol{n}+\mathbf{e}_k-\mathbf{e}_j)\hat{q}^{\mathrm{D}}_{kc}(\boldsymbol{n}+\mathbf{e}_k-\mathbf{e}_j,\boldsymbol{n}-\mathbf{e}_j)r_{kc,jc}\hat{p}^{\mathrm{A}}_{jc}(\boldsymbol{n}-\mathbf{e}_j,\boldsymbol{n})\\
&\qquad+\alpha_j^{-}\hat{\pi}(\boldsymbol{n}+\mathbf{e}_j)\hat{p}^{\mathrm{A}}_{jc^-}(\boldsymbol{n}+\mathbf{e}_j,\boldsymbol{n}).
\end{aligned}$$

After summing these equations over $j=1,\ldots,N$, using the identity

$$\sum_{j=1}^{N}(\alpha_j-\beta_j) = -\sum_{j=1}^{N}\alpha_j^{-} + \sum_{j=1}^{N}\lambda_j + \sum_{j=1}^{N}\lambda_j^{-} - \sum_{j=1}^{N}\beta_j r_{jc,0},$$

and rearranging the terms, we obtain

$$\begin{aligned}
&\sum_{j=1}^{N}\hat{\pi}(\boldsymbol{n})\hat{q}^{\mathrm{D}}_j(\boldsymbol{n},\boldsymbol{n}-\mathbf{e}_j) + \sum_{j=1}^{N}\hat{\pi}(\boldsymbol{n})\lambda_j\frac{\Psi(\boldsymbol{n})}{\Phi(\boldsymbol{n})} + \sum_{j=1}^{N}\hat{\pi}(\boldsymbol{n})\lambda_j^{-}\frac{\Psi(\boldsymbol{n})}{\Phi(\boldsymbol{n})}\hat{p}^{\mathrm{A}}_{jc^-}(\boldsymbol{n},\boldsymbol{n}-\mathbf{e}_j)\\
&= \sum_{j=0}^{N}\sum_{k=0}^{N}\hat{\pi}(\boldsymbol{n}+\mathbf{e}_k-\mathbf{e}_j)\hat{q}^{\mathrm{D}}_k(\boldsymbol{n}+\mathbf{e}_k-\mathbf{e}_j,\boldsymbol{n}-\mathbf{e}_j)r_{kc,jc}\hat{p}^{\mathrm{A}}_{jc}(\boldsymbol{n}-\mathbf{e}_j,\boldsymbol{n})\\
&\quad+\sum_{j=1}^{N}\alpha_j^{-}\hat{\pi}(\boldsymbol{n}+\mathbf{e}_j)\hat{p}^{\mathrm{A}}_{jc^-}(\boldsymbol{n}+\mathbf{e}_j,\boldsymbol{n}) + \sum_{j=1}^{N}(\alpha_j^{-}-\lambda_j^{-})\hat{\pi}(\boldsymbol{n})\frac{\Psi(\boldsymbol{n})-\Psi(\boldsymbol{n}-\mathbf{e}_j)}{\Phi(\boldsymbol{n})}.
\end{aligned}$$

This is exactly the global balance equation for the modified state-dependent network since, by the quasi-reversibility of the nodes in the original network, the last two terms on the right-hand side represent the rate going into state $\boldsymbol{n}$ due to the negative customers, i.e.,

$$\begin{aligned}
&\sum_{j=1}^{N}\alpha_j^{-}\hat{\pi}(\boldsymbol{n}+\mathbf{e}_j)\hat{p}^{\mathrm{A}}_{jc^-}(\boldsymbol{n}+\mathbf{e}_j,\boldsymbol{n})\\
&\quad= \sum_{k=0}^{N}\sum_{j=1}^{N}\hat{\pi}(\boldsymbol{n}+\mathbf{e}_k+\mathbf{e}_j)\hat{q}^{\mathrm{D}}_{jc}(\boldsymbol{n}+\mathbf{e}_k+\mathbf{e}_j,\boldsymbol{n}+\mathbf{e}_j)r_{kc,jc^-}\hat{p}^{\mathrm{A}}_{jc^-}(\boldsymbol{n}+\mathbf{e}_j,\boldsymbol{n}),
\end{aligned}$$

$$\sum_{j=1}^{N}(\alpha_j^- - \lambda_j^-)\hat{\pi}(\boldsymbol{n})\frac{\Psi(\boldsymbol{n}) - \Psi(\boldsymbol{n} - \mathbf{e}_j)}{\Phi(\boldsymbol{n})}$$

$$= \sum_{k=1}^{N}\sum_{j=1}^{N}\hat{\pi}(\boldsymbol{n} + \mathbf{e}_k)\hat{q}_{kc}^{\mathrm{D}}(\boldsymbol{n} + \mathbf{e}_k, \boldsymbol{n})r_{kc,jc^-}\big(1 - \hat{p}_{jc^-}^{\mathrm{A}}(\boldsymbol{n}, \boldsymbol{n} - \mathbf{e}_j)\big).$$

Hence $\hat{\pi}$ is the stationary measure of the network. This completes the second proof of Theorem 9.11. □

Note that in the second proof of Theorem 9.11 the quasi-reversibility of the original network, i.e., the case where $\Psi(\boldsymbol{n}) = 1[\boldsymbol{n} \geq \mathbf{0}]$ and $\Phi(\boldsymbol{n}) = 1$, plays a crucial role. It suggests that the cross local balance may be used for obtaining and proving stationary distributions of more general state-dependent networks with signals, as long as the quasi-reversibility is satisfied in the corresponding network with state-independent transitions. This is indeed true under certain restrictions on the arrival and departure effects, and in the next chapter we develop a general procedure for doing this. In the following two sections we present several results for such state-dependent networks.

9.4 NETWORKS WITH NEGATIVE SIGNALS AND STATE-DEPENDENT TRANSITIONS

Most networks with concurrent batch departures can be viewed as special cases of networks with multiple classes of negative signals, e.g., the networks discussed in Chapter 7. In this section we present the results for state-dependent networks with negative signals.

Consider a network with N nodes. There is a single class of customers, denoted by c, and I classes of negative signals, denoted by $\{u^-; u = 1, 2, \ldots, I\}$. When the state of the network is $\boldsymbol{n}$, customers arrive at node j from the outside according to a state-dependent process with rate

$$\lambda_j\frac{\Psi(\boldsymbol{n})}{\Phi(\boldsymbol{n})}, \qquad j = 1, \ldots, N,$$

and negative signals of class u^- arrive at node j from the outside according to a state-dependent process with rate

$$\lambda_{ju}^-\frac{\Psi(\boldsymbol{n})}{\Phi(\boldsymbol{n})}, \qquad j = 1, \ldots, N,\ u = 1, \ldots, I.$$

The service times at node j are exponentially distributed with rate

$$q_{jc}^{\mathrm{D}}(\boldsymbol{n}, \boldsymbol{n} - \mathbf{e}_j) = \mu_j\frac{\Psi(\boldsymbol{n} - \mathbf{e}_j)}{\Phi(\boldsymbol{n})}, \qquad j = 1, \ldots, N.$$

Upon a service completion, a customer at node j joins node k as a customer with probability $r_{jc,kc}$, as a class u^- negative signal with probability r_{jc,ku^-}, and the customer leaves the network with probability $r_{jc,0}$. The arrival of a customer at node j increases the number of customers by 1, and the arrival of a class u^-, $u = 1, 2, \ldots, I$, negative signal at node j induces a customer at node j to leave the network with probability

$$p^{\mathrm{A}}_{ju^-}(\boldsymbol{n}, \boldsymbol{n} - \mathbf{e}_j) = \frac{\Psi(\boldsymbol{n} - \mathbf{e}_j)}{\Psi(\boldsymbol{n})}, \qquad u = 1, \ldots, I. \tag{4.1}$$

This is possible because Ψ is assumed to be nondecreasing in each component. In this case the number of customers at node j decreases by 1, and the customer induced to leave goes to node k as a regular customer with probability $r_{ju^-,kc}$, as a class v^- negative signal with probability r_{ju^-,kv^-}, and it leaves the network with probability $r_{ju^-,0}$. With probability

$$1 - \frac{\Psi(\boldsymbol{n} - \mathbf{e}_j)}{\Psi(\boldsymbol{n})}$$

the arrival of a class u^- negative signal at node j has no effect at the node. Hence (4.1) is also referred to as the effective probability, i.e., with probability (4.1) the signal is effective. Clearly, when $\Psi(\boldsymbol{n}) = 1[\boldsymbol{n} \geq \mathbf{0}]$ and $\Phi(\boldsymbol{n}) = 1$, the arrival of a negative signal at node j is effective if and only if the queue is not empty, i.e., $n_j > 0$. This special case has been studied in Chapter 5.

The proof of the following result runs in parallel with the second proof of Theorem 9.11, and it is a special case of a more general result that will be proved in Chapter 10 (Theorem 10.15). It is presented here, therefore, without a proof.

Theorem 9.12 *The stationary measure of the network described above is*

$$\pi(\boldsymbol{n}) = \Phi(\boldsymbol{n}) \prod_{j=1}^{N} \rho_j^{n_j},$$

where

$$\rho_j = \frac{\alpha_j}{\sum_{u=1}^{I} \alpha^-_{ju} + \mu_j}, \qquad j = 1, \ldots, N,$$

and α_j and α^-_{ju} are the solutions to the traffic equations

$$\alpha_j = \lambda_j + \sum_{k=1}^{N} \mu_k \rho_k r_{kc,jc} + \sum_{k=1}^{N} \sum_{v=1}^{I} \alpha^-_{kv} \rho_k r_{kv^-,jc}, \quad j = 1, \ldots, N, \tag{4.2}$$

$$\alpha^-_{ju} = \lambda^-_{ju} + \sum_{k=1}^{N} \mu_k \rho_k r_{kc,ju^-} + \sum_{k=1}^{N} \sum_{v=1}^{I} \alpha^-_{kv} \rho_k r_{kv^-,ju^-}, \tag{4.3}$$

$$j = 1, \ldots, N, \; u = 1, \ldots, I.$$

In what follows we present two special cases of this result.

Example 9.13 (Henderson–Northcote–Taylor model) Consider a network with N nodes. When the state of the network is $\boldsymbol{n}$, customers arrive from the outside according to a state-dependent process with rate

$$\lambda \frac{\Psi(\boldsymbol{n})}{\Phi(\boldsymbol{n})},$$

and signals arrive from the outside according to a state-dependent process with rate

$$\lambda^{-} \frac{\Psi(\boldsymbol{n})}{\Phi(\boldsymbol{n})}.$$

When a customer arrives at the network, it is routed to node j with probability $r_{0c,jc}$, where

$$\sum_{j=1}^{N} r_{0c,jc} = 1,$$

and when a signal arrives, it is routed to node j as a type u signal with probability r_{00^-,ju^-}, $u = 0, 1, \ldots,$ where

$$\sum_{j=1}^{N} \sum_{u=0}^{\infty} r_{00^-,ju^-} = 1.$$

The service rate at node j is

$$\mu_j \frac{\Psi(\boldsymbol{n} - \mathbf{e}_j)}{\Phi(\boldsymbol{n})}, \qquad j = 1, \ldots, N.$$

Upon a service completion at node j, a customer goes to node k as a type u signal with probability r_{jc,ku^-}, $u = 0, 1, \ldots$, and as a customer with probability $r_{j,k}$, $j, k = 1, \ldots, N$, and the customer leaves the network with probability $r_{j,0}$. These probabilities satisfy

$$\sum_{k=0}^{N} r_{j,k} + \sum_{k=1}^{N} \sum_{u=0}^{\infty} r_{jc,ku^-} = 1, \qquad j = 1, \ldots, N.$$

An arriving signal at node j is effective with probability

$$\frac{\Psi(\boldsymbol{n} - \mathbf{e}_j)}{\Psi(\boldsymbol{n})}, \tag{4.4}$$

and it is not effective with probability

$$1-\frac{\Psi(\boldsymbol{n}-\mathbf{e}_j)}{\Psi(\boldsymbol{n})}. \tag{4.5}$$

An ineffective signal leaves the network in state $\boldsymbol{n}$, i.e., it has no effect on the network state.

A type 0 signal can trigger a customer to move. When a type 0 signal arrives at node j, it induces a customer to depart for node k as a customer with probability $r_{j0^-,kc}$, $j,k=1,\ldots,N$, and to leave the network with probability $r_{j0^-,0}$, where

$$\sum_{k=1}^{N} r_{j0^-,kc}+r_{j0^-,0}=1, \qquad j=1,2,\ldots,N.$$

A type u signal arrives with the intention of deleting a batch of u customers, $u=1,2,\ldots$. However, it may happen that fewer than u customers are removed. An effective type u signal, $u\geq 1$, that arrives at node j when the network is in state $\boldsymbol{n}$ decreases the queue length of node j by ℓ with probability

$$\frac{\Psi(\boldsymbol{n}-u\mathbf{e}_j)}{\Psi(\boldsymbol{n}-\mathbf{e}_j)}, \qquad \ell=u, \tag{4.6}$$

$$\frac{\Psi(\boldsymbol{n}-\ell\mathbf{e}_j)-\Psi(\boldsymbol{n}-(\ell+1)\mathbf{e}_j)}{\Psi(\boldsymbol{n}-\mathbf{e}_j)}, \qquad 1\leq \ell<u. \tag{4.7}$$

The above probabilities have to be multiplied by $\Psi(\boldsymbol{n}-\mathbf{e}_j)/\Psi(\boldsymbol{n})$ to obtain the probabilities that a type u ($u\geq 1$) signal that arrives at node j in state $\boldsymbol{n}$ is effective and decreases the number of customers at node j by ℓ. These indeed constitute a probability distribution since

$$\frac{\Psi(\boldsymbol{n}-u\mathbf{e}_j)}{\Psi(\boldsymbol{n}-\mathbf{e}_j)}+\sum_{\ell=1}^{u-1}\frac{\Psi(\boldsymbol{n}-\ell\mathbf{e}_j)-\Psi(\boldsymbol{n}-(\ell+1)\mathbf{e}_j)}{\Psi(\boldsymbol{n}-\mathbf{e}_j)}=1.$$

Note that when the network is in state $\boldsymbol{n}$, the probability that a signal of type u, $u\geq\ell\geq 1$, removes at least ℓ customers from node j is

$$\sum_{k=\ell}^{u-1}\frac{\Psi(\boldsymbol{n}-k\mathbf{e}_j)-\Psi(\boldsymbol{n}-(k+1)\mathbf{e}_j)}{\Psi(\boldsymbol{n}-\mathbf{e}_j)}+\frac{\Psi(\boldsymbol{n}-u\mathbf{e}_j)}{\Psi(\boldsymbol{n}-\mathbf{e}_j)}=\frac{\Psi(\boldsymbol{n}-\ell\mathbf{e}_j)}{\Psi(\boldsymbol{n}-\mathbf{e}_j)}. \qquad \square$$

To see that this network is a special case of the model in Theorem 9.12, construct a network with the classes of negative signals

$$\{u^-;u=0,1,\ldots\}.$$

The service rate at node j is

$$q^{\mathrm{D}}_{jc}(\boldsymbol{n}, \boldsymbol{n} - \mathbf{e}_j) = \mu_j \frac{\Psi(\boldsymbol{n} - \mathbf{e}_j)}{\Phi(\boldsymbol{n})}.$$

A class 0^- negative signal is defined in the same way as a type 0 signal. The effect of a class u^- ($u \geq 1$) negative signal in Theorem 9.12 is, however, different from that of a type u negative signal in the present example. When the state of the network is $\boldsymbol{n}$, the arrival of a class u^- negative signal at node j triggers *one* customer to move with probability (4.4), and nothing happens with probability (4.5). Assume that

$$f_{ju^-,u^-}(\boldsymbol{n}, \boldsymbol{n} - \mathbf{e}_j) = 1, \qquad u = 1, 2, \ldots,$$

and that the routing probabilities of class u^- negative signals are

$$r_{ju^-,j(u-1)^-} = 1, \qquad j = 1, \ldots, N, \ u \geq 2,$$
$$r_{j1^-,0} = 1, \qquad j = 1, \ldots, N.$$

Thus, if a class u^- negative signal is effective, then it triggers one customer at node j to leave and returns to node j as a class $(u-1)^-$ negative signal; the class $(u-1)^-$ negative signal, if effective, induces another customer to depart and returns to node j as a class $(u-2)^-$ negative signal, and so on. This process continues until either u customers are removed or $\ell < u$ customers are removed due to the ineffectiveness of a signal. It is left as an exercise for the reader to show that the resulting effect of a class u^- negative signal is exactly as defined in (4.6) and (4.7).

Applying Theorem 9.12 to this network we obtain the following result.

Proposition 9.14 *The stationary measure of the network described above is*

$$\pi(\boldsymbol{n}) = \Phi(\boldsymbol{n}) \prod_{j=1}^{N} (1 - \rho_j)\rho_j^{n_j},$$

where ρ_j *is given by*

$$\rho_j = \frac{\alpha_j}{\mu_j + \alpha_j^-}, \qquad j = 1, \ldots, N,$$

and α_j and α_j^- are the solutions to the traffic equations

$$\begin{aligned}
\alpha_j &= \lambda r_{0c,jc} + \sum_{k=1}^{N} \mu_k \rho_k r_{kc,jc} + \sum_{k=1}^{N} \lambda^- r_{00^-,k0^-} \rho_k r_{k0^-,jc} \\
&\quad + \sum_{i=1}^{N} \sum_{k=1}^{N} \mu_i \rho_i r_{ic,k0^-} \rho_k r_{k0^-,jc}, \qquad j = 1, \ldots, N, \\
\alpha_j^- &= \lambda^- \sum_{u=0}^{\infty} r_{00^-,ju^-} + \lambda^- \sum_{\ell=0}^{\infty} \rho_j^\ell \sum_{u=\ell+1}^{\infty} r_{00^-,ju^-} + \sum_{k=1}^{N} \mu_k \rho_k r_{kc,j0^-} \\
&\quad + \sum_{k=1}^{N} \sum_{\ell=0}^{\infty} \mu_k \rho_k \rho_j^\ell \sum_{u=\ell+1}^{\infty} r_{kc,ju^-}, \qquad j = 1, \ldots, N.
\end{aligned}$$

Example 9.15 (State-dependent coalescence networks) We now extend the queueing networks with coalescence of Chapter 7 to include state-dependent transition rates. When the state of the network is $\boldsymbol{n}$, customers arrive at node j according to a state-dependent process with rate

$$\lambda_j \frac{\Psi(\boldsymbol{n})}{\Phi(\boldsymbol{n})}, \qquad j = 1, \ldots, N.$$

Each node serves customers in batches with a state-dependent rate. The batch sizes are random with a probability mass function

$$P(B_j = n) = b_j(n), \qquad n = 1, 2, \ldots,$$

for $j = 1, 2, \ldots, N$. At node j, B_j customers are processed simultaneously and, at the service completion, they coalesce into a single customer. When the network is in state $\boldsymbol{n}$, node j processes a batch of u customers at rate

$$\mu_j b_j(u) \frac{\Psi(\boldsymbol{n} - u\mathbf{e}_j)}{\Phi(\boldsymbol{n})}. \tag{4.8}$$

The u customers coalesce into a single customer and this customer joins node k as a regular customer with probability $r_{jc,kc}$, as a type v negative signal with probability r_{jc,kv^-}, $v = 0, 1, \ldots$, and it leaves the network with probability $r_{jc,0}$. Clearly,

$$r_{jc,0} + \sum_{k=1}^{N} r_{jc,kc} + \sum_{k=1}^{N} \sum_{v=0}^{\infty} r_{jc,kv^-} = 1, \qquad j = 1, \ldots, N.$$

However, it may also occur that fewer than $B_j = u$ customers are processed in a batch, and the rate at which a batch of $\ell < u$ customers completes a service due to this event is

$$\mu_j b_j(u) \frac{\Psi(\boldsymbol{n} - \ell \mathbf{e}_j) - \Psi(\boldsymbol{n} - (\ell + 1)\mathbf{e}_j)}{\Phi(\boldsymbol{n})}, \qquad \ell = 1, \ldots, u - 1. \tag{4.9}$$

When this occurs, these ℓ customers coalesce into a partial batch, and leave the network.

Negative signals arrive at node j from the outside according to a state-dependent process with rate

$$\lambda_j^- \frac{\Psi(\boldsymbol{n})}{\Phi(\boldsymbol{n})}, \qquad j = 1, \ldots, N.$$

An arriving signal at node j is of type u with probability r_{0,ju^-}, where

$$\sum_{u=1}^{\infty} r_{0,ju^-} = 1, \qquad j = 1, \ldots, N.$$

The effects of signals are more general than in the previous example. When a signal arrives at node j, either from the outside or from another node, it is effective with probability (4.4), and it is ineffective with probability (4.5). An ineffective signal leaves the network in state $\boldsymbol{n}$. Type u, $u \geq 1$, signals arrive at node j with the intention of triggering u customers to move, but it may happen that only $\ell < u$ are removed; these events occur with probabilities (4.6) and (4.7), respectively. If u customers are triggered by a type u signal, they are called a full batch, and if $\ell < u$ customers are triggered by a type u signal, the ℓ customers are called a partial batch. A full batch coalesces into a single customer, and moves to node k as a customer with probability $r_{js^-,kc}$, as a type v negative signal with probability r_{js^-,kv^-}, and leaves the network with probability $r_{js^-,0}$, where

$$\sum_{k=1}^{N} r_{js^-,kc} + \sum_{k=1}^{N} \sum_{v=1}^{\infty} r_{js^-,kv^-} + r_{js^-,0} = 1, \qquad j = 1, \ldots, N.$$

A partial batch coalesces into an incomplete unit and leaves the network. □

Like the previous example, this network is also a special case of the model discussed in this section. To see this, construct a network with the class of negative signals

$$\{u^+; u = 1, 2, \ldots\} \cup \{u^-; u = 1, 2, \ldots\}.$$

Assume that when the network is in state $\boldsymbol{n}$, class u^+ negative signals arrive at node j from the outside according to a state-dependent process with rate

$$\mu_j b_j(u) \frac{\Psi(\boldsymbol{n})}{\Psi(\boldsymbol{n})}, \qquad u \geq 1.$$

As described earlier, the arrival of a class u^+ or class u^- negative signal at node j is effective with probability (4.4), and ineffective with probability (4.5). The arrival of an effective class u^+ ($u > 1$) negative signal at node j triggers one customer to leave as a class u^+ negative signal, and the arrival of an ineffective class u^+ negative signal has no impact on the network. A departure of class u^+ ($u > 1$) negative signal from node j returns to the same node as a class $(u-1)^+$ negative signal, i.e., $r_{ju^+,j(u-1)^+} = 1$. The departure of a class 1^+ positive signal has routing probabilities $r_{jc,kc}, r_{jc,kv^-}$, and $r_{jc,0}$. Thus a service completion at node j is described by a class u^+ negative signal with routing probabilities

$$\begin{aligned} r_{ju^+,j(u-1)^+} &= 1, && j,k = 1,\ldots,N,\ u \geq 2,\\ r_{j1^+,kc} &= r_{jc,kc}, && j,k = 1,\ldots,N,\\ r_{j1^+,kv^-} &= r_{jc,kv^-}, && j,k = 1,\ldots,N,\ v \geq 1. \end{aligned}$$

Clearly, the arrival of a class u^+ negative signal either triggers u customers to leave as a full batch and route according to probabilities $r_{jc,kc}$, r_{jc,kv^-}, and $r_{jc,0}$, or removes $\ell < u$ customers from the node. Thus these negative signals have the same effects as the service completions in the original model.

Similarly, when a class u^- ($u > 1$) negative signal arrives at node j it is effective with probability (4.4). An effective class u^- negative signal triggers a customer to leave as a class u^- negative signal and returns to the same node as a class $(u-1)^-$ negative signal, i.e., $r_{ju^-,j(u-1)^-} = 1$, $u = 2, 3, \ldots$. An effective class 1^- signal triggers one customer to move according to probabilities $r_{js^-,kc}$, r_{js^-,kv^-} and $r_{js^-,0}$. Thus we have

$$\begin{aligned} r_{ju^-,j(u-1)^-} &= 1, && j,k = 1,\ldots,N,\ u \geq 2,\\ r_{j1^-,kc} &= r_{js^-,kc}, && j,k = 1,\ldots,N,\\ r_{j1^-,kv^-} &= r_{js^-,kv^-}, && j,k = 1,\ldots,N,\ v \geq 1. \end{aligned}$$

Clearly, this again describes a batch triggering due to a type u negative signal.

Define

$$\begin{aligned} \bar{b}_j(n) &= \sum_{\ell=n}^{\infty} b_j(\ell),\\ \tilde{B}_j(\rho) &= \sum_{\ell=1}^{\infty} b_j(\ell)\rho^{\ell}. \end{aligned}$$

Applying Theorem 9.12 to this network yields the following result.

Proposition 9.16 *If the traffic equations*

$$\alpha_j = \lambda_j + \sum_{k=1}^{N} \mu_k \tilde{B}_k(\rho_k) r_{kc,jc} + \sum_{k=1}^{N}\sum_{u=1}^{\infty} \alpha_{ku}^- \rho_k^u r_{ku^-,jc}, \quad j = 1, \dots, N,$$

$$\alpha_{ju}^- = \lambda_j^- r_{0,ju^-} + \sum_{k=1}^{N} \mu_k \tilde{B}_k(\rho_k) r_{kc,ju^-} + \sum_{k=1}^{N}\sum_{v=1}^{\infty} \alpha_{kv}^- \rho_k^v r_{kv^-,ju^-},$$

$$j = 1, \dots, N, \; u \geq 1,$$

have solutions $\rho_1, \rho_2, \dots, \rho_N$, *where* ρ_j *is determined by*

$$\alpha_j = \mu_j \sum_{\ell=1}^{\infty} \rho_j^\ell \sum_{u=\ell}^{\infty} b_j(u) + \sum_{\ell=1}^{\infty} \rho_j^\ell \sum_{u=\ell}^{\infty} \alpha_{ju}^-,$$

then the stationary measure of the network is

$$\pi(\boldsymbol{n}) = \Phi(\boldsymbol{n}) \prod_{j=1}^{N} \rho_j^{n_j}.$$

We use the results of this section to analyze the predator–prey model discussed in Chapter 1.

Example 9.17 (The predator–prey model continued) Consider the predator–prey problem of Example 1.6. The ecosystem has N species. We represent each species by a single node in an N-node network, and an additional node, node 0, represents the outside of the ecosystem.

Let n_j be the population size of species j, and $\boldsymbol{n} = (n_1, n_2, \dots, n_N)$. Set

$$\Phi(\boldsymbol{n}) = \prod_{j=1}^{N} \prod_{\ell=1}^{n_j} \frac{1}{\xi_j(\ell)},$$

$$\Psi(\boldsymbol{n}) = f(\boldsymbol{n})\Phi(\boldsymbol{n}),$$

where ξ_j and f are arbitrarily given positive and nonnegative functions. We will impose some conditions on ξ_j and f later (see (4.11) below). The rate at which species j enters the ecosystem is

$$\lambda_j \frac{\Psi(\boldsymbol{n})}{\Phi(\boldsymbol{n})} = \lambda_j f(\boldsymbol{n}), \qquad j = 1, \dots, N.$$

The emission rate of individuals of species from node j is

$$\mu_j \frac{\Psi(\boldsymbol{n} - \mathbf{e}_j)}{\Phi(\boldsymbol{n})} = \mu_j f(\boldsymbol{n} - \mathbf{e}_j)\xi_j(n_j), \qquad j = 1, \ldots, N.$$

An emission represents a death if the emitted individual is transferred to node 0; it represents a member of species j hunting for a member of species k as a predator of species k if the emitted individual is transferred to node k, $k = 1, 2, \ldots, N$. The latter emission from node j can be considered as a class j signal arriving at node k. We denote the transferring probability by $r_{jc,ks(j)}$, with $s(j)$ representing a signal of class j. These signal transfer probabilities fully describe the diets of each species in the ecosystem, and they satisfy

$$\sum_{k=0}^{N} r_{jc,ks(j)} = 1, \qquad j = 1, \ldots, N.$$

The arrival of a signal at node 0 has no further effect on the network. The arrival of a signal at node k when the state of the network is $\boldsymbol{n}$ is effective with probability

$$\frac{\Psi(\boldsymbol{n} - \mathbf{e}_k)}{\Psi(\boldsymbol{n})} = \frac{f(\boldsymbol{n} - \mathbf{e}_k)}{f(\boldsymbol{n})}\xi_k(n_k); \tag{4.10}$$

and it is ineffective with probability

$$1 - \frac{f(\boldsymbol{n} - \mathbf{e}_k)}{f(\boldsymbol{n})}\xi_k(n_k).$$

This is the probability that a predator of species j fails to capture a member of species k to eat, and consequently dies of starvation. An effective class j signal at node k triggers the transfer of an individual from node k to node j with probability 1, reducing the population size at node k by one and increasing the population size at node j by one. This represents a member of species j eliminating a member of species k and returning to its base node. The net population size change at node j for such a transition is therefore 0. The routing probability of the class j signals is

$$r_{ks(j),jc} = 1.$$

Thus, the function f represents the dependency of the transition rates and the signal effectiveness probabilities on the state of the network, while the function $\xi_j(n_j)$ describes the dependency of the emission rates and the signal effectiveness probabilities on n_j. For the expression (4.10) to qualify for a probability, we need the condition

$$f(\boldsymbol{n} - \mathbf{e}_k)\xi_k(n_k) \leq f(\boldsymbol{n}), \qquad k = 1, \ldots, N. \tag{4.11}$$

For instance, simple choices for ξ_j and $f(\boldsymbol{n})$ are

$$\xi_j(\ell) = \frac{\ell}{\ell + C_j}$$

for some constant C_j, and $f(\boldsymbol{n}) = 1[\boldsymbol{n} \geq 0]$. The latter would ensure that when $n_j = 0$, the emission rate from node j becomes 0.

By Theorem 9.12, the stationary distribution of this model is

$$\pi(\boldsymbol{n}) = C \prod_{j=1}^{N} \frac{\rho_j^{n_j}}{\prod_{\ell=1}^{n_j} \xi_j(\ell)},$$

where C is a normalization constant, and ρ_j is determined through the traffic equations

$$\begin{aligned}
\alpha_j &= \lambda_j + \sum_{k=1}^{N} \alpha_{ks(j)} \rho_k \,, & j &= 1, 2, \ldots, N, \\
\alpha_{js(k)} &= \mu_k \rho_k r_{kc,js(k)} \,, & j, k &= 1, 2, \ldots, N, \\
\alpha_j^- &= \sum_{k=1}^{N} \alpha_{js(k)} \,, & j &= 1, 2, \ldots, N,
\end{aligned}$$

where

$$\rho_j = \frac{\alpha_j}{\mu_j + \alpha_j^-}, \qquad j = 1, 2, \ldots, N.$$

In these traffic equations, α_j is the arrival rate of species j, $\alpha_{ks(j)}$ is the arrival rate of species j hunting for species k, and α_j^- is the total arrival rate of predators for species j. Clearly, the first three equations can be simplified to

$$\begin{aligned}
\alpha_j &= \lambda_j + \mu_j \rho_j \sum_{k=1}^{N} \rho_k r_{jc,ks(j)}, & j &= 1, 2, \ldots, N, \\
\alpha_j^- &= \sum_{k=1}^{N} \mu_k \rho_k r_{kc,js(k)}, & j &= 1, 2, \ldots, N.
\end{aligned}$$

The interpretation of the first equation is that the arrival rate at a node is equal to the arrival rate at the node from the outside plus the portion of the departures from the node that can safely return.

To illustrate, consider an ecosystem of three species. Species 1 is the only predator of both species 2 and 3, and we are only interested in this level of predator–prey interaction, i.e., assume that ample food is available for species 2 and 3 so that they form the lowest level of the food chain. In this example we will only model departures from the system due to predator–prey interaction and due to starvation. Let μ_1 be the rate at which members of species 1 seek food, and set $\mu_2 = \mu_3 = 0$, $r_{1c,2s(1)} = p$, $r_{1c,3s(1)} = 1 - p$, so that the diet of species 1 is fully described by the parameter p. The functions Ψ, Φ, and ξ_j are as suggested above with $C_j = 0$. All other parameters are 0. A member of species 1 hunts a member of species 2 and 3 with probabilities p and $1 - p$, respectively, and the hunt is successful if the population of the hunted species is not 0.

Let α_1, α_2 and α_3 be the overall arrival rates of the three types of species, and let α_j^- be the arrival rate of predators for species j, $j = 1, 2, 3$. It follows from our assumption that $\alpha_1^- = 0$. The traffic equations for this model are

$$\begin{aligned}
\alpha_1 &= \lambda_1 + \mu_1\rho_1 p\rho_2 + \mu_1\rho_1(1-p)\rho_3, \\
\alpha_2 &= \lambda_2, \qquad \alpha_2^- = \rho_1\mu_1 p, \\
\alpha_3 &= \lambda_3, \qquad \alpha_3^- = \rho_1\mu_1(1-p), \\
\rho_1 &= \frac{\alpha_1}{\mu_1 + \alpha_1^-} = \frac{\alpha_1}{\mu_1}, \\
\rho_2 &= \frac{\alpha_2}{\mu_2 + \alpha_2^-} = \frac{\lambda_2}{\rho_1\mu_1 p}, \\
\rho_3 &= \frac{\alpha_3}{\mu_3 + \alpha_3^-} = \frac{\lambda_3}{\rho_1\mu_1(1-p)}.
\end{aligned}$$

Solving these equations yields

$$\begin{aligned}
\rho_1 &= \frac{\lambda_1 + \lambda_2 + \lambda_3}{\mu_1}, \\
\rho_2 &= \frac{\lambda_2}{(\lambda_1 + \lambda_2 + \lambda_3)p}, \\
\rho_3 &= \frac{\lambda_3}{(\lambda_1 + \lambda_2 + \lambda_3)(1-p)}.
\end{aligned}$$

The stationary distribution of the populations is

$$\pi(n_1, n_2, n_3) = \prod_{j=1}^{3}(1 - \rho_j)\rho_j^{n_j}.$$

For this model to be stable, we need $\rho_j < 1$ for $j = 1, 2$ and 3, which are equivalent to

$$\begin{aligned}
\mu_1 &> \lambda_1 + \lambda_2 + \lambda_3, \\
p &> \frac{\lambda_2}{\lambda_1 + \lambda_2 + \lambda_3}, \\
1 - p &> \frac{\lambda_3}{\lambda_1 + \lambda_2 + \lambda_3}.
\end{aligned}$$

Thus p must satisfy

$$\frac{\lambda_1 + \lambda_2}{\lambda_1 + \lambda_2 + \lambda_3} > p > \frac{\lambda_2}{\lambda_1 + \lambda_2 + \lambda_3}.$$

□

9.5 NETWORKS WITH POSITIVE AND NEGATIVE SIGNALS AND STATE-DEPENDENT TRANSITIONS

We observed in Chapter 8 that most networks with concurrent batch arrivals and concurrent batch departures are special cases of networks with multiple classes of positive and negative signals. In this section we present the state-dependent version of these networks. As in the case of state-independent transitions, we have to introduce additional departures in order to obtain product form stationary measures.

Consider a network with N nodes, one class of customers, denoted by c, I^+ classes of positive signals, denoted by $\{u^+; u = 1, 2, \ldots, I^+\}$, and I^- classes of negative signals, denoted by $\{u^-; u = 1, 2, \ldots, I^-\}$. When the network is in state $\boldsymbol{n}$, customers, positive signals of class u^+, and negative signals of class u^- arrive at node j from the outside with respective rates

$$\lambda_j \frac{\Psi(\boldsymbol{n})}{\Phi(\boldsymbol{n})}, \qquad \lambda_{ju}^+ \frac{\Psi(\boldsymbol{n})}{\Phi(\boldsymbol{n})}, \qquad \text{and} \qquad \lambda_{ju}^- \frac{\Psi(\boldsymbol{n})}{\Phi(\boldsymbol{n})}.$$

The service rate at node j is

$$q_{jc}^{\mathrm{D}}(\boldsymbol{n}, \boldsymbol{n} - \mathbf{e}_j) = \mu_j \frac{\Psi(\boldsymbol{n} - \mathbf{e}_j)}{\Phi(\boldsymbol{n})}.$$

Upon a service completion at node j, the customer joins node k as a customer with probability $r_{jc,kc}$, as a class u^+ positive signal with probability r_{jc,ku^+}, as a class u^- negative signal with probability r_{jc,ku^-}, and the customer leaves the system with probability $r_{jc,0}$. The arrival of a class u^+ positive signal at node j increases the number of customers by 1 and the positive signal immediately goes to node k as a customer with probability $r_{ju^+,kc}$, as a class v^+ positive signal with probability r_{ju^+,kv^+}, as a class v^- negative signal with probability r_{ju^+,kv^-}, and it leaves the system with probability $r_{ju^+,0}$. The arrival of a class u^- negative signal at node j is effective with probability

$$\frac{\Psi(\boldsymbol{n} - \mathbf{e}_j)}{\Phi(\boldsymbol{n})}.$$

An effective class u^- negative signal triggers a customer at node j to leave for node k as a customer with probability $r_{ju^-,kc}$, as a class v^+ positive signal with probability r_{ju^-,kv^+}, as a class v^- negative signal with probability r_{ju^-,kv^-}, and it leaves the system with probability $r_{ju^-,0}$. With probability

$$1 - \frac{\Psi(\boldsymbol{n} - \mathbf{e}_j)}{\Phi(\boldsymbol{n})}$$

a class u^- negative signal at node j is ineffective and the state of the node does not change.

If $\Psi(\boldsymbol{n}) = 1[\boldsymbol{n} \geq \mathbf{0}]$ and $\Phi(\boldsymbol{n}) = 1$, then the network is the one studied in Chapter 5 and its stationary measure is given by $\prod_{j=1}^{N} \rho_j^{n_j}$, where

$$\rho_j = \frac{\alpha_j + \sum_{u=1}^{I^+} \alpha_{ju}^+}{\mu_j + \sum_{u=1}^{I^-} \alpha_{ju}^-},$$

and α_j, α_{ju}^+ and α_{ju}^- are the solutions to the traffic equations

$$\alpha_j = \lambda_j + \sum_{k=1}^{N}\sum_{v=1}^{I} \mu_k \rho_k r_{kv,jc} + \sum_{k=1}^{N}\sum_{v=1}^{I^-} \rho_k \alpha_{kv}^- r_{kv^-,jc} + \sum_{k=1}^{N}\sum_{v=1}^{I^+} \alpha_{kv}^+ \rho_k^{-1} r_{kv^+,jc},$$
$$j = 1, \ldots, N,\ u = 1, \ldots, I, \tag{5.1}$$

$$\alpha_{ju}^+ = \lambda_{ju}^+ + \sum_{k=1}^{N}\sum_{v=1}^{I} \mu_k \rho_k r_{kv,ju^+} + \sum_{k=1}^{N}\sum_{v=1}^{I^-} \alpha_{kv}^- \rho_k r_{kv^-,ju^+}$$
$$+ \sum_{k=1}^{N}\sum_{v=1}^{I^+} \alpha_{kv}^+ \rho_k^{-1} r_{kv^+,ju^+}, \qquad j = 1, \ldots, N,\ u = 1, \ldots, I^+, \tag{5.2}$$

$$\alpha_{ju}^- = \lambda_{ju}^- + \sum_{k=1}^{N}\sum_{v=1}^{I} \mu_k \rho_k r_{kv,ju^-} + \sum_{k=1}^{N}\sum_{v=1}^{I^-} \rho_k \alpha_{kv}^- r_{kv^-,ju^-}$$
$$+ \sum_{k=1}^{N}\sum_{v=1}^{I^+} \alpha_{kv}^+ \rho_k^{-1} r_{kv^+,ju^-}, \qquad j = 1, \ldots, N,\ u = 1, 2, \ldots, I^-. \tag{5.3}$$

Theorem 9.18 *If there is an additional departure process of class u^+ positive signals from node j with rate*

$$q_{ju^+}^{\mathrm{D}}(\boldsymbol{n}, \boldsymbol{n}) = \alpha_{ju}^+ \rho_j^{-1} \frac{\Psi(\boldsymbol{n}) - \Psi(\boldsymbol{n} - \mathbf{e}_j)}{\Phi(\boldsymbol{n})}, \qquad u \geq 1, \tag{5.4}$$

then the stationary measure of the state-dependent network with positive and negative signals is

$$\pi(\boldsymbol{n}) = \Phi(\boldsymbol{n}) \prod_{j=1}^{N} \rho_j^{n_j}, \tag{5.5}$$

where ρ_j, α_j, α_{ju}^+ and α_{ju}^- are determined by the traffic equations (5.1) to (5.3).

Remark 9.19 Several remarks are in order.

(i) The additional condition (5.4) is concerned only with positive signals, so if the network does not have positive signals, e.g., the network of the previous section, then no additional departure is needed for the result to hold.

(ii) If $\Psi(\boldsymbol{n}) = 1[\boldsymbol{n} \geq \boldsymbol{0}]$, then it follows from (5.4) that the additional term is not zero if and only if $n_j = 0$. This explains why in the case of state-independent transition rates an additional term at a node is needed only when it is empty.

(iii) Similar to Corollary 4.18, if a positive signal cannot be transformed into either a negative signal or a customer that can be transformed into a negative signal, then the product form solution (5.5) is a stochastic upper bound for the original network without the additional departures (5.4).

Theorem 9.18 is, again, a special case of Theorem 10.15 of Chapter 10, so a proof is not given here. We present two special cases of this result. One is the state-dependent version of the assembly-transfer network, and the other is the state-dependent version of the concurrent batch movement network; both are discussed in Chapter 8.

Example 9.20 (State-dependent assembly-transfer networks) We extend the assembly-transfer network discussed in Chapter 8 to include state-dependent transition rates. Assume that when the state of the network is $\boldsymbol{n}$, a batch of u customers arrives at node j at rate

$$\lambda_j a_j(u) \frac{\Psi(\boldsymbol{n})}{\Phi(\boldsymbol{n})}, \tag{5.6}$$

where $\{a_j(u); u \geq 1\}$ is a probability mass function. The services at each node occur also in batches. The probability mass function of the service batch at node j is $\{b_j(u); u \geq 1\}$. The service rates of a full batch of u customers and a partial batch of $\ell < u$ customers are

$$\mu_j b_j(u) \frac{\Psi(\boldsymbol{n} - u\mathbf{e}_j)}{\Phi(\boldsymbol{n})}, \tag{5.7}$$

$$\mu_j b_j(u) \frac{\Psi(\boldsymbol{n} - \ell\mathbf{e}_j) - \Psi(\boldsymbol{n} - (\ell+1)\mathbf{e}_j)}{\Phi(\boldsymbol{n})}, \qquad \ell < u. \tag{5.8}$$

A full batch of u customers that departs from node j goes to node k as a batch of v customers with probability $r_{ju,kv}$ and leaves the network with probability $r_{ju,0}$. A partial batch leaves the network with probability 1.

We formulate this network as a network of Theorem 9.18. To that end, we introduce the class of positive signals $\{u^+; u = 1, 2, \ldots\}$, and the class of negative signals $\{u^-(\ell); \ell \le u, u = 1, 2, \ldots\}$. Class u^+ positive signals arrive at node j from the outside according to a state-dependent process with rate (5.6). The arrival of a class u^+ positive signal at node j adds one customer and then immediately leaves as a class u^+ positive signal; the routing probabilities of the positive signals are

$$r_{ju^+, j(u-1)^+} = 1, \qquad j = 1, \ldots, N, \; u \ge 2,$$
$$r_{j1^+,0} = 1, \qquad j = 1, \ldots, N.$$

Thus, a departing class u^+ positive signal returns to the same node as a class $(u-1)^+$ positive signal, adds another customer to the node, and then returns as a class $(u-2)^+$ positive signal, etc. This process continues until a total of u customers are added to the node, and a class 1^+ positive signal leaves the network.

Class $u^-(u)$ negative signals arrive at node j from the outside according to a state-dependent process with rate

$$\mu_j b_j(u) \frac{\Psi(\boldsymbol{n})}{\Phi(\boldsymbol{n})}.$$

The arrival of a class $u^-(\ell)$ $(\ell \le u)$ negative signal at node j induces a customer to move with probability

$$\frac{\Psi(\boldsymbol{n} - \mathbf{e}_j)}{\Psi(\boldsymbol{n})}.$$

This customer leaves node j as a class $u^-(\ell)$ negative signal and its routing probabilities are

$$r_{ju^-(v), ju^-(v-1)} = 1, \qquad j = 1, \ldots, N, \qquad u, v \ge 2,$$
$$r_{ju^-(1), kv^+} = r_{ju,kv}, \qquad j, k = 1, \ldots, N, \quad u, v \ge 1,$$
$$r_{ju^-(1),0} = r_{ju,0}, \qquad j = 1, \ldots, N, \qquad u \ge 1.$$

Thus a departing class $u^-(u)$ negative signal returns to the same node as a class $u^-(u-1)$ negative signal, and, if effective, induces another customer to leave as a class $u^-(u-1)$ negative signal and return as a class $u^-(u-2)$ negative signal, etc. This process continues until either u customers are removed, which happens with probability (5.7), and a class $u^-(1)$ negative signal is routed according to probabilities $r_{ju,kv}$, $k = 1, \ldots, N$, $v = 1, 2, \ldots$, and $r_{ju,0}$; or $\ell < u$ customers are removed due to the ineffectiveness of negative signals. It is left as an exercise for the reader to show that this happens with probability (5.8). Since the batch departures are described by the arrivals of negative signals, there is no service process of regular customers in this model. The arrival rates of positive and negative signals from the outside are

$$\lambda_{ju}^{+} = \lambda_j a_j(u), \quad \lambda_{ju}^{-} = \mu_j b_j(u), \qquad u = 1, 2, \ldots.$$

This then yields a queueing network model that is exactly the state-dependent, assembly-transfer network. □

Applying Theorem 9.18 we obtain the stationary distribution of the assembly-transfer network with state-dependent transitions.

Proposition 9.21 *Let ρ_j be the solution to the traffic equation (2.1) of Chapter 8. Assume that when the state of the network is $\mathbf{n}$, an additional batch arrival process with rate*

$$\lambda_j^* \frac{\Psi(\boldsymbol{n}) - \Psi(\boldsymbol{n} - \mathbf{e}_j)}{\Phi(\boldsymbol{n})}$$

is activated at node j, where λ_j^* and the batch size distribution are given by (2.2) and (2.3) of Chapter 8. Then the stationary measure of the state-dependent, assembly-transfer network is

$$\pi(\boldsymbol{n}) = \Phi(\boldsymbol{n}) \prod_{j=1}^{N} \rho_j^{n_j}.$$

Example 9.22 (Networks with state-dependent concurrent batch movements) We now extend the networks with concurrent batch arrivals and concurrent batch departures, discussed in Section 8.3, to include state-dependent transitions. When the network is in state $\boldsymbol{n}$, batches of customers arrive at node j from the outside according to a state-dependent process with rate

$$\lambda_j \frac{\Psi(\boldsymbol{n})}{\Phi(\boldsymbol{n})}, \tag{5.9}$$

and with batch size distribution $\{a_j(u); u \geq 1\}$. Type u positive signals and type u negative signals arrive from the outside according to state-dependent processes with respective rates

$$\lambda_j a_j^{+}(u) \frac{\Psi(\boldsymbol{n})}{\Phi(\boldsymbol{n})} \quad \text{and} \quad \lambda_j a_j^{-}(u) \frac{\Psi(\boldsymbol{n})}{\Phi(\boldsymbol{n})}. \tag{5.10}$$

The service at node j occurs also in batches, and the batch size distribution is $\{b_j(u); u \geq 1\}$. The service rate for a full batch of size u is

$$\mu_j b_j(u) \frac{\Psi(\boldsymbol{n} - u\mathbf{e}_j)}{\Phi(\boldsymbol{n})} .$$

It may happen that fewer than u customers are processed, and a partial batch of $\ell < u$ customers is processed at rate

$$\mu_j b_j(u) \frac{\Psi(\boldsymbol{n} - \ell\mathbf{e}_j) - \Psi(\boldsymbol{n} - (\ell+1)\mathbf{e}_j)}{\Phi(\boldsymbol{n})}, \qquad \ell < u.$$

When a full batch of u customers is processed at node j, it joins node k as a batch of v customers with probability $r_{ju,kv}$, as a type v positive signal with probability r_{ju,kv^+}, as a type v negative signal with probability r_{ju,kv^-}, and the batch leaves the network with probability $r_{ju,0}$. The arrival of a type u positive signal at node j adds u customers at node j and then leaves for node k as a batch of v customers with probability $r_{ju^+,kv}$, as a type v positive signal with probability r_{ju^+,kv^+}, as a type v negative signal with probability r_{ju^+,kv^-}, and it leaves the network with probability $r_{ju^+,0}$. The arrival of a type u negative signal at node j is effective in triggering u customers to leave with probability

$$\frac{\Psi(\boldsymbol{n} - u\mathbf{e}_j)}{\Psi(\boldsymbol{n})} .$$

If effective, it triggers u customers to depart for node k as a batch of v customers with probability $r_{ju^-,kv}$, as a type v positive signal with probability r_{ju^-,kv^+}, as a type v negative signal with probability r_{ju^-,kv^-}, and the u customers leave the network with probability $r_{ju^-,0}$. A noneffective type u negative signal removes $\ell < u$ customers with probability

$$\frac{\Psi(\boldsymbol{n} - \ell\mathbf{e}_j) - \Psi(\boldsymbol{n} - (\ell+1)\mathbf{e}_j)}{\Psi(\boldsymbol{n})}, \qquad 1 \leq \ell < u.$$

To see that this model is a special case of the network of this section, consider a network with positive and negative signals but without services of regular customers, similar to Example 9.20. The class of positive signals is

$$\{u^+; u \geq 1\} \cup \{u^+(\ell); \ell \leq u, u \geq 1\},$$

and the class of negative signals is

$$\{u^-(\ell); \ell \leq u, u \geq 1\}.$$

Class u^+ positive signals arrive from the outside at rate

$$\lambda_j a_j(u)\frac{\Psi(\boldsymbol{n})}{\Phi(\boldsymbol{n})}, \qquad u \geq 1.$$

Class $u^+(u)$ positive signals arrive from the outside at the rate (5.10), and class $u^-(u)$ negative signals arrive from the outside at rate

$$\mu_j b_j(u)\frac{\Psi(\boldsymbol{n})}{\Phi(\boldsymbol{n})}, \qquad u \geq 1.$$

The arrival of a class u^+ or $u^+(\ell)$ ($\ell \leq u$) positive signal at node j adds one customer and then leaves immediately as a class u^+ or $u^+(\ell)$ departure. The arrival of a class $u^-(\ell)$ ($1 \leq \ell \leq u$) negative signal is effective with probability

$$\frac{\Psi(\boldsymbol{n} - \mathbf{e}_j)}{\Phi(\boldsymbol{n})},$$

and if effective, it triggers one customer to leave as a class $u^-(\ell)$ negative signal. The routing probabilities are

$$\begin{aligned}
r_{ju^+,j(u-1)^+} &= 1, & j &= 1, \ldots, N, & u &\geq 2,\\
r_{j1^+,0} &= 1, & j &= 1, \ldots, N, & &\\
r_{ju^+(\ell),ju^+(\ell-1)} &= 1, & j &= 1, \ldots, N, & u &\geq 1, 2 \leq \ell \leq u,\\
r_{ju^+(1),kv^+} &= r_{ju^+,kv}, & j, k &= 1, \ldots, N, & u, v &\geq 1,\\
r_{ju^+(1),kv^+(v)} &= r_{ju^+,kv^+}, & j, k &= 1, \ldots, N, & u, v &\geq 1,\\
r_{ju^+(1),kv^-(v)} &= r_{ju^+,kv^-}, & j, k &= 1, \ldots, N, & u, v &\geq 1,\\
r_{ju^+(1),0} &= r_{ju^+,0}, & j &= 1, \ldots, N, & u &\geq 1,\\
r_{ju^-(\ell),ju^-(\ell-1)} &= 1, & j &= 1, \ldots, N, & u &\geq 1, 2 \leq \ell \leq u,\\
r_{ju^-(1),kv^+} &= r_{ju^-,kv}, & j, k &= 1, \ldots, N, & u, v &\geq 1,\\
r_{ju^-(1),kv^+(v)} &= r_{ju^-,kv^+}, & j, k &= 1, \ldots, N, & u, v &\geq 1,\\
r_{ju^-(1),kv^-(v)} &= r_{ju^-,kv^-}, & j, k &= 1, \ldots, N, & u, v &\geq 1,\\
r_{ju^-(1),0} &= r_{ju^-,0}, & j &= 1, \ldots, N, & u &\geq 1.
\end{aligned}$$

It is left as an exercise for the reader to show that this network with positive and negative signals is equivalent to the model with state-dependent concurrent movements. □

Applying Theorem 9.18 to the concurrent batch movement network of Example 9.22 yields the following result.

Proposition 9.23 *Assume that whenever node j is empty, an additional batch arrival process is activated with rate*

$$\lambda_j^* \frac{\Psi(\boldsymbol{n}) - \Psi(\boldsymbol{n} - \mathbf{e}_j)}{\Phi(\boldsymbol{n})},$$

and batch size distribution $a_j^*(n)$, where λ_j^* and $a_j^*(n)$ are as defined in Theorem 8.6. Furthermore, assume that there is an additional departure process of type $u > \ell$ positive signals from node j at rate

$$\alpha_{ju}^+ \rho_j^{-u} \frac{\Psi(\boldsymbol{n} - \ell\mathbf{e}_j) - \Psi(\boldsymbol{n} - (\ell+1)\mathbf{e}_j)}{\Phi(\boldsymbol{n})}, \tag{5.11}$$

where α_{ju}^+ and ρ_j are the solutions to the traffic equations for Theorem 8.6. Then the stationary measure of the network is

$$\pi(\boldsymbol{n}) = \Phi(\boldsymbol{n}) \prod_{j=1}^{N} \rho_j^{n_j}.$$

Note that, if $\Psi(\boldsymbol{n}) = 1[\boldsymbol{n} \geq \mathbf{0}]$, then we must have $\ell = n_j$ for the additional term (5.11) not to be 0. Thus in this case the result is reduced to that of Theorem 8.6.

9.6 NETWORKS WITH ROUTING DEPENDING UPON SERVICE REQUIREMENTS

The remaining part of this chapter focuses on networks in which the routing depends on historical information.

A common assumption in the models discussed so far is that the routing probabilities are Markovian, and depend only on the departing node. From a practical point of view it appears useful to generalize this framework and allow for dependencies between the routing probabilities and past service times or other information pertaining to the history of the customer. For example, a customer induced to move by a signal may be more likely to leave the system if the service just received is longer. This section considers a model with routing probabilities depending upon the service requirement received at the departing node as well as upon the cause of departure. The next section considers a model with the routing of a customer depending upon the number of times the customer has been induced to move by signals in the past.

Consider a network with N nodes, I classes of customers and a single class of signals. Each node consists of a single server. Customers of class u, $u = 1, \ldots, I$, arrive at node j, $j = 1, \ldots, N$, from the outside according to a Poisson process with rate λ_{ju}, and signals arrive at node j from the outside according to a Poisson process with rate λ_j^-. The service requirements of class u customers at node j have distribution F_{ju}. If the service discipline at a node is symmetric, as defined in Chapter 5, then the distribution F_{ju} may be arbitrary. If the service discipline is not symmetric, then we have to assume that the service requirements at that node are exponentially distributed with the same rate for each class of customers. In this case the service discipline as well as the effects of the signals on the node are allowed to be more general.

The routing probabilities are defined as follows. When a class u customer completes its service at a symmetric node j after completing x units of service requirement, it goes to node k as a customer of class v with probability $r_{ju,kv}(x)$, as a signal with probability $r_{ju,ks^-}(x)$, and the customer leaves the system with probability $r_{ju,0}(x)$. These transition probabilities satisfy

$$\sum_{k=1}^{N}\sum_{v=1}^{I} r_{ju,kv}(x) + \sum_{k=1}^{N} r_{ju,ks^-}(x) + r_{ju,0}(x) = 1,$$
$$\text{for } j = 1, \ldots, N,\ u = 1, \ldots, I, \text{ and } x \geq 0.$$

When a signal arrives at node j, either from the outside or from another node, it induces one of the customers, if one or more are present, to depart immediately. If this customer is of class u and it has completed x units of service requirement, it goes to node k as a customer of class v with probability $r_{ju^-,kv}(x)$, as a signal with probability $r_{ju^-,ks^-}(x)$, and the customer leaves the system with probability $r_{ju^-,0}(x)$. Again we have

$$\sum_{k=1}^{N}\sum_{v=1}^{I} r_{ju^-,kv}(x) + \sum_{k=1}^{N} r_{ju^-,ks^-}(x) + r_{ju^-,0}(x) = 1,$$
$$\text{for } j = 1, \ldots, N,\ u = 1, \ldots, I, \text{ and } x \geq 0.$$

When a signal arrives at an empty node, nothing happens.

Note that in the transition probabilities $r_{ju,kv}(x)$, $r_{ju,ks^-}(x)$, and $r_{ju,0}(x)$, the variable x does not denote the time the class u customer spends in service at node j. It denotes the amount of service requirement the customer at node j has received (see Chapter 5). Similarly, in $r_{ju^-,kv}(x)$, $r_{ju^-,ks^-}(x)$, and $r_{ju^-,0}(x)$, the variable x denotes the amount of service requirement already received by the customer at node j at the time of arrival of the signal.

For the results of this section to hold, we have to impose the following restriction on the non-symmetric nodes. If node j operates according to a non-symmetric service discipline (e.g., FCFS), then the routing probabilities from node j are independent of x. It is left as an exercise to show that the results may not hold otherwise.

Let

$$\boldsymbol{n} = (\boldsymbol{n}_1, \ldots, \boldsymbol{n}_N)$$

denote the state of the network, where

$$\boldsymbol{n}_j = (n_{j1}, \ldots, n_{jI}),$$

and n_{ju} represents the number of class u customers at node j. We define $n_j = \sum_{u=1}^{I} n_{ju}$ as the total number of customers at node j. Let α_{ju}, α_j^-, $j = 1, \ldots, N$ and $u = 1, \ldots, I$ be the average arrival rates of class u customers and signals at node j. They satisfy the traffic equations

$$\begin{aligned}\alpha_{ju} = \lambda_{ju} &+ \sum_{k=1}^{N}\sum_{v=1}^{I}\alpha_{kv}\int_0^\infty e^{-\alpha_k^- x} r_{kv,ju}(x)dF_{kv}(x) \\ &+ \sum_{k=1}^{N}\sum_{v=1}^{I}\alpha_{kv}\int_0^\infty \alpha_k^- e^{-\alpha_k^- x}(1 - F_{kv}(x))r_{kv^-,ju}(x)dx, \qquad (6.1)\end{aligned}$$

$$\begin{aligned}\alpha_j^- = \lambda_j^- &+ \sum_{k=1}^{N}\sum_{v=1}^{I}\alpha_{kv}\int_0^\infty e^{-\alpha_k^- x} r_{kv,js^-}(x)dF_{kv}(x) \\ &+ \sum_{k=1}^{N}\sum_{v=1}^{I}\alpha_{kv}\int_0^\infty \alpha_k^- e^{-\alpha_k^- x}(1 - F_{kv}(x))r_{kv^-,js^-}(x)dx. \qquad (6.2)\end{aligned}$$

In order to ensure stability we need the condition

$$\sum_{u=1}^{I} \alpha_{ju} E(\min\{S_j^-, S_{ju}\}) < 1, \qquad j = 1, \ldots, N,$$

where S_j^- is an exponential random variable with rate α_j^- which is independent of S_{ju}, and S_{ju} is a generic service requirement of a class u customer at node j with distribution function F_{ju}.

Theorem 9.24 *The stationary probability distribution of the network has the product form*

$$\pi(\boldsymbol{n}) = \prod_{j=1}^{N} \pi_j(\boldsymbol{n}_j),$$

where $\pi_j(\boldsymbol{n}_j)$ is the stationary probability of node j when it is in isolation and *subject to Poisson arrivals of class u customers at rate α_{ju} and Poisson arrivals of signals at rate α_j^-, i.e.,*

$$\pi_j(\boldsymbol{n}_j) = b_j^{-1}\frac{(n_{j1}+\cdots+n_{jI})!}{n_{j1}!\cdots n_{jI}!}\prod_{u=1}^{I}\Big(\alpha_{ju}E(\min\{S_j^-, S_{ju}\})\Big)^{n_{ju}},$$

where

$$b_j = \frac{1}{1-\sum_{u=1}^{I}\alpha_{ju}E(\min\{S_j^-, S_{ju}\})}.$$

The result presented in Theorem 9.24 is fairly general. Some special cases are the following.

Case 1. Consider a network with only regular customers and no signals, i.e., $\lambda_j^- = 0$ and $r_{ju,ks^-} = 0$ for all j, k and u. It represents a conventional queueing network with routing probabilities depending on the service requirement. In this case the traffic equations are reduced to the linear equations

$$\alpha_{ju} = \lambda_{ju} + \sum_{k=1}^{N}\sum_{v=1}^{I}\alpha_{kv}\bar{r}_{kv,ju}, \qquad j = 1,\ldots,N,\ u = 1,\ldots,I,$$

and $\lambda_j^- = 0$, for all j, where

$$\bar{r}_{ju,kv} = \int_0^\infty r_{ju,kv}(x)dF_{ju}(x),$$

$$\bar{r}_{ju,0} = \int_0^\infty r_{ju,0}(x)dF_{ju}(x)$$

are the average routing probabilities of class u customers leaving node j. The stationary distribution of the network is the product form with $\pi_j(\boldsymbol{n}_j)$ given by

$$\pi_j(\boldsymbol{n}_j) = b_j\frac{(n_{j1}+\cdots+n_{jI})!}{n_{j1}!\cdots n_{jI}!}\prod_{u=1}^{I}(\alpha_{ju}E(S_{ju}))^{n_{ju}}.$$

Moreover, if all the routing probabilities are constant, i.e., $r_{ju,iv}(x) = r_{ju,iv}$, and $r_{ju,0}(x) = r_{ju,0}$ for all $x \geq 0$, then $\bar{r}_{ju,iv} = r_{ju,iv}$ and $\bar{r}_{ju,0} = r_{ju,0}$. It then reduces to the result for multi-class Jackson networks with service positions discussed in Chapter 5.

Case 2. Consider the case where the routing probabilities are constant and do not depend on the service requirements, i.e.,

$$r_{ju,kv}(x) = r_{ju,kv}, \qquad r_{ju,ks^-}(x) = r_{ju,ks^-}, \qquad r_{ju,0}(x) = r_{ju,0},$$

$$r_{ju^-,kv}(x) = r_{ju^-,kv}, \quad r_{ju^-,ks^-}(x) = r_{ju\cdot ks^-}, \quad r_{ju^-,0}(x) = r_{ju^-,0},$$

for all x. In this case the traffic equations are simplified to

$$\alpha_{ju} = \lambda_{ju} + \sum_{k=1}^{N}\sum_{v=1}^{I} \alpha_{kv}\hat{F}_{kv}(\alpha_k^-)r_{kv,ju} + \sum_{k=1}^{N}\sum_{v=1}^{I} \alpha_{kv}(1 - \hat{F}_{kv}(\alpha_k^-))r_{kv^-,ju},$$

$$\alpha_j^- = \lambda_j^- + \sum_{k=1}^{N}\sum_{v=1}^{I} \alpha_{kv}\hat{F}_{kv}(\alpha_k^-)r_{kv,js^-} + \sum_{k=1}^{N}\sum_{v=1}^{I} \alpha_{kv}(1 - \hat{F}_{kv}(\alpha_k^-))r_{kv^-,js^-},$$

where

$$\hat{F}_{kv}(s) = \int_0^\infty e^{-sx} dF_{kv}(x)$$

is the Laplace transform of the service requirement distribution F_{kv}. This result is discussed in Chapter 6.

Case 3. Consider the case where the routing probabilities do not depend on the cause of departure, i.e., whether it is due to a regular service completion or due to a signal. So

$$r_{ju,kv}(x) = r_{ju^-,kv}(x), \quad r_{ju,ks^-}(x) = r_{ju^-,ks^-}(x),$$

for all x. In this case the traffic equations reduce to the linear equations

$$\alpha_{ju} = \lambda_{ju} + \sum_{k=1}^{N}\sum_{v=1}^{I} \alpha_{kv}\bar{r}_{kv,ju}, \qquad j = 1, \ldots, N, \ u = 1, \ldots, I,$$

$$\alpha_j^- = \lambda_j^- + \sum_{k=1}^{N}\sum_{v=1}^{I} \alpha_{kv}\bar{r}_{kv,js^-}, \qquad j = 1, \ldots, N,$$

where

$$\bar{r}_{kv,ju} = \int_0^\infty r_{kv,ju}(x)dF_{kv}^*(x),$$

$$\bar{r}_{kv,js^-} = \int_0^\infty r_{kv,js^-}(x)dF_{kv}^*(x)$$

are the average routing probabilities of class v customers departing from node i and $F_{kv}^*(x)$ is the cumulative distribution function of $\min\{S_{kv}, S_k^-\}$, i.e.,

$$F_{kv}^*(x) = 1 - e^{-\alpha_k^- x}(1 - F_{kv}(x)).$$

Observe that in this case the result is exactly the same as that of Case 1 with the service requirement of a class v customer at node k replaced by $\min\{S_{kv}, S_k^-\}$.

Case 4. Consider the case where a signal simply destroys a customer (negative customer), i.e., $r_{ju^-,0} = 1$ for all j and u. In this case the traffic equations are

$$\alpha_{ju} = \lambda_{ju} + \sum_{k=1}^{N}\sum_{v=1}^{I} \alpha_{kv} \int_0^\infty e^{-\alpha_k^- x} r_{kv,ju}(x) dF_{kv}(x),$$

$$\alpha_j^- = \lambda_j^- + \sum_{k=1}^{N}\sum_{v=1}^{I} \alpha_{kv} \int_0^\infty e^{-\alpha_k^- x} r_{kv,js^-}(x) dF_{kv}(x).$$

Furthermore, if $r_{kv,ju}(x) = r_{kv,ju}$ and $r_{kv,js^-}(x) = r_{kv,js^-}$ are constants for all $x \geq 0$, and the service times are exponentially distributed with rate μ_{ju} for all j and u, then the traffic equations become

$$\alpha_{ju} = \lambda_{ju} + \sum_{k=1}^{N}\sum_{v=1}^{I} \alpha_{kv} \hat{F}_{kv}(\alpha_k^-) r_{kv,ju},$$

$$\alpha_j^- = \lambda_j^- + \sum_{k=1}^{N}\sum_{v=1}^{I} \alpha_{kv} \hat{F}_{kv}(\alpha_k^-) r_{kv,js^-}.$$

PROOF OF THEOREM 9.24. We use Theorem 4.9 to prove this result. We first consider the case where all the routing probabilities $r_{ju,kv}(x)$ are lattice with respect to x, i.e., they are left continuous and change only at points $x_\ell = \epsilon\ell$ for $\ell = 1, \ldots$ and $\epsilon > 0$. For convenience let $x_0 = 0$. Arbitrary $r_{ju,kv}(x)$'s can be approximated by these routing functions as ϵ goes to zero. Call a class u customer that departs from node j after having completed greater than $x_{\ell-1}$ but not more than x_ℓ units of service requirements a class (u, x_ℓ) or a class $(u, x_\ell)^-$ customer, depending upon whether the departure is due to a regular service completion or due to the triggering of a signal.

We show that with such a classification each node is quasi-reversible. Clearly we only need to consider symmetric nodes since the quasi-reversibility of non-symmetric nodes, which do not have routing probabilities that depend on the service requirement, have already been established in Chapter 5.

Consider node j in isolation. Let class u customers arrive according to a Poisson process with rate α_{ju} and signals arrive according to a Poisson process with rate α_j^-. The service requirements of class u customers have distribution F_{ju}. Let S_j^- and S_{ju} be independent random variables having an exponential distribution with mean $1/\alpha_j^-$ and distribution F_{ju}, respectively. Then, the departure of a class u customer can be regarded as completing $S_j^- \wedge S_{ju}$ units of its service requirement. If

$$x_{\ell-1} < S_{ju} \leq x_\ell \quad \text{and} \quad S_{ju} \leq S_j^-,$$

then the customer is classified as a class (u, x_ℓ) departure, and if

$$x_{\ell-1} < S_j^- \le x_\ell \quad \text{and} \quad S_j^- < S_{ju},$$

then it is classified as a class $(u, x_\ell)^-$ departure, for $\ell = 1, 2, \ldots$.

Similar to the discussions in Sections 6.3 and 6.4, it can be shown that the node is quasi-reversible with respect to departure classes (u, x_ℓ) and $(u, x_\ell)^-$, where the departure rate of class (u, x_ℓ) customers is

$$\alpha_{ju} P(x_{\ell-1} < S_{ju} \le x_\ell, S_{ju} \le S_j^-) = \alpha_{ju} \int_{x_{\ell-1}}^{x_\ell} e^{-\alpha_j^- y} dF_{ju}(y),$$

and the departure rate of class $(u, x_\ell)^-$ is

$$\alpha_{ju} P(x_{\ell-1} < S_j^- \le x_\ell, S_j^- \le S_{ju}) = \alpha_{ju} \int_{x_{\ell-1}}^{x_\ell} (1 - F_{ju}(y))\alpha_j^- e^{-\alpha_j^- y} dy.$$

The routing probabilities of these customers are $r_{ju,kv}(x_\ell)$ and $r_{ju^-,kv}(x_\ell)$, respectively. An application of Theorem 4.9 yields the result for each $\epsilon > 0$. The general case is obtained by letting ϵ go to zero, since the stationary distribution of the network, supplemented by the attained service requirements of all the customers in the network with arbitrarily distributed services, is continuous with respect to service requirement distributions. The proof of this continuity argument is technical; the details are therefore not given here (see the reference notes). □

9.7 NETWORKS WITH ROUTING DEPENDING ON THE NUMBER OF INTERRUPTIONS BY SIGNALS

The network considered in the previous section has routing probabilities that depend on the historical information pertaining to each customer at the node of departure. We now present a network in which the routing probabilities depend on the customer's historical information with regard to the nodes previously visited, i.e., the number of service interruptions.

Consider the same network except for the routing probabilities. It is assumed now that when a customer completes its service at a node or is induced to leave by a signal, its routing probabilities depend on the number of times its services have been interrupted in the past by signals. Specifically, we assume that upon a service completion of a class u customer at node j, it goes to node k as a customer of class v with probability $r_{ju,kv}(m)$, as a signal with probability $r_{ju,ks^-}(m)$, and it leaves the system with probability $r_{ju,0}(m)$, where m denotes the number of service interruptions this customer has experienced in the past, $m = 0, 1, \ldots$. Similarly, when a signal arrives at node j, either from the outside or from another node, it induces a class u customer to leave for node k as a customer of class v with probability $r_{ju^-,kv}(m)$, as a signal with probability $r_{ju^-,ks^-}(m)$, and to leave the system with probability $r_{ju^-,0}(m)$. These transition probabilities satisfy

$$\sum_{k=1}^{N}\sum_{v=1}^{I} r_{ju,kv}(m) + \sum_{k=1}^{N} r_{ju,ks^-}(m) + r_{ju,0}(m) = 1,$$

$$\sum_{k=1}^{N}\sum_{v=1}^{I} r_{ju^-,kv}(m) + \sum_{k=1}^{N} r_{ju^-,ks^-}(m) + r_{ju^-,0}(m) = 1,$$

for $j = 1, \ldots, N,\ u = 1, \ldots, I$ and all nonnegative integer m.

Let $\alpha_{ju}(m)$ denote the overall arrival rate of class u customers at node j that have been interrupted m times by signals in the past; let α_j^- denote the average arrival rate of signals at node j. Let $n_{ju}(m)$ denote the number of class u customers at node j that have been interrupted m times by signals, and let

$$n_{ju} = \sum_{m=0}^{\infty} n_{ju}(m).$$

The state of the network is given by $\boldsymbol{n} = (\boldsymbol{n}_1, \ldots, \boldsymbol{n}_N)$, where $\boldsymbol{n}_j = (n_{j1}, \ldots, n_{jI})$. The traffic equations for this network are

$$\alpha_{ju}(0) = \lambda_{ju} + \sum_{k=1}^{N}\sum_{v=1}^{I} \alpha_{kv}(0)\hat{F}_{kv}(\alpha_k^-) r_{kv,ju}(0), \tag{7.1}$$

$$\begin{aligned}\alpha_{ju}(m) = &\sum_{k=1}^{N}\sum_{v=1}^{I} \alpha_{kv}(m)\hat{F}_{kv}(\alpha_k^-) r_{kv,ju}(m) \\ &+ \sum_{k=1}^{N}\sum_{v=1}^{I} \alpha_{kv}(m-1)(1-\hat{F}_{kv}(\alpha_k^-)) r_{kv^-,ju}(m-1), \quad m > 0,\end{aligned} \tag{7.2}$$

$$\begin{aligned}\alpha_j^- = \lambda_j^- &+ \sum_{m=0}^{\infty}\sum_{k=1}^{N}\sum_{v=1}^{I} \alpha_{kv}(m)\hat{F}_{kv}(\alpha_k^-) r_{kv,js^-}(m) \\ &+ \sum_{m=0}^{\infty}\sum_{k=1}^{N}\sum_{v=1}^{I} \alpha_{kv}(m)(1-\hat{F}_{kv}(\alpha_k^-)) r_{kv^-,js^-}(m).\end{aligned} \tag{7.3}$$

For this model we have a result similar to Theorem 9.24.

Theorem 9.25 *Assume*

$$\sum_{u=1}^{I} \alpha_{ju} E(\min\{S_j^-, S_{ju}\}) < \infty,$$

where S_j^- is an exponential random variable with mean $1/\alpha_j^-$, α_{ju} is the total arrival rate of class u customers at node j, i.e.,

$$\alpha_{ju} = \sum_{m=0}^{\infty} \alpha_{ju}(m),$$

and $\alpha_{ju}(m)$ and α_j^- are the solutions to the traffic equations (7.1), (7.2) and (7.3). The stationary distribution of the network is $\pi(\boldsymbol{n}) = \prod_{j=1}^{N} \pi_j(\boldsymbol{n}_j)$, where $\pi_j(\boldsymbol{n}_j)$ is given by

$$\pi_j(\boldsymbol{n}_j) = b_j^{-1} \frac{(n_{j1} + \cdots + n_{jI})!}{n_{j1}! \cdots n_{jI}!} \prod_{u=1}^{I} (\alpha_{ju} E(\min\{S_j^-, S_{ju}\}))^{n_{ju}},$$

where

$$b_j = \frac{1}{1 - \sum_{u=1}^{I} \alpha_{ju} E(\min\{S_j^-, S_{ju}\})}.$$

The proof of this result is similar to that of Theorem 9.24, and is left to the reader as an exercise.

9.8 REFERENCE NOTES

Linear traffic networks have been studied in depth, see for example, Chandy and Martin (1983), Boucherie and Van Dijk (1991, 1993) Henderson, Pearce, Taylor and Schassberger (1995), and Miyazawa (1994), and its structure is now well understood. In particular, a full characterization of linear traffic networks has been obtained by Miyazawa (1996). However, much less is known about the structure of nonlinear networks. The material presented in Sections 9.3, 9.4 and 9.5 is based on some more general results that are included in the next chapter, which, in turn, are taken from Miyazawa and Chao (1998). Sections 9.6 and 9.7 are based on Chao and Pinedo (1995*b*). For age-dependent routing in Generalized Semi-Markov Processes (GSMP), see Rumswicz (1987). Example 9.9 is taken from Kelly (1979), in which the clustering process is assumed to satisfy the conditions that (2.17) and (2.18) have a solution. For more discussion on clustering processes, see Kelly (1979) or Whittle (1986). The continuity of the network in the proof of Theorem 9.24 follows from the continuity of Markov processes, e.g., see Miyazawa (1991). Example 9.10 is an extension of the circuit switching network with Poisson arrivals, which traces back to the original work of Erlang (see Burman, Lehoczky and Lim (1984), Kelly (1991), and Ross (1995) for related references). Example 9.13 is taken from Henderson, Northcote and Taylor (1994*b*). For results on nonlinear networks with batch movements, see Henderson, Northcote and Taylor (1995), Serfozo and Yang (1998), Chao and Serfozo (1997), and Miyazawa and Chao (1998). Example 9.15 on state-dependent coalescence networks was first studied in Chao and Zheng (1998). Example 9.17 is taken from Henderson et al. (1994*b*), in which the predator–prey model is studied under the assumption that each species has at most one predator.

EXERCISES

9.1 Prove the statement of Remark 9.5.

9.2 Extend Example 9.9 to allow more general cluster merging processes (three into one or one into three, etc.).

9.3 Extend the model of Theorem 9.12 to allow the routing probabilities of a signal-triggered departure to depend on the source of the signal. That is, if a customer at node j is triggered to leave by a class u negative signal that arrives at node j from node i, then the customer goes to node k as a customer with probability $r_{ju^-,kc}(i)$, as a class v^- negative signal with probability $r_{ju^-,kv^-}(i)$, and it leaves the network with probability $r_{ju^-,0}(i)$, where

$$\sum_{k=1}^{N} r_{ju^-,kc}(i) + \sum_{k=1}^{N}\sum_{v=1}^{I} r_{ju^-,kv^-}(i) + r_{ju^-,0}(i) = 1,$$
$$i = 0, \ldots, \quad j = 1, \ldots, N, \quad v = 1, \ldots, I.$$

Show that the network has a product form stationary distribution. Write down the traffic equations for this model.

9.4 Extend Example 9.15 to allow the routing probabilities to depend on the service batch sizes and triggering batch sizes. Assume that, when a full batch of u customers completes its service at node j, it joins node k as a regular customer with probability $r_{ju,kc}$, as a type v negative signal with probability r_{ju,kv^-}, and it leaves the network with probability $r_{ju,0}$; similarly, when a full batch of u customers at node j is induced to move by a type u negative signal, it joins node k as a regular customer with probability $r_{ju^-,kc}$, as a type v negative signal with probability r_{ju^-,kv^-}, and it leaves the network with probability $r_{ju^-,0}$. State and prove the result for the model. What are the traffic equations?

9.5 Prove Proposition 9.16.

9.6 Prove Proposition 9.14.

9.7 Prove Proposition 9.21.

9.8 Prove Proposition 9.23 using Theorem 9.18.

9.9 The quasi-reversibility used in proving Theorem 9.24 can be intuitively defined as follows. Considering the arrival process as a marked point process with the marks being the service times required by the customers, then quasi-reversibility is defined as that the future arrivals of the marked point process, the current state of the system, and the past departures of the marked point process are independent. Show that First-Come First-Served queues do not satisfy this quasi-reversibility property by considering a stationary $M/M/1$ queueing system. Assume that the current time is $t = 10$, and that there are three departures in $[0, 10]$. Specifically, we assume that the marked point process of past departures between 0 and 10 are $(T_1, S_1) = (2, 3.2)$, $(T_2, S_2) = (7, 4.5)$, and $(T_3, S_3) = (9, 1.5)$, where T_j is the departure time of customer j, and S_j is its service time, $j = 1, 2, \ldots$. Since the third customer leaves the system at time 9, and its service time is 1.5, this customer started service at time 7.5. However, the second customer leaves the system at time 7, the system is therefore empty between 7 and 7.5. In other words, knowing the marked point process of past departures enables us to obtain specific information about the state of the system at certain times, e.g., there is no customer in the system at time 7.4. From the fact that the states of the system (number of customers in the system) at time $t = 7.4$ and time $t = 10$ are not independent, we conclude that the marked point process of past departures and the current state of the system are *not* independent.

9.10 Consider a network with two nodes in series. The first node is the $M/M/1$ queue just described in Exercise 9.9, and the second node is another $M/M/1$ queue. There is a Poisson arrival process with rate 1 at node 1. Let the transition probabilities from node 1 be $p_{12}(x) = 1$ for $x < 1$, and $p_{10}(x) = 1$ for $x \geq 1$, $p_{20}(x) = 1$ for all x. That is, a departure from node 1 goes to node 2 only if its service time at node 1 is less than 1, otherwise it leaves the system. Departures from node 2 always leave the system. Show that this network does not have a product form solution, i.e., the stationary distribution of the network cannot be written as the product of the marginal distributions.

9.11 Consider the network with routing depending upon the service requirements of Section 9.6. Change the routing probability $r_{ju^-,kv}(x)$ due to signal triggering so that x is the original amount of the service requirement for the triggered customer. Then, Theorem 9.24 holds true if we replace the traffic equations (6.1) and (6.2) by

$$\alpha_{ju} = \lambda_{ju} + \sum_{k=1}^{N}\sum_{v=1}^{I} \alpha_{kv} \int_0^\infty e^{-\alpha_k^- x} r_{kv,ju}(x) dF_{kv}(x)$$
$$+ \sum_{k=1}^{N}\sum_{v=1}^{I} \alpha_{kv} \int_0^\infty (1 - e^{-\alpha_k^- x}) r_{kv^-,ju}(x) dF_{kv}(x),$$
$$\alpha_j^- = \lambda_j^- + \sum_{k=1}^{N}\sum_{v=1}^{I} \alpha_{kv} \int_0^\infty e^{-\alpha_k^- x} r_{kv,js^-}(x) dF_{kv}(x)$$
$$+ \sum_{k=1}^{N}\sum_{v=1}^{I} \alpha_{kv} \int_0^\infty (1 - e^{-\alpha_k^- x}) r_{kv^-,js^-}(x) dF_{kv}(x).$$

10

Local Balance Revisited

The two most frequently used techniques for showing product form solutions are quasi-reversibility and local balance. In Chapters 3 and 4 we studied the theoretical structure, the properties, and the intuitive implications of quasi-reversibility. In this chapter we focus on the local balance. We discuss its relationship with quasi-reversibility, and explore the pivotal role it plays in extending the results for networks with state-independent transitions to networks with state-dependent transitions. We also present some of the important properties implied by the local balance, namely the preservation property under time reversal and the Arrival Theorem.

In this chapter we use a vector $\boldsymbol{x}$ to denote the state of a Markov chain; it usually represents the state of a network.

10.1 BIASED LOCAL BALANCE

In Chapter 2 we defined a full set of local balances for a Markov chain with transition rates $q(\boldsymbol{x}, \boldsymbol{x}')$, $\boldsymbol{x}, \boldsymbol{x}' \in \mathcal{S}$, as

$$\pi(\boldsymbol{x}) \sum_{\boldsymbol{x}' \in \mathcal{S}} q_j(\boldsymbol{x}, \boldsymbol{x}') = \sum_{\boldsymbol{x}' \in \mathcal{S}} \pi(\boldsymbol{x}') q'_j(\boldsymbol{x}', \boldsymbol{x}), \qquad \boldsymbol{x} \in \mathcal{S},\ j \in J, \tag{1.1}$$

where J is a countable set, and $\{q_j;\ j \in J\}$ and $\{q'_j;\ j \in J\}$ are decompositions of the transition rate q of the Markov chain. We noted that a full set of local balances is useful for obtaining the stationary distributions of each node as well as of the entire network. In Chapter 9, we showed that such local balances characterize linear traffic networks. In conventional queueing networks, q_j and q'_j typically correspond to the transition rates for arrivals and departures. This, however, is not possible in nonlinear networks such as networks with negative signals, since the arrival of a negative signal at a node has the same effect as a departure. It turns out that the following form of the local balance is particularly useful for the study of nonlinear networks.

Definition 10.1 (Biased local balance) Suppose the transition rate function $q(\boldsymbol{x}, \boldsymbol{x}')$ of a continuous time Markov chain with state space $\mathcal{S}$ can be decomposed into transition functions $q_{jk}(\boldsymbol{x}, \boldsymbol{x}')$ over a finite or countable index set $j, k \in J$, i.e.,

$$q(\boldsymbol{x}, \boldsymbol{x}') = \sum_{j\in J}\sum_{k\in J} q_{jk}(\boldsymbol{x}, \boldsymbol{x}'), \qquad \boldsymbol{x}, \boldsymbol{x}' \in \mathcal{S}. \tag{1.2}$$

The continuous time Markov chain is said to satisfy the biased local balance with respect to $\{q_{jk};\ j, k \in J\}$ if there exists a set of numbers γ_j, $j \in J$, such that

$$\sum_{j\in J} \gamma_j = 0, \tag{1.3}$$

and

$$\pi(\boldsymbol{x}) \left(\sum_{k\in J}\sum_{\boldsymbol{x}'\in\mathcal{S}} q_{jk}(\boldsymbol{x}, \boldsymbol{x}') + \gamma_j \right) = \sum_{k\in J}\sum_{\boldsymbol{x}'\in\mathcal{S}} \pi(\boldsymbol{x}') q_{kj}(\boldsymbol{x}', \boldsymbol{x}), \qquad j \in J, \boldsymbol{x} \in \mathcal{S}, \tag{1.4}$$

where π is the stationary distribution of the Markov chain. Clearly, (1.3) and (1.4) imply global balance.

Note that the biased local balance is different from the full set of local balances of Definition 2.14 in Chapter 2, since the left-hand side of (1.4) has a constant term γ_j. If $\gamma_j = 0$, then the biased local balance is indeed a special case of a full set of local balances because of (1.2).

In many queueing network applications, the indices $j, k \in J$ of the transition function q_{jk} represent either a pair of nodes or a pair of a node and a customer class, and they represent transitions for the corresponding arrivals and departures. For instance, if in the queueing network of Theorem 9.4 we set

$$q_{jk}(\boldsymbol{n} + \mathbf{e}_j, \boldsymbol{n} + \mathbf{e}_k) = \mu_j(\boldsymbol{n} + \mathbf{e}_j) r_{jk},$$

then the local balance (1.9) of Chapter 9 is identical to (1.4) with $\gamma_j = 0$. Similarly, if in the network of Theorem 9.7

$$q_{\bar{\boldsymbol{a}}\bar{\boldsymbol{a}}'}(\boldsymbol{n}, \boldsymbol{n} + \boldsymbol{a} - \boldsymbol{a}') = \mu(\boldsymbol{n}, \bar{\boldsymbol{a}}) r(\bar{\boldsymbol{a}}, \bar{\boldsymbol{a}}'),$$

then the local balance (2.4) of Chapter 9 is of the form (1.4) with $\gamma_j = 0$. In the latter case, the indices represent neither nodes nor customer classes; they represent pairs of arrival and departure vectors.

The following example illustrates the biased local balance equations with nonzero bias terms.

Example 10.2 (A batch service queue) Consider the batch service queue of Section 2.6. The queue is characterized by

$$q^{\mathrm{A}}(n, n+1) = \lambda,$$
$$q^{\mathrm{D}}(n+\ell, n) = \mu b(\ell),$$
$$q^{\mathrm{I}}(n, 0) = \mu \overline{b}(n+1).$$

As always, the unspecified components are assumed to be zero. Suppose the queue is stable, i.e., $\lambda < \mu E(Y)$. From Lemma 2.32 and the equivalence of (6.4) and (6.5) in Chapter 2, it follows that the stationary distribution π of the queue is

$$\pi(n) = (1-\rho)\rho^n, \qquad n = 0, 1, \ldots,$$

where $\rho < 1$ is the unique solution of the equation

$$\lambda + \mu = \lambda/\rho + \mu \sum_{\ell=1}^{\infty} \rho^{\ell} b(\ell). \tag{1.5}$$

Hence we have

$$\sum_{n'=0}^{\infty} q^{\mathrm{A}}(n, n') = \lambda,$$
$$\sum_{n'=0}^{\infty} \pi(n') q^{\mathrm{D}}(n', n) = \mu(1-\rho)\rho^n \sum_{\ell=1}^{\infty} \rho^{\ell} b(\ell),$$
$$\sum_{n'=0}^{\infty} \pi(n') q^{\mathrm{I}}(n', n) = 1[n=0]\mu(1-\rho) \sum_{\ell=1}^{\infty} \rho^{\ell} \overline{b}(\ell+1).$$

Multiplying both sides of (1.5) by $(1-\rho)\rho^n$ yields

$$\pi(n)\left(\sum_{n'=0}^{\infty} q^{\mathrm{A}}(n, n') + \mu - \lambda\rho^{-1}\right) = \sum_{n'=0}^{\infty} \pi(n') q^{\mathrm{D}}(n', n), \qquad n \geq 0. \tag{1.6}$$

Subtracting this equation from the global balance equation of the queue defined by

$$q(n, n') = q^{\mathrm{A}}(n, n') + q^{\mathrm{D}}(n, n') + q^{\mathrm{I}}(n, n'),$$

we obtain

$$\pi(n)\left(\sum_{n'=0}^{\infty}(q^{\mathrm{D}}(n,n')+q^{\mathrm{I}}(n,n'))-\mu+\lambda\rho^{-1}\right)$$
$$=\sum_{n'=0}^{\infty}\pi(n')(q^{\mathrm{A}}(n',n)+q^{\mathrm{I}}(n',n)),\qquad n\geq 0.\qquad (1.7)$$

Hence, if we define $q_{jk},\ j,k=0,1$, by setting

$$q_{01}=q^{\mathrm{A}},\quad q_{10}=q^{\mathrm{D}},\quad q_{00}\equiv 0,\quad q_{11}=q^{\mathrm{I}},$$

then (1.6) and (1.7) represent biased local balances with bias terms

$$\gamma_0=\mu-\lambda\rho^{-1}\quad\text{and}\quad\gamma_1=\lambda\rho^{-1}-\mu.$$

□

In a similar fashion as in Example 10.2, it can be shown that the batch arrival and batch service queue of Section 2.7 satisfies biased local balance. This suggests that any quasi-reversible queue satisfies biased local balance. This is indeed true. In the next section we extend the biased local balance of Example 10.2 to networks with quasi-reversible nodes.

The bias terms in biased local balance equations do not correspond to transition rates. There is another way to present the biased local equations. Consider, for instance, Example 10.2 again. From (1.5), it follows that

$$\lambda=\mu\frac{\rho\left(1-\sum_{\ell=1}^{\infty}\rho^{\ell}b(\ell)\right)}{1-\rho}$$
$$=\mu\sum_{\ell=1}^{\infty}\sum_{m=1}^{\ell}\rho^{m}b(\ell).$$

Multiplying both sides of this equation by ρ^{n-1} and substituting the geometric stationary distribution π, we obtain

$$\lambda\pi(n-1)=\mu\sum_{\ell=1}^{\infty}\sum_{m=1}^{\ell}\pi(n+m-1)b(\ell),\qquad n\geq 1.\qquad (1.8)$$

This represents a balance of flows between states $0,1,\ldots,n-1$ and states $n,n+1,\ldots$, which, in Section 2.1, is referred to as a *cross balance*. We will show in Section 10.4 that, under some additional conditions, a local form of a cross balance, referred to as a *cross local balance*, is equivalent to the biased local balance. In Chapters 7 and 9, we have already seen such cross local balances (see (1.5) of Chapter 7 and (3.4) of Chapter 9). Later in this chapter it will be shown that the cross local balance is a key in obtaining networks with state-dependent transitions.

10.2 QUASI-REVERSIBILITY AND BIASED LOCAL BALANCE

As we have seen in earlier chapters, quasi-reversibility and local balance are very important techniques for verifying product form solutions. These two properties concern different aspects of complex systems. Quasi-reversibility is a flow property that preserves the Poisson flows in a system. Local or biased local balance equations, on the other hand, are more detailed forms of balance equations than global balance. In this section we explore the relationship between quasi-reversibility and biased local balance.

We use the notation and framework of Chapter 4. The network consists of N nodes, numbered from 1 to N. Node 0 is added to represent the outside. The set of arrival and departure classes of node j is denoted by T_j. Node j has state space $\mathcal{S}_j$ and is characterized by p^{A}_{ju}, q^{D}_{ju} and q^{I}_j on state space $\mathcal{S}_j$. The N nodes are connected to one another through routing probabilities $r_{ju,kv}$, as described in Section 4.2. Exogenous class u arrivals, i.e., class u departures from node 0 (the outside), form a Poisson process with rate β_{0u}, and each arrival is routed to node j as a class v arrival with probability $r_{0u,jv}$. Let T_0 be the set of entity classes that departs from and arrives at node 0. For notational convenience, assume $\mathcal{S}_0 = \{0\}$. The Poisson arrivals of exogenous entities are described by

$$p^{\text{A}}_{0u}(0,0) = 1, \quad q^{\text{D}}_{0u}(0,0) = \beta_{0u}, \quad q^{\text{I}}_0(0,0) = 0, \qquad u \in T_0.$$

Define, for $j, k = 0, 1, \ldots, N$,

$$q_{jj}(\boldsymbol{x}, \boldsymbol{x}') = 1[x_\ell = x'_\ell;\, \ell \neq j]\left(\sum_{u,v}\sum_{y_j \in \mathcal{S}_j} q^{\text{D}}_{ju}(x_j, y_j) r_{ju,jv} p^{\text{A}}_{jv}(y_j, x'_j) + q^{\text{I}}_j(x_j, x'_j)\right),$$

$$q_{jk}(\boldsymbol{x}, \boldsymbol{x}') = 1[x_\ell = x'_\ell;\, \ell \neq j, k] \sum_{u,v} q^{\text{D}}_{ju}(x_j, x'_j) r_{ju,kv} p^{\text{A}}_{kv}(x_k, x'_k), \qquad j \neq k.$$

As we discussed in Section 4.2, the network is described by a Markov chain with state space $\mathcal{S} = \mathcal{S}_0 \times \mathcal{S}_1 \times \cdots \times \mathcal{S}_N$ and transition rates

$$q(\boldsymbol{x}, \boldsymbol{x}') = \sum_{j=0}^{N}\sum_{k=0}^{N} q_{jk}(\boldsymbol{x}, \boldsymbol{x}'), \qquad \boldsymbol{x}, \boldsymbol{x}' \in \mathcal{S}, \tag{2.1}$$

where $q_{jk}(\boldsymbol{x}, \boldsymbol{x})$ is the transition rate from state $\boldsymbol{x}$ to itself.

To analyze the network process we need to study each node in isolation. The transition rate function $q_j^{(\boldsymbol{\alpha}_j)}$ of node j on state space $\mathcal{S}_j$ is defined by

$$q_j^{(\boldsymbol{\alpha}_j)}(x_j, x'_j) = \sum_{u \in T_j} \alpha_{ju} p^{\text{A}}_{ju}(x_j, x'_j) + \sum_{v \in T_j} q^{\text{D}}_{jv}(x_j, x'_j) + q^{\text{I}}_j(x_j, x'_j), \quad x_j, x'_j \in \mathcal{S}_j,$$

for a set of nonnegative numbers $\boldsymbol{\alpha}_j = \{\alpha_{ju};\, u \in T_j\}$. Here, α_{ju} represents the arrival rate of class u entities at node j. Let π_j be the stationary distribution of $q_j^{(\boldsymbol{\alpha}_j)}$; its dependency on $\boldsymbol{\alpha}_j$ is made implicit. Assume that node j is quasi-reversible, i.e., there exists a set of nonnegative numbers $\{\beta_{ju};\ j = 0, 1, \ldots, N, u \in T_j\}$ such that

$$\sum_{x'_j \in \mathcal{S}_j} \pi_j(x'_j) q^{\mathrm{D}}_{ju}(x'_j, x_j) = \beta_{ju} \pi_j(x_j), \qquad x_j \in \mathcal{S}_j,$$

$$\text{for all } j = 1, \ldots, N, u \in T_j. \tag{2.2}$$

For $j = 0$, it is clearly satisfied. The β_{ju} is interpreted as the departure rate of class u entities from node j.

Assume that α_{ju} and β_{ju} satisfy the traffic equations

$$\alpha_{ju} = \sum_{k=0}^{N} \sum_{v \in T_k} \beta_{kv} r_{kv,ju}, \qquad j = 1, \ldots, N, u \in T_j. \tag{2.3}$$

It follows from Theorem 4.3 that the stationary distribution π of the network has the product form

$$\pi(\boldsymbol{x}) = \prod_{j=1}^{N} \pi_j(x_j), \qquad \boldsymbol{x} \equiv (x_1, x_2, \ldots, x_N) \in \mathcal{S}. \tag{2.4}$$

To show that the quasi-reversible network satisfies the biased local balance equations, first note that (2.3) implies

$$\sum_{j=0}^{N} \sum_{u \in T_j} \alpha_{ju} = \sum_{j=0}^{N} \sum_{u \in T_j} \beta_{ju}. \tag{2.5}$$

This equation indicates that the total flow rates are preserved, but it does *not* imply that the total arrival rate to the network is equal to the total departure rate from the network, i.e.,

$$\sum_{j=1}^{N} \sum_{u \in T_j} \alpha_{ju} = \sum_{j=1}^{N} \sum_{u \in T_j} \beta_{ju}\,. \tag{2.6}$$

This equation is referred to as the flow balance condition.

Set

$$\gamma_j = \sum_{u \in T_j} \alpha_{ju} - \sum_{u \in T_j} \beta_{ju}, \qquad j = 0, 1, \ldots, N, \tag{2.7}$$

which, by (2.5), satisfies $\sum_{j=0}^{N} \gamma_j = 0$. Note that the biased local balance (1.4) for $j = 0$ is equivalent to the definition of γ_0, i.e., (2.7) for $j = 0$.

The following theorem not only shows that a quasi-reversible network satisfies the biased local balance, but that the biased local balance property also allows us to present a simpler proof for the product form result.

Theorem 10.3 *A quasi-reversible network characterized by (2.2) and (2.3) satisfies the biased local balance with respect to* $\{q_{jk};\ j,k = 0,1,\ldots,N\}$ *and the product form distribution* π *of (2.4), with the bias term being* γ_j *of (2.7) for* $j = 0,1,\ldots,N$. *Hence the product form (2.4) is the stationary distribution of the network.*

PROOF. For notational convenience we define, for an N-dimensional vector $\boldsymbol{a}$ and the numbers b and c, the vector $\boldsymbol{a}_j(b)$ as a vector $\boldsymbol{a}$ with its jth component replaced by b. If $j = 0$, then $\boldsymbol{a}_j(b)$ is defined as $\boldsymbol{a}$. The vector $\boldsymbol{a}_{jk}(b,c)$ is obtained by applying this operator twice, i.e., the jth and kth components of $\boldsymbol{a}$ are replaced by b and c, i.e., $\boldsymbol{a}_{jk}(b,c) = (\boldsymbol{a}_j(b))_k(c)$. Thus $\boldsymbol{a}_{jj}(b,c) = \boldsymbol{a}_j(c)$. Also, we define $x_k[x'_j]$ for each $k,\ j = 0,1,\ldots,N$ such that $x_k[x'_j] = x_k$ for $j \neq k$ and $x_j[x'_j] = x'_j$.

From (2.2) and (2.3), it follows that

$$\alpha_{ju} = \sum_{k=0}^{N}\sum_{v\in T_k}\frac{1}{\pi_k(x_k)}\sum_{x'_k\in\mathcal{S}_k}\pi_k(x'_k)q^{\mathrm{D}}_{kv}(x'_k,x_k)r_{kv,ju}\,.$$

Multiplying both sides of this equation by $\pi(\boldsymbol{x}_j(x'_j))p^{\mathrm{A}}_{ju}(x'_j,x_j)$, where π is given by (2.4), yields

$$\begin{aligned}&\pi(\boldsymbol{x}_j(x'_j))\alpha_{ju}p^{\mathrm{A}}_{ju}(x'_j,x_j)\\ &= \sum_{k=0}^{N}\sum_{v\in T_k}\sum_{y_k\in\mathcal{S}_k}\pi(\boldsymbol{x}_{jk}(x'_j,y_k))q^{\mathrm{D}}_{kv}(y_k,x_k[x'_j])r_{kv,ju}p^{\mathrm{A}}_{ju}(x'_j,x_j).\end{aligned}\qquad(2.8)$$

After summing both sides of (2.8) over all $u \in T_j$ and $x'_j \in \mathcal{S}_j$ and using the fact that π_j is the stationary distribution of $q_j^{(\boldsymbol{\alpha}_j)}$, the left-hand side becomes

$$\begin{aligned}&\sum_{u\in T_j}\sum_{x'_j\in\mathcal{S}_j}\pi(\boldsymbol{x}_j(x'_j))\alpha_{ju}p^{\mathrm{A}}_{ju}(x'_j,x_j)\\ &= \pi(\boldsymbol{x})\Bigg(\sum_{x'_j\in\mathcal{S}_j}\Bigg(\sum_{u\in T_j}q^{\mathrm{D}}_{ju}(x_j,x'_j)+q^{\mathrm{I}}_j(x_j,x'_j)\Bigg)+\sum_{u\in T_j}\alpha_{ju}\Bigg)\\ &\quad-\sum_{x'_j\in\mathcal{S}_j}\sum_{u\in T_j}\pi(\boldsymbol{x}_j(x'_j))q^{\mathrm{D}}_{ju}(x'_j,x_j)-\sum_{x'_j\in\mathcal{S}_j}\pi(\boldsymbol{x}_j(x'_j))q^{\mathrm{I}}_j(x'_j,x_j)\\ &= \pi(\boldsymbol{x})\Bigg(\sum_{x'_j\in\mathcal{S}_j}\Bigg(\sum_{u\in T_j}q^{\mathrm{D}}_{ju}(x_j,x'_j)+q^{\mathrm{I}}_j(x_j,x'_j)\Bigg)+\gamma_j\Bigg)-\sum_{x'_j\in\mathcal{S}_j}\pi(\boldsymbol{x}_j(x'_j))q^{\mathrm{I}}_j(x'_j,x_j),\end{aligned}\qquad(2.9)$$

where the second equality follows from the quasi-reversibility (2.2) and the definition of γ_j (2.7). The right-hand side of (2.8) becomes

$$\sum_{k=0}^{N}\sum_{u\in T_j}\sum_{v\in T_k}\sum_{x'_j\in\mathcal{S}_j}\sum_{y_k\in\mathcal{S}_k}\pi(\boldsymbol{x}_{jk}(x'_j,y_k))q^{\mathrm{D}}_{kv}(y_k,x_k[x'_j])r_{kv,ju}p^{\mathrm{A}}_{ju}(x'_j,x_j)$$

$$= \sum_{k=0}^{N} \sum_{\boldsymbol{x}' \in \mathcal{S}} \pi(\boldsymbol{x}') q_{kj}(\boldsymbol{x}', \boldsymbol{x}) - \sum_{x'_j \in \mathcal{S}_j} \pi(\boldsymbol{x}_j(x'_j)) q_j^1(x'_j, x_j). \quad (2.10)$$

Comparing (2.9) and (2.10) yields (1.4). □

In the next chapter we will show that the reverse of Theorem 10.3 is also true, i.e., if a product form satisfies a biased local balance, then each node of the network is quasi-reversible (see Theorem 11.7). This, however, does not mean that a product form implies quasi-reversibility. Quasi-reversibility is stronger than the condition that the network has a product form stationary distribution (see Chapter 11).

The following result is immediate from Theorem 10.3.

Corollary 10.4 *If each node of a quasi-reversible network is internally balanced, i.e.,*

$$\sum_{u \in T_j} \alpha_{ju} = \sum_{u \in T_j} \beta_{ju}, \qquad j = 1, \dots, N,$$

then the network is locally balanced with respect to $\{q_{jk};\ j, k = 0, 1, \dots, N\}$.

Example 10.5 (Assembly-transfer network) Consider the assembly-transfer network discussed in Chapter 8. Node j has a single exponential server with rate μ_j. Customers arrive at node j from the outside according to a batch Poisson process with rate λ_j and each brings in a batch of ℓ customers with probability $a_j(\ell)$, $\ell \geq 1$. The services at node j are also in batches and the batch size distribution is $\{b_j(\ell);\ \ell \geq 1\}$. When a full batch of ℓ customers finishes service at node j, it joins node k as a batch of m customers with probability $r_{j\ell,km}$, and it leaves the network with probability $r_{j\ell,0}$. Partial batches leave the network with probability 1.

It follows from Theorem 8.2 that, under the condition that whenever node j is empty an additional arrival process is activated with rate (2.2) and batch size distribution (2.3) in Chapter 8, the stationary distribution of the network is

$$\pi(\boldsymbol{n}) = \prod_{j=1}^{N} (1 - \rho_j) \rho_j^{n_j},$$

where ρ_j is determined by the traffic equations (2.8) in Chapter 8.

Define a class ℓ arrival (departure) as a batch arrival (full batch departure) of size ℓ, $\ell = 1, 2, \dots$. This network is quasi-reversible and satisfies the biased local balance (1.4). Since the departure rate of class ℓ customers from node j is

$$\beta_{j\ell} = \mu_j \rho_j^{\ell} b_j(\ell),$$

it follows from the traffic equations that the bias term is (2.8) in Chapter 8 and that

$$\gamma_j = \sum_{\ell=1}^{\infty} (\alpha_{j\ell} - \beta_{j\ell})$$
$$= \lambda_j - \mu_j \sum_{\ell=1}^{\infty} \rho_j^{\ell} b_j(\ell) + \sum_{\ell=1}^{\infty} \sum_{k=1}^{N} \sum_{m=1}^{\infty} \mu_k \rho_k^m b_k(m) r_{km,j\ell}\,,$$
$$j = 1, 2, \ldots, N. \qquad \square$$

10.3 REFINED LOCAL BALANCES WITH RESPECT TO CUSTOMER CLASSES AND SERVICE POSITIONS

The biased local balance of the previous section is a balance with respect to each node. In some networks, e.g., the symmetric service networks of Chapters 5 and 6, there are multiple classes of entities, and a more refined form of local balance can be obtained, e.g., with respect to each class of customer in each position at each node. We first give an example of a symmetric service network with exponential servers, and then present a general result with regard to such a refined form of local balance.

Consider the symmetric service network with multiple classes of customers described in Section 5.1. Assume that all the processing requirements are exponentially distributed, with the service requirements of class u customers at node j being exponential with rate μ_{ju}. The state of the network is denoted by

$$\boldsymbol{x} = (\boldsymbol{x}_1, \boldsymbol{x}_2, \ldots, \boldsymbol{x}_N),$$

where

$$\boldsymbol{x}_j = (c_j(1), c_j(2), \ldots, c_j(n_j)), \qquad j = 1, \ldots, N,$$

n_j is the number of customers at node j, and $c_j(\ell)$ is the class of the customer in position ℓ, $\ell = 1, \ldots, n_j$. By Theorem 5.4, the stationary distribution of the network is

$$\pi(\boldsymbol{x}) = \prod_{j=1}^{N} \pi_j(\boldsymbol{x}_j),$$

and $\pi_j(\boldsymbol{x}_j)$ is the stationary distribution of node j when it is in isolation and subject to a Poisson arrival process of class u customers with rate α_{ju}, $u = 1, \ldots, I$, i.e.,

$$\pi_j(\boldsymbol{x}_j) = b_j^{-1} \prod_{\ell=1}^{n_j} \frac{\alpha_{jc_j(\ell)}}{\phi_j(\ell)\mu_{jc_j(\ell)}}, \tag{3.1}$$

where

$$b_j = \sum_{n=0}^{\infty} \frac{\left(\sum_{u=1}^{I} \alpha_{ju}/\mu_{ju}\right)^n}{\prod_{\ell=1}^{n} \phi_j(\ell)},$$

and the α_{ju}'s are determined by the traffic equations (1.1) in Chapter 5.

Before presenting the local balance equations for this network, we need some of the vector operators introduced in Chapter 5. While the operators in Section 5.2 are applied to the state of a single node, we apply the operators here to the state of the network. Let $\boldsymbol{x} \ominus \mathbf{e}_{j\ell}$ be the state of the network after a customer in position ℓ at node j leaves the network, let $\boldsymbol{x} \ominus \mathbf{e}_{j\ell} \oplus \mathbf{e}_{km}(v)$ be the state of the network after a customer in position ℓ at node j departs and joins position m of node k as a class v customer, and let $\boldsymbol{x} \oplus \mathbf{e}_{j\ell}(u)$ be the state of the network after an exogenous class u customer enters position ℓ at node j. Define the transition rate $q_{j\ell(u),km(v)}$ by

$$q_{0,j\ell(u)}(\boldsymbol{x}, \boldsymbol{x} \oplus \mathbf{e}_{j\ell}(u)) = \lambda_{ju}\delta_j(\ell, n_j),$$

$$\begin{aligned} &q_{j\ell(u),km(v)}(\boldsymbol{x}, \boldsymbol{x} \ominus \mathbf{e}_{j\ell} \oplus \mathbf{e}_{km}(v)) \\ &\qquad = \phi_j(n_j)\mu_{ju}\delta_j(\ell, n_j)r_{ju,kv}\delta_k(m, n_k+1)1[c_j(\ell) = u], \end{aligned}$$

$$q_{j\ell(u),0}(\boldsymbol{x}, \boldsymbol{x} \ominus \mathbf{e}_{j\ell}) = \phi_j(n_j)\mu_{ju}\delta_j(\ell, n_j)r_{ju,0}1[c_j(\ell) = u].$$

Theorem 10.6 *The product form distribution $\pi(\boldsymbol{x})$ satisfies the local balance equations*

$$\begin{aligned} &\pi(\boldsymbol{x}) \left[\sum_{k=1}^{N} \sum_{v \in T_k} \sum_{m=1}^{n_k+1} q_{j\ell(u),km(v)}(\boldsymbol{x}, \boldsymbol{x} \ominus \mathbf{e}_{j\ell} \oplus \mathbf{e}_{km}(v)) + q_{j\ell(u),0}(\boldsymbol{x}, \boldsymbol{x} \ominus \mathbf{e}_{j\ell}) \right] \\ &= \sum_{k=1}^{N} \sum_{v \in T_k} \sum_{m=1}^{n_k+1} \pi(\boldsymbol{x} \ominus \mathbf{e}_{j\ell} \oplus \mathbf{e}_{km}(v)) q_{km(v),j\ell(u)}(\boldsymbol{x} \ominus \mathbf{e}_{j\ell} \oplus \mathbf{e}_{km}(v), \boldsymbol{x}) \\ &+\pi(\boldsymbol{x} \ominus \mathbf{e}_{j\ell}) q_{0,j\ell(u)}(\boldsymbol{x} \ominus \mathbf{e}_{j\ell}, \boldsymbol{x}), \\ &\qquad \boldsymbol{x} \in \mathcal{S},\ j = 1, \ldots, N,\ \ell = 1, \ldots, n_j,\ u = c_j(\ell), \end{aligned} \tag{3.2}$$

or, equivalently,

$$\begin{aligned} &\pi(\boldsymbol{x})\phi_j(n_j)\mu_{ju}\delta_j(\ell, n_j) \\ &= \sum_{k=1}^{N} \sum_{v \in T_k} \sum_{m=1}^{n_k+1} \pi(\boldsymbol{x} \ominus \mathbf{e}_{j\ell} \oplus \mathbf{e}_{km}(v))\phi_k(n_k+1)\mu_{kv}\delta_k(m, n_k+1)r_{kv,ju}\delta_j(\ell, n_j) \\ &+\pi(\boldsymbol{x} \ominus \mathbf{e}_{j\ell})\lambda_{ju}\delta_j(\ell, n_j). \end{aligned} \tag{3.3}$$

Equations (3.2) and (3.3) hold for $j = 0$ if the subscript 0ℓ is interpreted as 0, and $\mathbf{e}_0$ is interpreted as the null vector with the convention that $\boldsymbol{x} \ominus \mathbf{e}_0 = \boldsymbol{x}$. Furthermore, these local balance equations constitute a full set of local balances, i.e., they imply global balance.

PROOF. After substituting the product form distribution π into the local balance equation (3.3), and canceling the common terms, we obtain

$$\begin{aligned}\alpha_{ju} &= \lambda_{ju} + \sum_{k=1}^{N} \sum_{v \in T_k} \sum_{m=1}^{n_k+1} \alpha_{kv} r_{kv,ju} \delta_k(m, n_k + 1) \\ &= \lambda_{ju} + \sum_{k=1}^{N} \sum_{v \in T_k} \alpha_{kv} r_{kv,ju}.\end{aligned}$$

This is exactly the traffic equation (1.1) in Chapter 5. Hence the local balance equations have to be satisfied. The proof is thus complete. □

Theorem 10.6 states that the rate going out of state $\boldsymbol{x}$ due to the departure of a class u customer from position ℓ of node j is equal to the rate going into state $\boldsymbol{x}$ due to the arrival of a class u customer at position ℓ of node j, and this holds for all $j = 1, \dots, N$ and all $\ell = 1, \dots, n_j$. This local balance is more refined than the local balance equations discussed in the previous section, since it is not only with respect to a node and a customer class, but it is also with respect to the position of a node.

To see why the network satisfies the refined local balance (3.2), first note that since the nodes in the symmetric service network are quasi-reversible and $\alpha_{ju} = \beta_{ju}$, it follows from Theorem 10.3 that the network satisfies local balances with respect to the nodes. A close look at the proof of Theorem 10.3 reveals that a crucial step in obtaining a biased local balance is the summation of (2.8) over all $u \in T_j$ and $x'_j \in \mathcal{S}_j$, which enables us to use the global balance equation for π_j in the derivation of (2.9). Hence, if we do not sum (2.8) over $u \in T_j$ but only over $x_j \in \mathcal{S}_j$, we would require a more detailed balance for π_j. For the symmetric service network, π_j is given by (3.1). So we have, for each $u \in T_j$ and $\boldsymbol{x}_j \in \mathcal{S}_j$,

$$\begin{aligned}&\pi_j(\boldsymbol{x}_j)\phi_j(n_j)\mu_{ju}\delta_j(\ell, n_j)1[c_j(\ell) = u] \\ &\qquad = \pi_j(\boldsymbol{x}_j \ominus \mathbf{e}_\ell)\alpha_{ju}\delta_j(\ell, n_j)1[c_j(\ell) = u], \qquad \ell \le n_j, \\ &\pi_j(\boldsymbol{x}_j)\alpha_{ju}\delta_j(\ell, n_j + 1) = \pi_j(\boldsymbol{x}_j \oplus \mathbf{e}_\ell)\phi(n_j + 1)\mu_{ju}\delta_j(\ell, n_j + 1), \qquad \ell \le n_j + 1.\end{aligned}$$

Thus

$$\pi_j(\boldsymbol{x}_j)\big(\alpha_{ju}\delta_j(\ell, n_j + 1) + \phi_j(n_j)\mu_{ju}\delta_j(\ell, n_j)1[c_j(\ell) = u]\big)$$

$$= \pi_j(\boldsymbol{x}_j \ominus \mathbf{e}_\ell)\alpha_{ju}\delta_j(\ell, n_j)1[c_j(\ell) = u]$$
$$+\pi_j(\boldsymbol{x}_j \oplus \mathbf{e}_\ell(u))\phi(n_j+1)\mu_{ju}\delta_j(\ell, n_j+1). \tag{3.4}$$

Summing (3.4) over $u \in T_j$ and ℓ yields the global balance for π_j. Note that (3.4) is a balance equation for each class, position, and node. This is exactly the balance equation we need.

We extend this local balance to the queueing network formulation in Section 10.2. Suppose the internal transition q_j^{I} can be decomposed for each class such that

$$\sum_{u \in T_j} q_{ju}^{\mathrm{I}}(x_j, x_j') = q_j^{\mathrm{I}}(x_j, x_j'), \qquad x_j, x_j' \in \mathcal{S}_j .$$

Define a rate function $q_{ju}^{(\boldsymbol{\alpha}_j)}$ for each $u \in T_j$ as

$$q_{ju}^{(\boldsymbol{\alpha}_j)}(x_j, x_j') = \alpha_{ju} p_{ju}^{\mathrm{A}}(x_j, x_j') + q_{ju}^{\mathrm{D}}(x_j, x_j') + q_{ju}^{\mathrm{I}}(x_j, x_j'), \quad x_j, x_j' \in \mathcal{S}_j, \tag{3.5}$$

and $q_{ju,kv}$ as

$$q_{ju,jv}(\boldsymbol{x}, \boldsymbol{x}') = 1[x_\ell = x_\ell'; \ell \neq j]$$
$$\times\left(\sum_{y_j \in \mathcal{S}_j} q_{ju}^{\mathrm{D}}(x_j, y_j) r_{ju,jv} p_{jv}^{\mathrm{A}}(y_j, x_j') + 1[u = v] q_{ju}^{\mathrm{I}}(x_j, x_j')\right),$$
$$q_{ju,kv}(\boldsymbol{x}, \boldsymbol{x}') = 1[x_\ell = x_\ell'; \ell \neq j, k] q_{ju}^{\mathrm{D}}(x_j, x_j') r_{ju,kv} p_{kv}^{\mathrm{A}}(x_k, x_k'), \qquad j \neq k.$$

Theorem 10.7 *Consider the quasi-reversible queueing network of Section 10.2 characterized by (2.2) and (2.3). Assume that π_j satisfies the balance equation for $q_{ju}^{(\boldsymbol{\alpha}_j)}$, for each $j = 1, \ldots, N$ and $u \in T_j$, i.e.,*

$$\pi_j(x_j) \sum_{x_j' \in \mathcal{S}_j} q_{ju}^{(\boldsymbol{\alpha}_j)}(x_j, x_j') = \sum_{x_j' \in \mathcal{S}_j} \pi_j(x_j') q_{ju}^{(\boldsymbol{\alpha}_j)}(x_j', x_j). \tag{3.6}$$

Then the network satisfies the biased local balance with respect to

$$\{q_{ju,kv}; \, j, k = 0, 1, \ldots, N, u, v \in T_j\}$$

and the product form π, with the bias term being $\gamma_{ju} = \alpha_{ju} - \beta_{ju}$, i.e.,

$$\pi(\boldsymbol{x})\left(\sum_{k=1}^{N} \sum_{v \in T_k} \sum_{\boldsymbol{x}' \in \mathcal{S}} q_{ju,kv}(\boldsymbol{x}, \boldsymbol{x}') + \gamma_{ju}\right) = \sum_{k=1}^{N} \sum_{v \in T_k} \sum_{\boldsymbol{x}' \in \mathcal{S}} \pi(\boldsymbol{x}') q_{kv,ju}(\boldsymbol{x}', \boldsymbol{x}),$$
$$j = 1, \ldots, N, u \in T_j, \boldsymbol{x} \in \mathcal{S}. \tag{3.7}$$

PROOF. The arguments run in parallel to the arguments in the proof of Theorem 10.3. The differences lie in the fact that we have to sum (3.4) over $x'_j \in \mathcal{S}_j$ but not over $u \in T_j$, i.e.,

$$\sum_{x'_j \in \mathcal{S}_j} \pi(\boldsymbol{x}_j(x'_j))\alpha_{ju} p^{\mathrm{A}}_{ju}(x'_j, x_j)$$
$$= \sum_{k=0}^{N} \sum_{v \in T_k} \sum_{x'_j \in \mathcal{S}_j} \sum_{y_k \in \mathcal{S}_k} \pi(\boldsymbol{x}_{jk}(x'_j, y_k)) q^{\mathrm{D}}_{kv}(y_k, x_k[x'_j]) r_{kv,ju} p^{\mathrm{A}}_{ju}(x'_j, x_j), \quad (3.8)$$

and that we do not use the global balance for π_j but rather the balance equation (3.6) and quasi-reversibility, i.e.,

$$\pi_j(x_j)\left(\alpha_{ju} + \sum_{x_j \in \mathcal{S}_j} (q^{\mathrm{D}}_{ju}(x_j, x'_j) + q^{\mathrm{I}}_{ju}(x_j, x'_j))\right)$$
$$= \pi_j(x_j)\beta_{ju} + \sum_{x'_j \in \mathcal{S}_j} \pi_j(x'_j)(\alpha_{ju} p^{\mathrm{A}}_{ju}(x'_j, x_j) + q^{\mathrm{I}}_{ju}(x'_j, x_j)). \qquad \square$$

The reader might ask whether this theorem is equivalent to Theorem 10.3 if we redefine a node for each pair j and u. This is true if state x_j of node j is a vector that consists of components that correspond to entity classes. This, however, is not required for Theorem 10.7 (see, for instance, the symmetric service network in Theorem 10.6).

10.4 CROSS LOCAL BALANCES WITH RESPECT TO POSITIVE AND NEGATIVE EFFECTS

In this section we introduce a structure of negative and positive effects of entities on the nodes in a queueing network. The network has three types of entities at each node: regular entities, negative entities, and positive entities. For simplicity, in this and the two subsequent sections we only consider a single class of regular entities, namely customers, which are denoted by c. The extension to multiple classes of regular entities is straightforward. Let T^{R}_j be the class of regular entities, i.e., $\{c\}$ in this case, and let T^-_j and T^+_j be the classes of negative and positive entities, respectively, for $j = 0, 1, \ldots, N$. Consequently, the set of classes T_j is a union of the disjoint subsets T^{R}_j, T^-_j and T^+_j, i.e.,

$$T_j = T^{\mathrm{R}}_j \cup T^-_j \cup T^+_j, \qquad j = 0, 1, \ldots, N.$$

Assume that the effects of negative entities are similar to regular customer departures, and the effects of positive entities are similar to regular customer arrivals.

Specifically, assume that when a class v negative entity arrives at node j, which is in state x_j, the state changes to x'_j with probability $p^{\mathrm{A}*}_{jv,c}(x_j, x'_j)$, which is similar to the change of state due to a class c regular departure. Similarly, when a class v positive entity arrives at node j, the state changes to x'_j with probability $p^{\mathrm{A}*}_{jv,c}(x_j, x'_j)$, which is similar to the change of state due to a class c regular arrival.

Example 10.8 (Effects of positive and negative entities) Assume node j is an $M/M/1$ queue with three classes of arrivals: regular customers c, positive entities c^+, and negative entities c^-. The state of node j is n_j, the number of regular customers. Assume that when a positive entity arrives at the node it increases the number of customers by 1, and when a negative entity arrives it decreases the number of customers by 1 provided the node is not empty. Clearly, the effect of a negative entity is similar to a regular service completion, and the effect of a positive entity is similar to the arrival of a regular customer. Using the notation just introduced, we have

$$p^{\mathrm{A}*}_{jc^+,c}(n_j, n_j + 1) = 1, \qquad n_j \geq 0,$$
$$p^{\mathrm{A}*}_{jc^-,c}(n_j, n_j - 1) = 1, \qquad n_j \geq 0. \qquad \square$$

Note that in this example $p^{\mathrm{A}*}_{jc^-,c}(n_j, n_j - 1)$ is defined only for $n_j \geq 1$. When $n_j = 0$, the effect of a negative entity cannot be similar to a regular service completion since there is nobody in service when the queue is empty. To take this into account in the general setting we define $p^{\mathrm{A}*}_{jv}(x_j, x_j)$ as the probability that an arriving class v positive or negative entity does not change the state of the node, i.e.,

$$p^{\mathrm{A}}_{jv}(x_j, x'_j) = p^{\mathrm{A}*}_{jv,c}(x_j, x'_j) + p^{\mathrm{A}*}_{jv}(x_j, x_j)1[x'_j = x_j], \quad x_j, x'_j \in \mathcal{S}_j, v \in T^-_j \cup T^+_j.$$

For convenience we sometimes also use the notation $p^{\mathrm{A}*}_{jv}(x_j, x_j)$, but it is understood as 0 when $x_j \neq x'_j$. Note that $\sum_{x'_j \in \mathcal{S}_j} p^{\mathrm{A}*}_{jv,c}(x_j, x'_j)$ may be less than 1, and may depend on x_j. We sometimes write $p^{\mathrm{A}*}_{jv,c}(x_j, x'_j)$ as $p^{\mathrm{A}-}_{jv,c}(x_j, x'_j)$ for $v \in T^-_j$ and as $p^{\mathrm{A}+}_{jv,c}(x_j, x'_j)$ for $v \in T^+_j$.

Assume that there exist nonnegative numbers α_{ju}'s and β_{ju}'s such that each node is quasi-reversible for all classes, i.e.,

$$\sum_{x'_j \in \mathcal{S}_j} \pi_j(x'_j) q^{\mathrm{D}}_{ju}(x'_j, x_j) = \beta_{ju}\pi_j(x_j), \quad x_j \in \mathcal{S}_j,$$
$$j = 0, 1, \ldots, N, \; u \in T_j = T^{\mathrm{R}}_j \cup T^-_j \cup T^+_j, \qquad (4.1)$$

where $\pi_j(x_j)$ is the stationary distribution of $q_j^{(\boldsymbol{\alpha}_j)}$ defined by

$$q_j^{(\boldsymbol{\alpha}_j)}(x_j, x_j') = \sum_{v \in T_j} \alpha_{jv} p_{jv}^{\mathrm{A}}(x_j, x_j') + q_{jc}^{\mathrm{D}}(x_j, x_j') + q_j^{\mathrm{I}}(x_j, x_j'), \quad (4.2)$$

$$j = 1, \ldots, N, \ x_j, x_j' \in \mathcal{S}_j,$$

and β_{ju}, $j = 1, \ldots, N$, $u \in T_j$, are determined by the traffic equations

$$\alpha_{ju} = \sum_{k=0}^{N} \sum_{v \in T_k} \beta_{kv} r_{kv,ju}, \qquad j = 1, \ldots, N, \ u \in T_j. \quad (4.3)$$

This network is referred to as a *quasi-reversible network with positive and negative entities.*

We assume that, for $v \in T_j^- \cup T_j^+$,

(a) $q_{jv}^{\mathrm{D}}(x_j, x_j') = 0$ for $x_j \neq x_j'$, i.e., positive and negative entities generated at a node do not change the state of that node.

In a sense, the departure rates of positive and negative entities $q_{jv}^{\mathrm{D}}(x_j, x_j)$ may be regarded as artificial since they have no direct effect on the node of departure. These rates are often introduced to ensure the quasi-reversibility of each node and to obtain a product form stationary distribution for the network. The additional departures in the assembly-transfer network and the concurrent batch movement network are examples of condition (a). Because of condition (a), (4.2) does not have the terms corresponding to q_{jv}^{D} for $v \in T_j^- \cup T_j^+$.

Theorem 10.9 *Consider a quasi-reversible network with positive and negative entities. Let π_j be the stationary distribution of $q_j^{(\boldsymbol{\alpha}_j)}$ defined in (4.2). The network has the stationary distribution π which is the product of the π_j and it satisfies the cross local balance equations*

$$\begin{aligned}
&\pi(\boldsymbol{x}) \sum_{x_j' \in \mathcal{S}_j} \left(q_{jc}^{\mathrm{D}}(x_j, x_j') + q_j^{\mathrm{I}}(x_j, x_j') \right) \\
&\quad + \sum_{k=0}^{N} \sum_{v \in T_j^-} \sum_{w \in T_k} \sum_{x_j' \in \mathcal{S}_j} \sum_{y_k \in \mathcal{S}_k} \pi(\boldsymbol{x}_k(y_k)) q_{kw}^{\mathrm{D}}(y_k, x_k) r_{kw,jv} p_{jv,c}^{\mathrm{A}-}(x_j, x_j') \\
&= \sum_{k=0}^{N} \sum_{v \in T_k} \sum_{x_j' \in \mathcal{S}_j} \sum_{y_k \in \mathcal{S}_k} \pi(\boldsymbol{x}_{jk}(x_j', y_k)) q_{kv}^{\mathrm{D}}(y_k, x_k[x_j']) r_{kv,jc} p_{jc}^{\mathrm{A}}(x_j', x_j)
\end{aligned}$$

$$+\sum_{k=0}^{N}\sum_{v\in T_j^+}\sum_{w\in T_k}\sum_{x_j'\in\mathcal{S}_j}\sum_{y_k\in\mathcal{S}_k}\pi(\boldsymbol{x}_{jk}(x_j',y_k))q_{kw}^{\mathrm{D}}(y_k,x_k[x_j'])r_{kw,jv}p_{jv,c}^{\mathrm{A}+}(x_j',x_j)$$
$$+\sum_{x_j'\in\mathcal{S}_j}\pi(\boldsymbol{x}_j(x_j'))q_j^{\mathrm{I}}(x_j',x_j),\qquad j=1,\ldots,N,\ x_j\in\mathcal{S}_j, \tag{4.4}$$

if and only if

$$\sum_{x_j'\in\mathcal{S}_j}\left(\sum_{v\in T_j^-}\pi_j(x_j')\alpha_{jv}p_{jv,c}^{\mathrm{A}-}(x_j',x_j)-\pi_j(x_j)\sum_{v\in T_j^+}\alpha_{jv}p_{jv,c}^{\mathrm{A}+}(x_j,x_j')\right)$$
$$=(\alpha_{jc}-\beta_{jc})\pi_j(x_j),\qquad j=1,\ldots,N,\ x_j\in\mathcal{S}_j. \tag{4.5}$$

On the other hand, for any measure π_j, $j=1,\ldots,N$, and their product π, if they satisfy (4.1), (4.4) and (4.5) together with (4.3), then π is the stationary measure of the network process. That is, they constitute a full set of local balance equations.

PROOF. Apply an argument similar to the one in Theorem 10.3. First note that, by Theorem 10.3, the stationary distribution of the network is the product form π. Summing (2.8) over $x_j'\in\mathcal{S}_j$, we obtain

$$\sum_{x_j'\in\mathcal{S}_j}\pi(\boldsymbol{x}_j(x_j'))\alpha_{jc}p_{jc}^{\mathrm{A}}(x_j',x_j)$$
$$=\sum_{k=0}^{N}\sum_{v\in T_k}\sum_{x_j'\in\mathcal{S}_j}\sum_{y_k\in\mathcal{S}_k}\pi(\boldsymbol{x}_{jk}(x_j',y_k))q_{kv}^{\mathrm{D}}(y_k,x_k[x_j'])r_{kv,jc}p_{jc}^{\mathrm{A}}(x_j',x_j),$$
$$j=0,1,\ldots,N.$$

Since $\pi_j(x_j)$ is the stationary distribution of

$$q_j^{(\boldsymbol{\alpha}_j)}(x_j,x_j')=\alpha_{jc}p_{jc}^{\mathrm{A}}(x_j,x_j')+\sum_{v\in T_j^-\cup T_j^+}\alpha_{jv}p_{jv,c}^{\mathrm{A}*}(x_j,x_j')$$
$$+q_{jc}^{\mathrm{D}}(x_j,x_j')+q_j^{\mathrm{I}}(x_j,x_j'),$$

substituting the balance equation for $q_j^{(\boldsymbol{\alpha}_j)}$ into the left-hand side and using the quasi-reversibility property (4.1) yields

$$\pi(\boldsymbol{x})\sum_{x_j'\in\mathcal{S}_j}\left(q_{jc}^{\mathrm{D}}(x_j,x_j')+q_j^{\mathrm{I}}(x_j,x_j')+\alpha_{jc}+\sum_{v\in T_j^-\cup T_j^+}\alpha_{jv}p_{jv,c}^{\mathrm{A}*}(x_j,x_j')\right)$$
$$-\beta_{jc}\pi(\boldsymbol{x})-\sum_{x_j'\in\mathcal{S}_j}\pi(\boldsymbol{x}_j(x_j'))\left(q_j^{\mathrm{I}}(x_j',x_j)+\sum_{v\in T_j^-\cup T_j^+}\alpha_{jv}p_{jv,c}^{\mathrm{A}*}(x_j',x_j)\right)$$

$$= \sum_{k=0}^{N} \sum_{v \in T_k} \sum_{x'_j \in \mathcal{S}_j} \sum_{y_k \in \mathcal{S}_k} \pi(\boldsymbol{x}_{jk}(x'_j, y_k)) q^{\mathrm{D}}_{kv}(y_k, x_k[x'_j]) r_{kv,jc} p^{\mathrm{A}}_{jc}(x'_j, x_j),$$

$$j = 0, 1, \ldots, N. \qquad (4.6)$$

From (4.1) and (4.3) we have

$$\pi(\boldsymbol{x}) \sum_{v \in T_j^-} \alpha_{jv} p^{\mathrm{A}-}_{jv,c}(x_j, x'_j)$$

$$= \sum_{k=0}^{N} \sum_{v \in T_j^-} \sum_{w \in T_k} \sum_{x'_j \in \mathcal{S}_j} \sum_{y_k \in \mathcal{S}_k} \pi(\boldsymbol{x}_k(y_k)) q^{\mathrm{D}}_{kw}(y_k, x_k) r_{kw,jv} p^{\mathrm{A}-}_{jv,c}(x_j, x'_j),$$

$$j = 0, 1, \ldots, N, \qquad (4.7)$$

and

$$\sum_{v \in T_j^+} \pi(\boldsymbol{x}_j(x'_j)) \alpha_{jv} p^{\mathrm{A}+}_{jv,c}(x'_j, x_j)$$

$$= \sum_{k=0}^{N} \sum_{v \in T_j^+} \sum_{w \in T_k} \sum_{x'_j \in \mathcal{S}_j} \sum_{y_k \in \mathcal{S}_k} \pi(\boldsymbol{x}_{jk}(x'_j, y_k)) q^{\mathrm{D}}_{kw}(y_k, x_k[x'_j]) r_{kw,jv} p^{\mathrm{A}+}_{jv,c}(x'_j, x_j),$$

$$j = 0, 1, \ldots, N. \qquad (4.8)$$

Hence, (4.6) is equivalent to the cross local balance equations (4.4) if and only if (4.5) holds.

We now prove that (4.4) and (4.5) imply global balance. Since (4.4) is equivalent to (4.6) under (4.5), we start with (4.6). Recall that node 0 has a single state $\mathcal{S}_0 = \{0\}$ and no internal transitions. So equation (4.6) also holds for $j = 0$. Actually (4.6) for $j = 0$ is equivalent to

$$\pi(\boldsymbol{x}) \left(\beta_{0c} + \alpha_{0c} + \sum_{v \in T_0^- \cup T_0^+} \alpha_{0v} - \beta_{0c} - \sum_{v \in T_0^- \cup T_0^+} \alpha_{0v} \right) = \pi(\boldsymbol{x}) \alpha_{0c} .$$

Summing (4.6) over $j = 0, 1, \ldots, N$ and using the fact that

$$\sum_{j=0}^{N} \sum_{u \in T_j} \alpha_{ju} = \sum_{j=0}^{N} \sum_{u \in T_j} \beta_{ju} \qquad (4.9)$$

and

$$\sum_{x'_j \in \mathcal{S}_j} \sum_{v \in T_j^- \cup T_j^+} \alpha_{jv} p^{\mathrm{A}*}_{jv,c}(x_j, x'_j) = \sum_{v \in T_j^- \cup T_j^+} \alpha_{jv} - \sum_{v \in T_j^- \cup T_j^+} \alpha_{jv} p^{\mathrm{A}*}_{jv}(x_j, x_j),$$

we obtain

$$\pi(\boldsymbol{x})\sum_{j=0}^{N}\left(\sum_{x_j'\in\mathcal{S}_j}(q_{jc}^{\mathrm{D}}(x_j,x_j')+q_j^{\mathrm{I}}(x_j,x_j'))+\sum_{v\in T_j^-\cup T_j^+}(\beta_{jv}-\alpha_{jv}p_{jv}^{\mathrm{A}*}(x_j,x_j))\right)$$
$$=\sum_{j=0}^{N}\sum_{k=0}^{N}\sum_{v\in T_k}\sum_{x_j'\in\mathcal{S}_j}\sum_{y_k\in\mathcal{S}_k}\pi(\boldsymbol{x}_{jk}(x_j',y_k))q_{kv}^{\mathrm{D}}(y_k,x_k[x_j'])r_{kv,jc}p_{jc}^{\mathrm{A}}(x_j',x_j)$$
$$+\sum_{j=0}^{N}\sum_{x_j'\in\mathcal{S}_j}\pi(\boldsymbol{x}_j(x_j'))\left(\sum_{v\in T_j^-\cup T_j^+}\alpha_{jv}p_{jv,c}^{\mathrm{A}*}(x_j',x_j)+q_j^{\mathrm{I}}(x_j',x_j)\right).$$

Using the fact that $p_{jv}^{\mathrm{A}*}(x_j,x_j')=0$ when $x_j\neq x_j'$, we can rewrite this equation and obtain

$$\pi(\boldsymbol{x})\sum_{j=0}^{N}\left(\sum_{x_j'\in\mathcal{S}_j}(q_{jc}^{\mathrm{D}}(x_j,x_j')+q_j^{\mathrm{I}}(x_j,x_j'))+\sum_{v\in T_j^-\cup T_j^+}\beta_{jv}\right)$$
$$=\sum_{j=0}^{N}\sum_{k=0}^{N}\sum_{v\in T_k}\sum_{x_j'\in\mathcal{S}_j}\sum_{y_k\in\mathcal{S}_k}\pi(\boldsymbol{x}_{jk}(x_j',y_k))q_{kv}^{\mathrm{D}}(y_k,x_k[x_j'])r_{kv,jc}p_{jc}^{\mathrm{A}}(x_j',x_j)$$
$$+\sum_{j=0}^{N}\sum_{x_j'\in\mathcal{S}_j}\pi(\boldsymbol{x}_j(x_j'))\left(\sum_{v\in T_j^-\cup T_j^+}\alpha_{jv}(p_{jv,c}^{\mathrm{A}*}(x_j',x_j)+p_{jv}^{\mathrm{A}*}(x_j',x_j))+q_j^{\mathrm{I}}(x_j',x_j)\right).$$

By rewriting α_{jv} using the traffic equations and observing that for $v\in T_j^-\cup T_j^+$,

$$\pi_j(x_j)q_{jv}^{\mathrm{D}}(x_j,x_j)=\beta_{jv}\pi_j(x_j),$$

we arrive at the global balance for the network. This completes the proof. □

Since $x_k[x_j']$ equals x_k for $k\neq j$ and equals x_j' for $k=j$, the state of the network just before the arrival of the negative entity at node j in the second term on the left-hand side of (4.4) is $\boldsymbol{x}$. Thus equations (4.4) imply that the rate going across $\boldsymbol{x}$ from above, due to regular class c departures and class c departures triggered by negative entities, is equal to the rate going into $\boldsymbol{x}$ from below, due to regular class c arrivals and class c arrivals triggered by positive entities. It is for this reason that equations (4.4) are called the *cross local balance equations*. The cross local balance equations (4.5) state that the net gain of arrivals over departures should be compensated by the difference in the gain due to negative and positive entities.

The cross local balances (4.4) and (4.5) will be used in Section 10.6 to extend the transition rates in such a way that they may also depend on the state of the entire network.

Remark 10.10 If the arrivals of positive entities have a similar effect as the arrivals of the regular entities and do not trigger instantaneous movements, then they may be considered as regular entities. However, we treat them differently here since such a formulation naturally leads to extensions that include positive and negative signals, which will be studied in the next section.

Example 10.11 (Negative and positive entities) Consider the queueing network with negative customers of Example 4.7. This network has two classes of entities, denoted by c and c^-, and they are referred to as regular and negative customers. Thus, $T_j^{\mathrm{R}} = \{c\}$, $T_j^- = \{c^-\}$ and $T_j^+ = \emptyset$ for $j = 1, \ldots, N$. This network has no internal transitions and is quasi-reversible for $T_j = \{c, c^-\}$. We must verify condition (4.5). Since

$$p^{\mathrm{A}-}_{jc^-,c}(n+1, n) = 1,$$

condition (4.5) can be written as

$$\pi_j(n+1)\alpha_j^- = (\alpha_j - \beta_j)\pi_j(n).$$

This is easily verified since, from Example 4.7, we have

$$\pi_j(n) = \left(1 - \frac{\alpha_j}{\mu_j + \alpha_j^-}\right)\left(\frac{\alpha_j}{\mu_j + \alpha_j^-}\right)^n,$$

$$\beta_j = \frac{\alpha_j \mu_j}{\mu_j + \alpha_j^-}.$$

Hence it follows from Theorem 10.9 that the cross local balance equations (4.4) are satisfied, i.e.,

$$\pi(\boldsymbol{n})\mu_j + \sum_{k=0}^{N} \pi(\boldsymbol{n}_k + \mathbf{e}_k)\mu_k r_{kc,jc^-} = \sum_{k=0}^{N} \pi(\boldsymbol{n} + \mathbf{e}_k - \mathbf{e}_j)\mu_k r_{kc,jc}, \quad (4.10)$$

for state $\boldsymbol{n} = (n_1, \ldots, n_N)$, $n_j \geq 1$. □

10.5 CROSS LOCAL BALANCE EQUATIONS IN NETWORKS WITH SIGNALS

We now consider cross local balance equations in networks with signals and instantaneous movements studied in Section 4.4. Recall that the dynamics of node j, when in state x_j, are characterized by the following rate and probability functions:

(i) $q^{\mathrm{D}}_{ju}(x_j, x'_j)$: the departure rate of class u entities from node j that changes the state from x_j to x'_j;

(ii) $p^{\mathrm{A}}_{ju}(x_j, x'_j)$: the probability that state x_j changes to x'_j when a class u entity arrives at node j;

(iii) $f_{ju,v}(x_j, x'_j)$: the probability that a class u arrival, which changed the state of node j from x_j to x'_j, triggers a class v departure;

(iv) $q^{\mathrm{I}}_j(x_j, x'_j)$: the internal transition rate of node j from state x_j to x'_j.

A class u departure from node j, either due to a service completion at node j or due to a signal, joins node k as a class v entity with probability $r_{ju,kv}$, $j = 0, 1, \dots, N$, $k = 0, 1, \dots, N$ and $v \in T_k$.

Similar to Section 10.2, define a family of transition rate functions $\{q_{jk}\}$ by

$$\begin{aligned} q_{jj}(\boldsymbol{x}, \boldsymbol{x}') &= \sum_{u,v} \sum_{y_j \in \mathcal{S}_j} q^{\mathrm{D}}_{ju}(x_j, y_j) \left[(rp^{\mathrm{A}} f)^n (rp^{\mathrm{A}}(1-f))(\boldsymbol{x}_j(y_j), \boldsymbol{x}')\right]_{ju,jv} \\ &\quad + 1[x_\ell = x'_\ell;\, \ell \neq j] q^{\mathrm{I}}_j(x_j, x'_j), \\ q_{jk}(\boldsymbol{x}, \boldsymbol{x}') &= \sum_{u,v} \sum_{n=0}^{\infty} \sum_{y_j \in \mathcal{S}_j} q^{\mathrm{D}}_{ju}(x_j, y_j) \left[(rp^{\mathrm{A}} f)^n (rp^{\mathrm{A}}(1-f))(\boldsymbol{x}_j(y_j), \boldsymbol{x}')\right]_{ju,kv}. \end{aligned}$$

The transition rates of the network with signals are then given by

$$q(\boldsymbol{x}, \boldsymbol{x}') = \sum_{j=0}^{N} \sum_{k=0}^{N} q_{jk}(\boldsymbol{x}, \boldsymbol{x}').$$

For notational convenience, we introduce the following vector and matrix notation. For each $\boldsymbol{x}, \boldsymbol{x}' \in \mathcal{S}$, $\pi q^{\mathrm{D}}(\boldsymbol{x}, \boldsymbol{x}')$ is a row vector whose ju component is

$$[\pi q^{\mathrm{D}}(\boldsymbol{x}, \boldsymbol{x}')]_{ju} = \pi(\boldsymbol{x}) q^{\mathrm{D}}_{ju}(x_j, x'_j),$$

and $rp^{\mathrm{A}} f(\boldsymbol{x}, \boldsymbol{x}')$ and $rp^{\mathrm{A}}(1-f)(\boldsymbol{x}, \boldsymbol{x}')$ are matrices the (ju, kv) components of which are, respectively,

$$\left[rp^{\mathrm{A}} f(\boldsymbol{x}, \boldsymbol{x}')\right]_{ju,kv} = \sum_{w \in T_j} r_{ju,kw} p^{\mathrm{A}}_{kw}(x_k, x'_k) f_{kw,v}(x_k, x'_k),$$

$$\left[rp^{\mathrm{A}}(1-f)(\boldsymbol{x}, \boldsymbol{x}')\right]_{ju,kv} = r_{ju,kv} p^{\mathrm{A}}_{kv}(x_k, x'_k) \left(1 - \sum_{w \in T_k} f_{kv,w}(x_k, x'_k)\right).$$

These components represent, respectively, the probability that a class u departure from node j goes to node k and triggers a class v departure, and the probability that a class u departure from node j goes to node k as a class v entity and does not trigger a departure. Multiplications of vector and matrices are defined as follows. Let

$$\boldsymbol{a}(\boldsymbol{x}, \boldsymbol{x}') = \{a_{ju}(\boldsymbol{x}, \boldsymbol{x}')\}$$

be a vector and let

$$A(\boldsymbol{x}, \boldsymbol{x}') = \{A_{ju,kv}(\boldsymbol{x}, \boldsymbol{x}')\},$$
$$B(\boldsymbol{x}, \boldsymbol{x}') = \{B_{ju,kv}(\boldsymbol{x}, \boldsymbol{x}')\},$$

be two matrices. Then, $\boldsymbol{a}A(\boldsymbol{x}, \boldsymbol{x}')$ and $AB(\boldsymbol{x}, \boldsymbol{x}')$ are defined by

$$\left[\boldsymbol{a}A(\boldsymbol{x}, \boldsymbol{x}')\right]_{ju} = \sum_{k=0}^{N} \sum_{v \in T_k} \sum_{y_k \in \mathcal{S}_k} a_{kv}(\boldsymbol{x}, \boldsymbol{x}_k(y_k)) A_{kv,ju}(\boldsymbol{x}_k(y_k), \boldsymbol{x}'),$$
$$\left[AB(\boldsymbol{x}, \boldsymbol{x}')\right]_{ju,kv} = \sum_{\ell=0}^{N} \sum_{w \in T_\ell} \sum_{y_\ell \in \mathcal{S}_\ell} A_{ju,\ell w}(\boldsymbol{x}, \boldsymbol{x}_\ell(y_\ell)) B_{\ell w,kv}(\boldsymbol{x}_\ell(y_\ell), \boldsymbol{x}').$$

We assume that $rp^{\mathrm{A}} f$ is transient in the sense that

$$\lim_{n \to \infty} \left[(rp^{\mathrm{A}} f)^n(\boldsymbol{x}, \boldsymbol{x}')\right]_{ju,kv} = 0, \quad j, k = 0, 1, \ldots, N, u \in T_j, v \in T_k, \quad (5.1)$$

which means that at any point in time there cannot be an infinite number of instantaneous transitions. Using

$$\sum_{x'_j \in \mathcal{S}_j} p^{\mathrm{A}}_{ju}(x_j, x'_j) = 1,$$
$$\sum_{k=0}^{N} \sum_{v \in T_k} r_{ju,kv} = 1,$$

and (5.1), we obtain

$$\sum_{j=0}^{N} \sum_{u \in T_j} \sum_{\boldsymbol{x}' \in \mathcal{S}} \sum_{n=0}^{\infty} \left[(\pi q^{\mathrm{D}})(rp^{\mathrm{A}} f)^n (rp^{\mathrm{A}}(1 - f))(\boldsymbol{x}, \boldsymbol{x}')\right]_{ju}$$
$$= \lim_{n \to \infty} \sum_{j=0}^{N} \sum_{u \in T_j} \sum_{\boldsymbol{x}' \in \mathcal{S}} \left[(\pi q^{\mathrm{D}}) \left(1 - \sum_{k=0}^{N} \sum_{v \in T_k} (rp^{\mathrm{A}} f)^n (\boldsymbol{x}, \boldsymbol{x}')\right)\right]_{ju}$$

$$= \sum_{j=0}^{N} \sum_{u \in T_j} \sum_{\boldsymbol{x}' \in \mathcal{S}} \left[(\pi q^{\mathrm{D}})(\boldsymbol{x}, \boldsymbol{x}') \right]_{ju}$$

$$= \pi(\boldsymbol{x}) \sum_{j=0}^{N} \sum_{u \in T_j} \sum_{x'_j \in \mathcal{S}_j} q^{\mathrm{D}}_{ju}(x_j, x'_j).$$

Thus, the global balance equations for the queueing network with signals can be written as

$$\pi(\boldsymbol{x}) \sum_{j=0}^{N} \sum_{x'_j \in \mathcal{S}_j} \left(\sum_{u \in T_j} q^{\mathrm{D}}_{ju}(x_j, x'_j) + q^{\mathrm{I}}_j(x_j, x'_j) \right)$$
$$= \sum_{j=0}^{N} \sum_{u \in T_j} \sum_{\boldsymbol{x}' \in \mathcal{S}} \sum_{n=0}^{\infty} \left[(\pi q^{\mathrm{D}})(r p^{\mathrm{A}} f)^n (r p^{\mathrm{A}}(1 - f))(\boldsymbol{x}', \boldsymbol{x}) \right]_{ju}$$
$$+ \sum_{j=0}^{N} \sum_{x'_j \in \mathcal{S}_j} \pi(\boldsymbol{x}_j(x'_j)) q^{\mathrm{I}}_j(x'_j, x_j). \tag{5.2}$$

Suppose all nodes of the network are quasi-reversible queues with signals, i.e., there exist nonnegative numbers α_{ju} and β_{ju} such that

$$\sum_{x'_j \in \mathcal{S}_j} \pi_j(x'_j) \left(q^{\mathrm{D}}_{ju}(x'_j, x_j) + \sum_{v \in T_j} \alpha_{jv} p^{\mathrm{A}}_{jv}(x'_j, x_j) f_{jv,u}(x'_j, x_j) \right)$$
$$= \beta_{ju} \pi_j(x_j), \qquad x_j \in \mathcal{S}_j, \tag{5.3}$$

for each $j = 0, 1, \ldots, N$ and $u \in T_j$, where π_j is the stationary distribution of $q_j^{(\boldsymbol{\alpha}_j)}$ defined by

$$q_j^{(\boldsymbol{\alpha}_j)}(x_j, x'_j) = \sum_{u \in T_j} (\alpha_{ju} p^{\mathrm{A}}_{ju}(x_j, x'_j) + q^{\mathrm{D}}_{ju}(x_j, x'_j)) + q^{\mathrm{I}}_j(x_j, x'_j). \tag{5.4}$$

We further assume that α_{ju} and β_{ju} satisfy the traffic equations

$$\alpha_{ju} = \sum_{k=0}^{N} \sum_{v \in T_k} \beta_{kv} r_{kv, ju}, \qquad j = 0, 1, \ldots, N, u \in T_j. \tag{5.5}$$

The biased local balance of Theorem 10.3 can now be extended to networks with signals. It also provides an alternative proof for the product form result for networks with signals, i.e., Theorem 4.9.

Theorem 10.12 *Consider a quasi-reversible network with signals characterized by (5.3) and (5.5), and let π_j be the stationary distribution of $q_j^{(\boldsymbol{\alpha}_j)}$ defined by (5.4). The product form π of (2.4) satisfies the biased local balance equations (1.4) with the bias term*

$$\gamma_j = \sum_{u \in T_j} (\alpha_{ju} - \beta_{ju}).$$

That is,

$$\begin{aligned}
&\pi(\boldsymbol{x})\left(\sum_{x'_j \in \mathcal{S}_j} \left(\sum_{u \in T_j} q^{\mathrm{D}}_{ju}(x_j, x'_j) + q^{\mathrm{I}}_j(x_j, x'_j) \right) + \gamma_j \right) \\
&= \sum_{u \in T_j} \sum_{\boldsymbol{x}' \in \mathcal{S}} \sum_{n=0}^{\infty} \left[(\pi q^{\mathrm{D}})(r p^{\mathrm{A}} f)^n (r p^{\mathrm{A}}(1-f))(\boldsymbol{x}', \boldsymbol{x}) \right]_{ju} \\
&\quad + \sum_{x'_j \in \mathcal{S}_j} \pi(\boldsymbol{x}_j(x'_j)) q^{\mathrm{I}}_j(x'_j, x_j), \qquad j = 0, 1, \ldots, N, \boldsymbol{x} \in \mathcal{S}. \qquad (5.6)
\end{aligned}$$

Hence, the product form (2.4) is the stationary distribution of the network.

PROOF. Again we use the same argument as in the proof of Theorem 10.3. From (5.3) and (5.5), we have

$$\alpha_{ju} = \sum_{k=0}^{N} \sum_{x'_k \in \mathcal{S}_k} \frac{\pi_k(x'_k)}{\pi_k(x_k)} \left(q^{\mathrm{D}}_{ku}(x'_k, x_k) + \sum_{v \in T_k} \alpha_{kv} p^{\mathrm{A}}_{kv}(x'_k, x_k) f_{kv,u}(x'_k, x_k) \right) r_{kv,ju}.$$

Multiplying both sides of this equation by $\pi(\boldsymbol{x}) = \prod_{j=1}^{N} \pi_j(x_j)$, we obtain

$$\begin{aligned}
\alpha_{ju}\pi(\boldsymbol{x}) &= \sum_{k=0}^{N} \sum_{x'_k \in \mathcal{S}_k} \pi(\boldsymbol{x}_k(x'_k)) \\
&\quad \times \left(q^{\mathrm{D}}_{ku}(x'_k, x_k) + \sum_{v \in T_k} \alpha_{kv} p^{\mathrm{A}}_{kv}(x'_k, x_k) f_{kv,u}(x'_k, x_k) \right) r_{kv,ju}, \\
&\qquad \text{for } j = 0, 1, \ldots, N \text{ and } u \in T_j.
\end{aligned}$$

Repeated substitution of the expression for α_{ju} into the α_{kv} on the right-hand side of this equation yields

$$\alpha_{ju}\pi(\boldsymbol{x}) = \sum_{k=0}^{N} \sum_{v \in T_k} \sum_{\boldsymbol{x}' \in \mathcal{S}} \sum_{n=0}^{\infty} \left[(\pi q^{\mathrm{D}})(r p^{\mathrm{A}} f)^n (\boldsymbol{x}', \boldsymbol{x}) \right]_{kv} r_{kv,ju}, \qquad (5.7)$$

where the convergence of the infinite sum on the right-hand side follows from the transience condition (5.1) and the fact that $\alpha_{ju} < \infty$. Since $\pi_j(x_j)$ is the stationary distribution for $q_j^{(\boldsymbol{\alpha}_j)}$ of (5.4), it follows from the quasi-reversibility condition (5.3) that

$$\begin{aligned}
\pi_j(x_j)\Bigg(\sum_{u\in T_j}\alpha_{ju} + \sum_{x'_j\in\mathcal{S}_j}\Bigg(\sum_{u\in T_j} q^{\mathrm{D}}_{ju}(x_j,x'_j) + q^{\mathrm{I}}_j(x_j,x'_j)\Bigg)\Bigg)\\
= \sum_{u\in T_j}\sum_{x'_j\in\mathcal{S}_j}\pi_j(x'_j)q^{\mathrm{D}}_{ju}(x'_j,x_j) + \sum_{u\in T_j}\sum_{x'_j\in\mathcal{S}_j}\alpha_{ju}\pi_j(x'_j)p^{\mathrm{A}}_{ju}(x'_j,x_j)\\
+ \sum_{x'_j\in\mathcal{S}_j}\pi_j(x'_j)q^{\mathrm{I}}_j(x'_j,x_j)\\
= \sum_{u\in T_j}\Bigg(\beta_{ju}\pi_j(x_j) - \sum_{x'_j\in\mathcal{S}_j}\sum_{v\in T_j}\alpha_{jv}\pi_j(x'_j)p^{\mathrm{A}}_{jv}(x'_j,x_j)f_{jv,u}(x'_j,x_j)\Bigg)\\
+ \sum_{u\in T_j}\sum_{x'_j\in\mathcal{S}_j}\alpha_{ju}\pi_j(x'_j)p^{\mathrm{A}}_{ju}(x'_j,x_j) + \sum_{x'_j\in\mathcal{S}_j}\pi_j(x'_j)q^{\mathrm{I}}_j(x'_j,x_j).
\end{aligned}$$

After rearranging terms and applying the definition of γ_j, we obtain

$$\begin{aligned}
\pi_j(x_j)\Bigg(\sum_{u\in T_j}\sum_{x'_j\in\mathcal{S}_j}\Big(q^{\mathrm{D}}_{ju}(x_j,x'_j) + q_j(x_j,x'_j)\Big) + \gamma_j\Bigg)\\
= \sum_{u\in T_j}\sum_{x'_j\in\mathcal{S}_j}\alpha_{ju}\pi_j(x'_j)p^{\mathrm{A}}_{ju}(x'_j,x_j)\Bigg(1 - \sum_{v\in T_j} f_{ju,v}(x'_j,x_j)\Bigg)\\
+ \sum_{x'_j\in\mathcal{S}_j}\pi_j(x'_j)q^{\mathrm{I}}_j(x'_j,x_j).
\end{aligned}$$

Multiplying both sides of this equation by $\pi(\boldsymbol{x})/\pi_j(x_j)$ yields

$$\begin{aligned}
\pi(\boldsymbol{x})\Bigg(\sum_{u\in T_j}\sum_{x'_j\in\mathcal{S}_j}\Big(q^{\mathrm{D}}_{ju}(x_j,x'_j) + q_j(x_j,x'_j)\Big) + \gamma_j\Bigg)\\
= \sum_{u\in T_j}\sum_{x'_j\in\mathcal{S}_j}\alpha_{ju}\pi(\boldsymbol{x}_j(x'_j))p^{\mathrm{A}}_{ju}(x'_j,x_j)\Bigg(1 - \sum_{v\in T_j} f_{ju,v}(x'_j,x_j)\Bigg)\\
+ \sum_{x'_j\in\mathcal{S}_j}\pi(\boldsymbol{x}_j(x'_j))q^{\mathrm{I}}_j(x'_j,x_j).
\end{aligned}$$

Substituting (5.7) into this equation gives (5.6). □

We now consider a more refined form of the local balance. As in the previous section, we again only consider the case of a single class of regular customers, denoted by c. Extension to multiple classes of regular customers is easy and is left to the reader. As in the previous section we introduce a structure for the effects of positive and negative signals. Let the class of entities T_j be partitioned into three disjoint sets $T_j^{\text{R}} = \{c\}$, T_j^- and T_j^+, which are the classes of regular entities, negative signals, and positive signals. For $v \in T_j^- \cup T_j^+$, the effects of a class v arrival, $p_{jv}^{\text{A}}(x_j, x_j')$, are decomposed into $p_{jv,c}^{\text{A}*}(x_j, x_j')$ and $p_{jv}^{\text{A}*}(x_j, x_j')$ in such a way that

$$p_{jv}^{\text{A}}(x_j, x_j') = p_{jv,c}^{\text{A}*}(x_j, x_j') + p_{jv}^{\text{A}*}(x_j, x_j)1[x_j' = x_j].$$

The $p_{jv,c}^{\text{A}*}(x_j, x_j')$ is the probability that an arrival of a class v signal, $v \in T_j^-$ ($v \in T_j^+$), has a similar effect as a departure (arrival) of a class c regular entity, and, as in the previous section, $p_{jv}^{\text{A}*}(x_j, x_j)$ is the probability that an arrival of a class v signal has no effect on the node. Also, as in Section 10.4, $p_{jv,c}^{\text{A}*}$ will be written as $p_{jv,c}^{\text{A}-}$ for $v \in T_j^-$ and $p_{jv,c}^{\text{A}+}$ for $v \in T_j^+$, and $p_{jv}^{\text{A}*}(x_j, x_j')$ is 0 if $x_j \neq x_j'$.

A class v positive or negative signal triggers a class u departure with probability $f_{jv,u}(x_j, x_j')$. A regular entity does not trigger any departure, so $f_{jc,w}(x_j, x_j') \equiv 0$, $w \in T_j$. Similar to Section 10.4, we assume that

$$q_{jv}^{\text{D}}(x_j, x_j') = 0, \qquad x_j \neq x_j', \quad v \in T_j^- \cup T_j^+.$$

Thus signal departures generated at node j do not change the state of node j. Assume that all the nodes are quasi-reversible queues with signals.

This network is referred to as a *quasi-reversible network with positive and negative signals*. Theorem 10.9 for networks with positive and negative entities can be extended to networks with positive and negative signals as follows. Its proof is similar to that of Theorem 10.9, and is therefore omitted.

Theorem 10.13 *Consider a quasi-reversible network with positive and negative signals satisfying (5.3) and (5.5). Suppose $q_j^{(\boldsymbol{\alpha}_j)}$ of (4.2) has a stationary distribution π_j, $j = 1, \ldots, N$. The stationary distribution π of the network is the product of π_j and it satisfies the cross local balance equations*

$$\pi(\boldsymbol{x}) \sum_{x_j' \in \mathcal{S}_j} \left(q_{jc}^{\text{D}}(x_j, x_j') + q_j^{\text{I}}(x_j, x_j')\right)$$

$$+ \sum_{k=0}^{N} \sum_{v \in T_j^-} \sum_{w \in T_k} \sum_{y_j \in \mathcal{S}_j} \sum_{\boldsymbol{x}' \in \mathcal{S}} \sum_{n=0}^{\infty} \left[(\pi q^{\text{D}})(r p^{\text{A}} f)^n(\boldsymbol{x}', \boldsymbol{x})\right]_{kw} r_{kw,jv} p_{jv,c}^{\text{A}-}(x_j, y_j)$$

$$= \sum_{\boldsymbol{x}' \in \mathcal{S}} \sum_{n=0}^{\infty} \left[(\pi q^{\text{D}})(r p^{\text{A}} f)^n (r p^{\text{A}})(\boldsymbol{x}', \boldsymbol{x})\right]_{jc}$$

$$+\sum_{k=0}^{N}\sum_{v\in T_j^+}\sum_{w\in T_k}\sum_{y_j\in\mathcal{S}_j}\sum_{\boldsymbol{x}'\in\mathcal{S}}\sum_{n=0}^{\infty}\left[(\pi q^{\mathrm{D}})(rp^{\mathrm{A}}f)^n(\boldsymbol{x}',\boldsymbol{x}_j(y_j))\right]_{kw}r_{kw,jv}p_{jv,c}^{\mathrm{A}+}(y_j,x_j)$$

$$+\sum_{x_j'\in\mathcal{S}_j}\pi(\boldsymbol{x}_j(x_j'))q_j^{\mathrm{I}}(x_j',x_j), \qquad j=1,\ldots,N,\ \boldsymbol{x}\in\mathcal{S}, \qquad (5.8)$$

if and only if (4.5) holds. On the other hand, for any measure π_j, $j=1,\ldots,N$, and their product π, if they satisfy (4.1), (4.5) and (5.8) together with (4.3), then π is the stationary measure of the network process.

As noted in the interpretation of Theorem 10.9, the left and right-hand sides of (5.8) represent the rates going across state $\boldsymbol{x}$ from above and from below due to departure effects (i.e., regular departures and departures triggered by negative signals) and arrival effects (i.e., regular arrivals and arrivals caused by positive signals). In (4.4), the transitions going across $\boldsymbol{x}$ from above and from below stop at $\boldsymbol{x}$ because there are no instantaneous movements. However, in (5.8) the transitions may continue after reaching state $\boldsymbol{x}$. For instance, in the second term on the left-hand side of (5.8), a class v negative signal, which arrives at node j and observes the network in state $\boldsymbol{x}$, induces the state of node j to change to y_j and may immediately trigger a departure (it is a departure of class u with probability $f_{jv,u}(x_j,y_j)$), and this triggering process may result in a sequence of instantaneous transitions at other nodes. Similarly, in the second term on the right-hand side, the process goes across state $\boldsymbol{x}$ from below when a class v positive signal arrives at node j, but this positive signal may trigger an immediate departure from node j and the process may continue and affect a number of nodes at the same time epoch.

Example 10.14 (Single class of positive and negative signals) Consider a queueing network with a single class of positive signals and a single class of negative signals as discussed in Section 4.5. The network has no internal transitions and entity class $T_j=\{c,s^-,s^+\}$. It is decomposed into $T_j^{\mathrm{R}}=\{c\}$, $T_j^-=\{s^-\}$ and $T_j^+=\{s^+\}$. This is a quasi-reversible network with signals, and the departure rate of regular customers from node j is

$$\beta_j=\frac{\alpha_j+\alpha_j^+}{\mu_j+\alpha_j^-}\mu_j, \qquad j=1,\ldots,N,$$

and the marginal distribution π_j is given by

$$\pi_j(n)=\left(1-\frac{\alpha_j+\alpha_j^+}{\mu_j+\alpha_j^-}\right)\left(\frac{\alpha_j+\alpha_j^+}{\mu_j+\alpha_j^-}\right)^n, \qquad n=0,1,\ldots.$$

Thus condition (4.5) is easily verified, and can be written as

$$\pi_j(n+1)\alpha_j^- - \pi_j(n)\alpha_j^+ = \left(\alpha_j - \beta_j\right)\pi_j(n).$$

By Theorem 10.13, we have the cross local balance equations

$$\begin{aligned}
\pi(\boldsymbol{n})\mu_j &+ \sum_{k=0}^{N}\sum_{\boldsymbol{n}'\in\mathcal{S}}\sum_{w\in T_k}\sum_{\ell=0}^{\infty}\left[(\pi q^{\mathrm{D}})(rp^{\mathrm{A}}f)^{\ell}(\boldsymbol{n}',\boldsymbol{n})\right]_{kw} r_{kw,js^-} \\
&= \sum_{\boldsymbol{n}'\in\mathcal{S}}\sum_{\ell=0}^{\infty}\left[(\pi q^{\mathrm{D}})(rp^{\mathrm{A}}f)^{\ell}(rp^{\mathrm{A}})(\boldsymbol{n}',\boldsymbol{n})\right]_{jc} \\
&\quad + \sum_{k=0}^{N}\sum_{\boldsymbol{n}'\in\mathcal{S}}\sum_{w\in T_k}\sum_{\ell=0}^{\infty}\left[(\pi q^{\mathrm{D}})(rp^{\mathrm{A}}f)^{\ell}(\boldsymbol{n}',\boldsymbol{n}-\mathbf{e}_j)\right]_{kw} r_{kw,js^+}, \qquad (5.9)
\end{aligned}$$

for $\boldsymbol{n} = (n_1, \ldots, n_N)$ with $n_j \geq 1$, $j = 1, \ldots, N$. □

10.6 CONSTRUCTION OF NETWORKS WITH STATE-DEPENDENT TRANSITIONS

It was shown in Chapter 9 that local balances and cross local balances play a key role in extending networks with state-independent transitions to networks with state-dependent transitions. In the case of linear networks the forms of the state-dependent transition rates are completely characterized by the local balance equations (see Theorems 9.4 and 9.7, as well as Theorem 12.38). The cross local balance equations obtained in the general setting in the previous sections enable us to construct network models with state-dependent transitions from those with state-independent transitions.

Consider the quasi-reversible network with positive and negative signals of Theorem 10.13. For simplicity we again only consider a single class of regular entities, denoted by c. We make the following assumptions with regard to each node $j = 1, 2, \ldots, N$ and $x_j \in \mathcal{S}_j$.

(i) There exists a unique state $x_j^+ \in \mathcal{S}_j$ for each $x_j \in \mathcal{S}_j$ such that

$$p_{jc}^{\mathrm{A}}(x_j, x_j^+) = 1,$$
$$q_{jc}^{\mathrm{D}}(x_j^+, x_j) > 0.$$

That is, when node j is in state x_j, the arrival of a class c entity changes the state to x_j^+ with probability 1. Let $\mathcal{S}_j^{\mathrm{D}} = \{x_j^+; x_j \in \mathcal{S}_j\}$ represent the set of states from which regular departures can occur, and let x_j^- be x_j' such that $(x_j')_j^+ = x_j$.

(ii) The probabilities $p^{\text{A}+}_{jv,c}(x_j, x'_j)$ and $p^{\text{A}-}_{jv,c}(x_j, x'_j)$ can be positive only if $p^{\text{A}}_{jc}(x_j, x'_j)$ and $q^{\text{D}}_{jc}(x_j, x'_j)$ are positive, respectively.

Intuitively, (i) implies that the state changes due to arrivals are the reverse of those due to departures, and (ii) is implied by the assumption that the effects of positive and negative signals are similar to the effects of arrivals and departures of regular entities, respectively. Note that even though x_j^+ is defined for every state $x_j \in \mathcal{S}_j$, x_j^- may not be defined for some x_j. Actually x_j^- is defined only for $x_j \in \mathcal{S}_j^{\text{D}}$. For example, if node j is an $M/M/1$ queue where x_j represents the number of customers at the node, then $x_j^+ = x_j + 1$, $x_j^- = x_j - 1$, $\mathcal{S}_j^{\text{D}} = \{n; n \geq 1\}$, and x_j^- is not defined for $x_j = 0$.

Let Φ be an arbitrary positive function on $\mathcal{S}$, and let $\Psi(\boldsymbol{x})$ be an arbitrary non-negative function satisfying

$$\Psi(\boldsymbol{x}) \leq \Psi(\boldsymbol{x}_j(x_j^+)), \qquad j = 1, \ldots, N, \quad \boldsymbol{x} \in \mathcal{S}. \tag{6.1}$$

We call the function Ψ satisfying (6.1) nondecreasing. For convenience, we use the notation

$$\begin{aligned}
&\mathcal{S}^{\text{D}}(jc) = \{\boldsymbol{x}_j(x_j^+); \boldsymbol{x} \in \mathcal{S}\}, \\
&\boldsymbol{x}_j^- = \boldsymbol{x}_j(x_j^-), \qquad \boldsymbol{x} \in \mathcal{S}^{\text{D}}(jc), \\
&\boldsymbol{x}_j^+ = \boldsymbol{x}_j(x_j^+), \qquad \boldsymbol{x} \in \mathcal{S}.
\end{aligned}$$

Construct a network with state-dependent transition rates using Ψ and Φ as follows. For $j = 1, \ldots, N$,

$$\hat{q}^{\text{D}}_{ju}(\boldsymbol{x}, \boldsymbol{x}') = \begin{cases} 1[\boldsymbol{x}' = \boldsymbol{x}_j^-] q^{\text{D}}_{jc}(x_j, x_j^-) \dfrac{\Psi(\boldsymbol{x}_j^-)}{\Phi(\boldsymbol{x})}, & u = c, \\ 1[\boldsymbol{x}' = \boldsymbol{x}] q^{\text{D}}_{ju}(x_j, x_j) \dfrac{\Psi(\boldsymbol{x})}{\Phi(\boldsymbol{x})}, & u \in T_j^+ \cup T_j^-, \end{cases}$$

$$\hat{q}^{\text{I}}_j(\boldsymbol{x}, \boldsymbol{x}') = 1[\boldsymbol{x}' = \boldsymbol{x}_j(x'_j)] q^{\text{I}}_j(x_j, x'_j) \frac{\Psi(\boldsymbol{x})}{\Phi(\boldsymbol{x})},$$

$$\hat{p}^{\text{A}-}_{jv,c}(\boldsymbol{x}, \boldsymbol{x}') = 1[\boldsymbol{x}' = \boldsymbol{x}_j^-] p^{\text{A}-}_{jv,c}(x_j, x_j^-) \frac{\Psi(\boldsymbol{x}_j^-)}{\Psi(\boldsymbol{x})}, \qquad v \in T_j^-,$$

$$\hat{p}^{\text{A}+}_{jv,c}(\boldsymbol{x}, \boldsymbol{x}') = 1[\boldsymbol{x}' = \boldsymbol{x}_j^+] p^{\text{A}+}_{jv,c}(x_j, x_j^+), \qquad v \in T_j^+,$$

$$\hat{p}^{\text{A}*}_{jv}(\boldsymbol{x}, \boldsymbol{x}) = \begin{cases} 1 - \hat{p}^{\text{A}-}_{jv,c}(\boldsymbol{x}, \boldsymbol{x}_j^-), & v \in T_j^-, \\ 1 - \hat{p}^{\text{A}+}_{jv,c}(\boldsymbol{x}, \boldsymbol{x}_j^+), & v \in T_j^+, \end{cases}$$

and for $v \in T_j^- \cup T_j^+$ and $w \in T_j$, the triggering probabilities are unchanged, i.e.,

$$\hat{f}_{jv,w}(\boldsymbol{x}, \boldsymbol{x}_j(x'_j)) = f_{jv,w}(x_j, x'_j).$$

Finally, node 0 is assumed to have a single state 0 and no internal transitions, and its departure functions, i.e., the arrival rates to the network, are

$$\hat{q}^{\mathrm{D}}_{0u}(\boldsymbol{x}, \boldsymbol{x}') \;=\; 1[\boldsymbol{x}' = \boldsymbol{x}]q^{\mathrm{D}}_{0u}(0,0)\frac{\Psi(\boldsymbol{x})}{\Phi(\boldsymbol{x})}, \qquad u \in T_0.$$

The modified transition functions $\hat{q}^{\mathrm{D}}_{ju}$, $\hat{p}^{\mathrm{A}-}_{jv,c}$, $\hat{p}^{\mathrm{A}+}_{jv,c}$, $\hat{p}^{\mathrm{A}*}_{jv}$ and $\hat{f}_{jv,w}$ determine a network that is referred to as a *network with state-dependent transitions.*

The motivation for this change is the following. Multiply both sides of (5.9) by $\Psi(\boldsymbol{x}_j^-)/\Phi(\boldsymbol{x})$ and substitute the departure and arrival effect functions of the modified network. The resulting formula becomes, under the conditions given below, the local balance for the modified network, which is expected to be (and is) a full local balance. Note that if $\Phi(\boldsymbol{x}) \equiv 1$ and $\Psi(\boldsymbol{x}) = 1[\boldsymbol{x} \in \mathcal{S}]$, we obtain the original network with state-independent transitions.

Theorem 10.15 *Suppose the quasi-reversible network of Theorem 10.13 has the stationary distribution* π. *Assume* $\Psi(\boldsymbol{x})$ *is nondecreasing and invariant with regard to the internal transitions, i.e., if* $q^{\mathrm{I}}_j(\boldsymbol{x}, \boldsymbol{x}') > 0$ *then* $\Psi(\boldsymbol{x}) = \Psi(\boldsymbol{x}')$. *If the departure rate functions of the queueing network with the state-dependent transitions above are modified in such a way that*

$$\hat{q}^{\mathrm{D}+}_{ju}(\boldsymbol{x}, \boldsymbol{x}') = \begin{cases} \hat{q}^{\mathrm{D}}_{ju}(\boldsymbol{x}, \boldsymbol{x}'), & u \in T_j^- \cup \{c\}, \\ \hat{q}^{\mathrm{D}}_{ju}(\boldsymbol{x}, \boldsymbol{x}') + 1[\boldsymbol{x}' = \boldsymbol{x}, \boldsymbol{x}_j^- \in \mathcal{S}]\dfrac{\Psi(\boldsymbol{x}) - \Psi(\boldsymbol{x}_j^-)}{\Phi(\boldsymbol{x})} & \\ \quad \times \dfrac{\pi(\boldsymbol{x}_j^-)}{\pi(\boldsymbol{x})} \displaystyle\sum_{v \in T_j^+} \alpha_{jv} \hat{p}^{\mathrm{A}+}_{jv,c}(\boldsymbol{x}_j^-, \boldsymbol{x}) \hat{f}_{jv,u}(\boldsymbol{x}_j^-, \boldsymbol{x}), & u \in T_j^+, \end{cases}$$

$$\boldsymbol{x}, \boldsymbol{x}' \in \mathcal{S},\ j = 0, 1, 2, \ldots, N, \tag{6.2}$$

then the modified network has the stationary measure $\hat{\pi}$ *given by*

$$\hat{\pi}(\boldsymbol{x}) = \Phi(\boldsymbol{x})\pi(\boldsymbol{x}), \qquad \boldsymbol{x} \in \mathcal{S}. \tag{6.3}$$

In particular, if

$$K \equiv \sum_{\boldsymbol{x} \in \mathcal{S}} \Phi(\boldsymbol{x})\pi(\boldsymbol{x}) < \infty,$$

then $\{K^{-1}\hat{\pi}(\boldsymbol{x});\ \boldsymbol{x} \in \mathcal{S}\}$ *is the stationary distribution of the modified network.*

Theorem 10.15 states that the state-dependent network generated by Ψ and Φ may not have a product form stationary distribution. However, if we introduce an additional departure process of positive signals at node j, which do not change the state of node j, with rate

$$1[\boldsymbol{x}' = \boldsymbol{x}, \boldsymbol{x}_j^- \in \mathcal{S}] \frac{\Psi(\boldsymbol{x}) - \Psi(\boldsymbol{x}_j^-)}{\Phi(\boldsymbol{x})} \frac{\pi(\boldsymbol{x}_j^-)}{\pi(\boldsymbol{x})} \sum_{v \in T_j^+} \alpha_{jv} \hat{p}^{\mathrm{A}+}_{jv,c}(\boldsymbol{x}_j^-, \boldsymbol{x}) \hat{f}_{jv,u}(\boldsymbol{x}_j^-, \boldsymbol{x}),$$

$$u \in T_j^+, \ j = 1, \ldots, N, \qquad (6.4)$$

then the network has a product form stationary distribution.

To gain some further insight into the newly introduced departures of positive signals, consider the case where in the original quasi-reversible network, $q^{\mathrm{D}}_{ju}(x_j, x_j)$, $u \in T_j^+$, is nonzero only when x_j^- does not exist, e.g., $x_j = 0$ when x_j is the number of customers at node j. In this case $q^{\mathrm{D}}_{ju}(x_j, x_j)$ represents the additional departure process when the node is empty. Indeed, this is the case in all applications in Chapters 4, 5 and 8. If the positive (negative) signals only trigger departures of positive (negative) signals, then the quasi-reversibility condition for $x_j^- \in \mathcal{S}_j$ implies that

$$\frac{\pi(\boldsymbol{x}_j^-)}{\pi(\boldsymbol{x})} \sum_{v \in T_j^+} \alpha_{jv} \hat{p}^{\mathrm{A}+}_{jv,c}(\boldsymbol{x}_j^-, \boldsymbol{x}) \hat{f}_{jv,u}(\boldsymbol{x}_j^-, \boldsymbol{x}) = \beta_{ju}, \quad u \in T_j^+.$$

The additional departure rate of positive signals (6.4) can thus be written as

$$1[\boldsymbol{x}' = \boldsymbol{x}, \boldsymbol{x}_j^- \in \mathcal{S}] \beta_{ju} \frac{\Psi(\boldsymbol{x}) - \Psi(\boldsymbol{x}_j^-)}{\Phi(\boldsymbol{x})}, \qquad u \in T_j^+. \qquad (6.5)$$

If $\Psi(\boldsymbol{x})$ is 0 for $\boldsymbol{x} \notin \mathcal{S}$, then the departure process of positive signals generated at each node $\hat{q}^{\mathrm{D}}_{ju}(\boldsymbol{x}, \boldsymbol{x})$, $u \in T_j^+$, can also be written as (6.5) with $\boldsymbol{x}_j^-$ removed from the indicator function. Therefore, we may simply assume that in the original network there are no departures of positive signals generated at any node. Theorem 10.15 then states that if there are additional departures of positive signals from node j, which do not change the state of node j, at rate

$$\beta_{ju} \frac{\Psi(\boldsymbol{x}) - \Psi(\boldsymbol{x}_j^-)}{\Phi(\boldsymbol{x})}, \qquad u \in T_j^+,$$

when the state of the network is $\boldsymbol{x}$, then the network has the stationary measure $\hat{\pi}(\boldsymbol{x}) = \Phi(\boldsymbol{x})\pi(\boldsymbol{x})$. This result includes all the models with additional arrivals (or departures) described in Chapters 4, 5 and 8, because by setting $\Phi(\boldsymbol{x}) \equiv 1$ and $\Psi(\boldsymbol{x}) = 1[\boldsymbol{x} \in \mathcal{S}]$, we obtain those results.

We need two lemmas to prove Theorem 10.15.

Lemma 10.16 *Under conditions (i) and (ii), equation (5.8) is equivalent to*

$$\pi(\boldsymbol{x})\sum_{x_j'\in\mathcal{S}_j}(q^{\mathrm{D}}_{jc}(x_j,x_j')+q^{\mathrm{I}}_j(x_j,x_j'))+\sum_{k=0}^{N}\sum_{v\in T_j^-}\sum_{w\in T_k}\beta_{kw}\pi(\boldsymbol{x})r_{kw,jv}p^{\mathrm{A}-}_{jv,c}(x_j,x_j^-)$$
$$=\sum_{k=0}^{N}\sum_{w\in T_k}\beta_{kw}\pi(\boldsymbol{x}_j^-)\left(r_{kw,jc}p^{\mathrm{A}}_{jc}(x_j^-,x_j)+\sum_{v\in T_j^+}r_{kw,jv}p^{\mathrm{A}+}_{jv,c}(x_j^-,x_j)\right)$$
$$+\sum_{x_j'\in\mathcal{S}_j}\pi(\boldsymbol{x}_j(x_j'))q^{\mathrm{I}}_j(x_j',x_j),\qquad j=1,\ldots,N,\ \boldsymbol{x}\in\mathcal{S}.\tag{6.6}$$

PROOF. It follows from the quasi-reversibility condition (5.3) and the traffic equations that

$$\beta_{ju}\pi(\boldsymbol{x})=\sum_{x_j'\in\mathcal{S}_j}\pi(\boldsymbol{x}_j(x_j'))q^{\mathrm{D}}_{ju}(x_j',x_j)+\sum_{\boldsymbol{x}'\in\mathcal{S}}\sum_{n=0}^{\infty}\left[(\pi q^{\mathrm{D}})(rp^{\mathrm{A}}f)^n rp^{\mathrm{A}}f(\boldsymbol{x}',\boldsymbol{x})\right]_{ju}$$
$$=\sum_{\boldsymbol{x}'\in\mathcal{S}}\sum_{n=0}^{\infty}\left[(\pi q^{\mathrm{D}})(rp^{\mathrm{A}}f)^n(\boldsymbol{x}',\boldsymbol{x})\right]_{ju}.\tag{6.7}$$

Substituting (6.7) into (5.8) completes the proof. □

Lemma 10.17 *If $\pi^*(\boldsymbol{x})=\Psi(\boldsymbol{x})\pi(\boldsymbol{x})$, then, for $j=1,2,\ldots,N$,*

$$\beta_{ju}\pi^*(\boldsymbol{x})=\sum_{\boldsymbol{x}'\in\mathcal{S}}\sum_{n=0}^{\infty}\left[(\hat{\pi}\hat{q}^{\mathrm{D}+})(r\hat{p}^{\mathrm{A}}\hat{f})^n(\boldsymbol{x}',\boldsymbol{x})\right]_{ju},\qquad u\in T_j,\ \boldsymbol{x}\in\mathcal{S},\tag{6.8}$$

$$\alpha_{ju}\pi^*(\boldsymbol{x})=\sum_{k=0}^{N}\sum_{w\in T_k}\sum_{\boldsymbol{x}'\in\mathcal{S}}\sum_{n=0}^{\infty}\left[(\hat{\pi}\hat{q}^{\mathrm{D}+})(r\hat{p}^{\mathrm{A}}\hat{f})^n(\boldsymbol{x}',\boldsymbol{x})\right]_{kw}r_{kw,ju},$$
$$u\in T_j,\ \boldsymbol{x}\in\mathcal{S}.\tag{6.9}$$

PROOF. Since (6.9) follows from (6.8) and the traffic equations (5.5), it suffices to prove (6.8). Let $\hat{\pi}(\boldsymbol{x})=\Phi(\boldsymbol{x})\pi(\boldsymbol{x})$. Multiplying both sides of the quasi-reversibility condition (5.3) by $\Psi(\boldsymbol{x})$ and applying the modified transitions, we obtain, for $j=1,2,\ldots,N$,

$$\beta_{ju}\pi^*(\boldsymbol{x})=\begin{cases}\hat{\pi}(\boldsymbol{x}_j^+)\hat{q}^{\mathrm{D}}_{jc}(\boldsymbol{x}_j^+,\boldsymbol{x}), & u=c,\\[2ex] \hat{\pi}(\boldsymbol{x})\hat{q}^{\mathrm{D}}_{ju}(\boldsymbol{x},\boldsymbol{x}) \\ \quad+\sum_{v\in T_j^-}\pi^*(\boldsymbol{x}_j^+)\alpha_{jv}\hat{p}^{\mathrm{A}-}_{jv,c}(\boldsymbol{x}_j^+,\boldsymbol{x})\hat{f}_{jv,u}(\boldsymbol{x}_j^+,\boldsymbol{x}) \\ \quad+\Psi(\boldsymbol{x})\sum_{v\in T_j^+}\pi(\boldsymbol{x}_j^-)\alpha_{jv}\hat{p}^{\mathrm{A}+}_{jv,c}(\boldsymbol{x}_j^-,\boldsymbol{x})\hat{f}_{jv,u}(\boldsymbol{x}_j^-,\boldsymbol{x}) \\ \quad+\sum_{v\in T_j^-\cup T_j^+}\pi^*(\boldsymbol{x})\alpha_{jv}\hat{p}^{\mathrm{A}*}_{jv}(\boldsymbol{x},\boldsymbol{x})\hat{f}_{jv,u}(\boldsymbol{x},\boldsymbol{x}), & u\in T_j^-\cup T_j^+.\end{cases}$$

Rewriting $\Psi(\boldsymbol{x})$ as

$$\left(\Psi(\boldsymbol{x}) - \Phi(\boldsymbol{x}_j^-)\right) + \Phi(\boldsymbol{x}_j^-)$$

and applying the modified departure rate functions (6.2) yields

$$\beta_{ju}\pi^*(\boldsymbol{x}) = \begin{cases} \hat{\pi}(\boldsymbol{x}_j^+)\hat{q}_{jc}^{\mathrm{D}}(\boldsymbol{x}_j^+, \boldsymbol{x}), & u = c, \\ \hat{\pi}(\boldsymbol{x})\hat{q}_{ju}^{\mathrm{D}+}(\boldsymbol{x}, \boldsymbol{x}) & \\ \quad + \sum_{v \in T_j^-} \pi^*(\boldsymbol{x}_j^+)\alpha_{jv}\hat{p}_{jv,c}^{\mathrm{A}-}(\boldsymbol{x}_j^+, \boldsymbol{x})\hat{f}_{jv,u}(\boldsymbol{x}_j^+, \boldsymbol{x}) & \\ \quad + \sum_{v \in T_j^+} \pi^*(\boldsymbol{x}_j^-)\alpha_{jv}\hat{p}_{jv,c}^{\mathrm{A}+}(\boldsymbol{x}_j^-, \boldsymbol{x})\hat{f}_{jv,u}(\boldsymbol{x}_j^-, \boldsymbol{x}) & \\ \quad + \sum_{v \in T_j^- \cup T_j^+} \pi^*(\boldsymbol{x})\alpha_{jv}\hat{p}_{jv}^{\mathrm{A}*}(\boldsymbol{x}, \boldsymbol{x})\hat{f}_{jv,u}(\boldsymbol{x}, \boldsymbol{x}), & u \in T_j^- \cup T_j^+. \end{cases} \tag{6.10}$$

Similarly, for $j = 0$, we have

$$\beta_{0u}\pi^*(\boldsymbol{x}) = \hat{\pi}(\boldsymbol{x})\hat{q}_{0u}^{\mathrm{D}+}(\boldsymbol{x}, \boldsymbol{x}). \tag{6.11}$$

Using the traffic equations (5.5) and repeatedly substituting β_{ju} of (6.10) and (6.11) into the right-hand side of (6.10), we obtain (6.8). □

PROOF OF THEOREM 10.15. All we need to do is to verify that the following global balance equation, corresponding to (5.2), holds:

$$\begin{aligned} &\hat{\pi}(\boldsymbol{x}) \sum_{j=0}^{N} \sum_{\boldsymbol{x}' \in \mathcal{S}} \left(\sum_{u \in T_j} \hat{q}_{ju}^{\mathrm{D}+}(\boldsymbol{x}, \boldsymbol{x}') + \hat{q}_j^{\mathrm{I}}(\boldsymbol{x}, \boldsymbol{x}') \right) \\ &= \sum_{j=0}^{N} \sum_{u \in T_j} \sum_{\boldsymbol{x}' \in \mathcal{S}} \sum_{n=0}^{\infty} \left[(\hat{\pi}\hat{q}^{\mathrm{D}+})(r\hat{p}^{\mathrm{A}}\hat{f})^n (r\hat{p}^{\mathrm{A}}(1 - \hat{f}))(\boldsymbol{x}', \boldsymbol{x}) \right]_{ju} \\ &\quad + \sum_{j=0}^{N} \sum_{\boldsymbol{x}' \in \mathcal{S}} \hat{\pi}(\boldsymbol{x}')\hat{q}_j^{\mathrm{I}}(\boldsymbol{x}', \boldsymbol{x}). \end{aligned} \tag{6.12}$$

Fix $\boldsymbol{x} \in \mathcal{S}$, and multiply both sides of (6.6) by $\Psi(\boldsymbol{x}_j^-)$. It follows from conditions (i), (ii) and the invariance property of Ψ with regard to the internal transitions that

$$\begin{aligned} &\hat{\pi}(\boldsymbol{x}) \left(\hat{q}_{jc}^{\mathrm{D}}(\boldsymbol{x}, \boldsymbol{x}_j^-) + \sum_{x_j' \in \mathcal{S}_j} \hat{q}_j^{\mathrm{I}}(\boldsymbol{x}, \boldsymbol{x}_j(x_j')) \right) \\ &+ \sum_{k=0}^{N} \sum_{v \in T_j^-} \sum_{w \in T_k} \beta_{kw}\pi^*(\boldsymbol{x}) r_{kw,jv}\hat{p}_{jv,c}^{\mathrm{A}-}(\boldsymbol{x}, \boldsymbol{x}_j^-) \end{aligned}$$

$$= \sum_{k=0}^{N} \sum_{w \in T_k} \beta_{kw} \pi^*(\boldsymbol{x}_j^-) \left(r_{kw,jc} \hat{p}_{jc}^{\mathrm{A}}(\boldsymbol{x}_j^-, \boldsymbol{x}) + \sum_{v \in T_j^+} r_{kw,jv} \hat{p}_{jv,c}^{\mathrm{A}+}(\boldsymbol{x}_j^-, \boldsymbol{x}) \right)$$
$$+ \sum_{x_j' \in \mathcal{S}_j} \hat{\pi}(\boldsymbol{x}_j(x_j')) \hat{q}_j^{\mathrm{I}}(\boldsymbol{x}_j(x_j'), \boldsymbol{x}), \qquad j = 1, \ldots, N. \tag{6.13}$$

Since in the following derivation the internal transition terms do not change, we drop them from both sides of the equations and restore them later.

Substituting β_{kw} of Lemma 10.17 in the first summation on the right-hand side of (6.13) and applying the traffic equations, we obtain

$$\hat{\pi}(\boldsymbol{x}) \hat{q}_{jc}^{\mathrm{D}}(\boldsymbol{x}, \boldsymbol{x}_j^-) + \sum_{v \in T_j^-} \alpha_{jv} \pi^*(\boldsymbol{x}) \hat{p}_{jv,c}^{\mathrm{A}-}(\boldsymbol{x}, \boldsymbol{x}_j^-)$$
$$= \sum_{\boldsymbol{x}' \in \mathcal{S}} \sum_{n=0}^{\infty} \left[(\hat{\pi} \hat{q}^{\mathrm{D}+})(r \hat{p}^{\mathrm{A}} \hat{f})^n r \hat{p}^{\mathrm{A}}(\boldsymbol{x}', \boldsymbol{x}) \right]_{jc} + \sum_{v \in T_j^+} \alpha_{jv} \pi^*(\boldsymbol{x}_j^-) \hat{p}_{jv,c}^{\mathrm{A}+}(\boldsymbol{x}_j^-, \boldsymbol{x}),$$
$$j = 1, \ldots, N. \tag{6.14}$$

Summing (6.14) over $j = 1, 2, \ldots, N$, together with (5.6) multiplied by $\Psi(\boldsymbol{x})$ for $j = 0$ yields

$$\hat{\pi}(\boldsymbol{x}) \left(\sum_{j=1}^{N} \hat{q}_{jc}^{\mathrm{D}}(\boldsymbol{x}, \boldsymbol{x}_j^-) + \sum_{u \in T_0} \hat{q}_{0u}^{\mathrm{D}}(\boldsymbol{x}, \boldsymbol{x}) \right)$$
$$+ \pi^*(\boldsymbol{x}) \sum_{j=1}^{N} \sum_{v \in T_j^-} \alpha_{jv} \hat{p}_{jv,c}^{\mathrm{A}-}(\boldsymbol{x}, \boldsymbol{x}_j^-) + \pi^*(\boldsymbol{x}) \sum_{u \in T_0} (\alpha_{0u} - \beta_{0u})$$
$$= \sum_{j=1}^{N} \sum_{\boldsymbol{x}' \in \mathcal{S}} \sum_{n=0}^{\infty} \left[(\hat{\pi} \hat{q}^{\mathrm{D}+})(r \hat{p}^{\mathrm{A}} \hat{f})^n r \hat{p}^{\mathrm{A}}(\boldsymbol{x}', \boldsymbol{x}) \right]_{jc}$$
$$+ \sum_{u \in T_0} \sum_{\boldsymbol{x}' \in \mathcal{S}} \sum_{n=0}^{\infty} \left[(\hat{\pi} \hat{q}^{\mathrm{D}+})(r \hat{p}^{\mathrm{A}} \hat{f})^n r \hat{p}^{\mathrm{A}}(\boldsymbol{x}', \boldsymbol{x}) \right]_{0u}$$
$$+ \sum_{j=1}^{N} \sum_{v \in T_j^+} \alpha_{jv} \pi^*(\boldsymbol{x}_j^-) \hat{p}_{jv,c}^{\mathrm{A}+}(\boldsymbol{x}_j^-, \boldsymbol{x}), \tag{6.15}$$

where the second term on the right-hand side is obtained through (6.7) and (6.9):

$$\Psi(\boldsymbol{x}) \sum_{u \in T_0} \sum_{\boldsymbol{x}' \in \mathcal{S}} \sum_{n=0}^{\infty} \left[(\pi q^{\mathrm{D}})(r p^{\mathrm{A}} f)^n r p^{\mathrm{A}}(\boldsymbol{x}', \boldsymbol{x}) \right]_{0u}$$
$$= \Psi(\boldsymbol{x}) \sum_{u \in T_0} \sum_{k=1}^{N} \sum_{v \in T_k} \beta_{kv} \pi(\boldsymbol{x}) r_{kv,0u} p_{0u}^{\mathrm{A}}(\boldsymbol{x}, \boldsymbol{x})$$

$$= \sum_{u \in T_0} \alpha_{0u} \pi^*(\boldsymbol{x})$$

$$= \sum_{u \in T_0} \sum_{\boldsymbol{x}' \in \mathcal{S}} \sum_{n=0}^{\infty} \sum_{k=0}^{N} \sum_{w \in T_k} \left[(\hat{\pi} \hat{q}^{\text{D}+})(r \hat{p}^{\text{A}} \hat{f})^n (\boldsymbol{x}', \boldsymbol{x}) \right]_{kw} r_{kw,0u} .$$

Similarly, multiplying both sides of (4.5) by $\Psi(\boldsymbol{x})$ yields

$$\pi^*(\boldsymbol{x}_j^+) \sum_{v \in T_j^-} \alpha_{jv} \hat{p}_{jv,c}^{\text{A}-}(\boldsymbol{x}_j^+, \boldsymbol{x}) - \pi^*(\boldsymbol{x}) \sum_{v \in T_j^+} \alpha_{jv} \hat{p}_{jv,c}^{\text{A}+}(\boldsymbol{x}, \boldsymbol{x}_j^+)$$
$$= \left(\alpha_{jc} - \beta_{jc} \right) \pi^*(\boldsymbol{x}). \tag{6.16}$$

It follows from identity (4.9) and (6.16) that

$$\sum_{j=1}^{N} \left(\beta_{jc} + \sum_{u \in T_j^- \cup T_j^+} \beta_{ju} \right) - \sum_{u \in T_0} (\alpha_{0u} - \beta_{0u})$$
$$= \sum_{j=1}^{N} \left(\alpha_{jc} + \sum_{v \in T_j^- \cup \in T_j^+} \alpha_{jv} \right)$$
$$= \sum_{j=1}^{N} \left(\alpha_{jc} + \sum_{v \in T_j^-} \alpha_{jv} \hat{p}_{jv,c}^{\text{A}-}(\boldsymbol{x}, \boldsymbol{x}_j^-) + \sum_{v \in T_j^+} \alpha_{jv} \hat{p}_{jv,c}^{\text{A}+}(\boldsymbol{x}, \boldsymbol{x}_j^+) \right)$$
$$+ \sum_{j=1}^{N} \sum_{v \in T_j^+ \cup T_j^-} \alpha_{jv} \hat{p}_{jv}^{\text{A}*}(\boldsymbol{x}, \boldsymbol{x})$$
$$= \sum_{j=1}^{N} \left(\beta_{jc} + \sum_{v \in T_j^-} \alpha_{jv} \hat{p}_{jv,c}^{\text{A}-}(\boldsymbol{x}, \boldsymbol{x}_j^-) + \frac{\pi^*(\boldsymbol{x}_j^+)}{\pi^*(\boldsymbol{x})} \sum_{v \in T_j^-} \alpha_{jv} \hat{p}_{jv,c}^{\text{A}-}(\boldsymbol{x}_j^+, \boldsymbol{x}) \right)$$
$$+ \sum_{j=1}^{N} \sum_{v \in T_j^+ \cup T_j^-} \alpha_{jv} \hat{p}_{jv}^{\text{A}*}(\boldsymbol{x}, \boldsymbol{x}),$$

where (6.16) is used in the last equality. Thus we obtain

$$\pi^*(\boldsymbol{x}) \sum_{j=1}^{N} \sum_{v \in T_j^-} \alpha_{jv} \hat{p}_{jv,c}^{\text{A}-}(\boldsymbol{x}, \boldsymbol{x}_j^-) + \pi^*(\boldsymbol{x}) \sum_{u \in T_0} (\alpha_{0u} - \beta_{0u})$$
$$= \sum_{j=1}^{N} \left(\pi^*(\boldsymbol{x}) \sum_{u \in T_j^- \cup T_j^+} \beta_{ju} - \pi^*(\boldsymbol{x}_j^+) \sum_{v \in T_j^-} \alpha_{jv} \hat{p}_{jv,c}^{\text{A}-}(\boldsymbol{x}_j^+, \boldsymbol{x}) \right)$$
$$- \pi^*(\boldsymbol{x}) \sum_{j=1}^{N} \sum_{v \in T_j^+ \cup T_j^-} \alpha_{jv} \hat{p}_{jv}^{\text{A}*}(\boldsymbol{x}, \boldsymbol{x}).$$

After substituting this into equation (6.15) and rearranging terms, we have

$$\hat{\pi}(\boldsymbol{x})\left(\sum_{j=1}^{N}\hat{q}^{\mathrm{D}}_{jc}(\boldsymbol{x},\boldsymbol{x}_j^-)+\sum_{u\in T_0}\hat{q}^{\mathrm{D}}_{0u}(\boldsymbol{x},\boldsymbol{x})\right)+\pi^*(\boldsymbol{x})\sum_{j=1}^{N}\sum_{u\in T_j^-\cup T_j^+}\beta_{ju}$$

$$=\sum_{j=1}^{N}\sum_{\boldsymbol{x}'\in\mathcal{S}}\sum_{n=0}^{\infty}\left[(\hat{\pi}\hat{q}^{\mathrm{D}+})(r\hat{p}^{\mathrm{A}}\hat{f})^n r\hat{p}^{\mathrm{A}}(\boldsymbol{x}',\boldsymbol{x})\right]_{jc}$$

$$+\sum_{u\in T_0}\sum_{\boldsymbol{x}'\in\mathcal{S}}\sum_{n=0}^{\infty}\left[(\hat{\pi}\hat{q}^{\mathrm{D}+})(r\hat{p}^{\mathrm{A}}\hat{f})^n r\hat{p}^{\mathrm{A}}(\boldsymbol{x}',\boldsymbol{x})\right]_{0u}$$

$$+\sum_{j=1}^{N}\sum_{v\in T_j^+}\alpha_{jv}\pi^*(\boldsymbol{x}_j^-)\hat{p}^{\mathrm{A}+}_{jv,c}(\boldsymbol{x}_j^-,\boldsymbol{x})+\sum_{j=1}^{N}\sum_{v\in T_j^-}\alpha_{jv}\pi^*(\boldsymbol{x}_j^+)\hat{p}^{\mathrm{A}-}_{jv,c}(\boldsymbol{x}_j^+,\boldsymbol{x})$$

$$+\sum_{j=1}^{N}\sum_{v\in T_j^+\cup T_j^-}\alpha_{jv}\pi^*(\boldsymbol{x})\hat{p}^{\mathrm{A}*}_{jv}(\boldsymbol{x},\boldsymbol{x}). \tag{6.17}$$

Substituting $\beta_{ju}\pi^*(\boldsymbol{x})$ of (6.10) into (6.17), we obtain

$$\hat{\pi}(\boldsymbol{x})\left(\sum_{j=1}^{N}\left(\hat{q}^{\mathrm{D}}_{jc}(\boldsymbol{x},\boldsymbol{x}_j^-)+\sum_{u\in T_j^-\cup T_j^+}\hat{q}^{\mathrm{D}+}_{ju}(\boldsymbol{x},\boldsymbol{x})\right)+\sum_{u\in T_0}\hat{q}^{\mathrm{D}}_{0u}(\boldsymbol{x},\boldsymbol{x})\right)$$

$$=\sum_{j=0}^{N}\sum_{\boldsymbol{x}'\in\mathcal{S}}\sum_{n=0}^{\infty}\left[(\hat{\pi}\hat{q}^{\mathrm{D}+})(r\hat{p}^{\mathrm{A}}\hat{f})^n r\hat{p}^{\mathrm{A}}(\boldsymbol{x}',\boldsymbol{x})\right]_{jc}$$

$$+\sum_{j=0}^{N}\sum_{v\in T_j^+}\pi^*(\boldsymbol{x}_j^-)\alpha_{jv}\hat{p}^{\mathrm{A}+}_{jv,c}(\boldsymbol{x}_j^-,\boldsymbol{x})\left(1-\sum_{w\in T_j}f_{jv,w}(\boldsymbol{x}_j^-,\boldsymbol{x})\right)$$

$$+\sum_{j=0}^{N}\sum_{v\in T_j^-}\pi^*(\boldsymbol{x}_j^+)\alpha_{jv}\hat{p}^{\mathrm{A}-}_{jv,c}(\boldsymbol{x}_j^+,\boldsymbol{x})\left(1-\sum_{w\in T_j}f_{jv,w}(\boldsymbol{x}_j^+,\boldsymbol{x})\right)$$

$$+\sum_{j=0}^{N}\sum_{v\in T_j^+\cup T_j^-}\pi^*(\boldsymbol{x})\alpha_{jv}\hat{p}^{\mathrm{A}*}_{jv}(\boldsymbol{x},\boldsymbol{x})\left(1-\sum_{w\in T_j}f_{jv,w}(\boldsymbol{x},\boldsymbol{x})\right).$$

Substituting α_{ju} of (6.9) into this equation yields, after adding the dropped internal transitions, the global balance equations (6.12) for $\boldsymbol{x}\in\mathcal{S}$. □

Remark 10.18 It follows from (6.9) that when the state of the network is $\boldsymbol{x}$, the average departure rate of class u entities from node j is $\beta_{ju}\Psi(\boldsymbol{x})/\Phi(\boldsymbol{x})$. Hence, if the stationary distribution exists, the total average departure rate from all nodes of the network is

$$K^{-1}\sum_{j=1}^{N}\sum_{u\in T_j}\beta_{ju}\sum_{\boldsymbol{x}\in\mathcal{S}}\Psi(\boldsymbol{x})\pi(\boldsymbol{x}).$$

This rate is a function of Ψ and may not be finite.

In Chapter 9, we deferred the proof of Theorem 9.18, which includes Theorems 9.11 and 9.12 as special cases. We now apply Theorem 10.15 to prove Theorem 9.18.

PROOF OF THEOREM 9.18. The network of Theorem 9.18 is a modification of the network of Section 4.5 with a single class of customers and multiple classes of positive and negative signals. The network of Section 4.5 is not quasi-reversible without the additional departures of positive signals. So it is modified in such a way that, whenever node j is empty, an additional Poisson arrival process of class u^+ positive signals is activated with rate

$$q^{\mathrm{D}}_{ju}(0,0)=\frac{\mu_j+\alpha_j^-}{\alpha_j+\alpha_j^+}\alpha_{ju}^+, \qquad u=1,2,\ldots,I^+,$$

where

$$\alpha_j^-=\sum_{u=1}^{I^-}\alpha_{ju}^-, \qquad \alpha_j^+=\sum_{u=1}^{I^+}\alpha_{ju}^+.$$

It follows from Theorem 4.17 that if

$$\rho_j\equiv\frac{\alpha_j+\alpha_j^+}{\mu_j+\alpha_j^-}<1, \qquad j=1,\ldots,N,$$

then the network is quasi-reversible and its stationary distribution is

$$\pi(\boldsymbol{n})=\prod_{j=1}^{N}(1-\rho_j)\rho_j^{n_j}. \tag{6.18}$$

In this network, conditions (i) and (ii) are clearly satisfied, and the triggering probability functions $f_{jv,u}$ are

$$f_{jv,u}(x_j,x_j')=\begin{cases}1[x_j'=x_j^-,\, v=u], & u\in T_j^-,\\ 1[x_j'=x_j^+,\, v=u], & u\in T_j^+.\end{cases}$$

Modify the network by introducing the state-dependent transitions using Φ and Ψ as described at the beginning of this section, which then results in the state-dependent network described in Chapter 9.

To apply Theorem 10.15, we need to verify (4.5), i.e.,

$$\sum_{v=1}^{I^-} \pi_j(n+1)\alpha_{jv} - \pi_j(n)\sum_{v=1}^{I^+} \alpha_{jv} = (\alpha_{jc} - \beta_{jc})\pi_j(n),$$
$$j = 1, \ldots, N, n \geq 0. \qquad (6.19)$$

This is equivalent to

$$\rho_j\alpha_j^- - \alpha_j^+ = \alpha_{jc} - \beta_{jc}, \qquad j = 1, 2, \ldots, N.$$

Since $\beta_{jc} = \rho_j\mu_j$, this reduces to the definition of ρ_j. Hence, condition (4.5) is indeed satisfied. Thus it follows from Theorem 10.15 that the modified network has the stationary distribution π of (6.3). To complete the proof of Theorem 9.18, we need to show that (5.4) of Chapter 9 is identical to the additional terms in (6.2). The additional terms in (6.2) can be classified into the following two types:

(a) those added in the modification to ensure quasi-reversibility, and

(b) those added to subsume state dependency.

The first type only appears at node j when $n_j = 0$, and the rate is $\alpha_{ju}^+\rho_j^{-1}$. Because of (b), the rate becomes

$$\alpha_{ju}^+\rho_j^{-1}\frac{\Psi(\boldsymbol{n})}{\Phi(\boldsymbol{n})}, \qquad u = 1, \ldots, I^+. \qquad (6.20)$$

On the other hand, the second type appears at node j when $n_j \geq 1$. Since negative signals do not trigger departures of positive signals, it follows from the quasi-reversibility of the original network that the additional term of (6.2) for $n_j \geq 1$ is identical to (5.4) of Chapter 9 (see the discussion after Theorem 10.15). This completes the proof of Theorem 9.18. □

Theorem 10.15 can also be applied to the network with arbitrary service time distributions and negative signals of Section 6.3. Assume that there are one class of customers, one class of negative signals, and no positive signals. No additional departures are needed because there are no positive signals. The network has internal transitions for the remaining service phases. However, these internal transitions are locally balanced among themselves (see (4.9)). Hence, if Φ and Ψ only depend on the number of customers at the nodes, Theorem 10.15 can be applied to obtain a network with state-dependent transitions. We need to verify (4.5), i.e., (6.19) with $\alpha_{jv} = 0$ for $v \in T_j^+$. Using the notation of Chapter 6, this is equivalent to

$$\frac{\alpha_j}{\bar{\mu}_j}\alpha_j^- = \alpha_j - \beta_j, \qquad\qquad j = 1, 2, \ldots, N,$$

which is clearly satisfied by (4.12) of Chapter 6. Thus we obtain the following result.

Theorem 10.19 *Let π be the stationary distribution of the network with symmetric service disciplines in Section 6.3. Modify the model to a network with state-dependent transitions using a nondecreasing function Ψ and a positive function Φ. Then, the modified network has the stationary measure $\{\Phi(\boldsymbol{x})\pi(\boldsymbol{x});\ \boldsymbol{x} \in \mathcal{S}\}$.*

It is possible to extend this result to networks with symmetric service disciplines, arbitrarily distributed service requirements, and both positive and negative signals. We would need, of course, additional departures of positive signals in order to obtain product form stationary distributions.

10.7 TIME-REVERSED PROCESSES AND THE ARRIVAL THEOREM

We have shown in Chapters 3 and 4 that, if a queueing network is quasi-reversible, its reversed process represents another quasi-reversible network. In other words, a quasi-reversible network is preserved under time reversal. A similar preservation property holds for local balances.

Proposition 10.20 *If the Markov chain with transition rate q satisfies the biased local balance with respect to $\{q_{jk}\}_{j,k \in J}$ and with bias terms $\{\gamma_j\}_{j \in J}$, then its time-reversed process represents another network that satisfies the biased local balance with respect to $\{\tilde{q}_{jk}\}_{j,k \in J}$ and with bias terms $\{-\gamma_j\}_{j \in J}$, where*

$$\tilde{q}_{jk}(\boldsymbol{x}, \boldsymbol{x}') = \frac{\pi(\boldsymbol{x}')}{\pi(\boldsymbol{x})} q_{kj}(\boldsymbol{x}', \boldsymbol{x}), \qquad j, k \in J,\ \ \boldsymbol{x}, \boldsymbol{x}' \in \mathcal{S}. \tag{7.1}$$

PROOF. Let $\tilde{q} = \sum_{j,k \in J} \tilde{q}_{jk}$. It is easy to see that

$$\pi(\boldsymbol{x})q(\boldsymbol{x}, \boldsymbol{x}') = \pi(\boldsymbol{x}')\tilde{q}(\boldsymbol{x}', \boldsymbol{x}), \qquad \boldsymbol{x}, \boldsymbol{x}' \in \mathcal{S},$$

and, substituting $\tilde{q}_{jk}$ into the biased local balance equation (1.4) yields

$$\sum_{k\in J}\sum_{x'\in\mathcal{S}}\pi(x')\tilde{q}_{jk}(x',x)+\pi(x)\gamma_j=\sum_{k\in J}\sum_{x'\in\mathcal{S}}\pi(x)\tilde{q}_{jk}(x,x').$$

This shows that the reversed process satisfies the biased local balance with respect to $\{\tilde{q}_{jk}\}$ and with bias term $\{-\gamma_j\}$. □

Note that this proposition does not require the condition that the stationary distribution π is product form. Indeed, the local balance and the biased local balance can be considered for networks more general than quasi-reversible networks.

We now consider the embedded stationary distribution at specific epochs in time. In this book, we have mainly studied the time stationary distribution π for queueing networks. We may be interested in the distribution of the network state observed by the arrivals at a given node. Such a distribution provides information that is important for the customers in the network (for example, to compute sojourn times). Let π^{A}_{ju} denote the embedded stationary distribution observed by class u arrivals at node j. This distribution is, in general, not equal to the time stationary distribution π. In Chapter 2 we have shown that if the observing epochs constitute a Poisson process, then $\pi^{\mathrm{A}}_{ju}=\pi$ (PASTA). The following theorem shows that in quasi-reversible networks any arrivals see time averages (ASTA). The arrivals may be customers, they may also be signals. This result is also known as the Arrival Theorem.

Theorem 10.21 (ASTA) *In a quasi-reversible network with signals and instantaneous movements, i.e., the model of Theorem 10.12, arrivals see time averages, i.e.,*

$$\pi^{\mathrm{A}}_{ju}(x)=\pi(x),\qquad \textit{for all } x\in\mathcal{S},\quad j=0,1,\ldots,N, u\in T_j.$$

PROOF. In the context of point processes, $\pi^{\mathrm{A}}_{ju}(x)$ is defined as the ratio of the rate going into state x due to the arrivals of class u entities at node j, where x is the state immediately before the arrival, and the total rate of class u arrivals at node j. That is,

$$\pi^{\mathrm{A}}_{ju}(x)=\frac{\sum_{k=0}^{N}\sum_{v\in T_k}\sum_{x'\in\mathcal{S}}\sum_{n=0}^{\infty}\left[(\pi q^{\mathrm{D}})(rp^{\mathrm{A}}f)^n(x',x)\right]_{kv}r_{kv,ju}}{\sum_{k=0}^{N}\sum_{v\in T_k}\sum_{x',x''\in\mathcal{S}}\sum_{n=0}^{\infty}[(\pi q^{\mathrm{D}})(rp^{\mathrm{A}}f)^n(x',x'')]_{kv}r_{kv,ju}}. \tag{7.2}$$

Thus it follows from (5.7) that $\pi^{\mathrm{A}}_{ju}=\pi$. □

The relationship (7.2) also holds for networks with state-dependent transitions. The next result follows immediately from equations (7.2) and (6.9).

Theorem 10.22 *Assume that the networks with state-dependent transitions discussed in Section 10.6 have stationary distributions. Then* $\pi^{\mathrm{A}}_{ju}(\boldsymbol{x}) = \pi(\boldsymbol{x})$ *for* $\boldsymbol{x} \in \mathcal{S}$ *and* $j = 0, 1, \ldots, N$, $u \in T_j$ *if and only if* $\Psi(\boldsymbol{x}) = c\Phi(\boldsymbol{x})$, $\boldsymbol{x} \in \mathcal{S}$, *for some constant* c, *i.e., the arrival process to the network from the outside is Poisson.*

10.8 REFERENCE NOTES

The biased local balance in Section 10.1 was introduced by Chao and Miyazawa (1998) for quasi-reversible networks without signals. Section 10.2 is also taken from Chao and Miyazawa (1998). The biased local balance for networks with signals, i.e., Theorem 10.15, is new. In contrast to the biased local balance of Section 10.2, which is with respect to each node, the refined local balance of Section 10.3 is with respect to each position at each node and each customer class; it may be referred to as a position–class local balance. This result for BCMP and symmetric service networks goes back to Baskett et al. (1975) and Kelly (1976). The result presented in Section 10.3 within the framework of quasi-reversible networks is new. The results of Sections 10.4, 10.5 and 10.6 are taken from Miyazawa and Chao (1998). The Arrival Theorem presented in Section 10.7 is new; however, from the quasi-reversibility point of view, it is immediate.

EXERCISES

10.1 Prove Theorem 10.13.

10.2 Consider the assembly-transfer network in Example 10.5 and Section 8.2.

(a) From equation (2.1) of Theorem 8.2, derive the balance equation

$$\sum_{\ell=1}^{\infty} \rho_j^{\ell} \mu_j \overline{b}_j(\ell) = \lambda_j^* \qquad j = 1, \ldots, N.$$

(b) Using (a), derive the balance

$$\sum_{\ell=1}^{\infty} \pi(\boldsymbol{n} + (\ell - 1)\mathbf{e}_j)\mu_j \overline{b}_j(\ell)$$

$$= \pi(\mathbf{0})\lambda_j^* \overline{a}_j^*(n_j) + \sum_{\ell=1}^{n_j} \pi(\boldsymbol{n} - \ell\mathbf{e}_j)\lambda_j \overline{a}_j(\ell)$$

$$+\sum_{k=1}^{N}\sum_{v=1}^{\infty}\sum_{\ell=1}^{n_j}\pi(\boldsymbol{n}+v\mathbf{e}_k-\ell\mathbf{e}_j)\mu_k b_k(v)\sum_{m=\ell}^{\infty}r_{kv,jm},$$
$$j=1,\ldots,N,\ \boldsymbol{n}\in\mathcal{S}.$$

(c) Show that the cross local balance equation (1.8) is obtained as a special case of this equation, and interpret (b) as a cross local balance.

10.3 Consider the network of Section 10.4 without positive signals, i.e., $T_j^+=\emptyset$. Assume that the network satisfies condition (4.5). Derive the balance equations

$$\sum_{x_j'\in\mathcal{S}_j}\pi(\boldsymbol{x}_j(x_j'))q_{jc}^{\mathrm{D}}(x_j',x_j)$$
$$+\sum_{k=0}^{N}\sum_{v\in T_j^-}\sum_{x_j'\in\mathcal{S}_j}\sum_{y_k\in\mathcal{S}_k}\pi(\boldsymbol{x}_{jk}(x_j',y_k))q_{kc}^{\mathrm{D}}(y_k,x_k[x_j'])r_{kc,jv}p_{jv,c}^{\mathrm{A}-}(x_j',x_j)$$
$$=\sum_{k=0}^{N}\sum_{v\in T_k}\sum_{y_k\in\mathcal{S}_k}\sum_{x_j'\in\mathcal{S}_j}\pi(\boldsymbol{x}_k(y_k))q_{kc}^{\mathrm{D}}(y_k,x_k)r_{kc,jc}p_{jc}^{\mathrm{A}}(x_j,x_j').$$

Interpret this equation as a cross local balance.

11

Characterization of Product Form and Stability Issues

This chapter focuses on theoretical issues concerning the characterization of product form solutions and the existence of stationary distributions of queueing networks. Section 11.1 introduces some basic notation, and Section 11.2 presents the necessary and sufficient conditions for a network to have a product form solution. Such a characterization yields a general procedure for verifying the existence of a product form distribution and obtaining such a distribution when it exists. Since quasi-reversibility is known to be a sufficient condition for product form, Section 11.3 explores how much stronger quasi-reversibility is than product form. Various scenarios are presented in which quasi-reversibility is also a necessary condition for product form. These results are, for simplicity, first presented for the case of single class transitions, and in Section 11.4 they are extended to the multiple-class case. Finally, existence and stability issues are discussed in Sections 11.5 and 11.6 via an iterative approach and a fixed point approach.

11.1 INTRODUCTION AND NOTATION

In the preceding chapters we discussed various network models that possess product form stationary distributions. Most of the results are obtained through quasi-reversibility. A natural question to ask is: Is quasi-reversibility also a necessary condition for product form ? The answer is negative. This chapter presents the necessary and sufficient conditions for product form. Such a characterization yields a general procedure for verifying whether a network has a product form solution and finding it in the case when it exists. Furthermore, the network has a product form stationary distribution and is *biased locally balanced* if and only if the network is quasi-reversible and certain traffic equations are satisfied. We also consider various scenarios in which quasi-reversibility is a necessary condition for product form.

We employ the framework of Section 4.2. The network consists of N nodes, indexed 1 to N, and the outside is labeled as node 0. However, unlike the formulation in Chapter 4, we assume here that the outside, i.e., node 0, has multiple states. Since departures from node 0 are arrivals to the network, such a formulation allows the arrival process from the outside to the network to be arbitrary. The state of the network is a vector of the states of the individual nodes and the outside world; its state space is

$$\mathcal{S} = \mathcal{S}_0 \times \mathcal{S}_1 \times \cdots \times \mathcal{S}_N.$$

As in Section 4.2, node j, $j = 0, 1, \ldots, N$, is subject to three types of state transitions, referred to as arrival, departure and internal transitions, denoted by

$$\{p_j^{\mathrm{A}}(x_j, y_j);\, x_j, y_j \in \mathcal{S}_j\},$$
$$\{q_j^{\mathrm{D}}(x_j, y_j);\, x_j, y_j \in \mathcal{S}_j\},$$
$$\{q_j^{\mathrm{I}}(x_j, y_j);\, x_j, y_j \in \mathcal{S}_j\}.$$

They represent, respectively, the transition probabilities due to arrivals, the transition rate due to departures, and the internal transition rate. Thus we must have

$$\sum_{y_j \in \mathcal{S}_j} p_j^{\mathrm{A}}(x_j, y_j) = 1, \qquad x_j \in \mathcal{S}_j.$$

The network is characterized by the following system dynamics:

(i) When node j is in state x_j, the departure transition rate that changes the state from x_j to y_j is $q_j^{\mathrm{D}}(x_j, y_j)$.

(ii) A departure from node j becomes an arrival at node k with probability r_{jk}, $j, k = 0, 1, \ldots, N$ (recall that node 0 represents the outside).

(iii) An arrival at node k changes the node's state from x_k to y_k with probability $p_k^{\mathrm{A}}(x_k, y_k)$, $y_k \in \mathcal{S}_k$.

(iv) The internal transition rate at node j is $q_j^{\mathrm{I}}(x_j, y_j)$ when the state is x_j.

For convenience in this chapter we include feedbacks at a node in the internal transition, so the new internal transition rate is

$$q_j^{\mathrm{I}*}(x_j, y_j) = q_j^{\mathrm{I}}(x_j, y_j) + \sum_{x_j'} q_j^{\mathrm{D}}(x_j, x_j') r_{jj} p_j^{\mathrm{A}}(x_j', y_j).$$

Note that the total departure transition rate from x_j to y_j at node j excluding feedback is

$$\sum_{k \neq j} q_j^{\mathrm{D}}(x_j, y_j) r_{jk} = (1 - r_{jj}) q_j^{\mathrm{D}}(x_j, y_j).$$

The network process has transition rates

$$q(\boldsymbol{x}, \boldsymbol{x}') = \sum_{j,k} q_{jk}(\boldsymbol{x}, \boldsymbol{x}'), \qquad \boldsymbol{x}, \boldsymbol{x}' \in \mathcal{S},$$

where

$$q_{jk}(\boldsymbol{x}, \boldsymbol{x}') = \begin{cases} q_j^{\mathrm{D}}(x_j, x_j') r_{jk} p_k^{\mathrm{A}}(x_k, x_k') 1[x_\ell = x_\ell', \ell \neq j, k)], & \text{if } j \neq k, \\ q_j^{\mathrm{I}*}(x_j, x_j') 1[x_\ell' = x_\ell, \ell \neq j], & \text{if } j = k. \end{cases}$$

Our objective is to find the necessary and sufficient conditions for the network to have a product form stationary distribution.

Remark 11.1 The network transition rates q and the thinned transition rates q_{jk} are identical to the networks considered in Sections 4.2 and 10.2. The dynamics of nodes, however, are different, since the internal transitions are altered. We will see that it changes the node transitions and arrival rates at nodes (see, for example, (2.1)). However, it does not change the quasi-reversibility property of a node, nor does it change the stationary distribution of the node when the node is quasi-reversible (see Remark 11.8).

The following notation will be used in our analysis. For a probability distribution π_j on $\mathcal{S}_j$, let

$$q_j^{\mathrm{D}}(x_j) = \sum_{y_j} q_j^{\mathrm{D}}(x_j, y_j),$$

$$q_j^{\mathrm{I}*}(x_j) = \sum_{y_j} q_j^{\mathrm{I}*}(x_j, y_j),$$

$$\tilde{p}_j^{\mathrm{A}}(x_j) = \frac{\sum_{y_j} \pi_j(y_j) p_j^{\mathrm{A}}(y_j, x_j)}{\pi_j(x_j)},$$

$$\tilde{q}_j^{\mathrm{D}}(x_j) = \frac{\sum_{y_j} \pi_j(y_j) q_j^{\mathrm{D}}(y_j, x_j)}{\pi_j(x_j)},$$

$$\tilde{q}_j^{\mathrm{I}*}(x_j) = \frac{\sum_{y_j} \pi_j(y_j) q_j^{\mathrm{I}*}(y_j, x_j)}{\pi_j(x_j)},$$

$$\beta_j = \sum_{x_j} \sum_{y_j} \pi_j(x_j) q_j^{\mathrm{D}}(x_j, y_j),$$
$$\nu_j = \sum_{x_j} \sum_{y_j} \pi_j(x_j) q_j^{\mathrm{I}*}(x_j, y_j).$$

Note that $q_j^{\mathrm{D}}(x_j)$ and $q_j^{\mathrm{I}*}(x_j)$ are different from the transition rate functions $q_j^{\mathrm{D}}(x_j, y_j)$ and $q_j^{\mathrm{I}*}(x_j, y_j)$. They are distinguished only by their arguments. When they are used without arguments, e.g., q_j^{D}, they represent the transition rate functions, e.g., $q_j^{\mathrm{D}}(x_j, y_j)$. Assume that β_j and ν_j are finite. Note that $\tilde{p}_j^{\mathrm{A}}(x_j), \tilde{q}_j^{\mathrm{D}}(x_j), \tilde{q}_j^{\mathrm{I}*}(x_j)$ as well as β_j, ν_j are functions of π_j. The following relationships can be easily verified:

$$\sum_{x_j} \pi_j(x_j) q_j^{\mathrm{D}}(x_j) = \sum_{x_j} \pi_j(x_j) \tilde{q}_j^{\mathrm{D}}(x_j) = \beta_j, \tag{1.1}$$
$$\sum_{x_j} \pi_j(x_j) q_j^{\mathrm{I}*}(x_j) = \sum_{x_j} \pi_j(x_j) \tilde{q}_j^{\mathrm{I}*}(x_j) = \nu_j. \tag{1.2}$$

11.2 NECESSARY AND SUFFICIENT CONDITIONS FOR PRODUCT FORM

We first consider the possible forms of marginal distributions when the network has a product form stationary distribution. Define the transition rate q_j for each node j by

$$q_j(x_j, y_j) = \alpha_j p_j^{\mathrm{A}}(x_j, y_j) + (1 - r_{jj}) q_j^{\mathrm{D}}(x_j, y_j) + q_j^{\mathrm{I}*}(x_j, y_j), \quad x_j, y_j \in \mathcal{S}_j, \tag{2.1}$$

where α_j is a parameter to be determined. Consider this process as node j operating in isolation. The first term on the right-hand side of (2.1) indicates that this isolated node has Poisson arrivals with rate α_j. The second and third terms are transition rates associated respectively with departures from node j and internal transitions at node j, where the internal transition includes feedback to node j (see (iv)). Note that the q_j is different from the corresponding transitions used in Sections 4.3 and 10.2.

Theorem 11.2 *If the network has the product form stationary distribution*

$$\pi(\boldsymbol{x}) = \prod_{j=0}^{N} \pi_j(x_j),$$

then each π_j is the stationary distribution for the q_j defined by (2.1). The coefficients α_j in (2.1) are the solutions to the traffic equations

$$\alpha_j = \sum_{k \neq j} \beta_k(\alpha_k) r_{kj}, \qquad j = 0, 1, \ldots, N, \tag{2.2}$$

where the $\beta_j(\alpha_j)$ is the β_j of (1.1) that depends on α_j through π_j.

PROOF. The global balance equations for the network are

$$\pi(\boldsymbol{x}) \sum_{\boldsymbol{y}} q(\boldsymbol{x}, \boldsymbol{y}) = \sum_{\boldsymbol{y}} \pi(\boldsymbol{y}) q(\boldsymbol{y}, \boldsymbol{x}), \qquad \boldsymbol{x} \in \mathcal{S}. \tag{2.3}$$

Since

$$\pi(\boldsymbol{y}) = \frac{\pi(\boldsymbol{x}) \pi_j(y_j) \pi_k(y_k)}{\pi_j(x_j) \pi_k(x_k)}, \tag{2.4}$$

for all $\boldsymbol{y}$ with $x_\ell = y_\ell$ for $\ell \neq j, k$, it follows from the definition of q that (2.3) is equivalent to

$$\begin{aligned} &\pi(\boldsymbol{x}) \sum_j \left(q_j^{\mathrm{I}*}(x_j) + q_j^{\mathrm{D}}(x_j) \sum_{k \neq j} r_{jk} \right) \\ &\qquad = \pi(\boldsymbol{x}) \sum_j \left(\tilde{q}_j^{\mathrm{I}*}(x_j) + \tilde{p}_j^{\mathrm{A}}(x_j) \sum_{k \neq j} r_{kj} \tilde{q}_k^{\mathrm{D}}(x_k) \right), \qquad \boldsymbol{x} \in \mathcal{S}. \end{aligned} \tag{2.5}$$

For a fixed j, we sum these equations over all x_ℓ for $\ell \neq j$. First, the left-hand side becomes

$$\begin{aligned} &\sum_{x_\ell : \ell \neq j} \pi(\boldsymbol{x}) \Bigg[q_j^{\mathrm{I}*}(x_j) + \sum_{j' \neq j} q_{j'}^{\mathrm{D}}(x_{j'}) r_{j'j} + q_j^{\mathrm{D}}(x_j) \sum_{k \neq j} r_{jk} \\ &\qquad + \sum_{j' \neq j} \left(q_{j'}^{\mathrm{I}*}(x_{j'}) + q_{j'}^{\mathrm{D}}(x_{j'}) \sum_{k \neq j, j'} r_{j'k} \right) \Bigg] \\ &= \pi_j(x_j) \left(q_j^{\mathrm{I}*}(x_j) + \sum_{j' \neq j} \beta_{j'} r_{j'j} + (1 - r_{jj}) q_j^{\mathrm{D}}(x_j) \right) + \sum_{j' \neq j} \left(\nu_{j'} + \beta_{j'} \sum_{k \neq j, j'} r_{j'k} \right) \\ &= \pi_j(x_j) (q_j^{\mathrm{I}*}(x_j) + \alpha_j + (1 - r_{jj}) q_j^{\mathrm{D}}(x_j)) + \sum_{j' \neq j} \left(\nu_{j'} + \beta_{j'} \sum_{k \neq j, j'} r_{j'k} \right). \end{aligned}$$

A similar manipulation on the right-hand side yields

$$\pi_j(x_j)(\tilde{q}_j^{\mathrm{I}*}(x_j) + \alpha_j \tilde{p}_j^{\mathrm{A}}(x_j) + (1 - r_{jj}) \tilde{q}_j^{\mathrm{D}}(x_j)) + \sum_{j' \neq j} \left(\nu_{j'} + \beta_{j'} \sum_{k \neq j, j'} r_{j'k} \right).$$

Thus it follows from (2.5) that

$$q_j^{1*}(x_j) + \alpha_j + (1 - r_{jj})q_j^{\mathrm{D}}(x_j) = \tilde{q}_j^{1*}(x_j) + \alpha_j \tilde{p}_j^{\mathrm{A}}(x_j) + (1 - r_{jj})\tilde{q}_j^{\mathrm{D}}(x_j). \quad (2.6)$$

These are the balance equations for the q_j divided by $\pi_j(x_j)$ with α_j given by (2.2). This completes the proof of Theorem 11.2. □

The next theorem presents the necessary and sufficient conditions for the network to have a product form distribution.

Theorem 11.3 *The network has the product form stationary distribution*

$$\pi(\boldsymbol{x}) = \prod_{j=0}^{N} \pi_j(x_j), \qquad \boldsymbol{x} \in \mathcal{S},$$

if and only if each π_j is the stationary distribution of q_j with coefficients α_j satisfying the traffic equations (2.2) and

$$(\tilde{q}_j^{\mathrm{D}}(x_j) - \beta_j)r_{jk}(\tilde{p}_k^{\mathrm{A}}(x_k) - 1) + (\tilde{q}_k^{\mathrm{D}}(x_k) - \beta_k)r_{kj}(\tilde{p}_j^{\mathrm{A}}(x_j) - 1) = 0, \quad (2.7)$$

for all $j \neq k$ and $x_j \in \mathcal{S}_j$, $x_k \in \mathcal{S}_k$.

Proof. Assume the product form is $\pi(\boldsymbol{x}) = \prod_{j=0}^{N} \pi_j(x_j)$. Since the conditions of Theorem 11.2 are satisfied, (2.5) holds. Dividing (2.5) by $\pi(\boldsymbol{x})$, and subtracting the summation of (2.6) over all j, yields

$$\sum_j (\alpha_j \tilde{p}_j^{\mathrm{A}}(x_j) + (1 - r_{jj})\tilde{q}_j^{\mathrm{D}}(x_j)) = \sum_j \left(\alpha_j + \tilde{p}_j^{\mathrm{A}}(x_j) \sum_{k \neq j} \tilde{q}_k^{\mathrm{D}}(x_k) r_{kj}\right). \quad (2.8)$$

For convenience define

$$D_{jk}(x_j, x_k) = (\tilde{q}_j^{\mathrm{D}}(x_j) - \beta_j)r_{jk}(\tilde{p}_k^{\mathrm{A}}(x_k) - 1).$$

Since the stationary distribution is product form, it follows from Theorem 11.2 that α_j and β_j satisfy (2.2). Hence, substituting α_j of (2.2) into (2.8) gives

$$\sum_j \sum_{k \neq j} D_{jk}(x_j, x_k) = 0. \quad (2.9)$$

Multiplying (2.9) by $\prod_{\ell \neq j,k} \pi_\ell(x_\ell)$, summing over x_ℓ for $\ell \neq j, k$, and observing that

$$\sum_{x_\ell} \pi_\ell(x_\ell) \left(D_{\ell k}(x_\ell, x_k) + D_{k\ell}(x_k, x_\ell)\right) = 0$$

yields

$$D_{jk}(x_j, x_k) + D_{kj}(x_k, x_j) = 0.$$

This is exactly (2.7).

Conversely, assume that each π_j is the stationary distribution of q_j and that (2.2) and (2.7) are satisfied. Since (2.7) implies (2.9), we obtain (2.8). Similarly, (2.6) follows from the fact that π_j is the stationary distribution of q_j. Hence, from the calculation of (2.8) we obtain (2.5). Thus the product form π satisfies the global balance (2.3). □

The following are three sufficient conditions for (2.7), so they are sufficient for the stationary distribution of the network to be product form.

(a) Both nodes j and k are quasi-reversible. Recall that node j is quasi-reversible if $\tilde{q}_j^{\mathrm{D}}(x_j)$ is independent of x_j; in this case it must equal β_j.

(b) Both nodes j and k are non-effective with respect to arrivals. Node j is said to be *non-effective with respect to arrivals* if $\tilde{p}_j^{\mathrm{A}}(x_j) = 1$ for all $x_j \in \mathcal{S}_j$.

(c) Either node j or node k is quasi-reversible as well as non-effective with respect to arrivals.

These sufficient conditions are further weakened if $r_{jk} = 0$ or $r_{kj} = 0$. These and other special cases will be discussed in the next section. Note that when the outside (node 0) is a Poisson source, then node 0 has only one state (say 0) which is non-effective with respect to arrivals, i.e., the state of the outside source is not changed when a customer leaves the network. On the other hand, the Poisson source is clearly quasi-reversible, so it belongs to case (c). Therefore, when a network is subject to Poisson arrivals from the outside, condition (2.7) only has to be verified for nodes other than 0. In this case the product form stationary distribution $\prod_{j=0}^N \pi_j(x_j)$ can be written as $\prod_{j=1}^N \pi_j(x_j)$.

Theorem 11.3 yields the following procedure for establishing the existence of a product form distribution for the network and obtaining the distribution when it exists.

Step 1. For the dummy parameter α_j compute the stationary distribution π_j of node j defined by q_j of (2.1).

Step 2. Compute β_j using (2.2), which is a function of α_j since π_j is a function of α_j. So write it as $\beta_j(\alpha_j)$.

Step 3. Solve the traffic equations (2.2).

Step 4. Check condition (2.7) for each pair j, k and all x_j, x_k.

If this four-step procedure is successful, then $\pi(\boldsymbol{x}) = \prod_{j=0}^{N} \pi_j(x_j)$ is the stationary distribution of the network.

Finding the vector $\boldsymbol{\alpha} = (\alpha_0, \alpha_1, \ldots, \alpha_N)$ that satisfies the traffic equations (2.2) is a fixed point problem whose solution is usually established by Brouwer's fixed point theorem (see Appendix C). It follows from the proof of Theorem 11.3 that such a fixed point always exists when the network has a product form distribution.

Theorem 11.4 *The network has a product form stationary distribution if and only if there exists a solution to the traffic equations (2.2) and it satisfies the condition of Step 4.*

Therefore, the procedure above, in principle, applies to any queueing network described in Section 11.1. If the procedure is successful it gives the product form distribution of the network; otherwise, i.e., if it does not lead to a solution that satisfies the condition of Step 4, the procedure concludes that the network does not have a product form solution. For a particular application, we may be able to construct an algorithm to compute the fixed point, e.g., an iterative algorithm. The existence and uniqueness of the solutions as well as stability issues are discussed in Sections 11.5 and 11.6.

The following examples illustrate the procedure.

Example 11.5 (A counterexample) Consider a tandem queueing network with nodes 0, 1, and 2, where 0 represents the outside. The state spaces of the three nodes are

$$\mathcal{S}_0 = \{0\}, \quad \mathcal{S}_1 = \{0, 1\}, \quad \mathcal{S}_2 = \{0, 1\}.$$

The three nodes are characterized by

$$\begin{aligned}
&p_0^{\mathrm{A}}(0,0) = 1, && q_0^{\mathrm{D}}(0,0) = 4, && \\
&p_j^{\mathrm{A}}(x_j, x_j') = 1/2, && q_j^{\mathrm{I}}(x_i, x_j') = 0, && j = 1, 2,\ x_j, x_j' = 0, 1, \\
&q_j^{\mathrm{D}}(0,0) = 4, && q_j^{\mathrm{D}}(0,1) = 2, && j = 1, 2, \\
&q_j^{\mathrm{D}}(1,0) = 2, && q_j^{\mathrm{D}}(1,1) = 1, && j = 1, 2.
\end{aligned}$$

The nonzero routing probabilities of the network are

$$r_{01} = r_{12} = r_{20} = 1.$$

Step 1. For dummy variables $(\alpha_0, \alpha_1, \alpha_2)$, the node transition rates q_j of (2.1) are

$$q_j(0,0) = \alpha_j/2 + 4, \quad q_j(0,1) = \alpha_j/2 + 2, \qquad j = 1,2,$$
$$q_j(1,0) = \alpha_j/2 + 2, \quad q_j(1,1) = \alpha_j/2 + 1, \qquad j = 1,2.$$

The balance equations for the two nodes are

$$\pi_j(0)\left(\alpha_j/2 + 2\right) = \pi_j(1)\left(\alpha_j/2 + 2\right),$$
$$\pi_j(0) + \pi_j(1) = 1.$$

Thus $\pi_j(0) = \pi_j(1) = 1/2$ for $j = 1, 2$.

Step 2. Computing the mean departure rates β_j yields

$$\beta_0 = \sum_{x_0, x_0'} \pi_0(x_0') q_0^{\mathrm{D}}(x_0', x_0) = 4,$$
$$\beta_1 = \sum_{x_1, x_1'} \pi_1(x_1') q_1^{\mathrm{D}}(x_1', x_1) = 9/2,$$
$$\beta_2 = \sum_{x_2, x_2'} \pi_2(x_2') q_2^{\mathrm{D}}(x_2', x_2) = 9/2.$$

Step 3. The calculation of the mean arrival rates α_j's is easy since the β_j's do not depend on the α_j's. That is,

$$\alpha_0 = \beta_2 r_{20} = 9/2, \qquad \alpha_1 = \beta_0 r_{01} = 4, \qquad \alpha_2 = \beta_1 r_{12} = 9/2.$$

Step 4. It can be checked easily that condition (2.7) is satisfied. Actually, nodes 1 and 2 are non-effective with respect to arrivals, i.e., $\tilde{p}_j^{\mathrm{A}}(x_j) = 1$ for all $j = 1, 2$ and all x_j. For instance,

$$\tilde{p}_1^{\mathrm{A}}(1) = \frac{\pi_1(0) q_1^{\mathrm{A}}(0,0) + \pi_1(1) q_1^{\mathrm{A}}(1,0)}{\pi_1(1))} = \frac{1/2}{1/2} = 1.$$

Other cases can be checked similarly.

Thus all the steps of the procedure are completed and satisfied. This shows that the network has a product form stationary distribution, which, in this case, is

$$\pi(0,0) = \pi(1,0) = \pi(0,1) = \pi(1,1) = 1/4. \qquad \square$$

Note that even though in this example each node is non-effective with respect to arrivals, neither node is quasi-reversible since

$$\tilde{q}_j^{\mathrm{D}}(0) = 6, \qquad \tilde{q}_j^{\mathrm{D}}(1) = 3, \qquad j = 1, 2.$$

The following example presents a case where each node is neither quasi-reversible nor non-effective with respect to arrivals.

Example 11.6 (A second counterexample) Consider a queueing network with nodes 0, 1, 2, and $\mathcal{S}_1 = \mathcal{S}_2 = \{0, 1\}$, $\mathcal{S}_0 = \{0\}$. Assume that the nonzero routing probabilities are

$$r_{jk} = 1/2, \qquad j, k = 0, 1, 2, \; j \neq k,$$

and the nodes are characterized by

$$\begin{aligned}
&p_0^{\mathrm{A}}(0,0) = 1, && q_0^{\mathrm{D}}(0,0) = 2, \\
&p_j^{\mathrm{A}}(0,0) = p_j^{\mathrm{A}}(1,1) = 0, && p_j^{\mathrm{A}}(0,1) = p_j^{\mathrm{A}}(1,0) = 1, \qquad j = 1, 2, \\
&q_1^{\mathrm{D}}(0,0) = q_1^{\mathrm{D}}(0,1) = 3/2, && q_1^{\mathrm{D}}(1,0) = q_1^{\mathrm{D}}(1,1) = 0, \\
&q_2^{\mathrm{D}}(0,0) = 1/6, && q_2^{\mathrm{D}}(0,1) = 1/2, \\
&q_2^{\mathrm{D}}(1,0) = 3/2, && q_2^{\mathrm{D}}(1,1) = 0, \\
&q_j^{\mathrm{I}}(x_i, x_j') = 0, && j = 1, 2, \; x_j, x_j' = 0, 1.
\end{aligned}$$

We again apply the four-step procedure to compute the stationary distribution of the network.

Step 1. For the dummy variables (α_1, α_2), the node transition rates are

$$\begin{aligned}
&q_1(0,0) = 3/2, \quad q_1(0,1) = \alpha_1 + 3/2, \quad q_1(1,0) = \alpha_1, \\
&q_2(0,0) = 1/6, \quad q_2(0,1) = \alpha_2 + 1/2, \quad q_2(1,0) = \alpha_2 + 3/2,
\end{aligned}$$

where $q_1(1,1) = q_2(1,1) = 0$. The stationary distributions of the nodes are

$$\begin{aligned}
\pi_1(0) &= \frac{\alpha_1}{2\alpha_1 + 3/2}, & \pi_1(1) &= \frac{\alpha_1 + 3/2}{2\alpha_1 + 3/2}, \\
\pi_2(0) &= \frac{\alpha_2 + 3/2}{2\alpha_2 + 2}, & \pi_2(1) &= \frac{\alpha_1 + 1/2}{2\alpha_2 + 2}.
\end{aligned}$$

Step 2. From the π_j it follows that

$$\beta_0 = 2, \quad \beta_1 = \frac{6\alpha_1}{4\alpha_1 + 3}, \quad \beta_2 = \frac{13/6\alpha_2 + 7/4}{2\alpha_2 + 2}.$$

Step 3. The traffic equations for this network are

$$\alpha_0 = \frac{1}{2}\beta_1 + \frac{1}{2}\beta_2, \qquad \alpha_1 = 1 + \frac{1}{2}\beta_2, \qquad \alpha_2 = 1 + \frac{1}{2}\beta_1.$$

The solutions of these equations are $\alpha_0 = 1, \alpha_1 = \alpha_2 = 3/2, \beta_1 = \beta_2 = 1$. Thus

$$\pi_1(0) = 1/3, \quad \pi_1(2) = 2/3, \qquad \pi_2(0) = 3/5, \quad \pi_2(1) = 2/5.$$

Step 4. In order to verify condition (2.7), first note that

$$\begin{aligned} &\tilde{q}_1^{\mathrm{D}}(0) = 3/2, \quad &&\tilde{q}_1^{\mathrm{D}}(1) = 3, \quad &&\tilde{q}_2^{\mathrm{D}}(0) = 7/6, \quad &&\tilde{q}_2^{\mathrm{D}}(1) = 3/4, \\ &\tilde{p}_1^{\mathrm{A}}(0) = 2, &&\tilde{p}_1^{\mathrm{A}}(1) = 1/2, &&\tilde{p}_2^{\mathrm{A}}(0) = 2/3, &&\tilde{p}_2^{\mathrm{A}}(1) = 2/3. \end{aligned}$$

A routine check verifies (2.7). For instance, when $j = 1$, $k = 2$ and $x_j = x_k = 0$,

$$\begin{aligned} &(\tilde{q}_1^{\mathrm{D}}(0) - \beta_1)r_{12}(\tilde{p}_2^{\mathrm{A}}(0) - 1) + (\tilde{q}_2^{\mathrm{D}}(0) - \beta_2)r_{21}(\tilde{p}_1^{\mathrm{A}}(0) - 1) \\ &\quad = \left(\frac{3}{2} - 1\right)\frac{1}{2}\left(\frac{2}{3} - 1\right) + \left(\frac{7}{6} - 1\right)\frac{1}{2}(2 - 1) = 0. \end{aligned}$$

It can be verified that other cases of (2.7) are also zero. Hence by Theorem 11.3, the stationary distribution of the network is given by the product form

$$\pi(x_1, x_2) = \pi_1(x_1)\pi_2(x_2), \qquad x_1, x_2 = 0, 1.$$

Note that neither node 1 nor node 2 is quasi-reversible since

$$\tilde{q}_1^{\mathrm{D}}(0) = 3/2 \neq \beta_1 = 1, \qquad \tilde{q}_2^{\mathrm{D}}(0) = 7/6 \neq \beta_2 = 1.$$

Furthermore, neither node is non-effective with respect to arrivals since

$$\begin{aligned} &\tilde{p}_1^{\mathrm{A}}(0) \neq 1, \qquad &&\tilde{p}_1^{\mathrm{A}}(1) \neq 1, \\ &\tilde{p}_2^{\mathrm{A}}(0) \neq 1, &&\tilde{p}_2^{\mathrm{A}}(1) \neq 1. \end{aligned}$$

□

11.3 QUASI-REVERSIBILITY REVISITED

As we observed in the previous section, quasi-reversibility is a sufficient condition but not a necessary condition for product form. In this section we are concerned with how much stronger this condition is than necessary. It turns out that quasi-reversibility is equivalent to a product form that satisfies the biased local balance equations we introduced in Chapter 10.

Recall that a Markov chain with transition rate q is said to satisfy *biased local balance* with respect to a positive probability measure π on $\mathcal{S}$ and real numbers $\boldsymbol{\gamma} = \{\gamma_j;\ j = 0, 1, \ldots, N\}$ if $\sum_j \gamma_j = 0$ and

$$\pi(\boldsymbol{x})\left(\sum_k \sum_{\boldsymbol{y}} q_{jk}(\boldsymbol{x}, \boldsymbol{y}) + \gamma_j\right) = \sum_k \sum_{\boldsymbol{y}} \pi(\boldsymbol{y}) q_{kj}(\boldsymbol{y}, \boldsymbol{x}), \qquad (3.1)$$
$$j = 0, 1, \ldots, N,\ \boldsymbol{x} \in \mathcal{S}.$$

The π must be the stationary distribution of the Markov chain since the global balance equations are the sum of these biased local balance equations over j. Also, we say that the Markov chain is *locally balanced* with respect to π when all the γ_j's are 0.

Theorem 11.7 *The following statements are equivalent.*

(i) The network satisfies the biased local balance with respect to a product form distribution $\pi(\boldsymbol{x}) = \prod_{j=0}^{N} \pi_j(x_j)$ *and* $\boldsymbol{\gamma} = \{\gamma_j;\ j = 0, 1, \ldots, N\}$.

(ii) Each node q_j *is quasi-reversible with respect to* π_j *for some* α_j *that satisfies*

$$\alpha_j = \sum_{k \neq j} \beta_k r_{kj}, \quad j = 0, 1, \ldots, N. \qquad (3.2)$$

If these statements hold, then

$$\gamma_j = \alpha_j - (1 - r_{jj})\beta_j, \quad j = 0, 1, \ldots, N. \qquad (3.3)$$

PROOF. Suppose (i) holds. Fix node j. Since π has product form, it follows from Theorem 11.2 that π_j is the stationary distribution of node q_j for some $\alpha_j \geq 0$ satisfying (3.2). Using the same argument that we used when we derived (2.5) from (2.3) and (2.4), we obtain the following equation from the biased local balance equations (3.1):

$$q_j^{\mathrm{I}*}(x_j) + (1 - r_{jj}) q_j^{\mathrm{D}}(x_j) + \gamma_j = \tilde{q}_j^{\mathrm{I}*}(x_j) + \tilde{p}_j^{\mathrm{A}}(x_j) \sum_{k \neq j} \tilde{q}_k^{\mathrm{D}}(x_k) r_{kj}. \qquad (3.4)$$

Multiplying (3.4) by $\pi_j(x_j)$, summing over x_j, and applying (1.1) and (1.2), we obtain

$$(1 - r_{jj})\beta_j + \gamma_j = \sum_{k \neq j} \tilde{q}_k^{\mathrm{D}}(x_k) r_{kj}.$$

Fix $\ell \neq j$. Multiplying this equation by $\prod_{k \neq j,\ell} \pi_k(x_k)$, summing over x_k for $k \neq j, \ell$, and applying (1.1) and (3.2) yields

$$\begin{aligned}(1 - r_{jj})\beta_j + \gamma_j &= \tilde{q}_\ell^{\mathrm{D}}(x_\ell) r_{\ell j} 1(\ell \neq j) + \sum_{k \neq j,\ell} \beta_k r_{kj} \\ &= \alpha_j + (\tilde{q}_\ell^{\mathrm{D}}(x_\ell) - \beta_\ell) r_{\ell j} 1(\ell \neq j).\end{aligned} \tag{3.5}$$

Summing over j and using (3.2) gives rise to

$$(\tilde{q}_\ell^{\mathrm{D}}(x_\ell) - \beta_\ell) \sum_{j \neq \ell} r_{\ell j} = 0.$$

This proves $\tilde{q}_\ell^{\mathrm{D}}(x_\ell) = \beta_\ell$. Thus, each node q_ℓ is quasi-reversible, so (ii) is proved.

Assume (ii) holds. Since quasi-reversibility implies (2.7), the conditions for the product form of Theorem 11.3 are satisfied, and therefore (2.6) holds. Substituting $p_j^{\mathrm{A}}(x_j) = 1$ and $\tilde{q}_j^{\mathrm{D}}(x_j) = \beta_j$ in (2.6), we obtain

$$q_j^{\mathrm{I}*}(x_j) + \alpha_j + (1 - r_{jj}) q_j^{\mathrm{D}}(x_j) = \tilde{q}_j^{\mathrm{I}*}(x_j) + \alpha_j \tilde{p}_j^{\mathrm{A}}(x_j) + (1 - r_{jj})\beta_j, \quad x_j \in \mathcal{S}_j.$$

Define γ_j by (3.3). Applying (3.3) to the expression above yields

$$q_j^{\mathrm{I}*}(x_j) + (1 - r_{jj}) q_j^{\mathrm{D}}(x_j) + \gamma_j = \tilde{q}_j^{\mathrm{I}*}(x_j) + \alpha_j \tilde{p}_j^{\mathrm{A}}(x_j), \quad x_j \in \mathcal{S}_j. \tag{3.6}$$

From (3.2) and the fact that $\beta_j = \tilde{q}_j^{\mathrm{D}}(x_j)$, it follows that

$$\alpha_j = \sum_{k \neq j} \tilde{q}_k^{\mathrm{D}}(x_k) r_{kj}.$$

Substituting this α_j into (3.6) yields (3.4), which implies (3.1). Hence q satisfies the biased local balance with respect to π and $\boldsymbol{\gamma}$. This completes the proof that (ii) implies (i). □

Remark 11.8 If node j is quasi-reversible, it is easy to see that (3.4) is identical to the global balance equations of the node transition rates (single-class case) used in Sections 4.3 and 10.2. Hence, the stationary distribution of (2.1) for each node is the same as the one that we used in the earlier chapters. Since the departure rate function only differs by the constant multiplier $1 - r_{jj}$, quasi-reversibility of each node with (2.1) is also equivalent to that of each node defined in earlier chapters. The arrival rates α_j, however, are different; the arrival rate in the previous setup is equal to $\alpha_j + r_{jj}\beta_j$ in the current notation.

Theorems 11.3 and 11.7 show that quasi-reversibility is a sufficient condition for product form. The remaining part of this section considers scenarios under which quasi-reversibility is also a necessary condition for product form.

Corollary 11.9 *If in a queueing network there are no immediate reverse moves, i.e., $r_{jk} \neq 0$ implies $r_{kj} = 0$, and $\tilde{p}_k^{\mathrm{A}}(x_k)$ is not identically 1 for all x_k, i.e.,*

$$\tilde{p}_k^{\mathrm{A}}(x_k) \neq 1, \quad \textit{for at least one } x_k \in \mathcal{S}_k,$$

then the product form implies that node j is quasi-reversible. In particular, if the discrete-time Markov chain on $\mathcal{S}_k$ with transition probability $\{p_k^{\mathrm{A}}(x_k, y_k); x_k, y_k \in \mathcal{S}_k\}$ is transient, then (3.7) is satisfied.

PROOF. Under the assumptions, equation (2.7) is reduced to

$$(\tilde{q}_j^{\mathrm{D}}(x_j) - \beta_j) r_{jk} (\tilde{p}_k^{\mathrm{A}}(x_k) - 1) = 0. \tag{3.7}$$

Fix an x_k that satisfies (3.7). Since (3.7) holds for every x_j and $r_{jk}(\tilde{p}_k^{\mathrm{A}}(x_k) - 1) \neq 0$, we conclude that $\tilde{q}_j^{\mathrm{D}}(x_j) = \beta_j$ for all x_j, i.e., node j is quasi-reversible.

To prove the second part, first note that $\tilde{p}_k^{\mathrm{A}}(x_k) = 1$ for all x_k is, by definition of $\tilde{p}_k^{\mathrm{A}}(x_k)$, equivalent to π_k being a positive stationary measure for the Markov chain with transition probability $\{p_k^{\mathrm{A}}(x_k, y_k); x_k, y_k \in \mathcal{S}_k\}$; this cannot be true if p_k^{A} is transient. Thus there must be at least one x_k such that (3.7) is satisfied. □

For a network with regular customers, if arrivals do not decrease the number of customers at a node, then p_k^{A} is clearly transient. This is the case for traditional networks, which are referred to as customer-based networks without signals. Here, signals are meant that arrivals do not trigger immediate departures, as used in the earlier chapters.

Definition 11.10 (Non-terminal node) Node j is said to be *non-terminal* if

$$1 - r_{jj} - r_{j0} > 0,$$

i.e., a departure from a non-terminal node arrives at another node in the network with a positive probability.

Feedforward networks clearly satisfy the first condition of Corollary 11.9 for all nodes that are non-terminal. Thus we obtain the following result.

Corollary 11.11 *A customer-based feedforward queueing network without signals has a product form distribution if and only if all non-terminal nodes are quasi-reversible.*

Of course, there are many queueing networks with feedback that satisfy the conditions of Corollary 11.9.

Example 11.12 (Necessity of quasi-reversibility) Consider the customer-based network with four nodes and no signals depicted in Figure 11.1. Customers arrive at nodes 1 and 2 according to Poisson processes. Departures from nodes 1 and 2 join node 3, and departures from node 4 may join node 1, node 2, or leave the network. Clearly, this network satisfies the conditions of Corollary 11.9, so the quasi-reversibility of each node is both necessary and sufficient for the stationary distribution of the network to be product form. □

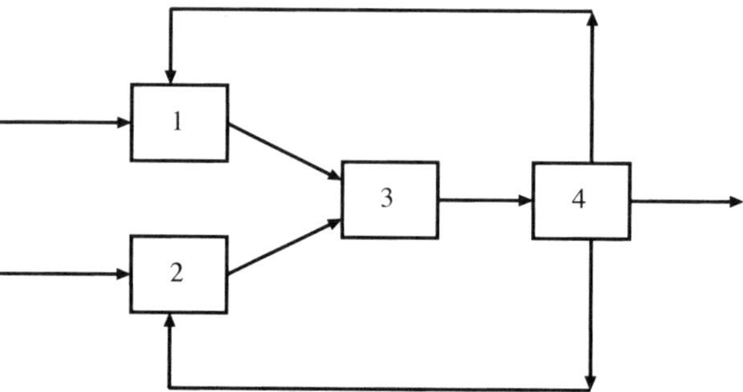

Figure 11.1 A four-node network

To present the next result, we need to introduce two new concepts. Node j is called a *conventional queue* if it has an empty state, denoted by 0, from which there can be no departures or internal transitions, and state 0 cannot be reached via an arrival or an internal transition, i.e.,

$$\tilde{p}_j^{\mathrm{A}}(0) = q_j^{\mathrm{D}}(0) = 0,$$
$$q_j^{\mathrm{I}*}(x_j, x_j') = 0, \qquad \text{if either } x_j = 0 \text{ or } x_j' = 0.$$

Clearly, if a network has an outside Poisson source, then node 0 is not conventional.

A queueing network is called conventional if its outside source is Poisson and all other nodes are conventional. Let α_j^+ denote the average arrival rate at node j including the feedback, i.e.,

$$\alpha_j^+ = \sum_{k=0}^{N} \beta_k r_{kj} = \alpha_j + r_{jj}\beta_j .$$

Node j is said to be *internally balanced* if $\alpha_j^+ = \beta_j$, i.e., the average arrival rate equals the average departure rate.

Theorem 11.13 *Suppose a queueing network has conventional nodes and the outside source, i.e., node 0, is non-effective with respect to arrivals. If the network has a product form stationary distribution, then a non-terminal node k is quasi-reversible if and only if either one of the following two conditions holds.*

(a) Node k has a path connecting it to an internally balanced node.

(b) Node k is directly connected to some node $j \neq 0$, but node j is not directly connected to node k, i.e, $r_{kj} > 0$ and $r_{jk} = 0$.

In these cases, all the non-terminal nodes are internally balanced. Note that condition (b) is satisfied if the destination node k is a terminal node.

PROOF. Suppose the network has a product form stationary distribution. By Theorem 11.2, the marginal distribution for each node satisfies the balance equations (2.6) with the coefficients determined by the traffic equations (2.2). Substituting $x_j = 0$ in (2.6) yields

$$\alpha_j + (1 - r_{jj})q_j^{\mathrm{D}}(0) + q_j^{\mathrm{I}*}(0) = \alpha_j \tilde{p}_j^{\mathrm{A}}(0) + (1 - r_{jj})\tilde{q}_j^{\mathrm{D}}(0) + \tilde{q}_j^{\mathrm{I}*}(0).$$

Thus, from the fact that node j is a conventional queue, it follows that

$$\alpha_j = (1 - r_{jj})\tilde{q}_j^{\mathrm{D}}(0). \tag{3.8}$$

Because of the product form, we also have (2.7) of Theorem 11.3. Letting $x_j = 0$ in (2.7) and substituting (3.8), we obtain

$$(\alpha_j^+ - \beta_j)\frac{r_{jk}}{1 - r_{jj}}(\tilde{p}_k^{\mathrm{A}}(x_k) - 1) - (\tilde{q}_k^{\mathrm{D}}(x_k) - \beta_k)r_{kj} = 0, \quad k \neq j,\ 0. \tag{3.9}$$

Substituting $x_k = 0$ in (3.9) yields

$$(\alpha_j^+ - \beta_j)\frac{r_{jk}}{1 - r_{jj}} + (\alpha_k^+ - \beta_k)\frac{r_{kj}}{1 - r_{kk}} = 0. \tag{3.10}$$

Since the network process is irreducible, any non-terminal node k has an arc directly connecting it to some other node j, i.e., $r_{kj} > 0$. From (3.9) it follows that node k is quasi-reversible if and only if

$$(\alpha_j^+ - \beta_j) r_{jk} (\tilde{p}_k^{\mathrm{A}}(x_k) - 1) = 0.$$

Letting $x_k = 0$ we conclude from this formula that either $\alpha_j^+ - \beta_j = 0$ or $r_{jk} = 0$. The latter is exactly (b), while the former is (a) since node j is internally balanced. Thus (a) and (b) are necessary for the quasi-reversibility of node k. Conversely, it follows from (3.9) that (b) implies that node k is quasi-reversible. If (a) is satisfied, then we need to use induction to show that node k is quasi-reversible. First, if node k is directly connected to node j, i.e., $r_{kj} > 0$, then it follows from (3.9) that node k is quasi-reversible. If node k is connected to j via $i_1, i_2, \ldots, i_\ell$, then applying (3.10) to nodes i_ℓ and j shows that node i_ℓ is internally balanced, and applying (3.10) to nodes $i_{\ell-1}$ and i_ℓ shows that node $i_{\ell-1}$ is internally balanced, etc. After showing that node i_1 is internally balanced, we apply (3.9) to nodes k and i_1 to obtain that node k is quasi-reversible.

To show that each non-terminal node has to be internally balanced, assume that node j is quasi-reversible. Then, from (3.8) and $\tilde{q}_j^{\mathrm{D}}(0) = \beta_j$ it follows that

$$\begin{aligned} \alpha_j^+ &= \alpha_j + r_{jj}\beta_j \\ &= \alpha_j + r_{jj}\tilde{q}_j^{\mathrm{D}}(0) \\ &= \tilde{q}_j^{\mathrm{D}}(0) \\ &= \beta_j \,. \end{aligned}$$

The proof of Theorem 11.13 is thus complete. □

From the sufficient condition (c) in Section 11.2, it follows that if the outside source is Poisson, then no condition is required on the terminal nodes for the product form to hold.

The next result follows immediately from Theorem 11.13.

Corollary 11.14 *Consider a queueing network with only conventional nodes and all non-terminal nodes satisfying either (a) or (b) of Theorem 11.13. The network has a product form distribution if and only if all the non-terminal nodes are quasi-reversible.*

Considering an even more special case we obtain the following result.

Corollary 11.15 *Consider a conventional queueing network with Poisson arrivals from the outside and each node internally balanced, i.e., the departure rate of each node is equal to its arrival rate. The network has a product form stationary distribution if and only if all non-terminal nodes are quasi-reversible. If the outside source node is also a conventional queue, then the network has a product form stationary distribution if and only if all nodes are quasi-reversible.*

Note the difference between Corollary 11.11 and Corollary 11.15. In Corollary 11.11 the non-terminal nodes do not have to be conventional or internally balanced, but the topological structure of the network is restricted. On the other hand, in Corollary 11.15, the network topology may be arbitrary, but each non-terminal node has to be conventional and internally balanced.

11.4 NETWORKS WITH MULTIPLE CLASSES OF TRANSITIONS

This section extends the results of the previous sections to the case with multiple classes of transitions. These extensions are quite straightforward, so their proofs are left as exercises.

Let T_j be the class of arrival and departure transitions of node j. As we discussed in Chapter 3, the arrival and departure classes are often different and T_j is then taken to be the union of all classes. In such cases null transitions have to be introduced. For each $u \in T_j$, let $q^{\text{D}}_{ju}(x_j, x'_j)$ and $p^{\text{A}}_{ju}(x_j, x'_j)$ denote, respectively, the departure transition rate of class u entities and the arrival transition probability of class u entities, and let $q^{\text{I}}_j(x_j, x'_j)$ denote the internal transition rate. The new internal transition rate $q^{\text{I}*}_j(x_j, x'_j)$, which includes feedback transitions, is

$$q^{\text{I}*}_j(x_j, y_j) = q^{\text{I}}_j(x_j, y_j) + \sum_{x'_j} \sum_{u,v} q^{\text{D}}_{ju}(x_j, x'_j) r_{ju,jv} p^{\text{A}}_{jv}(x'_j, y_j).$$

The routing probability r_{jk} is now extended to $r_{ju,kv}$. The transition rates of the network are defined as

$$q(\boldsymbol{x}, \boldsymbol{y}) = \sum_{j,k} q_{jk}(\boldsymbol{x}, \boldsymbol{y}), \quad \boldsymbol{x}, \boldsymbol{y} \in \mathcal{S}, \tag{4.1}$$

where

$$q_{jj}(\boldsymbol{x}, \boldsymbol{y}) = q^{\text{I}*}_j(x_j, y_j) 1(y_\ell = x_\ell, \ell \neq j),$$

$$q_{jk}(\boldsymbol{x}, \boldsymbol{y}) = \sum_{u \in T_j} \sum_{v \in T_k} q^{\text{D}}_{ju}(x_j, y_j) r_{ju,kv} p^{\text{A}}_{kv}(x_k, y_k) 1(y_\ell = x_\ell, \ell \neq j, k),$$

$$\text{for } j \neq k.$$

As in Section 11.2, for a distribution π_j on $\mathcal{S}_j$, define

$$\begin{aligned} q^{\mathrm{D}}_{ju}(x_j) &= \sum_{y_j} q^{\mathrm{D}}_{ju}(x_j, y_j), \\ \tilde{q}^{\mathrm{D}}_{ju}(x_j) &= \pi_j(x_j)^{-1} \sum_{y_j} \pi_j(y_j) q^{\mathrm{D}}_{ju}(y_j, x_j), \\ \tilde{p}^{\mathrm{A}}_{ju}(x_j) &= \pi_j(x_j)^{-1} \sum_{y_j} \pi_j(y_j) p^{\mathrm{A}}_{ju}(y_j, x_j), \\ \beta_{ju} &= \sum_{x_j} \sum_{y_j} \pi_j(x_j) q^{\mathrm{D}}_{ju}(x_j, y_j). \end{aligned} \tag{4.2}$$

The corresponding notation $\tilde{q}^{\mathrm{I}*}_j$ and ν_j for the internal transitions are defined as in Section 11.1. We assume that

$$\beta_j = \sum_{u \in T_j} \beta_{ju} < \infty \qquad \text{and} \qquad \nu_j < \infty$$

for all j. The transition rate function q_j for the local process of node j now becomes

$$q_j(x_j, y_j) = \sum_{u \in T_j} \left(\alpha_{ju} p^{\mathrm{A}}_{ju}(x_j, y_j) + \left(1 - \sum_{v \in T_j} r_{ju,jv}\right) q^{\mathrm{D}}_{ju}(x_j, y_j) \right) + q^{\mathrm{I}*}_j(x_j, y_j), \qquad x_j, y_j \in \mathcal{S}_j, \tag{4.3}$$

where the coefficients α_{ju} are determined by the traffic equations

$$\alpha_{ju} = \sum_{k \neq j} \sum_{v \in T_k} \beta_{kv} r_{kv,ju}, \qquad u \in T_j, \ j = 0, 1, \ldots, N. \tag{4.4}$$

Finally, D_{jk} is replaced by

$$D_{ju,kv}(x_j, x_k) = (\tilde{q}^{\mathrm{D}}_{ju}(x_j) - \beta_{ju}) r_{ju,kv} (\tilde{p}^{\mathrm{A}}_{kv}(x_k) - 1).$$

Theorem 11.3 can now be generalized to networks with multiple classes of transitions as follows.

Theorem 11.16 *The stationary distribution of the network with multiple classes of transitions has the product form* $\pi(\boldsymbol{x}) = \prod_{j=0}^{N} \pi_j(x_j)$ *if and only if* π_j *is the stationary distribution of* q_j *with the coefficients* α_{ju} *determined by (4.4), and*

$$\sum_{u \in T_j} \sum_{v \in T_k} (D_{ju,kv}(x_j, x_k) + D_{kv,ju}(x_k, x_j)) = 0, \qquad j \neq k, x_j \in \mathcal{S}_j, x_k \in \mathcal{S}_k. \tag{4.5}$$

Analogous to the networks with a single class of transitions, we have the following four-step procedure for establishing the existence of a product form stationary distribution and obtaining the distribution when it exists.

Step 1. For the dummy vector $\boldsymbol{\alpha}_j = (\alpha_{ju}; u \in T_j)$, compute the stationary distribution π_j of node j with the q_j of (4.3).

Step 2. Compute $\beta_{ju}, u \in T_j$, using (4.2), which is a function of $\boldsymbol{\alpha}_j$ since π_j is a function of $\boldsymbol{\alpha}_j$. Write it as $\beta_{ju}(\boldsymbol{\alpha}_j)$.

Step 3. Solve the traffic equations

$$\alpha_{ju} = \sum_{k \neq j} \sum_{v \in T_j} \beta_{kv}(\boldsymbol{\alpha}_k) r_{kv,ju}, \quad u \in T_j, \ j = 0, 1, \ldots, N. \tag{4.6}$$

Step 4. Verify condition (4.5) for each pair k, j and all x_j, x_k.

If this four-step procedure is successful, then $\pi(\boldsymbol{x}) = \prod_{j=0}^{N} \pi_j(x_j)$ is the stationary distribution of the network. Again, finding $\boldsymbol{\alpha} = (\boldsymbol{\alpha}_0, \boldsymbol{\alpha}_1, \ldots, \boldsymbol{\alpha}_N)$ that satisfies the traffic equations (4.6) is a fixed point problem, and theoretically, such a fixed point always exists if the network has a product form distribution. That is, the network has a product form stationary distribution if and only if there exists a solution to the traffic equations (4.6) such that condition (4.5) in Step 4 is satisfied.

Recall from Chapter 3 that in the case of multiple classes of transitions, q_j is said to be *quasi-reversible* if π_j is the stationary distribution of q_j and $\tilde{q}^{\mathrm{D}}_{ju}(x_j)$ is independent of x_j for each $u \in T_j$. As in the case of single-class transitions, the three sufficient conditions for (4.5) are

(a) both nodes j and k are quasi-reversible;

(b) both nodes j and k are non-effective with respect to arrivals, where node j is said to be *non-effective with respect to arrivals* if $\tilde{p}^{\mathrm{A}}_{ju}(x_j) = 1$ for all $u \in T_j$ and $x_j \in \mathcal{S}_j$; and

(c) either node j or node k is quasi-reversible as well as non-effective with respect to arrivals.

The biased local balance condition for the network with multiple classes of transitions also has the form (3.1) but with the q_{jk} now defined as in (4.1). It can be shown that this condition is satisfied if quasi-reversibility holds for all pairs of j and u. However, the biased local balance plus the product form stationary distribution do not imply quasi-reversibility in the multiple class case. For example, it is easy to see that if

$$\sum_{u \in T_j} (\tilde{\alpha}^{\mathrm{D}}_{ju}(x_j) - \beta_{ju}) r_{ju,kv} = 0, \tag{4.7}$$

then (4.5) is satisfied, and the network has a product form. We can also show, using the same argument as in the proof of Theorem 11.7, that (4.7) implies a biased local balance. However, (4.7) is clearly weaker than quasi-reversibility. Thus, when there are multiple classes of transitions, quasi-reversibility is sufficient but not necessary for a product form that satisfies a biased local balance.

11.5 SOLVING TRAFFIC EQUATIONS: AN ITERATIVE APPROACH

In the previous sections we presented the necessary and sufficient conditions for a queueing network to have a product form distribution, and developed a general procedure to verify these conditions. These results simultaneously establish the existence of the stationary distribution, which is often referred to as the *stability* of the network. A key step in verifying these conditions is to solve the traffic equations (2.2) or (4.6). Solving the traffic equations, in general, is not easy, except in the case of linear traffic equations, i.e., the mean departure rate β_j is a linear function of the mean arrival rate α_j. In fact, it is not even clear whether there exists a solution to the traffic equations when it is not known in advance if the network has a product form stationary distribution. So analyzing the traffic equations is the first step in analyzing a queueing network problem. Little work has been done in this area, and the analysis has been mainly on a case-by-case basis. We present several such results in this section and the next.

Consider a queueing network with multiple classes of entities and signals (see Section 4.4). Assume that the outside, i.e., node 0, is a Poisson source. In this respect, the network is less general than the network in Section 11.4, but it is more general in the dynamics of the network since it allows instantaneous movements. Because node 0 is a Poisson source, we can omit the equation for node 0 in the traffic equations.

Let the mean arrival and departure rates be denoted by the vectors $\boldsymbol{\alpha}$ and $\boldsymbol{\beta}$, i.e.,

$$\boldsymbol{\alpha} = (\alpha_{ju};\ j = 1, 2, \dots, N, u \in T_j),$$

$$\boldsymbol{\beta}(\boldsymbol{\alpha}) = (\beta_{ju};\ j = 1, 2, \dots, N, u \in T_j),$$

where the departure rate vector has the argument $\boldsymbol{\alpha}$ to emphasize that it is a function of $\boldsymbol{\alpha}$. The departure rate β_{ju} is

$$\beta_{ju} = \sum_{x_j, x_j' \in S_j} \pi(x_j') q_{ju}^{\mathrm{D}}(x_j, x_j'),$$

only if the stationary distribution π_j of $q_j^{(\boldsymbol{\alpha})}$ exists. However, if the node has a stationary measure and it is quasi-reversible with respect to this measure, then, as seen from the definition of quasi-reversibility in Chapter 3, the β_{ju}'s are defined for all $u \in T_j$. In this section we let $\beta_{ju} = \infty$ for all $u \in T_j$ if either $q_j^{(\boldsymbol{\alpha})}$ does not have a stationary measure satisfying quasi-reversibility, or $\alpha_{jv} = \infty$ for some $v \in T_j$. Under this convention, $\boldsymbol{\beta}(\boldsymbol{\alpha})$ is well defined for all $\boldsymbol{\alpha} \geq \mathbf{0}$, but some of its components may be infinity.

Similarly, the arrival rates from the outside are denoted by

$$\boldsymbol{\lambda} = (\lambda_{ju};\ j = 1, 2, \ldots, N, u \in T_j).$$

The routing probability matrix is

$$R = \{r_{ju,kv};\ j, k = 1, 2, \ldots, N, u \in T_j, v \in T_k\}.$$

Note that node 0, the outside source, is neither included in the matrix R nor in the vectors $\boldsymbol{\alpha}$ and $\boldsymbol{\beta}$. Using these notations, the traffic equations can be expressed as

$$\boldsymbol{\alpha} = \boldsymbol{\lambda} + \boldsymbol{\beta}(\boldsymbol{\alpha})R. \tag{5.1}$$

Clearly, solving the traffic equations may be regarded as a fixed point problem for the multivariate function

$$f(\boldsymbol{\alpha}) = \boldsymbol{\lambda} + \boldsymbol{\beta}(\boldsymbol{\alpha})R.$$

There are two approaches in establishing the existence of a fixed point. The first one is an algorithmic approach which is considered in this section; and the second is an analytic approach using the fixed point theorem, which will be discussed in the next section. The advantage of the first approach is that it gives a numerical solution, i.e., the fixed point.

Let $\boldsymbol{\alpha}^{(0)} = \mathbf{0}$. Inductively define a sequence $\boldsymbol{\alpha}^{(n)}$ by

$$\boldsymbol{\alpha}^{(n)} = \boldsymbol{\lambda} + \boldsymbol{\beta}(\boldsymbol{\alpha}^{(n-1)})R, \qquad n = 1, 2, \ldots. \tag{5.2}$$

In general, it is not obvious whether $\boldsymbol{\alpha}^{(n)}$ converges as n goes to infinity, although it does numerically do so in many cases. However, if it does, the converging solution satisfies the traffic equations (provided $\boldsymbol{\beta}(\boldsymbol{\alpha})$ is continuous in $\boldsymbol{\alpha}$, which is true in all our applications).

The following lemma provides sufficient conditions for such a convergence.

Lemma 11.17 *If the following two conditions are satisfied:*

(i) $\boldsymbol{\beta}(\boldsymbol{\alpha})R$ *is nondecreasing in* $\boldsymbol{\alpha}$*, and*

(ii) $\boldsymbol{\beta}(\boldsymbol{\alpha})$ *is continuous in* $\boldsymbol{\alpha}$*,*

then the traffic equations (5.1) have a minimum nonnegative solution, which may have some components that are infinity, and (5.2) converges to the minimum solution.

PROOF. Let $\boldsymbol{\alpha}^*$ be an arbitrary solution of (5.1), which exists since $\boldsymbol{\alpha}^* = \infty$ is a trivial solution. Since

$$\boldsymbol{\alpha}^{(0)} \leq \boldsymbol{\alpha}^{(1)}$$

and

$$\boldsymbol{\alpha}^{(n+1)} - \boldsymbol{\alpha}^{(n)} = (\boldsymbol{\beta}(\boldsymbol{\alpha}^{(n)}) - \boldsymbol{\beta}(\boldsymbol{\alpha}^{(n-1)}))R,$$

the monotonicity assumption (i) implies that $\boldsymbol{\alpha}^{(n)}$ is a nondecreasing sequence. On the other hand,

$$\mathbf{0} = \boldsymbol{\alpha}^{(0)} \leq \boldsymbol{\alpha}^*$$

implies

$$\boldsymbol{\alpha}^{(1)} = \boldsymbol{\lambda} + \boldsymbol{\beta}(\boldsymbol{\alpha}^{(0)}) \leq \boldsymbol{\lambda} + \boldsymbol{\beta}(\boldsymbol{\alpha}^*) = \boldsymbol{\alpha}^*, \tag{5.3}$$

and by induction we obtain

$$\boldsymbol{\alpha}^{(n)} \leq \boldsymbol{\alpha}^*, \qquad n = 0, 1, \ldots.$$

This shows that $\boldsymbol{\alpha}^{(n)}$ is a nondecreasing sequence that is bounded from above by $\boldsymbol{\alpha}^*$. Therefore it must have a limit. Let $\boldsymbol{\alpha}$ denote this limit. The continuity assumption (ii) implies that $\boldsymbol{\alpha}$ satisfies (5.1). Since $\boldsymbol{\alpha} \leq \boldsymbol{\alpha}^*$, $\boldsymbol{\alpha}$ is a minimum solution. □

Combining this lemma with the product form results of Chapter 4 yields the following result.

Proposition 11.18 *Consider the network in Section 4.4 with multiple classes of entities and signals. If all nodes are uniformly quasi-reversible and conditions (i) and (ii) of Lemma 11.17 are satisfied, then the network has a product form stationary measure if and only if the iterative algorithm (5.2) converges to a finite solution* $\boldsymbol{\alpha}$. *In particular, if* $q_j^{(\boldsymbol{\alpha}_j)}$, *for this* $\boldsymbol{\alpha}$, *has a stationary distribution, then the network has a product form stationary distribution.*

PROOF. The sufficiency follows immediately from Theorem 4.9. We consider the necessity. Since the product form stationary measure exists and the network is uniformly quasi-reversible, the traffic equation (5.1) has a finite solution. By Lemma 11.17, the iterative algorithm (5.2) converges to a minimum solution. Hence, the minimum solution must be finite. □

We now apply Proposition 11.18 to two of the networks discussed in Chapters 7 and 8, namely the coalescence network and the assembly-transfer network.

Example 11.19 (The coalescence network) Consider the network with customer coalescence discussed in Section 7.1. Each node in this network is uniformly quasi-reversible and its traffic equations are

$$\alpha_j = \lambda_j + \sum_{k=1}^{N} \sum_{\ell=1}^{\infty} a_k(\ell) \rho_k^{\ell} \mu_k r_{kj}, \quad j = 1, 2, \ldots, N, \tag{5.4}$$

$$\alpha_j = \mu_j \sum_{\ell=1}^{\infty} \bar{a}_j(\ell) \rho_j^{\ell}, \qquad j = 1, 2, \ldots, N. \tag{5.5}$$

It is easily seen from (5.5) that ρ_j is a nondecreasing function of α_j. Since

$$\beta_j(\boldsymbol{\alpha}) = \mu_j \sum_{\ell=1}^{\infty} a_j(\ell) \rho_j^{\ell},$$

$\beta_j(\boldsymbol{\alpha})$ is also a nondecreasing function of $\boldsymbol{\alpha}$. The continuity of $\beta_j(\boldsymbol{\alpha})$ with regard to $\boldsymbol{\alpha}$ is obvious. Thus it follows from Proposition 11.18 that there exists a solution to the traffic equations (5.4) and (5.5), and the network has the product form stationary distribution if the q_j for the solution α_j of (5.4) and (5.5) has a stationary distribution, i.e., $\rho_j < 1$ for $j = 1, 2, \ldots, N$. □

Example 11.20 (The assembly-transfer network) Consider the assembly-transfer network discussed in Section 8.1. Again, each node in this network is uniformly quasi-reversible. Its traffic equations are

$$\alpha_{ju} = \lambda_j a_j(u) + \sum_{k=1}^{N} \sum_{v=1}^{\infty} \mu_k \rho_k^v b_k(v) r_{kv,ju}, \tag{5.6}$$

$$\sum_{u=1}^{\infty} \alpha_{ju} \rho_j^{-u} + \sum_{u=1}^{\infty} \mu_j b_j(u) \rho^u = \sum_{u=1}^{\infty} \alpha_{ju} + \mu_j. \tag{5.7}$$

Write (5.7) as

$$\sum_{\ell=1}^{\infty} \sum_{u=\ell}^{\infty} \alpha_{ju} \rho_j^{-\ell+1} = \mu_j \sum_{\ell=1}^{\infty} \sum_{u=\ell}^{\infty} b_j(u) \rho_j^{\ell}.$$

It can be shown from this expression that its solution $\rho_j(\boldsymbol{\alpha}_j)$ is nondecreasing in $\boldsymbol{\alpha}_j = (\alpha_{ju}; u \geq 1)$ (see Exercise 11.6). Since

$$\beta_{jv}(\boldsymbol{\alpha}) = \mu_j \rho_j^v b_j(v),$$

$\beta_{jv}(\boldsymbol{\alpha})$ is nondecreasing in $\boldsymbol{\alpha}$. By Proposition 11.18, the traffic equations (5.6) and (5.7) always have a solution, and the network has the product form stationary distribution if and only if $\rho_j(\boldsymbol{\alpha}_j) < 1$ for $j = 1, 2, \ldots, N$. □

In some cases we can obtain more explicit conditions for stability. For instance, consider the assembly-transfer network in Example 11.20. Since α_{ju} is the average arrival rate of size u batches at node j, the following sufficient conditions are intuitively clear.

Corollary 11.21 *If the solution to the traffic equations (5.6) and (5.7) satisfies*

$$\sum_{u=1}^{\infty} u\alpha_{ju} < \mu_j E(Y_j), \qquad j = 1, 2, \ldots, N, \tag{5.8}$$

where Y_j is a generic service batch size at node j, then the network is stable and has a product form stationary distribution. In particular, if

$$\lambda_j E(X_j) + \sum_{k=1}^{N} \sum_{u=1}^{\infty} \sum_{v=1}^{\infty} \mu_k b_k(v) r_{kv,ju} u < \mu_j E(Y_j), \qquad j = 1, \ldots, N, \tag{5.9}$$

where X_j is a generic arrival batch size from the outside, then the network is stable.

PROOF. Define

$$f(\rho_j) = \sum_{u=1}^{\infty} \alpha_{ju} \rho_j^{-u} + \sum_{u=1}^{\infty} \mu_j b_j(u) \rho^u - \sum_{u=1}^{\infty} \alpha_{ju} - \mu_j.$$

It is left as an exercise for the reader to show that under condition (5.8) f has a solution in $[0, 1)$. From the fact that a solution to (5.6) and (5.7) always exists, it follows that under condition (5.8) the network has the product form stationary distribution. That condition (5.9) implies (5.8) is obvious. □

In applications, condition (ii) can be checked easily and is usually true. However, (i) is a much stronger condition and may not be satisfied. Nevertheless, in many cases the iterative procedure still generates a converging sequence. Thus the iterative procedure is very often used in practice for solving nonlinear traffic equations.

11.6 EXISTENCE OF SOLUTIONS: A FIXED POINT THEOREM APPROACH

Even though in many networks the monotonicity condition of Lemma 11.17 is not satisfied, there still exist solutions to the traffic equations. In these cases the existence of solutions to traffic equations can often be established by Brouwer's fixed point theorem (see Theorem C.1 in Appendix C). In this section, we apply this theorem to several of the networks discussed in earlier chapters.

We first consider a simple queueing network with negative customers, i.e., the model discussed in Example 4.7. Let the routing probabilities be such that the stochastic process of the network is irreducible. In Chapter 4 we assumed that there exists a solution to the traffic equations

$$\alpha_j = \lambda_j + \sum_{k=1}^{N} \rho_k \mu_k r_{kc,jc}, \qquad j = 1, \ldots, N, \tag{6.1}$$

$$\alpha_j^- = \lambda_j^- + \sum_{k=1}^{N} \rho_k \mu_k r_{kc,jc^-}, \qquad j = 1, \ldots, N, \tag{6.2}$$

where

$$\rho_j = \frac{\alpha_j}{\alpha_j^- + \mu_j}, \qquad j = 1, \ldots, N. \tag{6.3}$$

We are interested in the conditions under which there exists a solution to the traffic equations (6.1) to (6.3), and, if one exists, whether the solution is unique.

First, if (6.1) and (6.2) do have a solution α_j, α_j^- such that

$$\rho_j = \frac{\alpha_j}{\alpha_j^- + \mu_j} < 1, \qquad j = 1, \ldots, N,$$

then the solution has to be unique. This is because under these conditions

$$\prod_{j=1}^{N} (1 - \rho_j)\rho_j^{n_j}$$

is a stationary distribution of the network; hence by the uniqueness of the stationary distribution of an irreducible Markov process, it must be unique.

However, Lemma 11.17 is not applicable to this model since

$$\beta_j(\boldsymbol{\alpha}) = \frac{\alpha_j}{\mu_j + \alpha_j^-}\mu_j$$

is not nondecreasing in (α_j, α_j^-). A simple sufficient condition for the existence of a solution to traffic equations (6.1) and (6.2), which only requires network parameters, is given in the following theorem.

Theorem 11.22 *If the inequalities*

$$\lambda_j + \sum_{k=1}^{N} \mu_k r_{kc,jc} < \lambda_j^- + \mu_j, \quad j = 1, \ldots, N, \tag{6.4}$$

hold, the the traffic equations (6.1) and (6.2) have a solution $\{\alpha_j, \alpha_j^-; j = 1, 2, \ldots, N\}$ *such that* $\rho_j < 1$, $j = 1, 2, \ldots, N$.

PROOF. For $\boldsymbol{\rho} = (\rho_1, \ldots, \rho_N)$, define

$$f(\boldsymbol{\rho}) = \Big(\frac{\alpha_1}{\mu_1 + \alpha_1^-}, \frac{\alpha_2}{\mu_2 + \alpha_2^-}, \ldots, \frac{\alpha_N}{\mu_N + \alpha_N^-}\Big),$$

where α_j and α_j^- are given by (6.1) and (6.2). The α_j and α_j^- are both functions of $\boldsymbol{\rho}$. Condition (6.4) together with (6.1) implies

$$\alpha_j \le \lambda_j + \sum_{k=1}^{N} \mu_k r_{kc,jc} < \lambda_j^- + \mu_j \le \alpha_j^- + \mu_j.$$

Let

$$r_j = \frac{\lambda_j + \sum_{k=1}^{N} \mu_k r_{kc,jc}}{\lambda_j^- + \mu_j},$$

and $\boldsymbol{r} = (r_1, \ldots, r_N)$. Clearly $r_j < 1$ for $j = 1, \ldots, N$. The function $f(\cdot)$ is a continuous mapping from $[\mathbf{0}, \boldsymbol{r}]$ onto $[\mathbf{0}, \boldsymbol{r}]$. By Brouwer's fixed point theorem, there exists a $\boldsymbol{\rho} \in [\mathbf{0}, \boldsymbol{r}]$ such that $f(\boldsymbol{\rho}) = \boldsymbol{\rho} < \mathbf{1}$, where $\mathbf{1}$ is the column vector of 1's. The α_j and α_j^- are then determined by (6.1) and (6.2). This completes the proof of Theorem 11.22. □

We next show that the traffic equations (6.1) and (6.2) always have a solution. For convenience, we write the traffic equations in matrix form

$$\boldsymbol{\alpha} = \boldsymbol{\lambda} + \boldsymbol{\alpha} F R, \tag{6.5}$$

$$\boldsymbol{\alpha}^- = \boldsymbol{\lambda}^- + \boldsymbol{\alpha} F R^-, \tag{6.6}$$

where R is the matrix of the $r_{jc,kc}$'s, R^- is the matrix of the r_{jc,kc^-}'s, and F is the diagonal matrix with diagonal elements

$$\frac{\mu_1}{\mu_1 + \alpha_1^-}, \frac{\mu_2}{\mu_2 + \alpha_2^-}, \ldots, \frac{\mu_N}{\mu_N + \alpha_N^-}.$$

Note that all diagonal elements of F are between 0 and 1. Assume that every customer either ultimately leaves the network or changes into a signal, which implies that $(FR)^n$ converges to the zero matrix as n goes to infinity. In this case, $I - FR$ and $I - R$ are invertible, and $(I - FR)^{-1}$ and $(I - R)^{-1}$ are nonnegative matrices (see Exercise 11.2). So, from (6.5) we obtain

$$\boldsymbol{\alpha} = \boldsymbol{\lambda}(I - FR)^{-1}.$$

Substituting this into (6.6) yields

$$\boldsymbol{\alpha}^- = \boldsymbol{\lambda}^- + \boldsymbol{\lambda}(I - FR)^{-1} F R^-.$$

Define the function $g(\cdot)$ as

$$g(\boldsymbol{\alpha}^-) = \boldsymbol{\lambda}^- + \boldsymbol{\lambda}(I - FR)^{-1} F R^-.$$

Clearly, $g(\boldsymbol{\alpha}^-)$ is continuous and decreasing in $\boldsymbol{\alpha}^-$. Hence, if we let

$$\boldsymbol{u} = \boldsymbol{\lambda}^- + \boldsymbol{\lambda}(I - R)^{-1} R^-,$$

then

$$\boldsymbol{\lambda}^- \le \lim_{\boldsymbol{\alpha}^- \to \infty} g(\boldsymbol{\alpha}^-) \le g(\boldsymbol{\alpha}^-) \le g(\mathbf{0}) = \boldsymbol{u}.$$

Thus $g(\boldsymbol{\alpha}^-)$ is a continuous mapping from $[\boldsymbol{\lambda}^-, \boldsymbol{u}]$ to $[\boldsymbol{\lambda}^-, \boldsymbol{u}]$. Applying Brouwer's fixed point theorem shows that there exists an $\boldsymbol{\alpha}^-$ such that

$$g(\boldsymbol{\alpha}^-) = \boldsymbol{\alpha}^-.$$

This $\boldsymbol{\alpha}^-$ determines F and thus also $\boldsymbol{\alpha}$, by

$$\boldsymbol{\alpha} = \boldsymbol{\lambda}(I - FR)^{-1},$$

and ρ_j is obtained from (6.3). A stationary measure of the network is $\prod_{j=1}^{N} \rho_j^{n_j}$. Summarizing, we have the next theorem.

Theorem 11.23 *A queueing network with negative customers always has a solution to the traffic equations (6.1), (6.2), and (6.3). The network is stable, i.e., its stationary distribution exists, if and only if $\rho_j < 1$ for all j. If the network is stable, then its stationary distribution is given by the product form $\prod_{j=1}^{N}(1 - \rho_j)\rho_j^{n_j}$.*

The α_j and α_j^- in the traffic equations (6.1) and (6.2) represent the average arrival rates of regular customers and negative customers when $\rho_j \leq 1$. If $\rho_j > 1$ then the departure rate from node j is μ_j. To study the stability of the network it suffices to only consider ρ_j such that $\rho_j \leq 1$. Taking this into consideration we modify the traffic equations (6.1) and (6.2) to

$$\alpha_j = \lambda_j + \sum_{k=1}^{N} (\rho_k \wedge 1)\mu_k r_{kc,jc}, \qquad j = 1, \ldots, N, \tag{6.7}$$

$$\alpha_j^- = \lambda_j^- + \sum_{k=1}^{N} (\rho_k \wedge 1)\mu_k r_{kc,jc^-}, \qquad j = 1, \ldots, N, \tag{6.8}$$

where ρ_j is still given by (6.3) and $\rho_k \wedge 1 = \min\{\rho_k, 1\}$. To show the existence of solutions to (6.7) and (6.8), we define the multivariate function F as

$$F(\boldsymbol{\alpha}, \boldsymbol{\alpha}^-) = (f_1(\boldsymbol{\alpha}_1), g_1(\boldsymbol{\alpha}_1), \ldots, f_N(\boldsymbol{\alpha}_N), g_N(\boldsymbol{\alpha}_N)),$$

where

$$f_j(\boldsymbol{\alpha}_j) = \lambda_j + \sum_{k=1}^{N} (\rho_k \wedge 1)\mu_k r_{kc,jc}, \qquad j = 1, \ldots, N,$$

$$g_j(\boldsymbol{\alpha}_j^-) = \lambda_j^- + \sum_{k=1}^{N} (\rho_k \wedge 1)\mu_k r_{kc,jc^-}, \qquad j = 1, \ldots, N.$$

Clearly, $F(\boldsymbol{\alpha}, \boldsymbol{\alpha}^-)$ is bounded. So Brouwer's fixed point theorem can be applied to F to show that (6.7) and (6.8) always have a solution.

We now apply this approach to the queueing networks with positive and negative signals of Section 4.5.1. Modifying the traffic equations (5.1), (5.2) and (5.3) of Chapter 4 we obtain

$$\begin{aligned}
\alpha_j &= \lambda_j + \sum_{k=1}^{N} (\rho_k \wedge 1)\mu_k r_{kc,jc} + \sum_{k=1}^{N} (\rho_k \wedge 1)\alpha_k^- r_{ks^-,jc} + \sum_{k=1}^{N} \rho_k^{-1}\alpha_k^+ r_{ks^+,jc}, \\
\alpha_j^+ &= \lambda_j^+ + \sum_{k=1}^{N} (\rho_k \wedge 1)\mu_k r_{kc,js^+} + \sum_{k=1}^{N} (\rho_k \wedge 1)\alpha_k^- r_{ks^-,js^+} \\
&\quad + \sum_{k=1}^{N} \rho_k^{-1}\alpha_k^+ r_{ks^+,js^+}, \\
\alpha_j^- &= \lambda_j^- + \sum_{k=1}^{N} (\rho_k \wedge 1)\mu_k r_{kc,js^-} + \sum_{k=1}^{N} (\rho_k \wedge 1)\alpha_k^- r_{ks^-,js^-} \\
&\quad + \sum_{k=1}^{N} \rho_k^{-1}\alpha_k^+ r_{ks^+,js^-}, \qquad j = 1, \ldots, N, \qquad (6.9)
\end{aligned}$$

where

$$\rho_j = \frac{\alpha_j + \alpha_j^+}{\mu_j + \alpha_j^-}. \qquad (6.10)$$

As in Section 4.5.1, we modify the network by introducing additional departures of positive signals with rate $\rho_j^{-1}\alpha_j^+$ whenever node j is empty. Then, all nodes in this network are uniformly quasi-reversible, provided $\rho_j < 1$ for $j = 1, 2, \ldots, N$.

Define the multivariate function $F(\boldsymbol{\alpha}, \boldsymbol{\alpha}^+, \boldsymbol{\alpha}^-)$ by the right-hand side of equations (6.9). To apply the fixed point theorem, we need to consider the boundedness of F. First, it follows from (6.10) that

$$\begin{aligned}
\rho_k^{-1}\alpha_k^+ &= \frac{\alpha_k^+}{\alpha_k + \alpha_k^+}(\mu_k + \alpha_k^-) \\
&\le \mu_k + \alpha_k^-. \qquad (6.11)
\end{aligned}$$

Hence if $\boldsymbol{\alpha}^-$ is bounded, then $F(\boldsymbol{\alpha}, \boldsymbol{\alpha}^+, \boldsymbol{\alpha}^-)$ is also bounded. To that end, let $p_{jk} = r_{js^-,ks^-} + r_{js^+,ks^-}$, and let P be an $N \times N$ matrix whose jk component is p_{jk}. We make the following assumption.

(i) The matrix P^n converges to zero as n goes to infinity.

For example, if negative signals eventually leave the network with probability 1 and if positive signals never change to negative signals, then (i) is satisfied.

Theorem 11.24 *Consider the queueing network with positive and negative signals of Section 4.5.1 that satisfies condition (i). The traffic equations (6.9) always have a solution, and the network is stable if and only if $\rho_j < 1$ for $j = 1, 2, \ldots, N$. Under these conditions the stationary distribution of the network is*

$$\pi(\boldsymbol{n}) = \prod_{j=1}^{N}(1-\rho_j)\rho_j^{n_j}.$$

PROOF. It follows from the discussions above that it suffices to prove that $\boldsymbol{\alpha}^-$ is bounded, which implies that the multivariate function F is bounded. From the third equation of (6.9) and inequality (6.11), we obtain

$$\alpha_j^- - \sum_{k=1}^{N}\alpha_k^- p_{kj} \le \lambda_j^- + \sum_{k=1}^{N}\mu_k(r_{kc,js^-} + r_{ks^+,js^-}).$$

Using the matrix notation $R^{c-} \equiv \{r_{kc,js^-}\}$ and $R^{+-} \equiv \{r_{ks^+,js^-}\}$, this can be rewritten as

$$\boldsymbol{\alpha}^-(I-P) \le \boldsymbol{\lambda}^- + \boldsymbol{\mu}(R^{c-} + R^{+-}).$$

Condition (i) implies that $I-P$ is invertible, and $(I-P)^{-1}$ is a nonnegative matrix (see Exercise 11.2). Multiplying both sides of the above inequality by $(I-P)^{-1}$, we obtain an upper bound for $\boldsymbol{\alpha}^-$. Applying Brouwer's fixed point theorem to F yields that the traffic equations (6.9) always have a solution. □

Remark 11.25 Under an additional condition, Theorem 11.24 can also be extended to the networks of Section 4.5.2 with multiple classes of positive and negative signals. In this case the last summation on the right-hand side of (6.9) becomes

$$\sum_{k=1}^{N}\sum_{v=1}^{I^-}\frac{\alpha_{kv}^+}{\alpha_k + \alpha_k^+}(\mu_k + \alpha_k^-)r_{kv^+,ju^-}.$$

This term cannot be bounded because of the summation of $\alpha_{kv}^- r_{kv^+,ju^-}$. So we have to assume that such a term vanishes, i.e., positive signals never become negative signals. Thus, if we replace condition (i) by

(i′) negative signals of any class eventually either become regular customers or leave the network with probability one, i.e., if $P = \{r_{ju^-,kv^-}\}$, then $I-P$ is invertible. Furthermore, $r_{ju^+,kv^-} = 0$ for all $j, k = 1, 2, \ldots, N$, $u = 1, \ldots, I^+$, and $v = 1, \ldots, I^-$, we then obtain a result similar to Theorem 11.24 for networks with multiple classes of positive and negative signals. Note that the numbers of the classes of positive and negative signals, I^+ and I^-, may be infinity. So R^{c-} and P may be matrices of infinite dimension. However, this does not cause any problem in the bounding of $\boldsymbol{\alpha}^-$, $\boldsymbol{\alpha}$, and $\boldsymbol{\alpha}^+$.

11.7 REFERENCE NOTES

The necessary and sufficient conditions for product form distributions were first reported in Takada and Miyazawa (1997), and extended in Chao, Miyazawa, Serfozo and Takada (1998). The material in Sections 11.1 to 11.4 is taken from the second reference, in which similar results are obtained for more general stochastic networks. The two examples of Section 11.2 are taken from the first reference. Some other discussions on the necessity of quasi-reversibility for the product form are contained in Harrison and Williams (1992), Glynn (1990), and Malinkovsky (1990).

The study of stability of queues traces back to Loynes (1962). The stability of single-class Jackson networks with general arrival and service processes is well understood (see, for example, Sigman (1990), Baccelli and Foss (1994) and Chang, Thomas and Kiang (1994)). Because of the paradoxical unstability in multi-class queueing networks, the stability analysis has become an area of active research in recent years; see, for example, Kumar (1993), Whitt (1993), Dai (1995) and Chen and Zhang (1997), among others. The study of stability focuses typically on the conditions, in terms of system data, under which a queueing system is stable. The discussion of stability in this chapter has a much narrower focus. The iterative approach to solving traffic equations discussed in Section 11.5 is taken from Miyazawa (1996). The fixed point approach to the existence of traffic equations has been employed in Gelenbe and Schassberger (1992), Henderson et al. (1994*b*), and Miyazawa and Taylor (1997). Theorem 11.24 is new, though the idea is essentially the same. Exercises 11.3 and 11.4 are taken from Chao et al. (1996).

EXERCISES

11.1 Consider a tandem queue with two nodes and one class of customers, denoted by c, and one class of signals, denoted by s. Customers arrive at node 1 from the outside according to a Poisson process with rate λ, and signals arrive at node j from the outside according to a Poisson process with rate λ_j^-, $j = 1, 2$. The state of the network is $\boldsymbol{n} = (n_1, n_2)$, where n_j is the number of customers at node j. The arrival of a customer at a node increases the number of customers by 1, and the arrival of a signal at a node reduces the number of customers by one, i.e.,

$$p_{jc}^{A}(n_j, n_j + 1) = 1, \qquad j = 1, 2,$$

$$p_{js}^{A}(n_j, n_j - 1) = 1[n_j > 0], \qquad j = 1, 2.$$

Assume that $r_{1c,2c} = 1, r_{2c,0} = 1$, and all other routing probabilities are 0. Let

$$q_{1c}^{D}(n_1, n_1 - 1) = \min\{n_1, 2\}\mu_1,$$

$$q_{2c}^{D}(n_2, n_2 - 1) = \mu_2.$$

Thus the service times at both nodes are exponentially distributed. There are two servers at node 1 and one server at node 2. Clearly, this is a special case of Example 6.7.

(i) Show that this tandem system with signals is equivalent to another tandem system without signals and state-dependent service rates. This transforms the network with multiple class of entities into one with a single class of entities.

(ii) Use (i) and Theorem 11.3 to show that this network does not have a product form stationary distribution.

11.2 Let P be an $N \times N$ $(N < \infty)$ square matrix with nonnegative and finite components.

(a) Show the following identity for $n = 1, 2, \ldots$,

$$(I - P)(I + P + P^2 + \cdots + P^n) = I - P^{n+1}.$$

(b) Prove that, if P^n converges to the zero matrix, then $(I - P)$ is invertible and

$$(I - P)^{-1} = \sum_{n=0}^{\infty} P^n$$

is a nonnegative matrix all the components of which are finite.

11.3 Consider a queueing network with customer coalescence and constant batch sizes, i.e., the K_j is a fixed number instead of a random variable. Show that the traffic equations

$$\alpha_j = \lambda_j + \sum_{k=1}^{N} (\rho_k \wedge 1)^{K_k} \mu_k r_{kj}, \qquad j = 1, \ldots, N,$$

$$\alpha_j = \mu_j \sum_{\ell=1}^{K_j} \rho_j^{\ell}, \qquad j = 1, 2, \ldots, N,$$

always have a solution ρ_j (possibly larger than 1), $j = 1, \ldots, N$, and that this solution must be unique.

11.4 Consider Exercise 11.3 again. Assume that the routing matrix $(r_{jk};\ j, k = 0, 1, \ldots, N)$ is irreducible. Let U be the set of non-bottleneck stations, i.e.,

$$U = \{j | \rho_j < 1\}.$$

Show that

$$\lim_{t\to\infty} P\{X_j(t) = n_j;\ j \in U\} = \prod_{j\in U}(1 - \rho_j)\rho_j^{n_j},$$

for all integers $n_j \geq 0$, and that any bottleneck station will end up with an infinite number of jobs, i.e.,

$$\lim_{t\to\infty} P\{X_j(t) \leq n_j\} = 0, \qquad \text{for all } j \notin U,$$

for any integers $n_j \geq 0$.

11.5 Consider the network in Section 11.1. Let $\pi(\boldsymbol{x})$ be the stationary distribution of the network. Assume that there exists a set of numbers β_j such that

$$\sum_{x'_j} \pi(\boldsymbol{x}_j(x'_j))q_j^{\mathrm{D}}(x'_j, x_j) = \beta_j\pi(\boldsymbol{x}), \qquad j = 1, \ldots, N.$$

Show that each node is quasi-reversible and that $\pi(\boldsymbol{x})$ is product form.

11.6 Define the functions f and g as

$$f(\rho_j, \boldsymbol{\alpha}_j) = \sum_{\ell=1}^{\infty}\sum_{u=\ell}^{\infty} \alpha_{ju}\rho_j^{-\ell+1},$$

$$g(\rho_j) = \mu_j\rho_j \sum_{\ell=1}^{\infty}\sum_{u=\ell}^{\infty} b_j(u)\rho_j^{\ell},$$

where $\boldsymbol{\alpha}_j = (\alpha_{j\ell};\ \ell \geq 1)$. Prove the following results.

(a) The equation $f(\rho_j, \boldsymbol{\alpha}_j) = g(\rho_j)$ has a nonnegative solution ρ_j for each $\boldsymbol{\alpha}_j \geq \mathbf{0}$ such that $\boldsymbol{\alpha}_j \neq \mathbf{0}$.

(b) The solution ρ_j in (a) is nondecreasing in $\boldsymbol{\alpha}_j$.

12

Discrete Time Networks

The preceding chapters focused primarily on continuous time models of queueing networks. However, in some applications, arrival and service times are synchronized, and discrete time models may be more appropriate. This chapter concentrates on discrete time queueing networks. From a technical point of view, discrete time models can be obtained from continuous time models either by tracing the jump epochs, or by slotting the time axis. Conversely, continuous time models can be constructed from discrete time models by introducing sojourn times between state changes. Nevertheless, there are crucial differences in the terminologies, the interpretations, as well as in the technical details. In discrete time models the concurrent movement is the most significant feature. These issues will be addressed in this chapter.

12.1 DISCRETE TIME MODELING

In discrete time queueing networks time is slotted, and multiple arrivals and departures may take place in a time slot at the same node as well as at different nodes. Hence, when we define arrivals and departures, we have to specify the order of their occurrences in each time slot. We assume that time is partitioned into intervals of equal length, which we refer to as *slots*. If departures take place at the beginning of each slot and arrivals at the end of each slot, then the formulation is called *late arrival*. If the departures and arrivals are conversely ordered, then the formulation is called *early arrival*. In the early arrival formulation, arriving customers can be served in the same slot. However, this is not the case in the late arrival formulation.

If in a queueing network a departure from a node becomes an arrival elsewhere, then the late arrival formulation is convenient. On the other hand, for quasi-reversibility the state immediately after a departure is important, so the early arrival formulation is more appropriate. In this chapter we adopt the late arrival formulation, but incorporate the advantages of both formulations by introducing two observation epochs. For convenience the starting time of each slot is called the *slot boundary*, and the time period between two adjacent slot boundaries is called the *slot inside* (see Figure 12.1). The state of the network is observed at the slot boundaries as well as during the slot insides. These two observation epochs may serve as alternative descriptions for the late arrival and early arrival formulations. Thus the difference between the two formulations in our modeling is nominal.

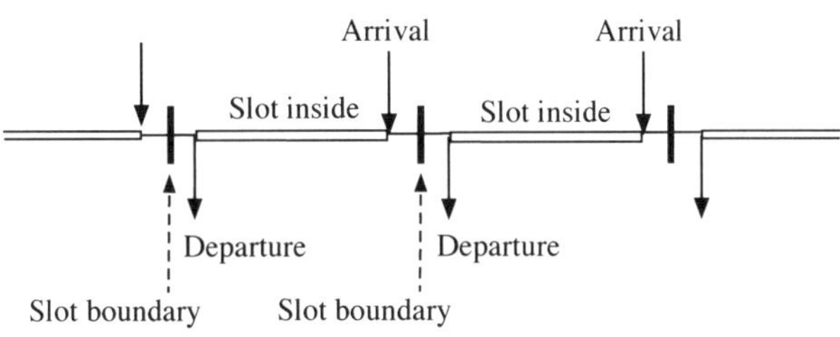

Figure 12.1 Time slot and boundary

Note that this modeling framework does not allow instantaneous triggerings by signals. On the one hand, the effects of signals can be included in a batch movement by aggregating all the effects of the signals within a time slot; on the other hand, the batch movement feature can also be considered in the framework of networks with signals using appropriate signaling mechanisms. Nevertheless, there are major differences since in the framework of networks with signals the events at all the nodes take place one after another in a sequential manner; as a result, the string of transitions may be terminated because of partial batches. Similarly, the batch movements cannot always catch the specifics of strings of sequential transitions. In this chapter we will not consider triggerings by signals. We focus mainly on the concurrency feature. For simplicity, we only consider batch movement networks whose states are vectors of customer populations over the nodes. Thus, the dynamics of the network are composed of batch arrivals and departures, as well as of internal transitions caused by batch movements. The first four sections, however, concentrate on networks without internal transitions.

The following example illustrates the concurrent movements in a discrete time network.

Example 12.1 (A synchronized production line) Suppose a production system is composed of N stages in series as shown in Figure 12.2.

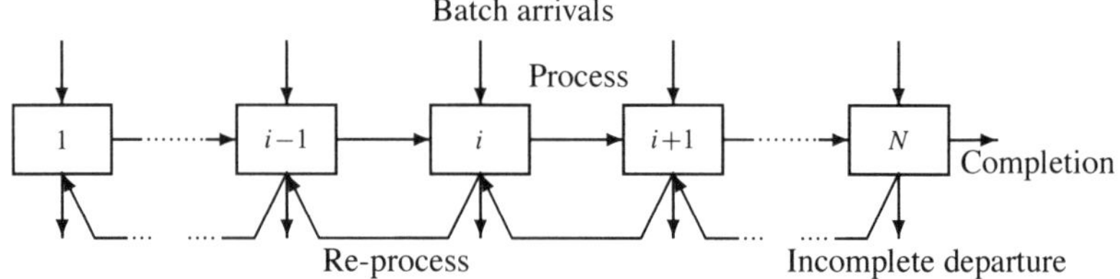

Figure 12.2 A synchronized production line

Each stage has a machine and a waiting buffer of infinite size. Actions at all the stages are synchronized, so time is slotted. In each time slot, new items arrive in a batch at stage 1 and have to be processed from stage 1 to stage N. In addition, items also arrive at stage j from outside suppliers, and have to be processed from stage j to stage N, $j = 2, \ldots, N$. Each machine is either up or down in a time slot. When a machine is up, it inspects the items in its upstream buffer, and processes the non-defective items. When a machine is down, it inspects the items but it cannot do any processing. In both cases, the defective items are either returned to the previous stage or removed from the system.

We assume that the departures are immediately transformed into arrivals at subsequent nodes. This model fits within the discrete time framework, and will be studied in Section 12.5. □

Consider a discrete time single queue with batch movements. Let X_ℓ be the number of customers in the system at slot boundary ℓ, $\ell = 0, 1, 2, \ldots$. At the beginning of slot ℓ, i.e., immediately after slot boundary ℓ, D_ℓ customers depart, changing the state to $Y_\ell = X_\ell - D_\ell$, where $D_\ell \leq X_\ell$. During slot inside ℓ, the number of customers in the system remains at Y_ℓ. At the end of this slot, A_ℓ customers arrive. Thus the state of the queue at slot boundary $\ell + 1$ is

$$X_{\ell+1} = X_\ell - D_\ell + A_\ell,$$

and, during slot inside $(\ell + 1)$, the state of the queue is

$$\begin{aligned} Y_{\ell+1} &= X_{\ell+1} - D_{\ell+1} \\ &= Y_\ell + A_\ell - D_{\ell+1}, \quad \ell = 0, 1, 2, \ldots. \end{aligned}$$

We assume that D_ℓ may depend on X_ℓ, but that it is independent of all other events before slot boundary ℓ. Thus there is a conditional probability function p^{D} such that

$$p^{\mathrm{D}}(n, i) = P(D_\ell = i | X_k, D_k, A_k, k \le \ell - 1, X_\ell = n), \quad i = 1, 2, \ldots, n. \quad (1.1)$$

This p^{D} is referred to as the *departure function*. Similarly, for A_ℓ, we assume that there is a probability function p^{A} such that

$$p^{\mathrm{A}}(n, i) = P(A_\ell = i | Y_k, A_k, D_{k+1}, k \le \ell - 1, Y_\ell = n), \qquad n, i \ge 0. \quad (1.2)$$

This p^{A} is referred to as the *arrival function*. Unlike in the continuous time models, we assume that there are no internal transitions. So the dynamics of the queue are completely governed by the arrival function p^{A} and the departure function p^{D}.

Remark 12.2 The conditional probabilities p^{A} and p^{D} correspond to the transition rate functions q^{A} and q^{D} in the continuous time setting. Their second arguments, however, are different. In q^{A} and q^{D} the second argument is the state of the queue after the corresponding transitions. Thus, if we would like to have consistency then $p^{\mathrm{A}}(n, i)$ and $p^{\mathrm{D}}(n, i)$ should be replaced by $p^{\mathrm{A}}(n, n + i)$ and $p^{\mathrm{D}}(n, n - i)$, respectively. Nonetheless it appears to be more convenient to use the present notation in the discrete time setting.

Example 12.3 (A discrete time queue) Consider the following departure dynamics of a queue. At slot boundary ℓ, independently of the state X_ℓ, S_ℓ customers are requested to leave the system. We assume that $\{S_\ell; \ell = 0, 1, 2, \ldots\}$ is a sequence of i.i.d. random variables. If $X_\ell < S_\ell$, then the system is emptied out, i.e., $D_\ell = X_\ell$. Otherwise, the queue length is changed to $X_\ell - S_\ell$, i.e., $D_\ell = S_\ell$. So

$$D_\ell = \min\{X_\ell, S_\ell\}, \qquad \ell = 1, 2, \ldots.$$

The departure function of this system is

$$p^{\mathrm{D}}(n, i) = \begin{cases} P(S_\ell = i), & i = 0, 1, \ldots, n-1, \\ P(S_\ell \ge n), & i = n. \end{cases}$$

The departure batch is called full if $D_\ell \ge S_\ell$, and partial otherwise. □

From assumptions (1.1) and (1.2), $\{X_\ell; \ell = 0, 1, 2, \ldots\}$ is a discrete time Markov chain with transition probability

$$p^{\mathrm{X}}(n, n') = \sum_{i=(n-n')^+}^{n} p^{\mathrm{D}}(n, i) p^{\mathrm{A}}(n - i, n' - (n - i)), \qquad n, n' \ge 0. \quad (1.3)$$

This process represents the late arrival formulation. In what follows we assume that p^{X} is irreducible. Similarly, $\{Y_\ell;\ \ell = 0, 1, 2, \ldots\}$ is also a discrete time Markov chain, which represents the early arrival formulation.

Assume that p^{X} has a stationary distribution π^{X}, and define π^{Y} by

$$\pi^{\mathrm{Y}}(n) = \sum_{i=0}^{\infty} \pi^{\mathrm{X}}(n+i) p^{\mathrm{D}}(n+i, i), \qquad n \geq 0. \tag{1.4}$$

It is left as an exercise for the reader to show that π^{Y} is the stationary distribution of the Markov chain $\{Y_\ell\}$. Furthermore, it follows from the relationship $X_{\ell+1} = Y_\ell + A_\ell$ that

$$\pi^{\mathrm{X}}(n) = \sum_{i=0}^{n} \pi^{\mathrm{Y}}(n-i) p^{\mathrm{A}}(n-i, i), \qquad n \geq 0. \tag{1.5}$$

Remark 12.4 Although $\{X_\ell\}$ is a Markov chain, it does not include information about the arrival and departure batch sizes. If we define the history $\mathcal{F}_\ell$, up to slot boundary ℓ, by

$$\mathcal{F}_\ell^{\mathrm{X}} = \sigma\{D_{k-1}, A_{k-1}, X_k;\ k \leq \ell\},$$

then $\{X_\ell\}$ is still a Markov chain with respect to $\{\mathcal{F}_\ell^{\mathrm{X}}\}$. Similarly, $\{Y_\ell\}$ is a Markov chain with respect to the history

$$\mathcal{F}_\ell^{\mathrm{Y}} = \sigma\{A_{k-1}, D_k, Y_k;\ k \leq \ell\}.$$

These modifications are similar to those in Section 3.1 for a continuous time Markov chain. Throughout this chapter it is implicitly assumed that Markov chains are defined with respect to all the relevant information. For instance, $\{X_\ell\}$ is defined with respect to the history $\{\mathcal{F}_\ell^{\mathrm{X}}\}$.

Another way of studying a discrete time queueing system is to consider the pair $(A_{\ell-1}, X_\ell)$ or (D_ℓ, Y_ℓ), which are also Markov chains. These formulations have been proven to be useful in the analysis of discrete time queues, see Section 12.8.

12.2 DISCRETE TIME QUASI-REVERSIBILITY

In a continuous time framework, the quasi-reversibility of each node plays a key role in networks with product form stationary distributions. This may not be the case in the discrete time setting because of the strong interactions between the different nodes enforced by the concurrent movements.

To see what could be expected for discrete time quasi-reversibility, consider a stationary, continuous time, quasi-reversible queue. It follows from Chapter 3 that the arrival and departure processes are Poisson, and the current state is independent of future arrivals and past departures. If we partition the time axis into slots of equal length, then the properties of the Poisson process imply that the numbers of arrivals during the slot insides are i.i.d. random variables with a common Poisson distribution, and the numbers of departures during the slot insides are also i.i.d. random variables with a common Poisson distribution.

Definition 12.5 (Batch Bernoulli process) A sequence of nonnegative i.i.d. integer-valued random variables is called a *batch Bernoulli process*. In particular, if the batch size distribution is Poisson, it is called a *Poisson batch Bernoulli process*.

Thus if we discretize a continuous time, quasi-reversible queue, the following properties hold.

(i) The arrival process is batch Bernoulli, and future arrival batches are independent of the current state of the system.

(ii) The departure process is batch Bernoulli, and the past departure batches are independent of the current state of the system.

(iii) The arrival batches and departure batches both have a Poisson distribution.

Indeed, as might be expect, networks of discrete time queues satisfying (i), (ii) and (iii) have a product form stationary distribution (see Theorem 12.17). However, the Poisson condition (iii), even though it has nice properties, is too restrictive in applications. Thus we define quasi-reversibility for discrete time queues using (i) and (ii), in parallel to Definition 3.4, as follows.

Definition 12.6 (Discrete time quasi-reversibility) A discrete time queue with arrival function p^{A} and departure function p^{D} is *quasi-reversible* if there exist two probability distributions $\{a(i)\}$ and $\{b(i)\}$ such that

$$p^{\mathrm{A}}(n, i) = a(i), \qquad n, i \geq 0, \tag{2.1}$$

and

$$\pi^{\mathrm{X}}(n+i)p^{\mathrm{D}}(n+i,i) = b(i)\pi^{\mathrm{Y}}(n), \qquad n, i \geq 0, \tag{2.2}$$

where π^{X} and π^{Y} are the stationary distributions of X_ℓ and Y_ℓ, respectively.

Remark 12.7 As we will see in Theorem 12.9 below, (2.1) and (2.2) hold only when $a(i) = b(i)$. So the queue can be quasi-reversible only when arriving batches and departing batches have the same probability mass function. This is because of the assumptions that any state change is counted as either an arrival or a departure and the effect of an arrival or a departure is simply the addition or the removal of customers. However, if these assumptions do not hold, e.g., if arrivals and departures are counted in a different way, then the batch size distributions may not be identical (see Exercise 12.3).

Because of (1.2), condition (2.1) implies the following properties:

(2.a) The arrival process $\{A_\ell\}$ is batch Bernoulli with batch size distribution $\{a(\ell)\}$.

(2.b) For any k, the future arrival batches $\{A_\ell; \ell \geq k\}$ are independent of the current state of the system Y_k.

On the other hand, since

$$\begin{aligned}\pi^{\mathrm{X}}(n+i)p^{\mathrm{D}}(n+i,i) &= P(X_\ell = n+i)P(D_\ell = i | X_\ell = n+i)\\ &= P(D_\ell = i, X_\ell = n+i)\\ &= P(D_\ell = i, Y_\ell = n)\\ &= P(D_\ell = i | Y_\ell = n)P(Y_\ell = n)\\ &= P(D_\ell = i | Y_\ell = n)\pi^{\mathrm{Y}}(n),\end{aligned}$$

condition (2.2) is equivalent to

$$P(D_\ell = i | Y_\ell = n) = b(i).$$

Hence, D_ℓ is independent of Y_ℓ and all events subsequent to slot inside ℓ because $\{Y_\ell; \ell = 0, 1, 2, \ldots\}$ is a Markov chain with respect to the history $\{\mathcal{F}_\ell^{\mathrm{Y}}\}$. This implies the following properties:

(2.c) The departure process $\{D_\ell\}$ is batch Bernoulli with batch size distribution $\{b(i)\}$.

(2.d) For any k, the past departure batches $\{D_\ell; \ell \leq k\}$ are independent of the current state of the system Y_k.

Clearly these four properties imply (2.1) and (2.2). Thus, we obtain the next lemma, which corresponds to Theorem 3.8 in the continuous time setting.

Lemma 12.8 *A stationary queue with arrival sequence $\{A_\ell\}$ and departure sequence $\{D_\ell\}$ is quasi-reversible if and only if (2.a)–(2.d) hold.*

The following result states that discrete time quasi-reversibility is completely characterized by the departure function.

Theorem 12.9 *Consider a stationary queue with no internal transitions, batch Bernoulli arrivals with batch size distribution $\{a(i)\}$, and departure function p^{D}. The queue is quasi-reversible if and only if $a(i) = b(i)$ and the departure function p^{D} is of the form*

$$p^{\mathrm{D}}(n, i) = \frac{\psi(n-i)}{\phi(n)} a(i), \tag{2.3}$$

where ψ is an arbitrary nonnegative function with

$$\sum_{n=0}^{\infty} \psi(n) < \infty, \tag{2.4}$$

and ϕ is a positive function determined by

$$\phi(n) = \sum_{i=0}^{n} \psi(n-i) a(i), \qquad n \geq 0. \tag{2.5}$$

In this case, the stationary distributions π^{X} and π^{Y} are given by

$$\begin{aligned} \pi^{\mathrm{X}}(n) &= c\phi(n), \qquad n \geq 0, \\ \pi^{\mathrm{Y}}(n) &= c\psi(n), \qquad n \geq 0, \end{aligned} \tag{2.6}$$

where c is the common normalization constant.

PROOF. We first prove necessity. Suppose the stationary distribution π^{X} exists, and the queue is quasi-reversible. Note that π^{X} is unique because p^{X} is irreducible.

If π^{Y} is defined by (1.4), then π^{Y} is the stationary distribution of Y_ℓ. Let $\psi = \pi^{\mathrm{Y}}$ and $\phi = \pi^{\mathrm{X}}$. The second quasi-reversibility condition (2.2) implies

$$p^{\mathrm{D}}(n,i) = \frac{\psi(n-i)}{\phi(n)}b(i). \tag{2.7}$$

Hence, from the fact that

$$\sum_{i=0}^{n} p^{\mathrm{D}}(n,i) = 1,$$

we obtain

$$\phi(n) = \sum_{i=0}^{n} \psi(n-i)b(i), \qquad n \geq 0.$$

It follows from (1.5) and the definitions of ϕ and ψ that

$$\phi(n) = \sum_{i=0}^{n} \psi(n-i)a(i), \qquad n \geq 0.$$

Comparing these two expressions for ϕ, we conclude by induction on n that

$$a(i) = b(i), \qquad \text{for } i = 0, 1, \ldots.$$

Since $\pi^{\mathrm{Y}} = \psi$ is a probability mass function, condition (2.4) is clearly satisfied. It follows from the definitions of ψ and ϕ that (2.6) holds with $c = 1$.

We now prove sufficiency. Assume that (2.3) holds for a probability mass function $\{a(i)\}$ and a nonnegative function ψ satisfying (2.4). Since (2.3) implies (2.5), it follows from (2.3) and $p^{\mathrm{A}}(n-i,n) = a(i)$ that

$$\phi(n+j-i)p^{\mathrm{D}}(n+j-i,j)p^{\mathrm{A}}(n-i,i) = \psi(n-i)a(j)a(i).$$

After summing both sides of the above equation over all i and j, and applying (2.5), we obtain

$$\begin{aligned}\sum_{j=0}^{\infty}\sum_{i=0}^{n} \phi(n+j-i)p^{\mathrm{D}}(n+j-i,j)p^{\mathrm{A}}(n-i,i) &= \sum_{j=0}^{\infty} \phi(n)a(j) \\ &= \phi(n).\end{aligned}$$

This shows that ϕ is an invariant measure for p^{X}. From (2.4) and (2.5) it follows that

$$\begin{aligned}\sum_{n=0}^{\infty}\phi(n) &= \sum_{n=0}^{\infty}\sum_{i=0}^{n}\psi(n-i)b(i)\\ &= \sum_{i=0}^{\infty}b(i)\sum_{n=i}^{\infty}\psi(n-i)\\ &= \sum_{n=0}^{\infty}\psi(n) < \infty.\end{aligned}$$

Thus ϕ can be scaled to a probability mass function, i.e., $\pi^{\mathrm{X}}(n) = c\phi(n)$ is the stationary distribution for p^{X}, where c is the normalization constant. Since the stationary distribution π^{Y} is obtained by

$$\pi^{\mathrm{Y}}(n) = \sum_{i=0}^{\infty}\pi^{\mathrm{X}}(n+i)p^{\mathrm{D}}(n+i,i),$$

equation (2.3) with $\phi(n) = \pi^{\mathrm{X}}(n)/c$ implies $\pi^{\mathrm{Y}}(n) = c\psi(n)$. This shows that (2.3) is equivalent to (2.2), and completes the proof of the theorem. □

It is of interest to investigate the possible departure functions of the form (2.3). They are not as general as they appear. By (2.4), we can assume without loss of generality that ψ is a probability mass function. The following three examples illustrate the most typical arrival batch size distributions, namely Bernoulli, geometric, and Poisson.

Example 12.10 (Bernoulli arrivals) Consider a single-server queue with single Bernoulli arrivals and geometrically distributed service times. At the beginning of each time slot there is an arrival with probability p and no arrival with probability $1-p$. At the end of the slot, if one or more customers are present, then there is a departure with probability q and no departure with probability $1-q$. Thus we have

$$\begin{aligned} a(0) &= 1-p, \qquad a(1) = p,\\ p^{\mathrm{D}}(n,0) &= 1-q, \qquad p^{\mathrm{D}}(n,1) = q, \qquad n \geq 1.\end{aligned}$$

We assume $0 < p$ and $q \leq 1$. It follows from Theorem 12.9 that this queue is quasi-reversible if and only if there exist distributions ϕ and ψ satisfying (2.3), i.e.,

$$\phi(0) = \psi(0)(1-p), \tag{2.8}$$

$$q\phi(n) = \psi(n-1)p, \qquad n \geq 1, \tag{2.9}$$

$$(1-q)\phi(n) = \psi(n)(1-p), \qquad n \geq 1. \tag{2.10}$$

Dividing (2.9) by (2.10) yields

$$\psi(n) = \frac{p}{1-p}\frac{1-q}{q}\psi(n-1), \qquad n \geq 1.$$

Inductively we obtain

$$\psi(n) = \left(\frac{p(1-q)}{q(1-p)}\right)^n \psi(0), \qquad n \geq 0,$$

$$\phi(n) = \begin{cases} (1-p)\psi(0), & n = 0, \\ \dfrac{p}{q}\left(\dfrac{p(1-q)}{q(1-p)}\right)^{n-1} \psi(0), & n \geq 1, \end{cases}$$

where

$$\psi(0) = \frac{1 - p/q}{1-p}.$$

These ϕ and ψ are proper distributions if and only if $p < q$. Hence, if the stability condition $p < q$ holds, the queue length distributions at the slot boundaries and during the slot insides are given by ϕ and ψ, respectively, and the queue is quasi-reversible. Since the mean arrival and service rates are p and q, the results are slightly different from those for the $M/M/1$ queue in continuous time.

A similar analysis shows that if the departure rates are state-dependent, i.e.,

$$p^{\mathrm{D}}(n,0) = 1 - q_n, \quad p^{\mathrm{D}}(n,1) = q_n, \qquad n \geq 1, \tag{2.11}$$

then the queue is also quasi-reversible with stationary distributions

$$\pi^{\mathrm{X}}(n) = \begin{cases} (1-p)\psi(0), & n = 0, \\ \dfrac{p}{q_n}\left(\dfrac{p}{1-p}\right)^n \displaystyle\prod_{\ell=1}^{n-1} \dfrac{1-q_\ell}{q_\ell}\psi(0), & n \geq 1, \end{cases} \tag{2.12}$$

$$\pi^{\mathrm{Y}}(n) = \left(\frac{p}{1-p}\right)^n \prod_{\ell=1}^{n} \frac{1-q_\ell}{q_\ell}\psi(0), \qquad n \geq 0, \tag{2.13}$$

where

$$\psi(0) = \frac{1}{\sum_{n=0}^{\infty}\left(\dfrac{p}{1-p}\right)^n \prod_{\ell=1}^{n} \dfrac{1-q_\ell}{q_\ell}},$$

which is assumed to be finite. Details are left as an exercise for the readers. Since for Bernoulli arrivals $a(n) = 0$ for $n \geq 2$, it follows from Theorem 12.9 that

$$p^{\mathrm{D}}(n,i) = 0, \qquad 2 \leq i \leq n.$$

Hence, the state-dependent departure function (2.11) is the only possible form for which the queue with Bernoulli arrivals is quasi-reversible. □

Example 12.11 (Geometric batch arrivals) Consider a queue with geometrically distributed batch arrivals, i.e.,

$$a(i) = (1-p)p^i, \qquad i \geq 0.$$

It follows from Theorem 12.9 that, if we substitute $\psi(i)$ and $\phi(i)$ in (2.3) by $\psi_0(i)p^i$ and $\phi_0(i)(1-p)p^i$, the queue is quasi-reversible if and only if its departure function is

$$p^{\mathrm{D}}(n,i) = \frac{\psi_0(n-i)}{\phi_0(n)}, \qquad i = 0, 1, \ldots, n, \tag{2.14}$$

for a nonnegative function ψ_0, with

$$\phi_0(n) = \sum_{i=0}^{n} \psi_0(i), \qquad n = 0, 1, \ldots.$$

Clearly, any given ψ_0 gives rise to a departure function and a quasi-reversible queue. In what follows we consider several choices of ψ_0 and derive the departure function and the stationary distribution of the queue.

First we consider the case where ψ_0 is a constant, i.e.,

$$\psi_0(i) \equiv c, \qquad i = 0, 1, \ldots,$$

for some number $c > 0$. Since ψ is a distribution and

$$\sum_{i=0}^{\infty} \psi(i) = \sum_{i=0}^{\infty} \psi_0(i)p^i = \frac{c}{1-p},$$

it follows that $c = 1 - p$, and

$$\phi_0(n) = (n+1)(1-p).$$

Its departure function is

$$p^{\mathrm{D}}(n,i) = \frac{\psi_0(n-i)}{\phi_0(n)} = \frac{1}{n+1}, \qquad i = 0, 1, \ldots, n.$$

Thus in this case the departure probability is uniform, i.e., the probability that i customers depart is the same for all i. The stationary distributions of the queue at the slot boundaries and during the slot insides are

$$\pi^{\mathrm{X}}(n) = \phi(n) = (n+1)(1-p)^2 p^n, \qquad n = 0, 1, \ldots,$$
$$\pi^{\mathrm{Y}}(n) = \psi(n) = (1-p)p^n, \qquad n = 0, 1, \ldots.$$

We next consider the case where ψ_0 is geometric, i.e.,

$$\psi_0(i) = cq^{-i}, \qquad i = 0, 1, \ldots,$$

for some constant c. Since

$$1 = \sum_{i=0}^{\infty} \psi(i) = \sum_{i=0}^{\infty} \psi_0(i) p^i = c\frac{1}{1 - p/q},$$

we must have $c = 1 - p/q$ and the queue is stable if and only if $q > p$. Thus

$$\begin{aligned}
\psi_0(i) &= (1 - p/q)q^{-i}, && i = 0, 1, \ldots, \\
\phi_0(n) &= \sum_{i=0}^{n} \psi_0(i) \\
&= \left(1 - \frac{p}{q}\right) q^{-n} \frac{1 - q^{n+1}}{1 - q}, && n = 0, 1, \ldots.
\end{aligned}$$

The departure function is

$$p^{\mathrm{D}}(n, i) = \frac{\psi_0(i)}{\phi_0(i)} = \frac{1-q}{1-q^{n+1}} q^i,$$

i.e., the service batch size has a truncated geometric distribution. The stationary distributions of the queue at the slot boundaries and during the slot insides are

$$\pi^{\mathrm{X}}(n) = \phi(n) = \left(1 - \frac{p}{q}\right) \frac{1-p}{1-q} \left(\frac{p}{q}\right)^n (1 - q^{n+1}), \quad n = 0, 1, \ldots,$$
$$\pi^{\mathrm{Y}}(n) = \psi(n) = \left(1 - \frac{p}{q}\right) \left(\frac{p}{q}\right)^n, \qquad n = 0, 1, \ldots.$$

We finally consider the case where $\psi_0(i)$ is chosen so that $\phi_0(i)$ is geometric. To that end we set $\psi_0(0) = c_0$ and $\psi_0(i) = c_1 q^{-i}$ for $i \geq 1$. Since

$$\begin{aligned}
\phi_0(n) &= \sum_{i=0}^{n} \psi_0(i) \\
&= c_0 + c_1 \frac{1 - q^n}{1 - q} q^{-n}, \qquad n \geq 1,
\end{aligned}$$

it follows that if we set

$$c_1 = (1-q)c_0,$$

then

$$\psi_0(n) = c_0(1-q)q^{-n}, \qquad n \geq 1,$$
$$\phi_0(n) = c_0 q^{-n}, \qquad n \geq 0.$$

This implies that the departure function is

$$p^{\mathrm{D}}(n,i) = (1-q)q^i, \qquad i = 0, 1, \ldots, n-1, n \geq 1,$$
$$p^{\mathrm{D}}(n,n) = q^n, \qquad n \geq 0.$$

That is, the requested departure batch size is geometrically distributed. However, if the requested batch size is more than the number of customers in the system, then the node is emptied out (see Example 12.3).

Since $\phi(n) = \phi_0(n)(1-p)p^n$ is the stationary distribution of the queue at the slot boundaries, the constant c_0 is

$$c_0 = \frac{1}{1-p}\left(1-\frac{p}{q}\right).$$

Thus the stationary distributions of the queue lengths at the slot boundaries and during the slot insides are

$$\pi^{\mathrm{X}}(n) = \left(1-\frac{p}{q}\right)\left(\frac{p}{q}\right)^n, \qquad n \geq 0,$$

$$\pi^{\mathrm{Y}}(n) = \begin{cases} \dfrac{1-p/q}{1-p}, & n = 0, \\ \dfrac{1-q}{1-p}\left(1-\dfrac{p}{q}\right)\left(\dfrac{p}{q}\right)^n, & n \geq 1. \end{cases}$$

□

Example 12.12 (Poisson batch arrivals) Assume that the arrival batch size distribution $\{a(i)\}$ is Poisson with mean λ, i.e.,

$$a(i) = e^{-\lambda}\frac{\lambda^i}{i!}, \qquad i = 0, 1, \ldots.$$

Let $\{\gamma(i); i \geq 0\}$ be a positively valued function with $\gamma(0) = 1$, and assume that the departure function is

$$p^{\mathrm{D}}(n,i) = \frac{c(n)}{i!}\gamma(n)\gamma(n-1)\cdots\gamma(n-i+1), \qquad i = 0, 1, \ldots, n, \quad (2.15)$$

where $c(n)$ is the normalization constant determined by

$$\sum_{i=0}^{n} p^{\mathrm{D}}(n,i) = 1 \ .$$

The queue is quasi-reversible because if ψ and ϕ are defined by

$$\psi(i) = \frac{\lambda^i}{\gamma(i)\gamma(i-1)\cdots\gamma(0)}, \qquad i \geq 0, \qquad (2.16)$$

$$\phi(i) = \frac{\lambda^i e^{-\lambda}}{c(i)\gamma(i)\gamma(i-1)\cdots\gamma(0)}, \qquad i \geq 0,$$

then (2.3) is satisfied. Actually, if ψ is an arbitrary positively valued function and the arrival batch size has a Poisson distribution, then the form of departure function (2.15) is equivalent to (2.3). This is because once ψ is given, γ is uniquely determined by ψ from (2.16).

From Theorem 12.9, the stationary distributions of the queue at the slot boundaries and during the slot insides are

$$\pi^{\mathrm{X}}(n) = K\frac{1}{\gamma(0)\cdots\gamma(n)}\frac{e^{-\lambda}\lambda^n}{c(n)}, \qquad n = 0, 1, \ldots, \qquad (2.17)$$

$$\pi^{\mathrm{Y}}(n) = K\frac{\lambda^n}{\gamma(0)\cdots\gamma(n)}, \qquad n = 0, 1, \ldots, \qquad (2.18)$$

where λ is the mean arrival batch size, and K is a normalization constant. A queue operating according to this departure rule is referred to as an *S-queue* (synchronized queue) by Walrand (1983*a*).

If

$$\gamma(i) = \frac{p}{1-p}i,$$

then $c(n) = (1-p)^n$. It follows from (2.15) that the departure function has the binomial distribution

$$p^{\mathrm{D}}(n,i) = \binom{n}{i} p^i (1-p)^{n-i}, \qquad i = 0, 1, \ldots, n.$$

In this case the stationary distributions of the queue are

$$\pi^{X}(n) = \phi(n) = e^{-\frac{\lambda}{p}} \frac{\left(\frac{\lambda}{p}\right)^n}{n!}, \qquad n = 0, 1, 2, \ldots,$$

$$\pi^{Y}(n) = \psi(n) = e^{-\frac{1-p}{p}\lambda} \frac{\left(\frac{1-p}{p}\lambda\right)^n}{n!}, \qquad n = 0, 1, 2, \ldots.$$

The stationary distributions of the numbers of customers at the slot boundaries and during the slot insides are therefore Poisson, with respective means λ/p and $(1-p)\lambda/p$. The interpretation of the service function is that each customer is served in a time slot, independently of the others, with probability p. So this service function corresponds to a queue with infinitely many servers and a geometric service time distribution.

Another special case is $\gamma(i) = \mu$, which implies that

$$p^{D}(n, i) = c(n)\frac{\mu^i}{i!}, \qquad i = 0, 1, \ldots, n.$$

Thus we have the truncated Poisson distribution. In this case we obtain from (2.17) and (2.18) that

$$\pi^{X}(n) = \left(1 - \frac{\lambda}{\mu}\right) e^{-\lambda} \sum_{i=0}^{n} \left(\frac{\lambda}{\mu}\right)^{n-i} \frac{\lambda^i}{i!}, \qquad n = 0, 1, \ldots,$$

$$\pi^{Y}(n) = \left(1 - \frac{\lambda}{\mu}\right)\left(\frac{\lambda}{\mu}\right)^n, \qquad n = 0, 1, \ldots. \qquad \square$$

Remark 12.13 Theorem 12.9 can be used in several different ways. In the three examples above we start out with the arrival batch size distribution, and study the various departure functions by varying ψ. In applications we may be given a queueing problem with arrival function $a(i)$ and departure function $p^{D}(n, i)$. We then can use Theorem 12.9 to check whether the queue is quasi-reversible. Another alternative is that, given the departure function $p^{D}(n, i)$, we can use Theorem 12.9 to find the arrival batch size distributions so that the queue is quasi-reversible. This can be achieved by identifying functions ψ such that

$$\frac{p^{D}(n, i)\phi(n)}{\psi(n-i)} = a(i)$$

is independent of n.

Remark 12.14 Note that all the departure functions in Examples 12.10, 12.11 and 12.12 are chosen independently of the arrival batch size distributions. This is important in applications since the service mechanism can be set up independently of the arrival mechanism. There are, of course, quasi-reversible queues where the departure functions do depend on the arrival batch size distributions.

12.3 NETWORKS OF DISCRETE TIME QUASI-REVERSIBLE QUEUES

We now connect discrete time quasi-reversible queues into a network. As in the single node case, departures from each node occur at the beginning of the time slot. These departures follow a routing mechanism, and arrive at the destination nodes at the end of the time slot.

By Lemma 12.8, if the arrivals at an isolated quasi-reversible queue follow a batch Bernoulli process, then the departures from the queue also follow a batch Bernoulli process, and the departure batch sizes have the same distribution as the arrival batch sizes. Furthermore, the past departure batches are independent of the current state of the queue. This implies that if queues of a similar type are connected with one another in tandem, then the arrivals at each queue are batch Bernoulli, and the states of the various queues are independent of each other. Actually, the last node in the tandem network may be an arbitrary queue. Thus we obtain the following result.

Theorem 12.15 *If we connect discrete time quasi-reversible queues with the same arrival batch size distribution in tandem, e.g., if all the nodes are of one of the types described in Examples 12.10, 12.11, or 12.12, then the stationary distribution of the tandem network is product form. The result still holds when the last node is an arbitrary queue.*

Example 12.16 (A discrete time tandem queue) Consider a tandem network with customers arriving at the first node according to a geometric batch Bernoulli process, i.e.,

$$a(i) = (1-p)p^i, \qquad i = 0, 1, \ldots,$$

and each node is quasi-reversible with respect to the geometric batch Bernoulli arrival process. The stationary distribution of the network at the slot boundaries is

$$\pi^{\mathrm{X}}(\boldsymbol{n}) = \prod_{j=1}^{N} \pi_j^{\mathrm{X}}(n_j),$$

where π_j is determined by the departure function p_j^{D} of node j. For instance, if the service batch size is uniform, then it follows from Example 12.10 that

$$\pi_j^{\mathrm{X}}(n_j) = (n_j + 1)(1 - p)^2 p^{n_j}, \qquad n_j \geq 0;$$

if the service batch size has a truncated geometric distribution with parameter $q_j > p$, then

$$\pi_j^{\mathrm{X}}(n_j) = \left(1 - \frac{p}{q_j}\right) \frac{1 - p}{1 - q_j} \left(\frac{p}{q_j}\right)^{n_j} \left(1 - q_j^{n_j+1}\right), \qquad n_j \geq 0;$$

and if node j has a geometric service batch size that empties the node if the requested batch size is greater than the number of customers present, then

$$\pi_j^{\mathrm{X}}(n_j) = \left(1 - \frac{p}{q_j}\right) \left(\frac{p}{q_j}\right)^{n_j}, \qquad n_j \geq 0. \qquad \square$$

However, we cannot mix different types of queues in the tandem network and obtain a product form distribution. This is because the quasi-reversibility of a queue depends on the arrival batch size distribution, and connecting different types of quasi-reversible nodes in tandem may cause some nodes to lose their quasi-reversibility. Furthermore, even if all the nodes are of the same type, we cannot extend the product form results to an arbitrary network topology, since the splitting and merging of batches may change the batch size distributions (see Exercise 12.8).

Thus, in general, we cannot expect networks of discrete time quasi-reversible queues to have product form distributions in the same way as in the continuous time framework. However, there is one exception, and that is the case when the arrival batch sizes are Poisson, i.e., S-queues. Since the Poisson distribution is preserved under splitting and merging, the product form result can be extended to arbitrary network topologies.

Consider a discrete time network with N nodes, numbered $1, 2, \ldots, N$. Node j is quasi-reversible with respect to Poisson batch Bernoulli arrivals. At the beginning of each time slot departures take place at each node. Each customer departing from node j proceeds independently to node k with probability r_{jk}, where r_{j0} is the probability that the departing customer leaves the network. Customers arrive at their destination at the end of the time slot. Customers arrive at node j from the outside according to a Poisson batch Bernoulli process with mean batch size λ_j, and they arrive at the end of the time slot. Let

$$\boldsymbol{n} = (n_1, \ldots, n_N)$$

be the state of the network, where n_j is the number of customers at node j at a slot boundary, and let $\mathcal{S}$ be the state space of the network. Let $\{P_\lambda(n)\}$ denote the Poisson distribution with mean λ, i.e.,

$$P_\lambda(n) = e^{-\lambda}\frac{\lambda^n}{n!}, \qquad n = 0, 1, \ldots.$$

Since batches from a node split according to a multinomial distribution, the network has transition probability

$$p^{\mathrm{X}}(\boldsymbol{n}, \boldsymbol{n}') = \sum_{\substack{n_i - a_i + |\boldsymbol{b}_i| = n_i' \\ i=1,2,\ldots,N}} \sum_{\substack{|\boldsymbol{b}_\ell| = a_\ell \le n_\ell \\ \ell=1,2,\ldots,N}} \prod_{k=1}^{N} P_{\lambda_k}(b_{0k}) \prod_{j=1}^{N} p_j^{\mathrm{D}}(n_j, a_j) c_j(a_j, \boldsymbol{b}_j),$$

where $\boldsymbol{b}_j = (b_{j0}, b_{j1}, \ldots, b_{jN})$, $|\boldsymbol{b}_j| = b_{j0} + b_{j1} + \cdots + b_{jN}$, and

$$c_j(a_j, \boldsymbol{b}_j) = \frac{a_j!}{b_{j0}! b_{j1}! \cdots b_{jN}!} \prod_{k=0}^{N} r_{jk}^{b_{jk}}.$$

Let

$$r_{0k} = \frac{\lambda_k}{\sum_{j=1}^{N} \lambda_j},$$

and assume that the routing matrix $\{r_{jk};\ j, k = 0, 1, \ldots, N\}$ is irreducible. The traffic equations

$$\alpha_j = \lambda_j + \sum_{k=1}^{N} \alpha_k r_{kj}, \qquad j = 1, \ldots, N, \tag{3.1}$$

have a unique positive solution α_j, $j = 1, \ldots, N$. The following result is a discrete time counterpart of Theorem 4.3.

Theorem 12.17 *If each node of the network is quasi-reversible with respect to Poisson batch Bernoulli arrivals, then the state of the network at the slot boundaries has the product form stationary distribution*

$$\pi^{\mathrm{X}}(\boldsymbol{n}) = \prod_{j=1}^{N} \pi_j^{\mathrm{X}}(n_j), \qquad \boldsymbol{n} \in \mathcal{S}, \tag{3.2}$$

where π_j^{X} is the stationary distribution at the slot boundaries of node j when it is in isolation and subject to Poisson batch Bernoulli arrivals with mean batch size

α_j. Similarly, the stationary distribution of the network during the slot insides is the product of the stationary distribution of each node during the slot insides, i.e.,

$$\pi^{Y}(\boldsymbol{n}) = \prod_{j=1}^{N} \pi_j^{Y}(n_j), \qquad \boldsymbol{n} \in \mathcal{S}. \tag{3.3}$$

PROOF. We show that if the process starts at slot boundary 1 with an initial distribution (3.2), then its distribution at any slot boundary ℓ is also (3.2).

First note that if node j starts at slot boundary ℓ with distribution $\pi_j^{X}(n_j)$ and is subject to Poisson batch Bernoulli arrivals with mean α_j, then its distribution at slot boundary $\ell+1$ is also $\pi_j^{X}(n_j)$, the departure batch size at the end of slot ℓ has a Poisson distribution with mean α_j, and the state of the node at slot boundary $\ell+1$ is independent of the departure batches from node j prior to slot boundary $\ell+1$.

We start the network process at slot boundary 1 with distribution $\pi^{X}(\boldsymbol{n})$. Each node has a departure batch at the beginning of the first time slot, and this batch is transformed into arrivals at the end of the first time slot. As we stated before, the departure batch size from node k ($k = 0, 1, \ldots, N$) has a Poisson distribution with mean α_k, and each departure joins node j with probability r_{kj}. If D_k is the number of departures from node k, then the number of arrivals at node j at the end of time slot 1 is

$$A_j = A_{0j} + \sum_{k=1}^{N} \sum_{\ell=1}^{D_k} 1_{kj}(\ell),$$

where $1_{jk}(\ell)$, $\ell = 1, 2, \ldots$, is a sequence of i.i.d. random variables taking value 1 with probability r_{kj} and 0 with probability $1-r_{kj}$, A_{0j} is a Poisson random variable with mean λ_j, and D_k is a Poisson random variable with mean α_k. All these random variables are independent of each other. It is left as an exercise for the reader to show that $A_1, A_2, \ldots, A_N$ are independent Poisson random variables with means $\alpha_1, \alpha_2, \ldots, \alpha_N$. Since the state of the network at slot boundary 2 is independent of the departures at slot boundary 1, it is also independent of $A_1, A_2, \ldots, A_N$. Therefore, the situation at slot boundary 2 is identical to that at slot boundary 1. This argument can be continued to show that the network has distribution $\pi^{X}(\boldsymbol{n})$ at any slot boundary.

A similar argument shows that the stationary distribution during the slot insides is given by (3.3). □

This theorem will be proved in a different way in the next section.

Remark 12.18 If the function ψ of (2.3) is positively valued, then each node in Theorem 12.17 is the S-queue of Example 12.12. Hence, as noted in Example 12.13, the departure functions can be chosen independently of the exogenous arrival batch size distributions using (2.15). This important feature is in general not true for the networks in the next section.

12.4 DISCRETE TIME LINEAR NETWORKS WITH CONCURRENT MOVEMENTS

We now consider arbitrarily distributed arrival batch sizes within the framework of linear traffic networks. The model in this section is essentially identical to the linear traffic network discussed in Chapter 9, but we reformulate it in the discrete time setting.

Consider a discrete time network with N nodes, numbered $1, \ldots, N$, and a single class of customers. At the beginning of each time slot ℓ, batches of customers are simultaneously released from the nodes as well as from the outside. Denote the departure vector $\bar{\boldsymbol{D}}_\ell$ by

$$\bar{\boldsymbol{a}} = (a_0, a_1, \ldots, a_N),$$

where a_j is the departing batch size from node j, and $j = 0$ represents the outside. This $\bar{\boldsymbol{a}}$ is called the *departure batch vector*. Note that a_0 is the total number of customers arriving at the network from the outside. As in Chapter 9, we attach an overbar to a vector to indicate that it includes node 0, i.e., the outside. We assume that these batches are transformed into arrival batches $\bar{\boldsymbol{A}}_\ell$, denoted by $\bar{\boldsymbol{a}}'$, at the end of the time slot with probability $r(\bar{\boldsymbol{a}}, \bar{\boldsymbol{a}}')$, where

$$\bar{\boldsymbol{a}}' = (a_0', a_1', \ldots, a_N'),$$

and a_0' represents the total number of customers leaving the network (arriving at node 0). This $\bar{\boldsymbol{a}}'$ is called the *arrival batch vector*, and the r is called the *routing function* or the *routing probability*. For linear networks we assume that the total number of customers remains unchanged with the transfer, i.e., $|\bar{\boldsymbol{a}}| = |\bar{\boldsymbol{a}}'|$, where $|\bar{\boldsymbol{a}}| = \sum_{i=0}^N a_i$. Let $\mathcal{A}$ be the set of possible departure batch vectors and arrival batch vectors, which is a subset of $\{0, 1, 2, \ldots\}^{N+1}$.

For instance, the independent routing mechanism described in the previous section can be obtained by setting

$$r(\bar{\boldsymbol{a}}, \bar{\boldsymbol{a}}') = \sum_{B\bar{\mathbf{1}}^t = \bar{\boldsymbol{a}}} \sum_{\bar{\mathbf{1}}B = \bar{\boldsymbol{a}}'} \prod_{j=0}^N a_j! \prod_{k=0}^N \frac{r_{jk}^{b_{jk}}}{b_{jk}!}, \tag{4.1}$$

for $|\bar{\boldsymbol{a}}| = |\bar{\boldsymbol{a}}'|$, and $\bar{\mathbf{0}}$, $r(\bar{\mathbf{0}}, \bar{\mathbf{0}}) = 1$, where B is the matrix $\{b_{ij}\}$, $\bar{\mathbf{1}}$ is the $(N+1)$-dimensional row vector $(1, 1, \ldots, 1)$, and $\bar{\mathbf{1}}^t$ is its transpose. Note that in this model there is a positive probability that no customer is released.

Consider r as the transition probability of a Markov chain with state space $\mathcal{A}$. Since $r(\bar{\boldsymbol{a}}, \bar{\boldsymbol{a}}') > 0$ only if $|\bar{\boldsymbol{a}}| = |\bar{\boldsymbol{a}}'|$, the state space $\mathcal{A}$ can be decomposed into disjoint closed sets with finite numbers of elements, where a subset of the state space is called *closed* if there are no transitions going out of it. Assume that r is irreducible on each closed set. Since each closed set is finite, r admits a stationary measure on each closed set, and it has a stationary measure ν satisfying

$$\nu(\bar{\boldsymbol{a}}) = \sum_{\bar{\boldsymbol{a}}'} \nu(\bar{\boldsymbol{a}}')r(\bar{\boldsymbol{a}}', \bar{\boldsymbol{a}}). \tag{4.2}$$

Note that $\nu(\bar{\boldsymbol{a}})$ is not unique, since it can be arbitrarily weighted on each closed set. In particular, $\nu(\bar{\boldsymbol{0}})$ can be chosen arbitrarily. We assume that

$$\sum_{\bar{\boldsymbol{a}}} \nu(\bar{\boldsymbol{a}}) < \infty. \tag{4.3}$$

The network dynamics are specified as follows. If at slot boundary ℓ the state of the network $\boldsymbol{X}_\ell$ is $\boldsymbol{n}$, then there is a departure of $\bar{\boldsymbol{D}}_\ell = \bar{\boldsymbol{a}} \in \mathcal{A}$ at the beginning of the time slot with probability

$$P(\bar{\boldsymbol{D}}_\ell = \bar{\boldsymbol{a}} | \boldsymbol{X}_\ell = \boldsymbol{n}) = p^{\mathrm{D}}(\boldsymbol{n}, \bar{\boldsymbol{a}}) = \begin{cases} \dfrac{\Psi(\boldsymbol{n} - \boldsymbol{a})}{\Phi(\boldsymbol{n})}\nu(\bar{\boldsymbol{a}}), & \bar{\boldsymbol{a}} \neq \bar{\boldsymbol{0}}, \\ C(\boldsymbol{n}), & \bar{\boldsymbol{a}} = \bar{\boldsymbol{0}}, \end{cases} \tag{4.4}$$

where $\boldsymbol{a} = (a_1, \ldots, a_N)$ is the vector $\bar{\boldsymbol{a}}$ with the first element removed, C is an arbitrary function such that

$$0 \leq C(\boldsymbol{n}) < 1,$$

Ψ is an arbitrary nonnegative function satisfying

$$\sum_{\boldsymbol{n}} \Psi(\boldsymbol{n}) < \infty,$$

and Φ is a positive function determined by

$$\Phi(\boldsymbol{n})(1 - C(\boldsymbol{n})) = \sum_{\bar{\boldsymbol{a}} \neq \bar{\boldsymbol{0}}} \Psi(\boldsymbol{n} - \boldsymbol{a})\nu(\bar{\boldsymbol{a}}). \tag{4.5}$$

We refer to p^{D} of (4.4) again as the departure function. However, the reader should note that p^{D}, through the batch size a_0, determines the batch arrival process from the outside. At the end of the time slot, the departure batch vector $\bar{\boldsymbol{a}}$ is transformed into an arrival batch vector $\bar{\boldsymbol{a}}'$ with probability $r(\bar{\boldsymbol{a}}, \bar{\boldsymbol{a}}')$. The network with departure function p^{D} of (4.4) and routing function r is called a *linear concurrent movement network*.

Let $\boldsymbol{X}_\ell$ and $\boldsymbol{Y}_\ell$ be the states of the network at slot boundary ℓ and slot inside ℓ, respectively, and let $\mathcal{S}$ be the common state space. Then Theorem 12.17 can be generalized as follows.

Theorem 12.19 *Consider the linear concurrent movement network with routing function r and departure function p^{D} of (4.4). If $C(\boldsymbol{n}) < c_0$ for all $\boldsymbol{n}$ for some $c_0 < 1$, then the network states $\boldsymbol{X}_\ell$ and $\boldsymbol{Y}_\ell$ have the stationary distributions π^{X} and π^{Y} given by*

$$\pi^{\mathrm{X}}(\boldsymbol{n}) = c\Phi(\boldsymbol{n}), \qquad \boldsymbol{n} \in \mathcal{S},$$

$$\pi^{\mathrm{Y}}(\boldsymbol{n}) = c\left(\Psi(\boldsymbol{n})\sum_{\bar{\boldsymbol{a}}\neq\bar{\boldsymbol{0}}}\nu(\bar{\boldsymbol{a}}) + \Phi(\boldsymbol{n})C(\boldsymbol{n})\right), \qquad \boldsymbol{n} \in \mathcal{S},$$

where c is the normalization constant.

PROOF. We first note that $\{\boldsymbol{X}_\ell\}$ is a Markov chain with transition probability

$$p^{\mathrm{X}}(\boldsymbol{n}, \boldsymbol{n}') = \sum_{\boldsymbol{n}-\boldsymbol{a}+\boldsymbol{a}'=\boldsymbol{n}'} p^{\mathrm{D}}(\boldsymbol{n}, \bar{\boldsymbol{a}})r(\bar{\boldsymbol{a}}, \bar{\boldsymbol{a}}').$$

Since $\boldsymbol{n} - \boldsymbol{a} = \boldsymbol{n}' - \boldsymbol{a}'$, multiplying both sides of (4.2) by $\Psi(\boldsymbol{n} - \boldsymbol{a})$ yields

$$\Phi(\boldsymbol{n})\frac{\Psi(\boldsymbol{n}-\boldsymbol{a})}{\Phi(\boldsymbol{n})}\nu(\bar{\boldsymbol{a}}) = \sum_{\bar{\boldsymbol{a}}'}\Phi(\boldsymbol{n}')\frac{\Psi(\boldsymbol{n}'-\boldsymbol{a}')}{\Phi(\boldsymbol{n}')}\nu(\bar{\boldsymbol{a}}')r(\bar{\boldsymbol{a}}', \bar{\boldsymbol{a}}).$$

This equation can be rewritten in terms of p^{D} as

$$\Phi(\boldsymbol{n})p^{\mathrm{D}}(\boldsymbol{n}, \bar{\boldsymbol{a}}) = \sum_{\boldsymbol{n}-\boldsymbol{a}=\boldsymbol{n}'-\boldsymbol{a}'}\Phi(\boldsymbol{n}')p^{\mathrm{D}}(\boldsymbol{n}', \bar{\boldsymbol{a}}')r(\bar{\boldsymbol{a}}', \bar{\boldsymbol{a}}). \tag{4.6}$$

Note that (4.6) is clearly satisfied if $\bar{\boldsymbol{a}} = \bar{\boldsymbol{0}}$. Summing both sides of (4.6) over all $\bar{\boldsymbol{a}}$, we obtain

$$\Phi(\boldsymbol{n}) = \sum_{\boldsymbol{n}'}\Phi(\boldsymbol{n}')p^{\mathrm{X}}(\boldsymbol{n}', \boldsymbol{n}).$$

Hence, Φ is the stationary measure of p^{X}. Since the total mass of Ψ is finite, it follows from (4.5) that

$$\begin{aligned}(1 - c_0)\sum_{\boldsymbol{n}}\Phi(\boldsymbol{n}) &\leq \sum_{\boldsymbol{n}}\Phi(\boldsymbol{n})(1 - C(\boldsymbol{n}))\\ &= \sum_{\boldsymbol{n}}\sum_{\bar{\boldsymbol{a}}\neq\bar{\boldsymbol{0}}}\Psi(\boldsymbol{n}-\boldsymbol{a})\nu(\bar{\boldsymbol{a}})\\ &= \sum_{\bar{\boldsymbol{a}}\neq\bar{\boldsymbol{0}}}\sum_{\boldsymbol{n}\geq\boldsymbol{a}}\Psi(\boldsymbol{n}-\boldsymbol{a})\nu(\bar{\boldsymbol{a}})\\ &= \sum_{\boldsymbol{n}}\Psi(\boldsymbol{n})\sum_{\bar{\boldsymbol{a}}\neq\bar{\boldsymbol{0}}}\nu(\bar{\boldsymbol{a}}) < \infty.\end{aligned}$$

Thus, Φ can be normalized to the stationary distribution with a constant c. Similar to (1.4), the stationary distribution π^{Y} during the slot insides is

$$\begin{aligned}\pi^{\mathrm{Y}}(\boldsymbol{n}) &= \sum_{\bar{\boldsymbol{a}}} \pi^{\mathrm{X}}(\boldsymbol{n}+\boldsymbol{a})p^{\mathrm{D}}(\boldsymbol{n}+\boldsymbol{a},\bar{\boldsymbol{a}}) \\ &= c\left(\Psi(\boldsymbol{n})\sum_{\bar{\boldsymbol{a}}\neq\bar{\boldsymbol{0}}} \nu(\bar{\boldsymbol{a}}) + \Phi(\boldsymbol{n})C(\boldsymbol{n})\right).\end{aligned}$$

This completes the proof. □

Note that in Theorem 12.19 the choice of $C(\boldsymbol{n})$ does not affect the stationary distribution π^{X} at the slot boundaries, but it may affect the stationary distribution π^{Y} during the slot insides. For this reason, we will not specify $C(\boldsymbol{n})$ when only the stationary distribution π^{X} is of concern.

Remark 12.20 It is easy to see that the argument in Theorem 12.19 still holds if we always set the first elements of the batch vectors equal to 0. In this case there are no arrivals from and departures to the outside. Because the number of customers remains unchanged in the transfers, the resulting network is called closed. Indeed, there are no essential differences between open and closed linear networks.

Corollary 12.21 *If the network process $\{\boldsymbol{X}_\ell\}$ is stationary, then under the assumptions of Theorem 12.19,*

$$P(\bar{\boldsymbol{A}}_\ell = \bar{\boldsymbol{a}} | \boldsymbol{X}_{\ell+1} = \boldsymbol{n}) = p^{\mathrm{D}}(\boldsymbol{n}, \bar{\boldsymbol{a}}). \tag{4.7}$$

In particular, if the arrivals from the outside follow a Poisson batch Bernoulli process and future arrivals from the outside are independent of the current state of the network, then the departures to the outside are also Poisson batch Bernoulli and the past departures to the outside are independent of the current state of the network. That is, the network is quasi-reversible with respect to arrivals from the outside and departures to the outside.

PROOF. The departure function is given by (4.4). If $\bar{\boldsymbol{a}} \neq \bar{\boldsymbol{0}}$, then the first statement follows from

$$\begin{aligned}P(\bar{\boldsymbol{A}}_\ell = \bar{\boldsymbol{a}} | \boldsymbol{X}_{\ell+1} = \boldsymbol{n}) &= \frac{P(\bar{\boldsymbol{A}}_\ell = \bar{\boldsymbol{a}}, \boldsymbol{X}_{\ell+1} = \boldsymbol{n})}{P(\boldsymbol{X}_{\ell+1} = \boldsymbol{n})} \\ &= \frac{\sum_{\bar{\boldsymbol{a}}'} P(\boldsymbol{X}_\ell = \boldsymbol{n}+\boldsymbol{a}'-\boldsymbol{a}, \bar{\boldsymbol{D}}_\ell = \bar{\boldsymbol{a}}')}{c\Phi(\boldsymbol{n})}\end{aligned}$$

$$
\begin{aligned}
&= \frac{\sum_{\bar{a}'} \Psi(\boldsymbol{n}-\boldsymbol{a})\nu(\bar{\boldsymbol{a}}')r(\bar{\boldsymbol{a}}', \bar{\boldsymbol{a}})}{\Phi(\boldsymbol{n})} \\
&= \frac{\Psi(\boldsymbol{n}-\boldsymbol{a})\nu(\bar{\boldsymbol{a}})}{\Phi(\boldsymbol{n})} \\
&= p^{\mathrm{D}}(\boldsymbol{n}, \bar{\boldsymbol{a}});
\end{aligned}
$$

if $\bar{\boldsymbol{a}} = \bar{\boldsymbol{0}}$, then it follows from $r(\bar{\boldsymbol{0}}, \bar{\boldsymbol{0}}) = 1$ that

$$
\begin{aligned}
P(\bar{\boldsymbol{A}}_\ell = \bar{\boldsymbol{0}} | \boldsymbol{X}_{\ell+1} = \boldsymbol{n}) &= \frac{P(X_\ell = \boldsymbol{n}, \bar{\boldsymbol{D}}_\ell = \bar{\boldsymbol{0}})}{P(\boldsymbol{X}_{\ell+1} = \boldsymbol{n})} \\
&= \frac{P(\bar{\boldsymbol{D}}_\ell = \bar{\boldsymbol{0}} | X_\ell = \boldsymbol{n}) P(X_\ell = \boldsymbol{n})}{c\Phi(\boldsymbol{n})} \\
&= \frac{C(\boldsymbol{n}) c\Phi(\boldsymbol{n})}{c\Phi(\boldsymbol{n})} \\
&= p^{\mathrm{D}}(\boldsymbol{n}, \bar{\boldsymbol{0}}).
\end{aligned}
$$

If exogenous arrivals are Poisson batch Bernoulli and future arrivals are independent of the current state of the network, then it follows from (4.7) that departures to the outside, i.e., the first element of $\bar{\boldsymbol{A}}_\ell$, $\ell \geq 0$, are also Poisson batch Bernoulli and past departures are independent of the current state of the network. □

Equation (4.7) and the fact that under time reversal arrivals become departures and departures become arrivals imply that the departure function of the reversed process is exactly the same as the arrival function of the original process. This property turns out to be an important characterization of linear networks, and will be discussed in more detail in Section 12.8.

Remark 12.22 Assume that the network is stationary and

$$
C(\boldsymbol{n}) = c \frac{\Psi(\boldsymbol{n})}{\Phi(\boldsymbol{n})} \nu(\bar{\boldsymbol{0}}). \tag{4.8}
$$

The state of the network during the slot insides is

$$
\boldsymbol{Y}_\ell = \boldsymbol{X}_\ell - \boldsymbol{D}_\ell,
$$

where $\boldsymbol{D}_\ell$ is $\bar{\boldsymbol{D}}_\ell$ with the first element, i.e., the arrivals from outside, deleted. Thus we have

$$
\begin{aligned}
P(\bar{\boldsymbol{D}}_\ell = \bar{\boldsymbol{a}}, \boldsymbol{Y}_\ell = \boldsymbol{n}) &= P(\boldsymbol{X}_\ell = \boldsymbol{n} + \boldsymbol{a}, \bar{\boldsymbol{D}}_\ell = \bar{\boldsymbol{a}}) \\
&= P(\bar{\boldsymbol{D}}_\ell = \bar{\boldsymbol{a}} | \boldsymbol{X}_\ell = \boldsymbol{n} + \boldsymbol{a}) P(\boldsymbol{X}_\ell = \boldsymbol{n} + \boldsymbol{a}) \\
&= \Psi(\boldsymbol{n}) \nu(\bar{\boldsymbol{a}}).
\end{aligned}
$$

Consequently,

$$P(\bar{\boldsymbol{D}}_\ell = \bar{\boldsymbol{a}} | \boldsymbol{Y}_\ell = \boldsymbol{n}) = \frac{\nu(\bar{\boldsymbol{a}})}{\sum_{\bar{\boldsymbol{a}}} \nu(\bar{\boldsymbol{a}})}. \tag{4.9}$$

Similarly, it can be shown that

$$P(\boldsymbol{Y}_\ell = \boldsymbol{n} | \bar{\boldsymbol{D}}_\ell = \bar{\boldsymbol{a}}) = P(\boldsymbol{Y}_\ell = \boldsymbol{n}).$$

This appears similar to the quasi-reversibility condition of the single queue. However, $\bar{\boldsymbol{D}}_\ell, \ell = 0, 1, \ldots$, in general do not constitute a sequence of i.i.d. random vectors, i.e., a vector valued Bernoulli process. This is because in the batch transfer network the departure batches $\bar{\boldsymbol{D}}_\ell$ are transformed at the end of the time slot into the arrival batches $\bar{\boldsymbol{A}}_\ell$, which correlate with $\boldsymbol{X}_{\ell+1}$ and therefore also with $\bar{\boldsymbol{D}}_{\ell+1}$, $\ell \geq 0$.

We now show that Theorem 12.17 is a special case of Theorem 12.19.

AN ALTERNATIVE PROOF FOR THEOREM 12.17 The nodes in the network of Theorem 12.17 are assumed to be quasi-reversible, and the batch sizes have Poisson distributions. The departure function of the entire network is

$$p^{\mathrm{D}}(\boldsymbol{n}, \bar{\boldsymbol{a}}) = P_\lambda(a_0) \prod_{j=1}^{N} \frac{\psi_j(n_j - a_j)}{\phi_j(n_j)} P_{\alpha_j}(a_j),$$

where $\lambda = \sum_{j=1}^{N} \lambda_j$, ϕ_j and ψ_j are the functions for the departures from node j satisfying the quasi-reversibility conditions in Theorem 12.9, and $P_{\alpha_j}(\cdot)$ is the probability mass function of a Poisson distribution with mean α_j. Without loss of generality we can assume that ϕ_j and ψ_j are distributions. Let

$$\Psi(\boldsymbol{n}) = \prod_{j=1}^{N} \psi_j(n_j),$$

$$\Phi(\boldsymbol{n}) = \prod_{j=1}^{N} \phi_j(n_j),$$

$$\nu(\bar{\boldsymbol{a}}) = P_\lambda(a_0) \prod_{j=1}^{N} P_{\alpha_j}(a_j) = \prod_{j=0}^{N} e^{-\alpha_j} \frac{\alpha_j^{a_j}}{a_j!}, \tag{4.10}$$

where for convenience we denote λ by α_0. The departure function p^{D} takes the form (4.4). Therefore, by Theorem 12.19, the product form $\Phi(\boldsymbol{n})$ is the stationary distribution of the network if we can verify that ν is the stationary measure of the routing function r, i.e.,

$$\nu(\bar{\boldsymbol{a}}) = \sum_{\bar{\boldsymbol{a}}'} \nu(\bar{\boldsymbol{a}}') r(\bar{\boldsymbol{a}}', \bar{\boldsymbol{a}}), \tag{4.11}$$

where the routing function r is given by (4.1). Since

$$r_{0j} = \frac{\lambda_j}{\lambda}, \qquad j = 1, \ldots, N,$$

the traffic equations (3.1) can be written as

$$\alpha_j = \sum_{k=0}^{N} \alpha_k r_{kj}, \qquad j = 1, \ldots, N,$$

where $r_{00} = 0$. From the binomial formula it follows that

$$a_k! \sum_{\sum_{i=0}^{N} b_{ik} = a_k} \prod_{j=0}^{N} \frac{(\alpha_j r_{jk})^{b_{jk}}}{b_{jk}!} = \left(\sum_{j=0}^{N} \alpha_j r_{jk} \right)^{a_k} = \alpha_k^{a_k}.$$

Using this identity, we obtain

$$\begin{aligned}
\sum_{\bar{\boldsymbol{a}}'} \nu(\bar{\boldsymbol{a}}') r(\bar{\boldsymbol{a}}', \boldsymbol{a}) &= \sum_{\bar{\boldsymbol{a}}'} \prod_{i=0}^{N} e^{-\alpha_i} \frac{\alpha_i^{a_i'}}{a_i'!} \sum_{B\bar{\mathbf{1}}' = \bar{\boldsymbol{a}}'} \sum_{\bar{\mathbf{1}} B = \bar{\boldsymbol{a}}} \prod_{j=0}^{N} a_j'! \prod_{k=0}^{N} \frac{r_{jk}^{b_{jk}}}{b_{jk}!} \\
&= \sum_{\bar{\boldsymbol{a}}'} \prod_{i=0}^{N} e^{-\alpha_i} \alpha_i^{a_i'} \sum_{B\bar{\mathbf{1}}' = \bar{\boldsymbol{a}}'} \prod_{k=0}^{N} \sum_{\bar{\mathbf{1}} B = \bar{\boldsymbol{a}}} \prod_{j=0}^{N} \frac{r_{jk}^{b_{jk}}}{b_{jk}!} \\
&= \prod_{i=0}^{N} e^{-\alpha_i} \sum_{\bar{\boldsymbol{a}}'} \sum_{B\bar{\mathbf{1}}' = \bar{\boldsymbol{a}}'} \prod_{m=0}^{N} \alpha_i^{b_{im}} \prod_{k=0}^{N} \sum_{\bar{\mathbf{1}} B = \bar{\boldsymbol{a}}} \prod_{j=0}^{N} \frac{r_{jk}^{b_{jk}}}{b_{jk}!} \\
&= \prod_{i=0}^{N} e^{-\alpha_i} \prod_{k=0}^{N} \sum_{\bar{\mathbf{1}} B = \bar{\boldsymbol{a}}} \prod_{j=0}^{N} \frac{(\alpha_j r_{jk})^{b_{jk}}}{b_{jk}!} \\
&= \prod_{i=0}^{N} e^{-\alpha_i} \prod_{k=0}^{N} \frac{\alpha_k^{a_k}}{a_k!} \\
&= \nu(\bar{\boldsymbol{a}}).
\end{aligned}$$

Thus it follows from Theorem 12.19 that the stationary distributions of $\boldsymbol{X}_\ell$ and $\boldsymbol{Y}_\ell$ are given by Φ and Ψ, respectively. This proves Theorem 12.17. □

There is an intuitive argument for proving (4.11). If a single customer circulates in a network with nodes $\{0, 1, \ldots, N\}$ according to routing probabilities r_{jk}, $j, k = 0, 1, \ldots, N$, spending one time unit at a node during a visit, then the steady state probability of the customer residing at node j is γ_j, determined by

$$\gamma_j = r_{0j} + \sum_{k=1}^{N} \gamma_k r_{kj}, \qquad j = 1, \ldots, N. \tag{4.12}$$

Now assume that there are $\sum_{j=0}^{N} a_j$ customers circulating in the network and they all circulate independently of each other. Each customer moves according to routing probabilities r_{jk}. Since each customer has probability γ_j of residing at node j, the probability that there are a_j customers at node j, $j = 0, 1, \ldots, N$, is

$$(a_0 + a_1 + \cdots + a_N)! \prod_{j=0}^{N} \frac{(\gamma_j)^{a_j}}{a_j!}. \tag{4.13}$$

Since (4.13) is the distribution of the customer population, it satisfies (4.11). Furthermore, because γ_j is proportional to α_j, we obtain ν of (4.10) as the solution to equation (4.11).

Linear networks with concurrent movements constitute a much larger class of networks than the class of networks with Poisson batch Bernoulli arrivals and independent routing. We conclude this section with two examples that illustrate Theorem 12.19. The first example extends Theorem 12.17 by allowing state-dependent service probabilities.

Example 12.23 (A network with state-dependent service rates) Consider a discrete time network with N nodes. When the network is in state $\boldsymbol{n}$, the departure function is

$$p^{\mathrm{D}}(\boldsymbol{n}, \bar{\boldsymbol{a}}) = K \frac{\Psi(\boldsymbol{n} - \boldsymbol{a})}{\Phi(\boldsymbol{n})} \frac{1}{\prod_{j=0}^{N} a_j!},$$

where K is some constant. Clearly, a special choice of Φ and Ψ yields the network of Theorem 12.17. A customer leaving a node selects its route independently of the other customers, and each customer leaving node j joins node k with probability r_{jk}, $j, k = 0, 1, \ldots, N$. Let α_j be the solution of

$$\alpha_j = \sum_{k=0}^{N} \alpha_k r_{kj}, \qquad j = 1, \ldots, N.$$

We assume $\alpha_0 = 1$.

We have already shown that the solution $\nu(\bar{\boldsymbol{a}})$ of the traffic equations for networks with independent routing is given by (4.10). Since

$$p^{\mathrm{D}}(\boldsymbol{n}, \bar{\boldsymbol{a}}) = K \frac{\Psi(\boldsymbol{n}-\boldsymbol{a})}{\Phi(\boldsymbol{n})} \frac{1}{\prod_{j=0}^{N} a_j!}$$

$$= \left[\frac{K \prod_{k=0}^{N} e^{\alpha_k} \left(\Psi(\boldsymbol{n}-\boldsymbol{a}) \prod_{j=1}^{N} \alpha_j^{n_j - a_j} \right)}{\Phi(\boldsymbol{n}) \prod_{j=1}^{N} \alpha_j^{n_j}} \right] \nu(\bar{\boldsymbol{a}}),$$

the departure function can be written in the form (4.4) if we regard the numerator and denominator in the square brackets above as $\Phi(\boldsymbol{n}-\boldsymbol{a})$ and $\Psi(\boldsymbol{n})$. Applying Theorem 12.19 we obtain the stationary distribution of the network

$$\pi^{\mathrm{X}}(\boldsymbol{n}) = c\Phi(\boldsymbol{n}) \prod_{j=1}^{N} \alpha_j^{n_j},$$

where c is the normalization constant. □

Example 12.24 (A discrete time Jackson network) Consider a linear concurrent movement network with at most one customer making a move in each time slot. In this case the routing function $r(\bar{\boldsymbol{a}}, \bar{\boldsymbol{a}}')$ is positive only if $\bar{\boldsymbol{a}}, \bar{\boldsymbol{a}}'$ are either the unit vectors $\bar{\mathbf{e}}_j, \bar{\mathbf{e}}_k$ or the null vector $\bar{\mathbf{0}}$. Denote $r(\bar{\mathbf{e}}_k, \bar{\mathbf{e}}_j)$ by r_{jk} for $j, k = 0, 1, \ldots, N$, where $r_{00} = 0$. It is assumed that $r(\bar{\mathbf{0}}, \bar{\mathbf{0}}) = 1$ (note that $\bar{\mathbf{e}}_0$ is different from $\bar{\mathbf{0}}$). Hence, the traffic equations (4.2) reduce to

$$\nu(\bar{\mathbf{e}}_j) = \sum_{k=0}^{N} \nu(\bar{\mathbf{e}}_k) r_{kj}, \qquad j = 1, \ldots, N, \tag{4.14}$$

where $\nu(\bar{\mathbf{0}})$ may be chosen arbitrarily. These traffic equations are exactly the same as those of the Jackson network in continuous time. Let $\alpha_j = \nu(\bar{\mathbf{e}}_j)$. It follows from (4.4) that the departure function p^{D} takes the form

$$p^{\mathrm{D}}(\boldsymbol{n}, \bar{\boldsymbol{a}}) = \begin{cases} \dfrac{\Psi(\boldsymbol{n}-\mathbf{e}_j)}{\Phi(\boldsymbol{n})} \nu(\bar{\mathbf{e}}_j), & \bar{\boldsymbol{a}} = \bar{\mathbf{e}}_j,\ j = 0, 1, \ldots, N, \\ C(\boldsymbol{n}), & \bar{\boldsymbol{a}} = \bar{\mathbf{0}}, \end{cases} \tag{4.15}$$

where $C(\boldsymbol{n})$ is arbitrary and is determined later. Clearly, (4.3) is satisfied. Hence, if the stationary distribution π^{X} exists, then by Theorem 12.19 π^{X} is proportional to $\Phi(\boldsymbol{n})$. In particular, set

$$\Psi(\boldsymbol{n}) = \Phi(\boldsymbol{n}) = \prod_{j=1}^{N} \left(\frac{\alpha_j}{q_j} \right)^{n_j},$$

where q_j are positive numbers satisfying $q_0 = \alpha_0$ and

$$\sum_{j=0}^{N} q_j = 1.$$

If we choose C such that

$$C(\boldsymbol{n}) = \sum_{j=1}^{N} q_j 1[n_j = 0],$$

then, when the state of the network is $\boldsymbol{n} = (n_1, \ldots, n_N)$, node j completes the service of a customer at the beginning of a time slot with probability q_j, provided there are customers present, and a customer arrives from the outside with probability q_0. It follows from Theorem 12.19 that π^{X} exists only if $\alpha_j < q_j$ for $j = 1, 2, \ldots, N$, and is given by

$$\pi^{\mathrm{X}}(\boldsymbol{n}) = \prod_{j=1}^{N} \left(1 - \frac{\alpha_j}{q_j}\right) \left(\frac{\alpha_j}{q_j}\right)^{n_j}.$$

We refer to this model as *a discrete time Jackson network.*

In general, (4.15) allows much more general departure functions. For instance, if, for a positive vector $\boldsymbol{M} = (M_1, M_2, \ldots, M_N)$,

$$\Psi(\boldsymbol{n}) = 1[\boldsymbol{n} \le \boldsymbol{M}] \prod_{j=1}^{N} \left(\frac{\alpha_j}{q_j}\right)^{n_j},$$

and

$$C(\boldsymbol{n}) = \sum_{j=1}^{N} q_j 1[n_j = 0] + \sum_{j=1}^{N} 1[n_j = M_j + 1] \sum_{k=0}^{N} q_k 1[k \ne j],$$

then the departure function is

$$p^{\mathrm{D}}(\boldsymbol{n}, \bar{\mathbf{e}}_j) = \begin{cases} q_j, & \boldsymbol{n} \le \boldsymbol{M},\ j = 0, 1, \ldots, N, \\ q_j, & n_j = M_j + 1,\ \boldsymbol{n} \le \boldsymbol{M} + \mathbf{e}_j,\ j = 1, 2, \ldots, N, \\ 0, & \text{otherwise.} \end{cases}$$

This describes a network with a finite buffer at each node and an additional common buffer. The common buffer can be used by any node when its own buffer is full. However, when the common buffer is used, all exogenous arrivals are blocked, and only a customer at the node that has a customer in the common buffer is served. Other service protocols can also be obtained by varying $\Psi(\boldsymbol{n})$, $\Psi(\boldsymbol{n})$, and $C(\boldsymbol{n})$. □

12.5 A DISCRETE TIME NONLINEAR NETWORK WITH CONCURRENT MOVEMENTS

The linear networks with concurrent batch movements studied in the previous sections have a restricted form of departure function (4.4). For instance, it does not include the departure function of Example 12.3. In this and the subsequent sections we consider the latter form of departure functions for concurrent batch movements, which correspond to a class of nonlinear networks.

12.5.1 Model Description

Consider a network with N nodes, numbered $1, 2, \dots, N$. For convenience, the set of nodes is denoted by $\mathcal{N}$, i.e.,

$$\mathcal{N} \equiv \{1, 2, \dots, N\}.$$

We only consider the state of the network at the slot boundaries, denoted by $\boldsymbol{n} = (n_1, n_2, \dots, n_N)$, where n_j is the customer population at node j, $j = 1, \dots, N$. The state space is denoted by $\mathcal{S}$.

Let $p^{\mathrm{T}}(\boldsymbol{a}, \boldsymbol{a}')$, $\boldsymbol{a}, \boldsymbol{a}' \in \mathcal{S}$, be a joint probability distribution, i.e.,

$$\sum_{\boldsymbol{a}, \boldsymbol{a}' \in \mathcal{S}} p^{\mathrm{T}}(\boldsymbol{a}, \boldsymbol{a}') = 1, \tag{5.1}$$

where $\boldsymbol{a} = (a_1, \dots, a_N)$ and $\boldsymbol{a}' = (a'_1, \dots, a'_N)$. The dynamics of the network are specified as follows. When the state of the network at a slot boundary is $\boldsymbol{n}$, the state of the network changes with probability $p^{\mathrm{T}}(\boldsymbol{a}, \boldsymbol{a}')$ for each pair $\boldsymbol{a}, \boldsymbol{a}' \in \mathcal{S}$.

(i) At the beginning of a time slot, a_j customers are requested to depart from node j, $j = 1, 2, \dots, N$. These N requests are independent of each other.

(ii) If every node has a sufficient number of customers, then a_j customers depart from node j at the beginning of the time slot, $j = 1, \dots, N$. These departures are transformed into arrivals at the end of the time slot and result in batch arrivals of size a'_j at node $j = 1, 2, \dots, N$.

(iii) If at some nodes there are insufficient customers present to satisfy the requests for departures, then these nodes are emptied out, and all departing customers, including those from nodes with sufficient numbers of customers, leave the network. In this case, there is no arrival at any node at the end of the time slot.

There are two major notational differences between this model and the one of the previous section. First, the batch vectors we use in this section do not include the component for node 0 (the outside), i.e., these vectors have no bars. Thus the total number $|\boldsymbol{a}|$ of departing customers may not be equal to the total number $|\boldsymbol{a}'|$ of arriving customers; the difference may be regarded as departures from (arrivals to) the outside of the network. Second, as seen from (5.1), p^{T} is not a conditional routing probability as defined for other networks. It is a joint probability function. For this reason we impose a superscript T to distinguish it from others. We refer to p^{T} as the *batch transfer function*. Clearly, if the state of the network at a slot boundary is $\boldsymbol{n}$, then with probability $p^{\mathrm{T}}(\boldsymbol{a}, \boldsymbol{a}')$ the state of the network at the next slot boundary is

$$\xi(\boldsymbol{n}, \boldsymbol{a}, \boldsymbol{a}') = \begin{cases} \boldsymbol{n} - \boldsymbol{a} + \boldsymbol{a}', & \boldsymbol{n} - \boldsymbol{a} \geq \boldsymbol{0}, \\ \boldsymbol{n} - \boldsymbol{a} \wedge \boldsymbol{n}, & \text{otherwise}, \end{cases}$$

where $\boldsymbol{a} \wedge \boldsymbol{n}$ is the vector whose jth element is the minimum of a_j and n_j. Let $\bar{\boldsymbol{a}} = (a_0, \boldsymbol{a})$ and $\bar{\boldsymbol{a}}' = (a_0', \boldsymbol{a}')$. Then, under a formulation similar to the one in Section 12.1 (see also Exercise 12.3), the dynamics of the present network can be specified by the transitions

$$p^{\mathrm{D}}(\boldsymbol{n}, \bar{\boldsymbol{a}}) r((\boldsymbol{n}, \bar{\boldsymbol{a}}), (\boldsymbol{n}', \bar{\boldsymbol{a}}')) = p^{\mathrm{T}}(\boldsymbol{a}, \boldsymbol{a}') 1[\boldsymbol{a} \leq \boldsymbol{n}],$$
$$p^{\mathrm{I}}(\boldsymbol{n}, \boldsymbol{n}') = 1[\boldsymbol{n}' = \boldsymbol{n} - \boldsymbol{a} \wedge \boldsymbol{n}] \sum_{\boldsymbol{a}'} p^{\mathrm{T}}(\boldsymbol{a}, \boldsymbol{a}') 1[\boldsymbol{a} > \boldsymbol{n}],$$

for $\boldsymbol{n}' = \xi(\boldsymbol{n}, \boldsymbol{a}, \boldsymbol{a}')$, where p^{I} represents an internal transition, and

$$\sum_{\boldsymbol{n}} \left(\sum_{\bar{\boldsymbol{a}}} p^{\mathrm{D}}(\boldsymbol{n}, \bar{\boldsymbol{a}}) + \sum_{\boldsymbol{n}'} p^{\mathrm{I}}(\boldsymbol{n}, \boldsymbol{n}') \right) = 1.$$

The following example illustrates the notation.

Example 12.25 (A three-node discrete time network) Consider a network with three nodes (see Figure 12.3), and batch transfer functions

$$p^{\mathrm{T}}((2, 3, 0), (0, 0, 1)) = 1/3,$$
$$p^{\mathrm{T}}((0, 0, 0), (1, 1, 0)) = 1/3,$$
$$p^{\mathrm{T}}((0, 0, 1), (0, 0, 0)) = 1/3.$$

All other $p^{\mathrm{T}}(\boldsymbol{a}, \boldsymbol{a}')$ are zero. Clearly, this is a joint probability since (5.1) is satisfied.

The interpretation of this network is that at the beginning of each time slot there are three possible events, and each occurs with probability 1/3. These events are:

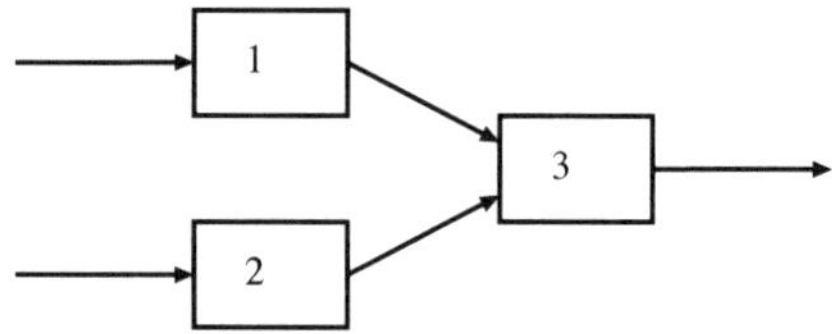

Figure 12.3 A three-node network

(a) a departure of two customers from node 1 and a departure of three customers from node 2 and these departures move to node 3 as a single customer;

(b) an arrival of one customer at node 1 and an arrival of one customer at node 2, both from the outside; and

(c) a departure of one customer from node 3 to the outside. □

The next example formulates the discrete time Jackson network of Example 12.24 within the framework of this section.

Example 12.26 (Reformulating the Jackson network) In the discrete time Jackson network, at the beginning of each time slot there is either an exogenous customer arrival at node j, which occurs with probability q_0, or one service completion at node j with probability q_j and the customer moves to node k with probability r_{jk}, where

$$\sum_{j=0}^{N} q_j = 1.$$

The dynamics under the present formulation are expressed by

$$p^{\mathrm{T}}(\boldsymbol{a}, \boldsymbol{a}') = \begin{cases} q_0 r_{j0}, & \boldsymbol{a} = \mathbf{0},\, \boldsymbol{a}' = \mathbf{e}_j, \\ q_j r_{jk}, & \boldsymbol{a} = \mathbf{e}_j,\, \boldsymbol{a}' = \mathbf{e}_k, \\ q_j \left(1 - \sum_{k=1}^{N} r_{jk}\right), & \boldsymbol{a} = \mathbf{e}_j,\, \boldsymbol{a}' = \mathbf{0}, \\ 0, & \text{otherwise}, \end{cases}$$

where $\mathbf{e}_j \in \mathcal{S}$ is the N-dimensional unit vector with the jth element equal to 1 and all other elements 0. □

12.5.2 Modification of the Network

The network with concurrent batch movements just defined is analytically intractable. Thus we modify the network by introducing additional arrivals in the same way as we did in Chapter 8. The following technical condition is needed.

(5.a) There exist positive numbers $\delta_1, \ldots, \delta_N$ such that

$$\sum_{a' \in \mathcal{S}} p^{\mathrm{T}}(\boldsymbol{a}, \boldsymbol{a}') \prod_{j=1}^{N} (1+\delta_j)^{a'_j} < \infty, \qquad \boldsymbol{a} \in \mathcal{S}.$$

(5.b) For all $j, k = 1, 2, \ldots, N$, with $j \neq k$, there is with a positive probability a path of batch transitions from node j to node k under the transition probability p^{T}.

Assumption (5.b) ensures the irreducibility of the Markov chain for the network process. If (5.b) is not satisfied, then we can decompose the network into subnetworks that satisfy (5.b) and analyze each subnetwork separately. So this is not an essential assumption.

We now consider a continuous time version of this network. The discrete time network can be considered to be a continuous time model in which events occur with intensity 1, and each event changes the state of the network according to the dynamics defined by p^{T}.

This continuous time network is now modified as follows. Introduce an additional batch arrival process from the outside the intensity of which depends on the state of the network. Let $\Lambda_{\boldsymbol{n}}(\boldsymbol{a})$ be the intensity of the additional arrival process with batch size $\boldsymbol{a}$ when the network is in state $\boldsymbol{n}$. This intensity is also a function of a positive vector $\boldsymbol{\rho} = (\rho_1, \ldots, \rho_N) < \mathbf{1} \equiv (1, \ldots, 1)$. Using the notation

$$\boldsymbol{\rho}^{\boldsymbol{n}} = \prod_{j=1}^{N} \rho_j^{n_j},$$

the intensity is

$$\Lambda_{\boldsymbol{n}}(\boldsymbol{a}) = \sum_{\boldsymbol{b} \in \mathcal{S}} \sum_{\boldsymbol{a}' \in \mathcal{S}} \boldsymbol{\rho}^{\boldsymbol{b}-\boldsymbol{a}'+\boldsymbol{a}} p^{\mathrm{T}}(\boldsymbol{b}, \boldsymbol{a}) \prod_{j \in \mathcal{N}_+(\boldsymbol{n})} 1[a'_j = a_j]$$
$$\times \left(\prod_{j \in \mathcal{N}_0(\boldsymbol{n})} 1[a'_j \geq a_j] - \prod_{j \in \mathcal{N}_0(\boldsymbol{n})} 1[a'_j = a_j] \right), \qquad (5.2)$$

where

$$\mathcal{N}_0(\boldsymbol{n}) = \{ j \in \mathcal{N} \mid n_j = 0 \},$$
$$\mathcal{N}_+(\boldsymbol{n}) = \mathcal{N} - \mathcal{N}_0(\boldsymbol{n}).$$

That is, $\mathcal{N}_0(\boldsymbol{n})$ and $\mathcal{N}_+(\boldsymbol{n})$ are the sets of empty and non-empty nodes, respectively, for each $\boldsymbol{n} \in \mathcal{S}$. It follows from (5.2) that the additional batch arrivals are activated only when one or more nodes are empty, i.e., $\mathcal{N}_0(\boldsymbol{n}')$ is not empty, since otherwise the term in the parentheses becomes zero. This modified network is considered to be a continuous time model, so it is referred to as a *continuous time network with additional arrivals*.

It follows from (5.b) that this continuous time Markov chain is irreducible. Suppose the Markov chain has a stationary distribution that is denoted by $\pi(\boldsymbol{n})$. The global balance equations are

$$\pi(\boldsymbol{n})\left(1 + \sum_{\boldsymbol{a}' \in \mathcal{S}} \Lambda_{\boldsymbol{n}}(\boldsymbol{a}')\right) = \sum_{\substack{\boldsymbol{n}', \boldsymbol{a}, \boldsymbol{a}' \in \mathcal{S} \\ \xi(\boldsymbol{n}', \boldsymbol{a}, \boldsymbol{a}') = \boldsymbol{n}}} \pi(\boldsymbol{n}') p^{\mathrm{T}}(\boldsymbol{a}, \boldsymbol{a}') + \sum_{\substack{\boldsymbol{n}' \in \mathcal{S} \\ \boldsymbol{n}' \leq \boldsymbol{n}}} \pi(\boldsymbol{n}') \Lambda_{\boldsymbol{n}'}(\boldsymbol{n} - \boldsymbol{n}'),$$

for each $\boldsymbol{n} \in \mathcal{S}$.

12.5.3 Characterization of the Product Form

Assume that the stationary distribution of the modified network with additional arrivals is of a geometric product form, i.e.,

$$\pi(\boldsymbol{n}) = \prod_{j=1}^{N} (1 - \rho_j) \rho_j^{n_j}, \qquad \boldsymbol{n} = (n_1, n_2, \ldots, n_N) \in \mathcal{S}.$$

The global balance equations can be rewritten as

$$1 + \sum_{\boldsymbol{a}' \in \mathcal{S}} \Lambda_{\boldsymbol{n}}(\boldsymbol{a}') = \sum_{\substack{\boldsymbol{n}', \boldsymbol{a}, \boldsymbol{a}' \in \mathcal{S} \\ \xi(\boldsymbol{n}', \boldsymbol{a}, \boldsymbol{a}') = \boldsymbol{n}}} \boldsymbol{\rho}^{\boldsymbol{n}' - \boldsymbol{n}} p^{\mathrm{T}}(\boldsymbol{a}, \boldsymbol{a}') + \sum_{\substack{\boldsymbol{n}' \in \mathcal{S} \\ \boldsymbol{n}' \leq \boldsymbol{n}}} \boldsymbol{\rho}^{\boldsymbol{n}' - \boldsymbol{n}} \Lambda_{\boldsymbol{n}'}(\boldsymbol{n} - \boldsymbol{n}'). \quad (5.3)$$

Equations (5.3) may not have a solution $\boldsymbol{\rho} < \mathbf{1}$. The following theorem gives the necessary and sufficient conditions, in terms of $p^{\mathrm{T}}(\boldsymbol{a}, \boldsymbol{a}')$, for (5.3) to have a solution. These conditions ensure that the stationary distribution of the modified network with additional arrivals is of the geometric product form.

We need the notation

$$\tilde{P}^{\mathrm{T}}(\boldsymbol{s}, \boldsymbol{t}) = \sum_{\boldsymbol{a} \in \mathcal{S}} \sum_{\boldsymbol{a}' \in \mathcal{S}} \boldsymbol{s}^{\boldsymbol{a}} \boldsymbol{t}^{\boldsymbol{a}'} p^{\mathrm{T}}(\boldsymbol{a}, \boldsymbol{a}') \quad (5.4)$$

for $\boldsymbol{s} = (s_1, \ldots, s_N)$ and $\boldsymbol{t} = (t_1, \ldots, t_N)$. Also, define the N-dimensional vector $\boldsymbol{\rho}(\mathcal{U})$ for the subset $\mathcal{U}$ of $\mathcal{N}$ by setting the jth element of $\boldsymbol{\rho}(\mathcal{U})$ equal to ρ_j for $j \in \mathcal{U}$, and equal to 1 for $j \in \mathcal{N} - \mathcal{U}$.

Theorem 12.27 *Under assumptions (5.a) and (5.b), the stationary distribution of the modified continuous time network with additional arrivals (5.2) exists and is of a geometric product form if and only if there exists a positive vector* $\boldsymbol{\rho} < \mathbf{1}$ *such that*

$$\tilde{P}^{\mathrm{T}}(\boldsymbol{\rho}(\mathcal{N}), \boldsymbol{\rho}^{-1}(\mathcal{U})) = \tilde{P}^{\mathrm{T}}(\boldsymbol{\rho}(\mathcal{N} - \mathcal{U}), \mathbf{1}), \quad \textit{for all } \mathcal{U} \subset \mathcal{N}. \tag{5.5}$$

Furthermore, if such a $\boldsymbol{\rho}$ *exists, then the stationary distribution is unique, and is a stochastic upper bound for the stationary distribution of the original discrete time network without the additional arrivals. That is, if* $\pi^0(\boldsymbol{n})$ *is the stationary distribution of the original network, then*

$$\sum_{\boldsymbol{n}' \geq \boldsymbol{n}} \pi^0(\boldsymbol{n}') \leq \prod_{j=1}^{N} \rho_j^{n_j}, \qquad \boldsymbol{n} \in \mathcal{S}. \tag{5.6}$$

PROOF. We first compute the second term on the right-hand side of (5.3). Fixing an $\boldsymbol{n} \in \mathcal{S}$, we can write this term as

$$\begin{aligned}
&\sum_{\boldsymbol{n}' \leq \boldsymbol{n}} \boldsymbol{\rho}^{\boldsymbol{n}'-\boldsymbol{n}} \Lambda_{\boldsymbol{n}'}(\boldsymbol{n} - \boldsymbol{n}') \\
&= \sum_{\boldsymbol{n}' \leq \boldsymbol{n}} \boldsymbol{\rho}^{\boldsymbol{n}'-\boldsymbol{n}} \sum_{\boldsymbol{a}} \sum_{\boldsymbol{a}'} \boldsymbol{\rho}^{\boldsymbol{a}-\boldsymbol{a}'+\boldsymbol{n}-\boldsymbol{n}'} p^{\mathrm{T}}(\boldsymbol{a}, \boldsymbol{a}') \prod_{j \in \mathcal{N}_+(\boldsymbol{n}')} 1[a'_j = n_j - n'_j] \\
&\quad \times \left(\prod_{j \in \mathcal{N}_0(\boldsymbol{n}')} 1[a'_j \geq n_j - n'_j] - \prod_{j \in \mathcal{N}_0(\boldsymbol{n}')} 1[a'_j = n_j - n'_j] \right) \\
&= \sum_{\boldsymbol{a}} \sum_{\boldsymbol{a}'} \boldsymbol{\rho}^{\boldsymbol{a}-\boldsymbol{a}'} \sum_{\boldsymbol{n}' \leq \boldsymbol{n}} p^{\mathrm{T}}(\boldsymbol{a}, \boldsymbol{a}') \\
&\quad \times \left(\prod_{j \in \mathcal{N}_+(\boldsymbol{n}')} 1[a'_j = n_j - n'_j] \prod_{j \in \mathcal{N}_0(\boldsymbol{n}')} 1[a'_j \geq n_j] - \prod_{j \in \mathcal{N}} 1[a'_j = n_j - n'_j] \right) \\
&= \sum_{\boldsymbol{a}} \sum_{\boldsymbol{a}'} \boldsymbol{\rho}^{\boldsymbol{a}-\boldsymbol{a}'} p^{\mathrm{T}}(\boldsymbol{a}, \boldsymbol{a}') - \sum_{\boldsymbol{a}} \sum_{\boldsymbol{a}'=\mathbf{0}}^{\boldsymbol{n}} \boldsymbol{\rho}^{\boldsymbol{a}-\boldsymbol{a}'} p^{\mathrm{T}}(\boldsymbol{a}, \boldsymbol{a}'),
\end{aligned}$$

where we have used the identity that, for any $\boldsymbol{a}'$ and $\boldsymbol{n}$,

$$\sum_{\boldsymbol{n}' \leq \boldsymbol{n}} \prod_{j \in \mathcal{N}_+(\boldsymbol{n}')} 1[a'_j = n_j - n'_j] \prod_{j \in \mathcal{N}_0(\boldsymbol{n}')} 1[a'_j \geq n_j] = 1,$$

which follows from the fact that the left-hand side is positive for $a'_j \geq n_j$ if and only if $n'_j = 0$. Hence, we have

$$\begin{aligned}
&\sum_{\boldsymbol{a},\boldsymbol{a}'} \sum_{\substack{\boldsymbol{n}'\geq \boldsymbol{a}\\ \boldsymbol{n}'=\boldsymbol{a}-\boldsymbol{a}'+\boldsymbol{n}}} \boldsymbol{\rho}^{\boldsymbol{n}'-\boldsymbol{n}} p^{\mathrm{T}}(\boldsymbol{a},\boldsymbol{a}') + \sum_{\boldsymbol{n}'\leq \boldsymbol{n}} \boldsymbol{\rho}^{\boldsymbol{n}'-\boldsymbol{n}} \Lambda_{\boldsymbol{n}'}(\boldsymbol{n}-\boldsymbol{n}')\\
&= \sum_{\boldsymbol{a}} \sum_{\boldsymbol{a}'=\boldsymbol{0}}^{\boldsymbol{n}} \boldsymbol{\rho}^{\boldsymbol{a}-\boldsymbol{a}'} p^{\mathrm{T}}(\boldsymbol{a},\boldsymbol{a}') + \sum_{\boldsymbol{a}} \sum_{\boldsymbol{a}'} \boldsymbol{\rho}^{\boldsymbol{a}-\boldsymbol{a}'} p^{\mathrm{T}}(\boldsymbol{a},\boldsymbol{a}') - \sum_{\boldsymbol{a}} \sum_{\boldsymbol{a}'=\boldsymbol{0}}^{\boldsymbol{n}} \boldsymbol{\rho}^{\boldsymbol{a}-\boldsymbol{a}'} p^{\mathrm{T}}(\boldsymbol{a},\boldsymbol{a}')\\
&= \sum_{\boldsymbol{a}} \sum_{\boldsymbol{a}'} \boldsymbol{\rho}^{\boldsymbol{a}-\boldsymbol{a}'} p^{\mathrm{T}}(\boldsymbol{a},\boldsymbol{a}')\\
&= \tilde{P}^{\mathrm{T}}(\boldsymbol{\rho},\boldsymbol{\rho}^{-1}).
\end{aligned} \tag{5.7}$$

We first prove necessity. Consider the cases $\mathcal{N}_0(\boldsymbol{n}) = \emptyset$ and $\mathcal{N}_0(\boldsymbol{n}) \neq \emptyset$ separately.

Case 1: $\mathcal{N}_0(\boldsymbol{n}) = \emptyset$, i.e., $\boldsymbol{n} > \boldsymbol{0}$.

In this case, $\xi(\boldsymbol{n}', \boldsymbol{a}, \boldsymbol{a}') = \boldsymbol{n}$ is equivalent to $\boldsymbol{n}' = \boldsymbol{a} - \boldsymbol{a}' + \boldsymbol{n}$ and $\boldsymbol{n}' \geq \boldsymbol{a}$. Thus it follows from (5.7) that

$$\sum_{\substack{\boldsymbol{n}',\boldsymbol{a},\boldsymbol{a}'\\ \xi(\boldsymbol{n}',\boldsymbol{a},\boldsymbol{a}')=\boldsymbol{n}}} \boldsymbol{\rho}^{\boldsymbol{n}'-\boldsymbol{n}} p^{\mathrm{T}}(\boldsymbol{a},\boldsymbol{a}') + \sum_{\boldsymbol{n}'\leq \boldsymbol{n}} \boldsymbol{\rho}^{\boldsymbol{n}'-\boldsymbol{n}} \Lambda_{\boldsymbol{n}'}(\boldsymbol{n}-\boldsymbol{n}') = \tilde{P}^{\mathrm{T}}(\boldsymbol{\rho},\boldsymbol{\rho}^{-1}).$$

Since no additional arrivals occur when $\boldsymbol{n} > \boldsymbol{0}$, the left-hand side of (5.3) is reduced to 1. So (5.3) is equivalent to

$$1 = \tilde{P}^{\mathrm{T}}(\boldsymbol{\rho},\boldsymbol{\rho}^{-1}). \tag{5.8}$$

Case 2: $\mathcal{N}_0(\boldsymbol{n}) \neq \emptyset$.

To compute the right-hand side of (5.3), we first consider the two identities

$$\begin{aligned}
&\sum_{\boldsymbol{a},\boldsymbol{a}'} \sum_{\substack{\boldsymbol{n}'\\ (\boldsymbol{n}'-\boldsymbol{a})^+=\boldsymbol{n}}} \boldsymbol{\rho}^{\boldsymbol{n}'-\boldsymbol{n}} p^{\mathrm{T}}(\boldsymbol{a},\boldsymbol{a}') = \sum_{\boldsymbol{a},\boldsymbol{a}'} p^{\mathrm{T}}(\boldsymbol{a},\boldsymbol{a}') \prod_{i\in\mathcal{N}_+(\boldsymbol{n})} \rho_i^{a'_i} \prod_{j\in\mathcal{N}_0(\boldsymbol{n})} \frac{1-\rho_j^{a'_j+1}}{1-\rho_j},\\
&\sum_{\boldsymbol{a},\boldsymbol{a}'} \sum_{\substack{\boldsymbol{n}'\geq \boldsymbol{a}\\ (\boldsymbol{n}'-\boldsymbol{a})^+=\boldsymbol{n}}} \boldsymbol{\rho}^{\boldsymbol{n}'-\boldsymbol{n}} p^{\mathrm{T}}(\boldsymbol{a},\boldsymbol{a}') = \sum_{\boldsymbol{a},\boldsymbol{a}'} \boldsymbol{\rho}^{\boldsymbol{a}} p^{\mathrm{T}}(\boldsymbol{a},\boldsymbol{a}'),
\end{aligned}$$

where $(\boldsymbol{n}' - \boldsymbol{a})^+ = \max\{\boldsymbol{n}' - \boldsymbol{a}, \boldsymbol{0}\}$ and the max operator is componentwise. Using these identities and (5.7), the right-hand side of (5.3) becomes

$$\sum_{\substack{\boldsymbol{n}',\boldsymbol{a},\boldsymbol{a}'\\ \xi(\boldsymbol{n}',\boldsymbol{a},\boldsymbol{a}')=\boldsymbol{n}}} \boldsymbol{\rho}^{\boldsymbol{n}'-\boldsymbol{n}} p^{\mathrm{T}}(\boldsymbol{a},\boldsymbol{a}') + \sum_{\boldsymbol{n}'\leq \boldsymbol{n}} \boldsymbol{\rho}^{\boldsymbol{n}'-\boldsymbol{n}} \Lambda_{\boldsymbol{n}'}(\boldsymbol{n}-\boldsymbol{n}')$$

$$= \sum_{a,a'} \sum_{\substack{n' \geq a \\ n'=a-a'+n}} \rho^{n'-n} p^{\mathrm{T}}(a, a') + \sum_{a,a'} \sum_{\substack{n' \\ (n'-a)^+=n}} \rho^{n'-n} p^{\mathrm{T}}(a, a')$$

$$- \sum_{a,a'} \sum_{\substack{n' \geq a \\ (n'-a)^+=n}} \rho^{n'-n} p^{\mathrm{T}}(a, a') + \sum_{n' \leq n} \rho^{n'-n} \Lambda_{n'}(n - n')$$

$$= \tilde{P}^{\mathrm{T}}(\rho, \rho^{-1}) + \sum_{a,a'} p^{\mathrm{T}}(a, a') \prod_{i \in \mathcal{N}_+(n)} \rho_i^{a'_i} \prod_{j \in \mathcal{N}_0(n)} \frac{1 - \rho_j^{a'_j+1}}{1 - \rho_j}$$

$$- \sum_{a,a'} \rho^{a} p^{\mathrm{T}}(a, a'). \tag{5.9}$$

We next consider the left-hand side of (5.3). Substituting (5.2) into (5.3) yields

$$1 + \sum_{a'} \Lambda_n(a')$$

$$= 1 + \sum_{a',b,a} \rho^{b-a+a'} p^{\mathrm{T}}(b, a) \prod_{j \in \mathcal{N}_+(n)} 1[a_j = a'_j]$$

$$\times \left(\prod_{i \in \mathcal{N}_0(n)} 1[a_j \geq a'_j] - \prod_{j \in \mathcal{N}_0(n)} 1[a_j = a'_j] \right)$$

$$= 1 + \sum_{b,a} \rho^{b-a} p^{\mathrm{T}}(b, a) \sum_{a'} \rho^{a'} \prod_{j \in \mathcal{N}_+(n)} 1[a'_j = a_j] \prod_{j \in \mathcal{N}_0(n)} 1[a_j \geq a'_j]$$

$$- \sum_{b,a} \rho^{b-a} p^{\mathrm{T}}(b, a) \sum_{a'} \rho^{a'} \prod_{j \in \mathcal{N}} 1[a'_j = a_j] \tag{5.10}$$

$$= 1 + \sum_{b,a} \rho^{b-a} p^{\mathrm{T}}(b, a) \prod_{i \in \mathcal{N}_+(n)} \rho_i^{a_i} \prod_{j \in \mathcal{N}_0(n)} \frac{1 - \rho_j^{a_j+1}}{1 - \rho_j} - \sum_{b,a} \rho^{b} p^{\mathrm{T}}(b, a).$$

From (5.8), (5.9) and (5.11), we obtain

$$\sum_{b,a} \rho^{b-a} p^{\mathrm{T}}(b, a) \prod_{i \in \mathcal{N}_+(n)} \rho_i^{a_i} \prod_{j \in \mathcal{N}_0(n)} \left(1 - \rho_j^{a_j+1}\right)$$

$$= \sum_{b,a} p^{\mathrm{T}}(b, a) \prod_{i \in \mathcal{N}_+(n)} \rho_i^{b_i} \prod_{j \in \mathcal{N}_0(n)} \left(1 - \rho_j^{b_j+1}\right). \tag{5.11}$$

Note that (5.11) depends on n only through $\mathcal{N}_0(n)$, which may be regarded as a subset of $\mathcal{N}$. If $\mathcal{N}_0(n) = \emptyset$, (5.11) is an identity. If $\mathcal{N}_0(n) = \{j\}$ for $j \in \mathcal{N}$, we have

$$\sum_{b,a} \rho^{b} p^{\mathrm{T}}(b, a) \prod_{i \in \{j\}} \rho_i^{-a_i} = \sum_{b,a} p^{\mathrm{T}}(b, a) \prod_{i \in \mathcal{N}-\{j\}} \rho_i^{b_i}.$$

By repeating these arguments, we obtain (5.5) inductively. We now show that (5.5) is sufficient for the global balance equations (5.3) to hold. It is easy to see that (5.5) implies (5.11) for $\mathcal{U} \neq \emptyset$, and (5.5) is identical to (5.8) when $\mathcal{U} = \emptyset$. It follows from the calculations above that (5.11) and (5.8) are equivalent to (5.3). Thus, (5.3) holds if and only if (5.5) holds. If there exists a $\boldsymbol{\rho} < \mathbf{1}$ satisfying (5.5), then the π of (5.3) is, by assumption (5.b), a unique stationary distribution. The inequality (5.6) is immediate, since the additional arrivals increase the customer population in the network. Thus the theorem is proved. □

Remark 12.28 In Theorem 12.27 we specified the arrival intensity of extra batches by (5.2). The reason why we choose this intensity is the following. The second term on the right-hand side of (5.3) can be written as

$$\sum_{\boldsymbol{n}' \leq \boldsymbol{n}} \boldsymbol{\rho}^{\boldsymbol{n}'-\boldsymbol{n}} \Lambda_{\boldsymbol{n}'}(\boldsymbol{n} - \boldsymbol{n}')$$

$$= \sum_{\boldsymbol{a},\boldsymbol{a}'} \boldsymbol{\rho}^{\boldsymbol{a}-\boldsymbol{a}'} p^{\mathrm{T}}(\boldsymbol{a}, \boldsymbol{a}') - \sum_{\boldsymbol{a}} \sum_{\boldsymbol{a}'=\boldsymbol{0}}^{\boldsymbol{n}} \boldsymbol{\rho}^{\boldsymbol{a}-\boldsymbol{a}'} p^{\mathrm{T}}(\boldsymbol{a}, \boldsymbol{a}') + d(\boldsymbol{n}), \tag{5.12}$$

where d is a function of $\boldsymbol{n}$. Note that $d(\boldsymbol{n})$ has to be constant for $\boldsymbol{n} > \mathbf{0}$, since the right-hand side of (5.3) is constant. Because the difference between the first and second terms on the right-hand side of (5.12) converges to zero when all components of $\boldsymbol{n}$ go to infinity, the function d that minimizes the additional arrivals has to satisfy $d(\boldsymbol{n}) = 0$ for $\boldsymbol{n} > \mathbf{0}$. Since the arrival intensity (5.2) is obtained by choosing d such that $d(\boldsymbol{n}) = \mathbf{0}$ for all $\boldsymbol{n} \in \mathcal{N}$, (5.2) is expected to be the one with the minimum additional arrivals.

Condition (5.5) is composed of $2^N - 1$ equations, and N unknowns, where N is the number of nodes, so (5.5) may not have a solution (see Exercise 12.13). It is therefore interesting to see if (5.5) can be reduced to N equations. This is, of course, not true in general. However, there are special network structures where (5.5) is reduced to a system of N equations that has a solution. This is the topic considered in the next section.

12.6 BATCH TRANSFER NETWORKS

In this section we consider two special classes of networks with concurrent movements for which condition (5.5) is reduced to N equations. In the first model, batches are transferred one at a time, while in the second one, multiple batches are transferred simultaneously but independently. We refer to these models as the *single batch transfer network* and the *multiple batch transfer network.*, respectively.

Consider a discrete time network with N nodes. At the beginning of each time slot there is a batch departure from one and only one node. More specifically, we assume that at the beginning of each time slot there is, with probability $p_{j,k}(a_j, a'_k)$, a request for a batch departure of a_j customers from node j, and this batch arrives at node k at the end of the time slot as a batch of a'_k customers, $j, k = 1, 2, \ldots, N$, where

$$\sum_{j=1}^{N}\sum_{k=1}^{N}\sum_{a=0}^{\infty}\sum_{a'=0}^{\infty} p_{j,k}(a, a') = 1. \tag{6.1}$$

Note that an exogenous batch of size a' arrives at node j with probability $p_{jj}(0, a')$ and a batch of size a leaves the network from node j with probability $p_{jj}(a, 0)$. Thus arrivals from and departures to the outside are defined differently from the other models that use an outside node 0. Since at any time slot there is at most a single batch departure from only one node and all customers in this batch join another node, we call this model a single batch transfer network. Using the notation of the previous section, the single batch transfer network has the transfer probability function

$$p^{\mathrm{T}}(\boldsymbol{a}, \boldsymbol{a}') = p_{j,k}(a_j, a'_k)1[(\boldsymbol{a}, \boldsymbol{a}') = (a_j\mathbf{e}_j, a'_k\mathbf{e}_k)], \qquad a_j, a'_k \geq 0.$$

It should be noted that $p_{j,k}(a_j, a'_k)$ is not a conditional probability, and should not be confused with the routing probabilities of customers. It is a joint probability of j, k, a_j, and a'_k. Condition (6.1) implies that in one time slot only one node can release a batch and all the customers in the batch proceed to the same node.

Example 12.29 (A single batch transfer network) Consider a tandem queue with two nodes, and batch transfer probabilities

$$p_{1,1}(0, 1) = 1/4, \quad p_{1,1}(0, 2) = 1/4,$$
$$p_{1,2}(2, 3) = 1/4, \quad p_{2,2}(1, 0) = 1/4.$$

The interpretation is that at the beginning of each time slot one and only one of the following four events occurs and each one occurs with the same probability:

(a) the arrival of one customer at node 1 from the outside;

(b) the arrival of two customers at node 1 from the outside;

(c) a departure of two customers from node 1 and these customers join node 2 as a batch of three customers; and

(d) a departure of one customer from node 2 that leaves the network. □

The continuous time version of the single batch transfer network is the assembly-transfer network studied in Chapter 8. It can be shown that the additional batch arrival condition needed for the product form is also equivalent to the one introduced in Section 8.2. We derive here the result of Theorem 8.2 again, but now we use the framework of the previous section.

We first note that $\tilde{P}^{\mathrm{T}}$ in (5.4) is given by

$$\begin{aligned}\tilde{P}^{\mathrm{T}}(\boldsymbol{s},\boldsymbol{t}) &= \sum_{j=1}^{N}\sum_{k=1}^{N}\sum_{a_j=1}^{\infty}\sum_{a'_k=1}^{\infty} p_{j,k}(a_j,a'_k)s_j^{a_j}t_k^{a'_k}\\ &= \sum_{j=1}^{N}\sum_{k=1}^{N}\tilde{P}_{j,k}(s_j,t_k), \end{aligned}\tag{6.2}$$

for all possible $\boldsymbol{s}=(s_1,\ldots,s_N)$ and $\boldsymbol{t}=(t_1,\ldots,t_N)$, where

$$\tilde{P}_{j,k}(s_j,t_k) = \sum_{a=1}^{\infty}\sum_{a'=1}^{\infty} p_{j,k}(a,a')s_j^{a}t_k^{a'} .$$

Lemma 12.30 *In the single batch transfer network with additional arrivals (continuous time), condition (5.5) is reduced to the N equations*

$$\sum_{i=1}^{N}\tilde{P}_{i,k}(\rho_i,\rho_k^{-1}) + \sum_{j=1}^{N}\tilde{P}_{k,j}(\rho_k,1) = \sum_{i=1}^{N}\tilde{P}_{i,k}(\rho_i,1) + \sum_{j=1}^{N}\tilde{P}_{k,j}(1,1),\qquad k=1,2,\ldots,N. \tag{6.3}$$

PROOF. Let $\mathcal{A}$ be an arbitrary subset of $\mathcal{N}$. Summing (6.3) over all $k\in\mathcal{A}$ yields

$$\begin{aligned}&\sum_{i\in\mathcal{N}}\sum_{k\in\mathcal{A}}\tilde{P}_{i,k}(\rho_i,\rho_k^{-1}) + \sum_{k\in\mathcal{A}}\sum_{j\in\mathcal{N}}\tilde{P}_{k,j}(\rho_k,1)\\ &\quad= \sum_{i\in\mathcal{N}}\sum_{k\in\mathcal{A}}\tilde{P}_{i,k}(\rho_i,1) + \sum_{k\in\mathcal{A}}\sum_{j\in\mathcal{N}}\tilde{P}_{k,j}(1,1) .\end{aligned}$$

Since

$$\begin{aligned}\sum_{k\in\mathcal{A}}\sum_{j\in\mathcal{N}}\tilde{P}_{k,j}(\rho_k,1) &= \sum_{k\in\mathcal{N}}\sum_{j\in\mathcal{N}}\tilde{P}_{k,j}(\rho_k,1) - \sum_{k\in\mathcal{N}-\mathcal{A}}\sum_{j\in\mathcal{N}}\tilde{P}_{k,j}(\rho_k,1),\\ \sum_{i\in\mathcal{N}}\sum_{k\in\mathcal{A}}\tilde{P}_{i,k}(\rho_i,1) &= \sum_{i\in\mathcal{N}}\sum_{k\in\mathcal{N}}\tilde{P}_{i,k}(\rho_i,1) - \sum_{i\in\mathcal{N}}\sum_{k\in\mathcal{N}-\mathcal{A}}\tilde{P}_{i,k}(\rho_i,1),\end{aligned}$$

we obtain

$$\sum_{i\in\mathcal{N}}\sum_{k\in\mathcal{A}}\tilde{P}_{i,k}(\rho_i,\rho_k^{-1})+\sum_{i\in\mathcal{N}}\sum_{k\in\mathcal{N}-\mathcal{A}}\tilde{P}_{i,k}(\rho_i,1)$$
$$=\sum_{k\in\mathcal{N}-\mathcal{A}}\sum_{j\in\mathcal{N}}\tilde{P}_{k,j}(\rho_k,1)+\sum_{k\in\mathcal{A}}\sum_{j\in\mathcal{N}}\tilde{P}_{k,j}(1,1)\,.$$

This is equivalent to condition (5.5) for $\mathcal{A}$, which completes the proof. □

It is readily seen that (6.3) is equivalent to the traffic equations of the assembly-transfer networks discussed in Chapter 8. The existence of a solution to equations (6.3) as well as the stability conditions have been established in Chapter 11 (see Example 11.20).

The single batch transfer network described above has the restriction that, in any time slot, there may be a batch departure from only one node. We now consider a network in which there may be batch departures from multiple nodes in any time slot.

Consider a network with N nodes. For each pair of nodes j and k, there is, with probability $p_{j,k}(a,a')$, a request for a batch of size a to depart from node j at the beginning of the slot and that arrives at node k as a batch of size a' at the end of the slot, where

$$\sum_{a=0}^{\infty}\sum_{a'=0}^{\infty}p_{j,k}(a,a')=1,\qquad j,k=1,2,\ldots,N.$$

Note that the batch transfer probability $p_{j,k}(a,a')$ here is different from that in the single batch transfer case, even though we used the same notation. In this model $p_{j,k}(a,a')$ is a joint probability of (a,a') *for each pair* j *and* k, and it implies that there may be simultaneous departures and arrivals at any set of nodes in the network. Similar to the single batch transfer model, $p_{jj}(0,a')$ and $p_{jj}(a,0)$ describe arrivals from the outside and departures to the outside.

Clearly, there are altogether N requests for departures at each node, and a total of N^2 requests for batch departures at the beginning of each time slot in the entire network. Similarly, there are a total of N^2 batch arrivals at the end of each time slot (N for each node) in the entire network. All these requests are assumed to be independent of each other. Thus each node releases N batches of customers independently and each batch may go to another node as a batch of a different size.

Example 12.31 (A multiple batch transfer network) Consider a network with four nodes. At the beginning of each time slot, one exogenous customer arrives at node 1 with probability p_1. At node 1 one customer is released for node 2 with probability p_{12}, and no customer is released at node 1 with probability $1 - p_{12}$. Independently, node 1 has probability p_{13} of releasing one customer for node 3 and probability $1 - p_{13}$ of not releasing a customer for node 3. Also, nodes 2 and 3 each release one customer for node 4 with probabilities p_2 and p_3, respectively. Node 4 releases one customer with probability p_4 for departure to the outside. All these events are independent of each other. See Figure 12.4.

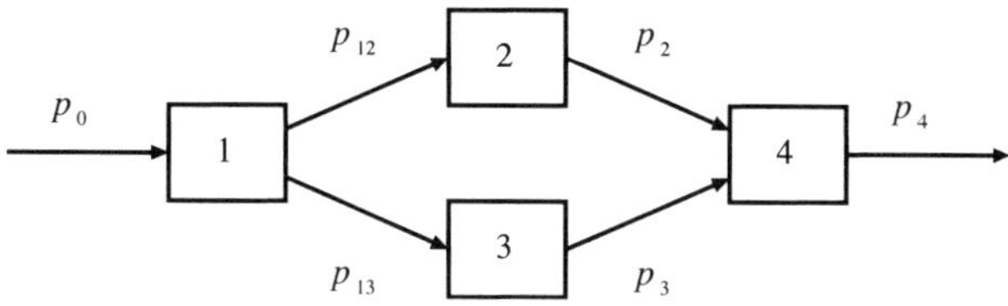

Figure 12.4 A multiple batch transfer network

This is a multiple batch transfer network with transfer probabilities

$$\begin{aligned}
p_{1,1}(0,1) &= p_1, & p_{1,1}(0,0) &= 1 - p_1, \\
p_{1,2}(1,1) &= p_{12}, & p_{1,2}(0,0) &= 1 - p_{12}, \\
p_{1,3}(1,1) &= p_{13}, & p_{1,3}(0,0) &= 1 - p_{13}, \\
p_{2,4}(1,1) &= p_2, & p_{2,4}(0,0) &= 1 - p_2, \\
p_{3,4}(1,1) &= p_3, & p_{3,4}(0,0) &= 1 - p_3, \\
p_{4,4}(1,0) &= p_4, & p_{4,4}(0,0) &= 1 - p_4.
\end{aligned}$$

Note that in this network six events may occur at the beginning of each time slot, and they are independent of each other:

(a) an arrival at node 1 from the outside;

(b) the request for a departure from node 1 for node 2;

(c) the request for a departure from node 1 for node 3;

(d) the request for a departure from node 2 for node 4;

(e) the request for a departure from node 3 for node 4; and

(f) the request for a departure from node 4 for the outside.

For example, the probability of not a single event occurring in one time slot is

$$(1-p_1)(1-p_{12})(1-p_{13})(1-p_2)(1-p_3)(1-p_4),$$

while the probability that all events occur, changing the state of the network from $\boldsymbol{n} = (n_1, n_2, n_3, n_4)$ at the beginning of the time slot to

$$\boldsymbol{n} + \mathbf{e}_1 - \mathbf{e}_1 + \mathbf{e}_2 - \mathbf{e}_1 + \mathbf{e}_3 - \mathbf{e}_2 + \mathbf{e}_4 - \mathbf{e}_3 + \mathbf{e}_4 - \mathbf{e}_4 = \boldsymbol{n} - \mathbf{e}_1 + \mathbf{e}_4$$

at the end of the time slot (assuming that a sufficient number of customers are present at each node) is $p_1 p_{12} p_{13} p_2 p_3 p_4$. The probability for other events can be computed similarly. □

Let $(X_{j,k}, Y_{j,k})$ be a pair of random variables with probability mass function $p_{j,k}(\cdot, \cdot)$, where $(X_{j,k}, Y_{j,k})$ are independent of other pairs with different subscripts. The total number of departures requested from node j at the beginning of the time slot, denoted by D_j, and the total number of arrivals at node j at the end of a time slot, denoted by A_j, are

$$D_j = \sum_{k=1}^{N} X_{j,k}, \qquad j = 1, 2, \ldots, N,$$

$$A_j = \sum_{k=1}^{N} Y_{k,j}, \qquad j = 1, 2, \ldots, N.$$

Let

$$\boldsymbol{D} = (D_1, \ldots, D_N),$$

$$\boldsymbol{A} = (A_1, \ldots, A_N)$$

denote the departure and arrival vectors of the network. Using the notation of the previous section, the batch transfer probability function p^{T} is

$$p^{\mathrm{T}}(\boldsymbol{a}, \boldsymbol{a}') = \sum_{\boldsymbol{X}\mathbf{1}=\boldsymbol{a}} \sum_{\mathbf{1}'\boldsymbol{Y}=\boldsymbol{a}'} \prod_{j=1}^{N} \prod_{k=1}^{N} p_{j,k}(x_{j,k}, y_{j,k}), \qquad \boldsymbol{a}, \boldsymbol{a}' \geq \mathbf{0}, \tag{6.4}$$

where the summation is taken over all $N \times N$ matrices $\boldsymbol{X} = \{x_{j,k}\}$ and $\boldsymbol{Y} = \{y_{j,k}\}$ all components of which are nonnegative integers.

After taking the z-transform of the batch transfer probability (6.4) for $\boldsymbol{s} = (s_1, \ldots, s_N)$ and $\boldsymbol{t} = (t_1, \ldots, t_N)$ we obtain

$$\tilde{P}^{\mathrm{T}}(\boldsymbol{s}, \boldsymbol{t}) = \prod_{j=1}^{N} \prod_{k=1}^{N} \tilde{P}_{j,k}(s_j, t_k),$$

where $\tilde{P}_{j,k}(s_j, t_k)$ is the z-transform of $p_{j,k}(a, a')$, i.e.,

$$\tilde{P}_{j,k}(s_j, t_k) = \sum_{a=0}^{\infty}\sum_{a'=0}^{\infty} p_{j,k}(a, a') s_j^a t_k^{a'}.$$

Let

$$\tilde{Q}_{j,k}(s_j, t_k) = -\log\left(\tilde{P}_{j,k}(s_j, t_k)\right),$$

and take the logarithm on both sides of condition (5.5). From the fact that $\tilde{Q}_{j,k}(1, 1) = 0$, it follows that

$$\sum_{j\in\mathcal{N}}\sum_{k\in\mathcal{A}} \tilde{Q}_{j,k}(\rho_j, \rho_k^{-1}) + \sum_{j\in\mathcal{N}}\sum_{k\in\mathcal{N}-\mathcal{A}} \tilde{Q}_{j,k}(\rho_j, 1) = \sum_{j\in\mathcal{N}-\mathcal{A}}\sum_{k\in\mathcal{N}} \tilde{Q}_{j,k}(\rho_j, 1) \quad (6.5)$$

for any subset $\mathcal{A}$ of $\mathcal{N}$. Equation (6.5) has the same form as equation (6.4). Thus we have the following lemma.

Lemma 12.32 *In the multiple batch transfer networks with additional arrivals, condition (5.5) is reduced to the N equations*

$$\sum_{j=1}^{N} \tilde{Q}_{j,k}(\rho_j, \rho_k^{-1}) + \sum_{j=1}^{N} \tilde{Q}_{k,j}(\rho_k, 1) = \sum_{j=1}^{N} \tilde{Q}_{j,k}(\rho_j, 1), \quad k = 1, 2, \ldots, N, \quad (6.6)$$

or, equivalently,

$$\prod_{i=1}^{N} \tilde{P}_{i,k}(\rho_i, \rho_k^{-1}) \prod_{j=1}^{N} \tilde{P}_{k,j}(\rho_k, 1) = \prod_{i=1}^{N} \tilde{P}_{i,k}(\rho_i, 1), \qquad k = 1, 2, \ldots, N. \quad (6.7)$$

We next consider algorithms for computing the stationary distributions of batch transfer networks. Since the single batch transfer network is equivalent to the assembly-transfer network, the iterative algorithm discussed in Chapter 11 can be used to solve it. In what follows we focus on the networks with multiple batch transfers. We first consider the following equation involving $\boldsymbol{\rho} \equiv (\rho_1, \rho_2, \ldots, \rho_N) \geq \mathbf{0}$ for a fixed $\boldsymbol{y} \equiv (y_1, y_2, \ldots, y_N) \in (0, 1]^N$:

$$\prod_{i=1}^{N} \frac{\tilde{P}_{i,k}(y_i, \rho_k^{-1})}{\tilde{P}_{i,k}(y_i, 1)} \prod_{j=1}^{N} \tilde{P}_{k,j}(\rho_k, 1) = 1, \qquad k = 1, 2, \ldots, N. \quad (6.8)$$

Lemma 12.33 *In the multiple batch transfer network with additional arrivals, equations (6.8) have at most one solution $\rho_k \in (0, 1)$, and such a solution exists for all $k = 1, 2, \ldots, N$ if $\mathbf{y} = \mathbf{1}$ and if the following two conditions hold:*

(6.a) $m'_j < m_j$ for all $j \in \mathcal{N}$, where

$$m'_j = \sum_{\boldsymbol{a}' \in \mathcal{S}} a'_j p^{\mathrm{T}}(\mathbf{1}, \boldsymbol{a}') \quad \text{and} \quad m_j = \sum_{\boldsymbol{a} \in \mathcal{S}} a_j p^{\mathrm{T}}(\boldsymbol{a}, \mathbf{1}).$$

(6.b) Not all distributions corresponding to the z-transforms $\tilde{P}_{j,k}(1, z)$ and $\tilde{P}_{k,j}(z, 1)$ are deterministic.

PROOF. For a real number θ, let

$$g_{k,j}(\theta) = \tilde{P}_{k,j}(e^\theta, 1),$$
$$h_{k,j}(\theta) = \log g_{k,j}(\theta)(\equiv -\tilde{Q}_{k,j}(e^\theta, 1)),$$

provided they exist. Then $h_{k,j}(\theta)$ is a convex function of θ in the domain where $h_{k,j}$ is defined. This can be seen by taking the second derivative of $h_{k,j}$, which is

$$\begin{aligned} h''_{k,j}(\theta) &= \frac{g''_{k,j}(\theta) g_{k,j}(\theta) - (g'_{k,j}(\theta))^2}{g^2_{k,j}(\theta)} \\ &= \frac{1}{2 g^2_{k,j}(\theta)} \int_0^\infty \int_0^\infty (x - y)^2 e^{\theta(x+y)} dG_{k,j}(x) dG_{k,j}(y) \geq 0, \end{aligned}$$

where $G_{k,j}$ is the distribution function of $g_{k,j}$. Note that $h_{k,j}$ is strictly convex unless $G_{k,j}$ is deterministic. Hence $\tilde{Q}_{k,j}(e^\theta, 1)$ is a concave function of θ. Similarly, $\tilde{Q}_{j,k}(y_j, e^\theta)$ is a concave function of θ for each fixed y_j. Let $\boldsymbol{\rho} = \mathbf{y}$ and $\rho_k = e^{-\theta}$ in (6.6). Rewriting the resulting formula yields

$$\sum_{j=1}^N \tilde{Q}_{j,k}(y_j, 1) - \sum_{j=1}^N \tilde{Q}_{j,k}(y_j, e^\theta) = \sum_{j=1}^N \tilde{Q}_{k,j}(e^{-\theta}, 1), \quad k = 1, 2, \ldots, N. \tag{6.9}$$

The right-hand side of (6.9) is increasing concave in θ, while the left-hand side is increasing convex in θ. Since both functions are equal to zero at $\theta = 0$, equations (6.9) have at most one solution $\theta > 0$. Thus the first part of the lemma follows from the fact that equations (6.9) and (6.8) are equivalent. The remaining part of the lemma follows from the strict convexity and the fact that the derivatives of the left-hand side and right-hand side of (6.9) at the origin are, under the given conditions, equal to m'_k and m_k, respectively. □

Remark 12.34 If condition (6.b) does not hold, equations (6.8) are reduced to $\rho_k^\ell = 1$ for some integer ℓ. So $\rho_k = 1$ is the only possible solution of (6.8) when $\ell \neq 0$. Note that $\ell = 0$ if and only if $m_k' = m_k$, which clearly implies that the number of customers at node k remains equal to the initial number.

We are now ready to present an algorithm for solving the multiple batch transfer network. If equations (6.8) are reduced to the identity $\rho_k^0 = 1$, it is understood that $\rho_k = 1$ is its only solution. Let ϵ be a pre-specified small number.

Step 1. Set $n = 0$, and $\boldsymbol{\rho}^{(0)} = \mathbf{1}$.

Step 2. Set $\mathbf{y} = \boldsymbol{\rho}^{(n)}$ in (6.8). Set $\rho_k^{(n+1)}$ equal to the minimal nonnegative solution ρ_k of (6.8) for any k that satisfies $\rho_k < 1$, and set $\rho_k^{(n+1)}$ equal to 1 otherwise. Note that the minimal solution is at least unity, which is a trivial solution of (6.8).

Step 3. If $\boldsymbol{\rho} = \mathbf{1}$, then STOP. Otherwise go to Step 4.

Step 4. If

$$\max_{j=1,2,\dots,N} |\rho_j^{(n+1)} - \rho_j^{(n)}| < \epsilon,$$

then set $\boldsymbol{\rho}^{(*)}$ equal to $\boldsymbol{\rho}^{(n+1)}$, and STOP. Otherwise go to Step 2.

Theorem 12.35 *If the iterative algorithm above has a limit $\boldsymbol{\rho}^{(\infty)}$ for the multiple batch transfer network with additional arrivals satisfying conditions (6.a) and (6.b) of Lemma 12.33, then the network has a stationary distribution of the geometric product form if and only if $\boldsymbol{\rho}^{(\infty)} < \mathbf{1}$.*

Unlike the algorithm for solving the single batch transfer networks, the algorithm above is not guaranteed to converge, even though it does converge in many numerical applications (see Section 12.7).

Remark 12.36 In the iterative algorithm above, equations (6.8) may be replaced by

$$\prod_{i=1}^{N} \frac{\tilde{P}_{i,k}(y_i, \rho_k^{-1})}{\tilde{P}_{i,k}(y_i, 1)} \prod_{j=1}^{N} \tilde{P}_{k,j}(y_k, 1) = 1, \qquad k = 1, 2, \dots, N. \qquad (6.10)$$

In this case, observe that (6.10) always has a unique solution ρ_k between 0 and 1 for all $k = 1, 2, \ldots, N$ provided $0 < y_i < 1$ for all $i \in \mathcal{N}$, since the first product on the left-hand side of (6.10) is greater than one. However, the speed of convergence seems to be slower than with equations (6.8) since we update the ρ_k of (6.8) in (6.10). Another possibility is to replace (6.8) by

$$\prod_{i=1}^{N} \frac{\tilde{P}_{i,k}(y_i, y_k^{-1})}{\tilde{P}_{i,k}(y_i, 1)} \prod_{j=1}^{N} \tilde{P}_{k,j}(\rho_k, 1) = 1, \qquad k = 1, 2, \ldots, N \,. \qquad (6.11)$$

Similar to (6.10), in this case we have a unique solution ρ_k between 0 and 1. However, we do not recommend (6.11) for the iteration process, because it may converge to the trivial solution **1** or **0**. These phenomena will be observed in the examples of the next section.

12.7 EXAMPLES OF BATCH TRANSFER NETWORKS

In this section we apply the multiple batch transfer network to two examples. The iterative algorithm is used to solve the second problem numerically.

Example 12.37 (The synchronized production line continued) The serial production line with synchronization discussed in Example 12.1 can be formulated as a multiple batch transfer network. We make the following assumptions.

(i) The number of arrivals at stage j is a_j with probability $A_j(a_j)$.

(ii) During each time slot the machine at stage j is up with probability ζ_j. If the machine is up it inspects the items at stage j and processes η_j non-defective items; otherwise it inspects the items but it does not process any items.

(iii) The items that are found defective are sent back to the previous stage (or to the outside supplier in the case of machine 1) for rework. The distribution of the number of items found defective during a cycle is denoted by $\{B_j(b_j); b_j \geq 0\}$.

Note that in this example as well as in the next, we use capital letters A_j and B_j to denote probability mass functions.

Assume that arrivals, inspections and processings are independent of each other and that all departing items are removed when the number of items at a stage is not sufficient for either inspection or processing. This model is then a multiple batch transfer network with

$$p_{j,j}(b_j, a_j) = A_j(a_j)1[b_j = 0],$$

$$p_{j,j+1}(b_j, a_{j+1}) = \begin{cases} \zeta_j, & b_j = \eta_j, a_{j+1} = \eta_j 1[j \neq N], \\ 1 - \zeta_j, & b_j = a_{j+1} = 0, \\ 0, & \text{otherwise,} \end{cases}$$
$$p_{j,j-1}(b_j, a_{j-1}) = B_j(b_j)1[a_{j-1} = b_j 1[j \neq 1]],$$
$$p_{j,k}(b_j, a_k) = 1[b_j = 0, a_k = 0], \qquad \text{for } k \neq j-1, j, j+1,$$

where $j - 1$ is interpreted as N when $j = 1$ and $j + 1$ is interpreted as 1 when $j = N$. Applying Theorem 12.27 and Lemma 12.32 we obtain that, under the modification of additional arrivals and conditions (6.a) and (6.b), the stationary distribution of the network is of a geometric product form if there exist ρ_j's satisfying

$$\zeta_j \rho_j^{\eta_j} \tilde{A}_j(\rho_j^{-1}) \tilde{B}_{j+1}(\rho_{j+1}, \rho_j^{-1}) \tilde{B}_j(\rho_j, 1) = \tilde{B}_{j+1}(\rho_{j+1}, 1), \quad j = 1,$$
$$\zeta_{j-1} \rho_{j-1}^{\eta_{j-1}} \rho_j^{-\eta_{j-1}} \zeta_j \rho_j^{\eta_j} \tilde{A}_j(\rho_j^{-1}) \tilde{B}_{j+1}(\rho_{j+1}, \rho_j^{-1}) \tilde{B}_j(\rho_j, 1)$$
$$= \zeta_{j-1} \rho_{j-1}^{\eta_{j-1}} \tilde{B}_{j+1}(\rho_{j+1}, 1), \qquad j = 2, \ldots, N-1,$$
$$\zeta_{j-1} \rho_{j-1}^{\eta_{j-1}} \rho_j^{-\eta_{j-1}} \zeta_j \rho_j^{\eta_j} \tilde{A}_j(\rho_j^{-1}) \tilde{B}_j(\rho_j, 1) = \zeta_{j-1} \rho_{j-1}^{\eta_{j-1}}, \qquad j = N,$$

where $\tilde{A}_j$ and $\tilde{B}_j$ are the z-transforms of probability distributions A_j and B_j. □

Example 12.38 (A time-slotted cyclic queue) Suppose N nodes, numbered 1 to N, are cyclically connected as shown in Figure 12.5. At each time slot, b_j customers depart from node j, and of these, a_j customers go to node $j + 1$ with probability $B_j(b_j, a_j)$. From the outside a'_j customers arrive at node j with probability $A_j(a'_j)$, where $j + 1$ is interpreted as 1 when $j = N$. These two movements are assumed to be independent of each other, and a_j, d_j and a'_j may be zero. Assume that all departing batches are removed when there are partial batch departures. If $B_j(1, 0) + B_j(1, 1) = 1$ for all j, then the model can be considered as a time-slotted ring with infinite buffers. This cyclic discrete time queue can be formulated as a multiple batch transfer network by letting

$$p_{j,j}(b_j, a_j) = A_j(a_j)1[b_j = 0],$$
$$p_{j,j+1}(b_j, a_{j+1}) = B_j(b_j, a_{j+1}),$$
$$p_{j,k}(b_j, a_k) = 1[b_j = 0, a_k = 0], \qquad \text{for } k \neq j, j+1,$$

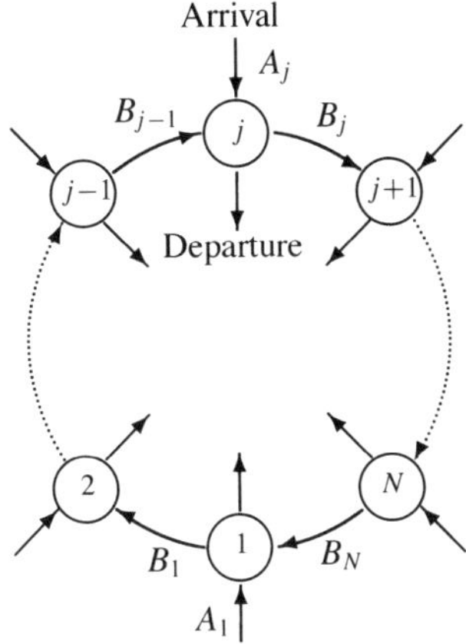

Figure 12.5 A time-slotted cyclic queue

where $j - 1$ is interpreted as N when $j = 1$. Under the modification of the additional arrivals and (6.a) and (6.b), the stationary distribution of the network has a geometric product form if there exist ρ_j's satisfying

$$\tilde{A}_j(\rho_j^{-1})\tilde{B}_{j-1}(\rho_{j-1}, \rho_j^{-1})\tilde{B}_j(\rho_j, 1) = \tilde{B}_{j-1}(\rho_{j-1}, 1), \quad j = 1, 2, \ldots, N, \tag{7.1}$$

where $\tilde{A}_j$ and $\tilde{B}_j$ are the z-transforms of probability distributions A_j and B_j.

Consider the case where at each time slot single customers are released simultaneously from all nodes. Since $\tilde{B}_{j-1}(\rho_{j-1}, 1) = \rho_{j-1}$, equation (7.1) is now reduced to

$$\tilde{A}_j(\rho_j^{-1})(1 - \beta_{j-1} + \beta_{j-1}\rho_j^{-1}) = \rho_j^{-1}, \qquad j = 1, 2, \ldots, N, \tag{7.2}$$

where β_j is the probability that a departing customer from node j goes to node $j + 1$. Since $\tilde{A}_j(\rho_j^{-1})$ is a convex function of ρ_j, it is not hard to see that (7.2) has solutions $\rho_j < 1$ for all j if and only if

$$\tilde{A}_j'(1) + \beta_{j-1} < 1, \qquad j = 1, 2, \ldots, N. \tag{7.3}$$

Since $\tilde{A}_j'(1)$ is the average number of exogenous arrivals at node j, it is intuitively clear that (7.3) is a stability condition. To obtain the decay rate ρ_j, we need to solve (7.2) numerically. For the case where A_j is geometrically distributed with mean α_j, we have

$$\rho_j = \frac{\alpha_j}{1 - \beta_{j-1}}, \qquad j = 1, 2, \ldots, N. \tag{7.4}$$

We apply the iterative algorithm to solve (7.1). Consider four different sets of parameters, which we refer to as cases (i), (ii), (iii) and (iv). In each case, $N = 3$ and the arrival batch size A_j is geometrically distributed with mean α_j, whose values are $\alpha_1 = \alpha_3 = 0.6$ and $\alpha_2 = \alpha_4 = 0.4$. The nonzero values of the B_j's are presented below.

(i) $B_1(1,0) = 0.9,\ B_1(1,1) = 0.1,$
$B_2(1,0) = 0.8,\ B_2(1,1) = 0.2,$
$B_3(1,0) = 0.8,\ B_3(1,1) = 0.2,$

(ii) $B_1(1,0) = 0.9,\ B_1(1,1) = 0.1,$
$B_2(1,0) = 0.8,\ B_2(1,1) = 0.2,$
$B_3(1,0) = 0.5,\ B_3(1,1) = 0.5,$

(iii) $B_1(0,0) = 0.01,\ B_1(1,0) = 0.899,\ B_1(1,1) = 0.1,$
$B_2(0,0) = 0.01,\ B_2(1,0) = 0.799,\ B_2(1,1) = 0.2,$
$B_3(0,0) = 0.01,\ B_3(1,0) = 0.799,\ B_3(1,1) = 0.2,$

(iv) $B_1(0,0) = 0.01,\ B_1(1,0) = 0.897,\ B_1(1,1) = 0.1,$
$B_1(2,1) = 0.01,\ B_1(2,2) = 0.01,$
$B_2(0,0) = 0.01,\ B_2(1,0) = 0.797,\ B_2(1,1) = 0.2,$
$B_2(2,1) = 0.01,\ B_2(2,2) = 0.01,$
$B_3(0,0) = 0.01,\ B_3(1,0) = 0.797,\ B_3(1,1) = 0.2,$
$B_3(2,1) = 0.01,\ B_3(2,2) = 0.01.$

To calculate the decay rates, we need to solve equation (7.1) for ρ_j given ρ_{j-1}. We apply a convergence criterion of 10^{-8} using double-precision calculations.

The results for the decay rates in each example are given in Table 12.1. The decay rates in case (i) are identical to these obtained from (7.4). For case (ii), based on the observation that $\rho_1 > 1$ in (7.4), it is expected that node 1 does not have a stationary distribution. In this case, ρ_1 converges to 1 as discussed in the previous section. For cases (iii) and (iv), it is difficult to obtain closed form expressions for the decay rates. The B_j's in (iii) and (iv) are slightly different from the B_j's in case (i), and so are the decay rates. Though ρ_1 is equal to ρ_3 in (i), ρ_1 is larger than ρ_3 in (iii) and (iv). This is because the probabilities that customers are transferred to node 1 from the previous node is greater than the corresponding probability for node 3.

Table 12.1 Numerical results of decay rates

	ρ_1	ρ_2	ρ_3
(i)	0.750000	0.444444	0.750000
(ii)	1.000000	0.444444	0.750000
(iii)	0.750939	0.444984	0.750767
(iv)	0.751877	0.446094	0.750867

12.8 STATE-DEPENDENT ROUTING AND STRUCTURAL REVERSIBILITY PROPERTIES

As noted in Chapter 4, if we reverse the time in a quasi-reversible network, we obtain another quasi-reversible network. In Corollary 12.21 it is shown that the departure function remains unchanged under time reversal. These results suggest that the networks discussed in Sections 12.3 and 12.4 may have more general structural properties. In this section we study the structural reversibility properties of queueing networks. It turns out that for a large class of networks with state-dependent routing probabilities, the structural reversibility of the departure function is equivalent to the network having linear traffic equations.

We basically work with the network discussed in Section 12.4. The differences are in the dynamics of the departures and the routing, which are now much more general. Let $\boldsymbol{X}_\ell$ be the state of the network at slot boundary ℓ. At the beginning of time slot ℓ, batches of customers

$$\bar{\boldsymbol{D}}_\ell = (D_\ell(0), D_\ell(1), \ldots, D_\ell(N))$$

are released, where $D_\ell(j)$ is the number of customers departing from node j, $j = 0, 1, \ldots, N$, and $D_\ell(0)$ represents the number of arrivals at the network (departure from node 0). The departure vector $\bar{\boldsymbol{D}}_\ell$ is transformed at the end of time slot ℓ into batches of arrivals

$$\bar{\boldsymbol{A}}_\ell = (A_\ell(0), A_\ell(1), \ldots, A_\ell(N)),$$

where $A_\ell(j)$ is the number of arrivals at node j, $j = 0, 1, \ldots, N$, and $A_j(0)$ represents the number of departures from the network (arrivals at node 0). Let $\boldsymbol{D}_\ell$ and $\boldsymbol{A}_\ell$ denote the vectors $\bar{\boldsymbol{D}}_\ell$ and $\bar{\boldsymbol{A}}_\ell$ with the first elements deleted, i.e.,

$$\begin{aligned}\boldsymbol{A}_\ell &= (A_\ell(1), A_\ell(2), \ldots, A_\ell(N)),\\ \boldsymbol{D}_\ell &= (D_\ell(1), D_\ell(2), \ldots, D_\ell(N)).\end{aligned}$$

The relationship

$$\boldsymbol{X}_{\ell+1} = \boldsymbol{X}_\ell - \boldsymbol{D}_\ell + \boldsymbol{A}_\ell$$

is immediate. Assume that customers are preserved when being transferred, i.e.,

$$|\bar{\boldsymbol{D}}_\ell| = |\bar{\boldsymbol{A}}_\ell|.$$

Let $\mathcal{S}$ denote the state space of the network at the slot boundaries, and let $\mathcal{A}$ be the set of possible arrival and departure vectors. The dynamics of the network are specified as follows. Recall that $\boldsymbol{a}$ is the vector $\bar{\boldsymbol{a}}$ with the first element deleted.

(8.a) The departure vector $\bar{\boldsymbol{D}}_\ell$ depends on the history up to slot boundary ℓ only through $\boldsymbol{X}_\ell$. That is, there is a conditional distribution p^{D} such that

$$p^{\mathrm{D}}(\boldsymbol{n}, \bar{\boldsymbol{a}}) = P(\bar{\boldsymbol{D}}_\ell = \bar{\boldsymbol{a}} | \boldsymbol{X}_k, \bar{\boldsymbol{D}}_k, \bar{\boldsymbol{A}}_k, k \leq \ell - 1, \boldsymbol{X}_\ell = \boldsymbol{n}), \quad \bar{\boldsymbol{a}} \in \mathcal{A}, \boldsymbol{n} \in \mathcal{S}.$$

We assume that $p^{\mathrm{D}}(\boldsymbol{n}, \bar{\boldsymbol{a}}) > 0$ only if $\boldsymbol{a} \leq \boldsymbol{n}$.

(8.b) The arrival vector $\bar{\boldsymbol{A}}_\ell$ depends on the history just after slot boundary ℓ only through $\boldsymbol{X}_\ell$ and $\bar{\boldsymbol{D}}_\ell$. This conditional probability function is denoted by

$$r((\boldsymbol{n}, \bar{\boldsymbol{a}}), (\boldsymbol{n}', \bar{\boldsymbol{a}}')) = P(\boldsymbol{X}_{\ell+1} = \boldsymbol{n}', \bar{\boldsymbol{A}}_\ell = \bar{\boldsymbol{a}}' | \boldsymbol{X}_\ell = \boldsymbol{n}, \bar{\boldsymbol{D}}_\ell = \bar{\boldsymbol{a}})$$

for $(\boldsymbol{n}, \bar{\boldsymbol{a}}), (\boldsymbol{n}', \bar{\boldsymbol{a}}') \in \mathcal{S} \times \mathcal{A}$. Clearly, this r vanishes unless $\boldsymbol{n}' = \boldsymbol{n} - \boldsymbol{a} + \boldsymbol{a}'$. We assume that $r((\boldsymbol{n}, \bar{\boldsymbol{a}}), (\boldsymbol{n} - \boldsymbol{a} + \boldsymbol{a}', \bar{\boldsymbol{a}}')) > 0$ if and only if $p^{\mathrm{D}}(\boldsymbol{n}, \bar{\boldsymbol{a}}) > 0$ and $|\bar{\boldsymbol{a}}| = |\bar{\boldsymbol{a}}'|$. That is, a departing customer either joins another node within the network or leaves the network.

Again, p^{D} is called the *departure function*, and for obvious reasons, r is called the *state-dependent routing* function. Note that, in (8.b), $\bar{\boldsymbol{a}}'$ is uniquely determined by $\boldsymbol{n}$, $\bar{\boldsymbol{a}}$ and $\boldsymbol{n}'$ as follows:

$$|\bar{\boldsymbol{a}}'| = |\bar{\boldsymbol{a}}| \quad \text{and} \quad \boldsymbol{a}' = \boldsymbol{n}' - \boldsymbol{n} + \boldsymbol{a}.$$

Hence, there is a function h such that

$$\bar{\boldsymbol{a}}' = h(\boldsymbol{n}', \bar{\boldsymbol{a}}, \boldsymbol{n}).$$

Obviously, $\bar{\boldsymbol{a}} = h(\boldsymbol{n}, \bar{\boldsymbol{a}}', \boldsymbol{n}')$. The routing function r is said to be *state-independent* if there exists a transition probability function $r_{\mathcal{A}}$ on $\mathcal{A}$ such that

$$r((\boldsymbol{n}, \bar{\boldsymbol{a}}), (\boldsymbol{n}', \bar{\boldsymbol{a}}')) = r_{\mathcal{A}}(\bar{\boldsymbol{a}}, \bar{\boldsymbol{a}}')1[\bar{\boldsymbol{a}}' = h(\boldsymbol{n}', \bar{\boldsymbol{a}}, \boldsymbol{n})], \quad \boldsymbol{a} \leq \boldsymbol{n}.$$

For instance, the network model considered in Section 12.4 has state-independent routing.

It follows from assumptions (8.a) and (8.b) that $\{\boldsymbol{X}_\ell\}$ is a Markov chain with transition probability function p^{X} given by

$$p^{\mathrm{X}}(\boldsymbol{n}, \boldsymbol{n}') = \sum_{\bar{\boldsymbol{a}}, \bar{\boldsymbol{a}}' \in \mathcal{A}} p^{\mathrm{D}}(\boldsymbol{n}, \bar{\boldsymbol{a}}) r((\boldsymbol{n}, \bar{\boldsymbol{a}}), (\boldsymbol{n}', \bar{\boldsymbol{a}}')).$$

However, we shall consider another Markov chain $\{(\bar{A}_{\ell-1}, X_\ell)\}$, that has a transition probability function p given by

$$p((\bar{a}, n), (\bar{a}', n')) = p^{\mathrm{D}}(n, \bar{a}'')r((n, \bar{a}''), (n', \bar{a}')),$$

where $\bar{a}'' = h(n, \bar{a}', n')$. We shall see that this Markov chain is useful for studying the dynamics of the time-reversed process of the network.

Suppose the Markov chain $\{(\bar{A}_{\ell-1}, X_\ell)\}$ has a stationary distribution π, i.e.,

$$\pi(\bar{a}, n) = \sum_{(\bar{a}', n') \in \mathcal{A} \times \mathcal{S}} \pi(\bar{a}', n') p((\bar{a}', n'), (\bar{a}, n)). \tag{8.1}$$

If we choose a state space $\mathcal{K}_\pi$ for the Markov chain $\{(\bar{A}_{\ell-1}, X_\ell)\}$ such that

$$\mathcal{K}_\pi = \{(\bar{a}, n) \in \mathcal{A} \times \mathcal{S};\ \pi(\bar{a}, n) > 0\},$$

then there exists a time-reversed process for $\{(\bar{A}_{\ell-1}, X_\ell)\}$ that is a discrete time Markov chain with transition probability function $\tilde{p}$ given by

$$\tilde{p}((\bar{a}, n), (\bar{a}', n')) = \frac{\pi(\bar{a}', n')}{\pi(\bar{a}, n)} p((\bar{a}', n'), (\bar{a}, n)),$$

for $(\bar{a}, n), (\bar{a}', n') \in \mathcal{K}_\pi$. We next define a transition probability function p^* by

$$p^*((n, \bar{a}), (n', \bar{a}')) = \tilde{p}((\bar{a}, n), (\bar{a}', n')). \tag{8.2}$$

Let $\{(X^*_\ell, \bar{D}^*_\ell)\}$ be a Markov chain with this transition probability function. Clearly, this Markov chain has the stationary distribution π^* given by

$$\pi^*(n, \bar{a}) = \pi(\bar{a}, n).$$

Let π^{X} be the stationary distribution of the network state X_ℓ, i.e.,

$$\pi^{\mathrm{X}}(n) = \sum_{\bar{a} \in \mathcal{A}} \pi(\bar{a}, n).$$

The following lemma shows that the network is always structurally reversible when it has a stationary distribution.

Lemma 12.39 *For the network described above, there exist a departure function $p^{\mathrm{D}*}$ and a routing function r^* such that*

$$\pi^{\mathrm{X}}(\boldsymbol{n})p^{\mathrm{D}}(\boldsymbol{n},\bar{\boldsymbol{a}})r((\boldsymbol{n},\bar{\boldsymbol{a}}),(\boldsymbol{n}',\bar{\boldsymbol{a}}')) = \pi^{\mathrm{X}}(\boldsymbol{n}')p^{\mathrm{D}*}(\boldsymbol{n}',\bar{\boldsymbol{a}}')r^*((\boldsymbol{n}',\bar{\boldsymbol{a}}'),(\boldsymbol{n},\bar{\boldsymbol{a}})), \tag{8.3}$$

$$p^*((\boldsymbol{n},\bar{\boldsymbol{a}}),(\boldsymbol{n}',\bar{\boldsymbol{a}}')) = r^*((\boldsymbol{n},\bar{\boldsymbol{a}}),(\boldsymbol{n}',\bar{\boldsymbol{a}}''))p^{\mathrm{D}*}(\boldsymbol{n}',\bar{\boldsymbol{a}}'), \tag{8.4}$$

for all $(\bar{\boldsymbol{a}},\boldsymbol{n}),(\bar{\boldsymbol{a}}',\boldsymbol{n}') \in \mathcal{K}_\pi$ and $\bar{\boldsymbol{a}}'' = h(\boldsymbol{n}',\bar{\boldsymbol{a}},\boldsymbol{n})$. Hence, the Markov chain $\{(\boldsymbol{X}^*_\ell,\bar{\boldsymbol{D}}^*_\ell)\}$ represents a network process with departure function $p^{\mathrm{D}*}$ and routing function r^*.

PROOF. Define a probability measure ν^* on $\mathcal{S}\times\mathcal{A}$ by

$$\nu^*(\boldsymbol{n}',\bar{\boldsymbol{a}}') = \sum_{\boldsymbol{n}\in\mathcal{S}} \pi^{\mathrm{X}}(\boldsymbol{n})p^{\mathrm{D}}(\boldsymbol{n},\bar{\boldsymbol{a}})r((\boldsymbol{n},\bar{\boldsymbol{a}}),(\boldsymbol{n}',\bar{\boldsymbol{a}}')),$$

for each $(\boldsymbol{n}',\bar{\boldsymbol{a}}') \in \mathcal{S}\times\mathcal{A}$, where $\bar{\boldsymbol{a}} = h(\boldsymbol{n},\bar{\boldsymbol{a}}',\boldsymbol{n}')$. It follows from (8.b) that $\nu^*(\boldsymbol{n},\bar{\boldsymbol{a}}) > 0$ if and only if $(\bar{\boldsymbol{a}},\boldsymbol{n}) \in \mathcal{K}_\pi$. Define r^* and $p^{\mathrm{D}*}$ by

$$r^*((\boldsymbol{n}',\bar{\boldsymbol{a}}'),(\boldsymbol{n},\bar{\boldsymbol{a}})) = \frac{1}{\nu^*(\boldsymbol{n}',\bar{\boldsymbol{a}}')}\pi^{\mathrm{X}}(\boldsymbol{n})p^{\mathrm{D}}(\boldsymbol{n},\bar{\boldsymbol{a}})r((\boldsymbol{n},\bar{\boldsymbol{a}}),(\boldsymbol{n}',\bar{\boldsymbol{a}}')),$$
$$p^{\mathrm{D}*}(\boldsymbol{n},\bar{\boldsymbol{a}}) = \frac{\nu^*(\boldsymbol{n},\bar{\boldsymbol{a}})}{\pi^{\mathrm{X}}(\boldsymbol{n})},$$

for $(\bar{\boldsymbol{a}},\boldsymbol{n}),(\bar{\boldsymbol{a}}',\boldsymbol{n}') \in \mathcal{K}_\pi$. Since π^{X} is the marginal distribution of ν^* with respect to the first component, $p^{\mathrm{D}*}(\boldsymbol{n},\bar{\boldsymbol{a}})$ is a conditional probability given $\boldsymbol{n}$, while from the definition of ν^*, r^* is a transition function. Obviously (8.3) is satisfied from the definition of $p^{\mathrm{D}*}$ and r^*. On the other hand, it follows from the definition of p that

$$\pi^{\mathrm{X}}(\boldsymbol{n})p^{\mathrm{D}}(\boldsymbol{n},\bar{\boldsymbol{a}})r((\boldsymbol{n},\bar{\boldsymbol{a}}),(\boldsymbol{n}',\bar{\boldsymbol{a}}')) = \pi^{\mathrm{X}}(\boldsymbol{n})p((\bar{\boldsymbol{a}}'',\boldsymbol{n}),(\bar{\boldsymbol{a}}',\boldsymbol{n}')),$$

for any $\bar{\boldsymbol{a}}'' \in \mathcal{A}$. Substituting this into the left-hand side of (8.3), and using (8.2), we obtain

$$\begin{aligned}&\pi^{\mathrm{X}}(\boldsymbol{n})\pi(\bar{\boldsymbol{a}}',\boldsymbol{n}')p^*((\boldsymbol{n}',\bar{\boldsymbol{a}}'),(\boldsymbol{n},\bar{\boldsymbol{a}}''))\\ &\quad= \pi^{\mathrm{X}}(\boldsymbol{n}')p^{\mathrm{D}*}(\boldsymbol{n}',\bar{\boldsymbol{a}}')r^*((\boldsymbol{n}',\bar{\boldsymbol{a}}'),(\boldsymbol{n},\bar{\boldsymbol{a}}))\pi(\bar{\boldsymbol{a}}'',\boldsymbol{n}),\end{aligned} \tag{8.5}$$

where $\bar{\boldsymbol{a}}' = h(\boldsymbol{n}',\bar{\boldsymbol{a}},\boldsymbol{n})$. Fix $\bar{\boldsymbol{a}}'$ and $\boldsymbol{n}'$ such that $(\bar{\boldsymbol{a}}',\boldsymbol{n}') \in \mathcal{K}_\pi$. Summing (8.5) over all $\bar{\boldsymbol{a}}'' \in \mathcal{A}$, and canceling $\pi^{\mathrm{X}}(\boldsymbol{n})$ on both sides yields

$$\pi(\bar{\boldsymbol{a}}',\boldsymbol{n}')\sum_{\bar{\boldsymbol{a}}''\in\mathcal{A}} p^*((\boldsymbol{n}',\bar{\boldsymbol{a}}'),(\boldsymbol{n},\bar{\boldsymbol{a}}'')) = \pi^{\mathrm{X}}(\boldsymbol{n}')p^{\mathrm{D}*}(\boldsymbol{n}',\bar{\boldsymbol{a}}')r^*((\boldsymbol{n}',\bar{\boldsymbol{a}}'),(\boldsymbol{n},\bar{\boldsymbol{a}})).$$

Summing both sides of the above equation over all possible $\boldsymbol{n}$ yields

$$\pi(\bar{\boldsymbol{a}}', \boldsymbol{n}') = \pi^{\mathrm{X}}(\boldsymbol{n}')p^{\mathrm{D}*}(\boldsymbol{n}', \bar{\boldsymbol{a}}'). \tag{8.6}$$

Substituting this into (8.5) and using it again for $\bar{\boldsymbol{a}}' = \bar{\boldsymbol{a}}''$ and $\boldsymbol{n}' = \boldsymbol{n}$, we obtain (8.4). This completes the proof. □

Let

$$\nu(\boldsymbol{n}, \bar{\boldsymbol{a}}) = \pi^{\mathrm{X}}(\boldsymbol{n})p^{\mathrm{D}}(\boldsymbol{n}, \bar{\boldsymbol{a}}),$$
$$\nu^*(\boldsymbol{n}, \bar{\boldsymbol{a}}) = \pi^{\mathrm{X}}(\boldsymbol{n})p^{\mathrm{D}*}(\boldsymbol{n}, \bar{\boldsymbol{a}}).$$

It follows from Lemma 12.39 that

$$\nu^*(\boldsymbol{n}, \bar{\boldsymbol{a}}) = \sum_{(\bar{\boldsymbol{a}}', \boldsymbol{n}') \in \mathcal{K}_\pi} \nu(\boldsymbol{n}', \bar{\boldsymbol{a}}')r((\boldsymbol{n}', \bar{\boldsymbol{a}}'), (\boldsymbol{n}, \bar{\boldsymbol{a}})), \qquad (\boldsymbol{n}, \bar{\boldsymbol{a}}) \in \mathcal{K}_\pi. \tag{8.7}$$

This equation is referred to as the *traffic equation*. We interpret $\nu^*(\boldsymbol{n}, \bar{\boldsymbol{a}})$ here as the rate at which the network process enters state $\boldsymbol{n}$ due to arrivals of batch $\bar{\boldsymbol{a}}$, while $\nu(\boldsymbol{n}, \bar{\boldsymbol{a}})$ is the rate at which the process leaves state $\boldsymbol{n}$ due to departures of batch $\bar{\boldsymbol{a}}$. In particular, if $\nu = \nu^*$, the traffic equation (8.7) becomes

$$\nu(\boldsymbol{n}, \bar{\boldsymbol{a}}) = \sum_{(\bar{\boldsymbol{a}}', \boldsymbol{n}') \in \mathcal{K}_\pi} \nu(\boldsymbol{n}', \bar{\boldsymbol{a}}')r((\boldsymbol{n}', \bar{\boldsymbol{a}}'), (\boldsymbol{n}, \bar{\boldsymbol{a}})). \tag{8.8}$$

This traffic equation is said to be *linear*. Note that (8.8) can be viewed as a local balance with respect to the batch movements in state $\boldsymbol{n}$.

We are concerned with the possible forms of the departure function p^{D} that yield closed form stationary distributions for an arbitrary routing function r. A partial answer to this problem for the case of state-independent routing is given in Section 12.4. Observe that if the traffic equations are linear, then, because of equation (8.8), $\nu(\boldsymbol{n}, \bar{\boldsymbol{a}}) = \pi^{\mathrm{X}}(\boldsymbol{n})p^{\mathrm{D}}(\boldsymbol{n}, \bar{\boldsymbol{a}})$ is a stationary measure of r, which can be used to identify the stationary distribution and possible departure functions.

Thus, we first consider the stationary measure for the transition probability function r. To this end, we impose an additional condition.

(8.c) The Markov chain defined by transition function r does not have any transient states.

We refer to (8.c) as a *recurrence condition*. This condition may not be satisfied for some networks, e.g., networks with blocking. However, it is in general not a very restrictive assumption. Let

$$\mathcal{K}_r = \{(\boldsymbol{n}, \bar{\boldsymbol{a}}) \in \mathcal{S} \times \mathcal{A}; r((\boldsymbol{n}', \bar{\boldsymbol{a}}'), (\boldsymbol{n}, \bar{\boldsymbol{a}})) > 0 \text{ for some } (\boldsymbol{n}', \bar{\boldsymbol{a}}') \in \mathcal{S} \times \mathcal{A}\}$$

be the state space of the Markov chain with transition function r. From conditions (8.a) and (8.b) it follows that $\mathcal{K}_r$ is a subset of $\{(\boldsymbol{n}, \bar{\boldsymbol{a}}) \in \mathcal{S} \times \mathcal{A}; \bar{\boldsymbol{a}} \leq \boldsymbol{n}\}$, and that it includes $\mathcal{K}_\pi$ in the sense that $(\bar{\boldsymbol{a}}, \boldsymbol{n}) \in \mathcal{K}_\pi$ implies $(\boldsymbol{n}, \bar{\boldsymbol{a}}) \in \mathcal{K}_r$.

As seen in the case of the state-independent routing, r is in general not irreducible, and its state space $\mathcal{K}_r$ may be decomposable into closed sets. To see what these closed sets look like, we consider the following discrete time Jackson network.

Example 12.40 (Irreducible sets for a Jackson network) Consider the discrete time Jackson network of Example 12.24. In this network, at most one customer is allowed to move in each time slot. Thus $\mathcal{A} = \{\bar{\mathbf{e}}_j\ ;\ j = 0, 1, \ldots, N\}$, where $\bar{\mathbf{e}}_j$ is an $(N + 1)$-dimensional unit vector whose jth element is 1. At the beginning of a time slot a customer departs from node j with probability q_j, and joins node k with probability r_{jk}, where $j = 0$ and $k = 0$ represent the outside. Clearly, conditions (8.a) and (8.b) are satisfied. This network can be described in the present framework by

$$p^{\mathrm{D}}(\boldsymbol{n}, \bar{\mathbf{e}}_j) = q_j,$$

$$r((\boldsymbol{n}, \bar{\mathbf{e}}_j), (\boldsymbol{n}', \bar{\mathbf{e}}_k)) = r_{jk} 1[\boldsymbol{n}' = \boldsymbol{n} - \mathbf{e}_j + \mathbf{e}_k],$$

for $\boldsymbol{n} \geq \mathbf{e}_j$. Thus, r is state-independent. Assume that $\{r_{jk}\}$ is irreducible. Then, the r satisfies (8.c). For this network there is, with a positive probability, a transition path from $(\boldsymbol{n}, \bar{\mathbf{e}}_j)$ to $(\boldsymbol{n}', \bar{\mathbf{e}}_k)$ if and only if

$$\boldsymbol{n} - \mathbf{e}_j = \boldsymbol{n}' - \mathbf{e}_k.$$

Hence the set $\mathcal{K}_r$ can be partitioned into irreducible subsets of the form

$$\{(\boldsymbol{n}, \bar{\mathbf{e}}_j) \in \mathcal{K}_r\ ;\ \boldsymbol{n} - \mathbf{e}_j = \boldsymbol{m},\ j = 0, 1, \ldots, N\}$$

for each $\boldsymbol{m} \in \mathcal{S}$, or

$$\{(\boldsymbol{m} + \mathbf{e}_j, \bar{\mathbf{e}}_j)\ ;\ j = 0, 1, \ldots, N\}$$

for each $\boldsymbol{m} \in \mathcal{S}$. □

Similar to Example 12.40, we can decompose the state space $\mathcal{K}_r$ into the disjoint closed sets

$$U_{\boldsymbol{m},\ell} = \{(\boldsymbol{n}, \bar{\boldsymbol{a}}) \in \mathcal{K}_r\ ;\ \boldsymbol{n} - \boldsymbol{a} = \boldsymbol{m},\ |\bar{\boldsymbol{a}}| = \ell\}$$

for each $\boldsymbol{m} \in \mathcal{S}$ and nonnegative integer ℓ. These disjoint sets can also be expressed as

$$U_{\boldsymbol{m},\ell} = \{(\boldsymbol{m} + \boldsymbol{a}, \bar{\boldsymbol{a}}) \in \mathcal{K}_r \,;\, |\bar{\boldsymbol{a}}| = \ell\}$$

for each $\boldsymbol{m} \in \mathcal{S}$. Note that $U_{\boldsymbol{n},\ell}$ is a finite set. Let $\langle \boldsymbol{n}, \bar{\boldsymbol{a}} \rangle$ denote a representative element in each irreducible set containing $(\boldsymbol{n}, \bar{\boldsymbol{a}}) \in \mathcal{K}_r$. This irreducible set is denoted by $V_{\langle \boldsymbol{n}, \bar{\boldsymbol{a}} \rangle}$. Since $V_{\langle \boldsymbol{n}, \bar{\boldsymbol{a}} \rangle}$ is included in $U_{\boldsymbol{m},\ell}$ which contains $(\boldsymbol{n}, \bar{\boldsymbol{a}})$, $V_{\langle \boldsymbol{n}, \bar{\boldsymbol{a}} \rangle}$ is finite. Hence, there exists a unique stationary distribution with respect to r on $V_{\langle \boldsymbol{n}, \bar{\boldsymbol{a}} \rangle}$. From these distributions we construct a stationary measure ν_0 on $\mathcal{K}_r$ such that $\nu_0(\boldsymbol{n}, \bar{\boldsymbol{a}}) > 0$ for all $(\boldsymbol{n}, \bar{\boldsymbol{a}}) \in \mathcal{K}_r$, and $\nu_0(\boldsymbol{n}, \bar{\boldsymbol{a}}) = 0$ for $(\boldsymbol{n}, \bar{\boldsymbol{a}}) \in \mathcal{S} \times \mathcal{A} \setminus \mathcal{K}_r$. Note that ν_0 is not unique, but we arbitrarily select one of them. Thus, ν_0 satisfies

$$\nu_0(\boldsymbol{n}, \bar{\boldsymbol{a}}) = \sum_{(\boldsymbol{n}', \bar{\boldsymbol{a}'}) \in \mathcal{K}_r} \nu_0(\boldsymbol{n}', \bar{\boldsymbol{a}'}) r((\boldsymbol{n}', \bar{\boldsymbol{a}'}), (\boldsymbol{n}, \bar{\boldsymbol{a}})) > 0\,, \quad (\boldsymbol{n}, \bar{\boldsymbol{a}}) \in \mathcal{K}_r.$$

Define a transition function $\tilde{r}$ on $\mathcal{K}_r$ by

$$\tilde{r}((\boldsymbol{n}, \bar{\boldsymbol{a}}), (\boldsymbol{n}', \bar{\boldsymbol{a}'})) = \frac{\nu_0(\boldsymbol{n}', \bar{\boldsymbol{a}'})}{\nu_0(\boldsymbol{n}, \bar{\boldsymbol{a}})} r((\boldsymbol{n}', \bar{\boldsymbol{a}'}), (\boldsymbol{n}, \bar{\boldsymbol{a}})), \quad (\boldsymbol{n}, \bar{\boldsymbol{a}}) \in \mathcal{K}_r.$$

The function $\tilde{r}$ is a transition function of the time-reversed Markov chain. Note that $\tilde{r}$ is uniquely determined on $\mathcal{K}_r$, even though ν_0 is not unique. The next theorem characterizes the networks with linear traffic equations.

Theorem 12.41 *Assume that the discrete time queueing network has stationary distribution π. The following three statements are equivalent.*

(i) The departure function is structurally reversible, i.e., it remains unchanged under time reversal, $p^{\mathrm{D}*} = p^{\mathrm{D}}$.

(ii) The traffic equations are linear, i.e., equations (8.8) hold.

(iii) The recurrence condition (8.c) holds, and there exist a nonnegative function Ψ on $\mathcal{K}_r$ and a positive function Φ on $\mathcal{S}$ such that the departure function p^{D} is given by

$$p^{\mathrm{D}}(\boldsymbol{n}, \bar{\boldsymbol{a}}) = \frac{\Psi(\langle \boldsymbol{n}, \bar{\boldsymbol{a}} \rangle) \nu_0(\boldsymbol{n}, \bar{\boldsymbol{a}})}{\Phi(\boldsymbol{n})}, \qquad (\boldsymbol{n}, \bar{\boldsymbol{a}}) \in \mathcal{K}_r, \tag{8.9}$$

where ν_0 is a stationary measure of r.

Conversely, instead of assuming the existence of π, suppose that (iii) is satisfied and

$$c^{-1} \equiv \sum_{(\boldsymbol{n},\bar{\boldsymbol{a}})\in\mathcal{K}_r} \Psi(\langle \boldsymbol{n}, \bar{\boldsymbol{a}}\rangle)\nu_0(\boldsymbol{n}, \bar{\boldsymbol{a}}) < \infty.$$

Then, the stationary distributions π^{X} and π of $\boldsymbol{X}_\ell$ and $(\bar{\boldsymbol{A}}_{\ell-1}, \boldsymbol{X}_\ell)$ are, respectively, given by

$$\pi^{\mathrm{X}}(\boldsymbol{n}) = c\Phi(\boldsymbol{n}), \qquad \boldsymbol{n} \in \mathcal{S},$$

$$\pi(\bar{\boldsymbol{a}}, \boldsymbol{n}) = c\Psi(\langle \boldsymbol{n}, \bar{\boldsymbol{a}}\rangle)\nu_0(\boldsymbol{n}, \bar{\boldsymbol{a}}), \qquad (\bar{\boldsymbol{a}}, \boldsymbol{n}) \in \mathcal{K}_r,$$

and $r^* = \tilde{r}$ holds on $\mathcal{K}_r$.

PROOF. Note that either (i) or (ii) implies (iii). From (8.7) it follows that (i) is equivalent to (ii) under (iii). Since ν_0 is unique up to a constant multiple on the irreducible sets, (ii) implies

$$\frac{\pi^{\mathrm{X}}(\boldsymbol{n})p^{\mathrm{D}}(\boldsymbol{n}, \bar{\boldsymbol{a}})}{\nu_0(\boldsymbol{n}, \bar{\boldsymbol{a}})} = \frac{\pi^{\mathrm{X}}(\boldsymbol{n}')p^{\mathrm{D}}(\boldsymbol{n}', \bar{\boldsymbol{a}'})}{\nu_0(\boldsymbol{n}', \bar{\boldsymbol{a}'})}, \tag{8.10}$$

for $(\boldsymbol{n}, \bar{\boldsymbol{a}}) \in V_{\langle \boldsymbol{n}', \bar{\boldsymbol{a}'}\rangle}$ satisfying $r((\boldsymbol{n}, \bar{\boldsymbol{a}}), (\boldsymbol{n}', \bar{\boldsymbol{a}'})) > 0$. Since $V_{\langle \boldsymbol{n}', \bar{\boldsymbol{a}'}\rangle}$ is irreducible, the left-hand side of (8.10) is the same for all $(\boldsymbol{n}, \bar{\boldsymbol{a}}) \in V_{\langle \boldsymbol{n}', \bar{\boldsymbol{a}'}\rangle} = V_{\langle \boldsymbol{n}, \bar{\boldsymbol{a}}\rangle}$. Denote this value by $\Psi(\langle \boldsymbol{n}, \bar{\boldsymbol{a}}\rangle)$. We obtain (8.9) by choosing $\Phi = \pi^{\mathrm{X}}$. Hence (ii) implies (iii).

On the other hand, assume that (iii) is satisifed. Let $\pi^{\mathrm{X}} = \Phi$. Then we have (8.8), since $\Psi(\langle \boldsymbol{n}, \bar{\boldsymbol{a}}\rangle)\nu_0(\boldsymbol{n}, \bar{\boldsymbol{a}})$ is also a stationary measure of r. Thus, we have (ii). Similarly, we have (8.1) by setting $\pi(\boldsymbol{n}, \bar{\boldsymbol{a}}) = \Psi(\langle \boldsymbol{n}, \bar{\boldsymbol{a}}\rangle)\nu_0(\bar{\boldsymbol{a}}, \boldsymbol{n})$. Note that π may not be a probability distribution. However, if c^{-1} is finite, then π can be normalized and the stationary distribution is obtained. Since π^{X} exists, it follows from the first part of the proof that $r^* = \tilde{r}$ on $\mathcal{K}_r$. □

Remark 12.42 Since $V_{\langle \boldsymbol{n}, \bar{\boldsymbol{a}}\rangle}$ is a subset of $U_{\boldsymbol{n}-\boldsymbol{a}, |\bar{\boldsymbol{a}}|}$, a simple example of Ψ is

$$\Psi(\langle \boldsymbol{n}, \bar{\boldsymbol{a}}\rangle) = f(\boldsymbol{n} - \boldsymbol{a}, |\bar{\boldsymbol{a}}|), \qquad (\boldsymbol{n}, \bar{\boldsymbol{a}}) \in \mathcal{K}_r,$$

where f is an arbitrary nonnegative function on $\mathcal{S} \times \{0, 1, 2, \ldots, \}$. This form is further simplified to

$$\Psi(\langle \boldsymbol{n}, \bar{\boldsymbol{a}}\rangle) = g(\boldsymbol{n} - \boldsymbol{a}), \qquad (\boldsymbol{n}, \bar{\boldsymbol{a}}) \in \mathcal{K}_r,$$

where g is an arbitrary nonnegative function on $\mathcal{S}$. The departure function of this form has been used in Section 12.4.

In view of Remark 12.42, Theorem 12.41 generalizes Theorems 12.17 and 12.19, in which condition (iii) is satisfied and a state-independent routing is assumed. In the latter case, (ii) implies that $r_{\mathcal{A}}(\bar{\boldsymbol{a}}, \bar{\boldsymbol{a}}') > 0$ only if $|\bar{\boldsymbol{a}}| = |\bar{\boldsymbol{a}}'|$. Hence, there exists a stationary measure ν of $r_{\mathcal{A}}$, and ν_0 can be given by $\nu_0(\boldsymbol{n}, \bar{\boldsymbol{a}}) = \nu(\bar{\boldsymbol{a}})$. This, together with (8.3) and (8.9), yields

$$r^*((\boldsymbol{n}', \bar{\boldsymbol{a}}'), (\boldsymbol{n}, \bar{\boldsymbol{a}})) = \frac{\nu(\bar{\boldsymbol{a}})}{\nu(\bar{\boldsymbol{a}}')} r_{\mathcal{A}}(\bar{\boldsymbol{a}}, \bar{\boldsymbol{a}}') 1[\bar{\boldsymbol{a}}' = h(\boldsymbol{n}, \bar{\boldsymbol{a}}, \boldsymbol{n}')].$$

Thus, we have the next corollary.

Corollary 12.43 *For the network with state-independent routing of Section 12.4, the time-reversed process is also a network with state-independent routing r^*, and the departure function of the reversed network process is the same as that of the original network, i.e., p^{D}.*

The structure of the reversed process of Corollary 12.43 is particularly useful when studying departure processes from the nodes. For instance, in the network of Theorem 12.17, batches arrive at the nodes from the outside according to independent Poisson batch Bernoulli processes, so it follows from the structure of the reversed process that the departure batches from the nodes to the outside are also independent Poisson batch Bernoulli processes.

12.9 REFERENCE NOTES

The early and late arrival formulations are discussed in Hunter (1983) and Takagi (1993). However, there does not seem to be a clear distinction in the literature between slot boundaries and slot insides, although the terminology slot boundary is standard. Related topics can be found in Woodward (1994). A single-server discrete time queue with batch Bernoulli arrivals is thoroughly studied in Takagi (1993). The discrete time quasi-reversibility of Section 12.4 was first introduced for the Poisson batch Bernoulli arrivals by Walrand (1983*a*), and extended by Ōsawa (1994) for general batch size distributions. However, their definitions use the independent properties (2.a)–(2.d). The definition of quasi-reversibility in this chapter is new. Theorem 12.9 slightly extends the results of Ōsawa (1994). Example 12.10 and the tandem network of Bernoulli arrival geometric service queues are due to Hsu and Burke (1976), Bharath and Kumar (1980), and Ōsawa (1989). The truncated geometric batch size for departures in Example 12.11 was originally studied in Pujolle, Claude and Seret (1985). The S-queue of Example 12.12 and Theorem 12.17 are due to Walrand (1983*a*). Theorem 12.19 extends the results of

Henderson and Taylor (1990), in which $p^{\mathrm{D}}(\boldsymbol{n}, \boldsymbol{n})$ is (implicitly) assumed to be (4.8). Similar results are also obtained in Boucherie and Van Dijk (1991), and extended in Miyazawa (1993). The present formulation of Theorem 12.19 and its proof are slightly different from these references. Sections 12.5–12.7 are based on Yamashita and Miyazawa (1998), and Section 12.8 is taken from Miyazawa (1997). Special cases of the results in Section 12.8 can be found in Boucherie and Van Dijk (1991), Towsley (1980), and Serfozo (1989*a*, 1993). Exercise 12.7 is taken from Boucherie (1992), and Exercise 12.9 from Henderson and Taylor (1990).

EXERCISES

12.1 Derive the transition probability function p^{Y} for the Markov chain $\{Y_\ell\}$, and show that π^{Y} defined by (1.4) is the stationary distribution of p^{Y}.

12.2 Find the transition probability function for the Markov chain $\{(A_{\ell-1}, X_\ell)\}$ for the single queue considered in Section 12.1.

12.3 Consider the single node queueing model in Section 12.1. Assume that the arrival and departure functions, denoted by $p^{\mathrm{A}+}$ and $p^{\mathrm{D}-}$, are decomposed as follows:

$$p^{\mathrm{A}+}(n, i) = p^{\mathrm{A}}(n, i) + p^{\mathrm{I}+}(n, i), \qquad n, i \geq 0,$$
$$p^{\mathrm{D}-}(n, i) = p^{\mathrm{D}}(n, i) + p^{\mathrm{I}-}(n, i), \qquad n \geq i \geq 0,$$

where $p^{\mathrm{I}+}$ and $p^{\mathrm{I}-}$ are considered internal transitions. Assume that the queue has the stationary distributions π^{X} and π^{Y} of the numbers of customers in the system at the slot boundaries and during the slot insides, respectively. The queue is called quasi-reversible if there exist positive numbers λ and μ and batch size distributions $\{a(n)\}$ and $\{b(n)\}$ such that

$$p^{\mathrm{A}}(n, i) = \lambda a(i), \qquad n, i \geq 0,$$
$$\pi^{\mathrm{X}}(n + i) p^{\mathrm{D}}(n + i, i) = \mu b(i) \pi^{\mathrm{Y}}(n), \qquad n, i \geq 0.$$

(a) Consider the queue with

$$p^{\mathrm{A}}(n, 1) = p^{\mathrm{A}+}(n, 1) = p, \qquad n \geq 0,$$
$$p^{\mathrm{I}}(n, 0) = p^{\mathrm{I}+}(n, 0) = 1 - p, \qquad n \geq 0,$$
$$p^{\mathrm{D}}(n, i) = p^{\mathrm{D}+}(n, i) = b(i), \qquad n \geq i \geq 0,$$
$$p^{\mathrm{I}-}(n, 0) = \bar{b}(n + 1), \qquad n \geq 0,$$

where $\{b(n)\}$ is a probability mass function on $\{1, 2, \ldots\}$, and $\bar{b}(n) = \sum_{\ell=n}^{\infty} b(\ell)$. Show that this model has, during the slot insides, the stationary measure

$$\pi^{\mathrm{Y}}(n) = \rho^n, \qquad n \geq 0,$$

where ρ is a solution of the equation

$$((1-p)\rho + p)\tilde{B}(\rho) = \rho,$$

and $\tilde{B}(z) = \sum_{\ell=1}^{\infty} z^{\ell} b(\ell)$.

(b) Find conditions under which this queue is stable.

(c) Under the condition obtained in (b), show that the queue is quasi-reversible, and that the arrival batch sizes and the departure batch sizes have different distributions.

12.4 Show that quasi-reversible queues with Poisson batch Bernoulli processes do not include First–Come First–Served (FCFS) queues. Give an example of a quasi-reversible queue with a non-Poisson Bernoulli arrival process and FCFS service discipline.

12.5 Consider a single queue with Bernoulli arrivals and state-dependent service rates (2.11). Verify that the queue is quasi-reversible and its stationary distributions are (2.12) and (2.13).

12.6 Consider a single queue where the arrival functions are uniformly distributed, i.e., for some fixed $N \geq 0$,

$$a(i) = \frac{1}{N+1}, \qquad i = 0, 1, \ldots, N.$$

Identify a departure function for this queue so that it is quasi-reversible. Find the corresponding stationary distributions.

12.7 Consider Example 12.12. Assume that the departure functions are now

$$p^{\mathrm{D}}(n, i) = c(n) \frac{\gamma(n) \cdots \gamma(n-i+1)}{i!} \frac{(1-\gamma(n-i)) \cdots (1-\gamma(1))}{(n-i)!},$$

where $c(n)$ is determined by $\sum_{i=0}^{n} p^{\mathrm{D}}(n, i) = 1$, or

$$p^{\mathrm{D}}(n, i) = c(n) \frac{\gamma(n) \cdots \gamma(n-i+1)}{i!} (1-\gamma(n-i)),$$

where $c(n)$ is determined by $\sum_{i=0}^{n} p^{\mathrm{D}}(n, i) = 1$. Show that in these cases the queue is also quasi-reversible. Give an interpretation for each departure function.

12.8 Show that the single queue of Section 12.2 is quasi-reversible if and only if its departure function p^{D} can be written as

$$p^{\mathrm{D}}(n,i) = P(A = i \mid A + B = n), \qquad i = 0, 1, \dots, n,\ n \geq 0,$$

where A and B are some independent nonnegative integer-valued random variables.

12.9 Show that the result of Theorem 12.15 cannot be extended to networks with splitting or merging by considering an example of a network with two queues. At each time slot, there is an arrival from the outside with probability 0.5 and there are no arrivals with probability 0.5. If an arrival occurs, it is routed to node 1 with probability 0.5 and it is routed to node 2 with probability 0.5. If there are customers present at queue i, $i = 1, 2$, then there is a departure with probability 0.5, and no departure with probability 0.5. Any departure from queues 1 or 2 leaves the system. Show that the stationary probability of the queue does not have a product form.

12.10 Consider a quasi-reversible queue subject to a batch Bernoulli arrival process, and suppose the stationary queue length distribution $\pi^{\mathrm{Y}}(n)$ is geometric. Find the classes of departure functions with the arrival batch size distributions being

(a) geometric;

(b) negative binomial, i.e., for $\eta > 0$ and $0 < \lambda < 1$,

$$a(n) = \binom{\eta + n - 1}{n}(1-\lambda)^{\eta}\lambda^{n}, \qquad n = 0, 1, \dots.$$

12.11 For each $\mathcal{A} \subset \mathcal{N}$, let $\boldsymbol{a}[\mathcal{U}] \in \mathcal{S}$ be defined in such a way that the jth element of $\boldsymbol{a}[\mathcal{U}]$ is a_j if $j \in \mathcal{U}$ and 0 otherwise. Let $\mathcal{S}[\mathcal{A}] = \{\boldsymbol{a}[\mathcal{A}]\,;\ \boldsymbol{a} \in \mathcal{S}\}$, and

$$\gamma_{\mathcal{A}}(\boldsymbol{a}) = \sum_{\boldsymbol{c}\in\mathcal{S}} \sum_{\boldsymbol{b}\in\mathcal{S}[\mathcal{N}-\mathcal{A}]} \rho^{\boldsymbol{c}} p^{\mathrm{T}}(\boldsymbol{c}, \boldsymbol{b}+\boldsymbol{a}), \qquad \boldsymbol{a} \in \mathcal{S}[\mathcal{A}],$$

which can be interpreted as the total arrival rate of the batches whose sizes over $\mathcal{A}$ equal $\boldsymbol{a} \in \mathcal{S}[\mathcal{A}]$. Show that the above equation can be written as

$$\sum_{\boldsymbol{a}\in\mathcal{S}[\mathcal{A}]} \gamma_{\mathcal{A}}(s)\rho^{-\boldsymbol{a}} = \sum_{\boldsymbol{b}\in\mathcal{S}[\mathcal{A}]} \sum_{\boldsymbol{c}\in\mathcal{S}[\mathcal{N}-\mathcal{A}]} \sum_{\boldsymbol{a}\in\mathcal{S}} \rho^{\boldsymbol{c}+\boldsymbol{b}} p^{\mathrm{T}}(\boldsymbol{c}+\boldsymbol{b}, \boldsymbol{a})\rho^{-\boldsymbol{b}}.$$

Give an interpretation of this expression as a traffic equation.

12.12 Consider a three-node network the transfer probability functions of which are given by

$$p^{\mathrm{T}}(\boldsymbol{a}, \boldsymbol{a}') = \begin{cases} \alpha_1, & \boldsymbol{a} = (0,0,0),\ \boldsymbol{a}' = (1,0,0), \\ \alpha_2, & \boldsymbol{a} = (0,0,0),\ \boldsymbol{a}' = (0,1,0), \\ \alpha_3, & \boldsymbol{a} = (1,1,0),\ \boldsymbol{a}' = (0,0,1), \\ \alpha_4, & \boldsymbol{a} = (0,0,1),\ \boldsymbol{a}' = (0,0,0), \\ 0, & \text{otherwise,} \end{cases}$$

where $\alpha_1 + \alpha_2 + \alpha_3 + \alpha_4 = 1$. This departure function implies that items from nodes 1 and 2 are combined into one item, and transferred to node 3 from which the item leaves the network. Write down condition (5.5), and show that it does not have a solution $\boldsymbol{\rho}$.

Appendix A

Basic Concepts in Probability Theory

This book assumes an introductory level of probability theory, and uses some basic notions such as mean and distribution function, mainly for discrete random variables. For the major part we do not need a measure-theoretic level of probability theory. However, in two places in Chapters 3 and 12 we do use conditional probabilities based on σ-fields, which we call *history*. In this appendix we present a formal framework of probability theory that covers this notion as well as some other related concepts.

A.1 PROBABILITY SPACE

Let Ω be a set, and $\mathcal{F}$ be a family, i.e., a set of subsets of Ω, such that

(i) $\Omega \in \mathcal{F}$,

(ii) if $A \in \mathcal{F}$, then $A^c \in \mathcal{F}$, where $A^c = \Omega \setminus A$,

(iii) if $A_n \in \mathcal{F}$ for $n = 1, 2, \ldots$, then $\cup_{n=1}^{\infty} A_n \in \mathcal{F}$.

This $\mathcal{F}$ is called a *σ-field* on Ω, and the pair $(\Omega, \mathcal{F})$ is called a *measurable space*. A measurable space $(\Omega, \mathcal{F})$ represents a random phenomenon, e.g., an experiment. An element $\omega \in \Omega$ is called a *sample*, which represents a possible outcome, and Ω is known as the *sample space* representing the set of all possible outcomes. An element of $\mathcal{F}$ is called an *event*, which is a collection of samples.

Let P be a function from the σ-field $\mathcal{F}$ of the measurable space $(\Omega, \mathcal{F})$ to $\mathbb{R} = (-\infty, +\infty)$ such that (i′) $0 \le P(A) \le 1$ for any $A \in \mathcal{F}$,

(ii′) $P(\Omega) = 1$,

(iii′) for a sequence of disjoint events $A_n \in \mathcal{F}$, $n = 1, 2, \ldots$, i.e., $A_n \cap A_m = \emptyset$ for $n \neq m$,

$$P\left(\bigcup_{n=1}^{\infty} A_n\right) = \sum_{n=1}^{\infty} P(A_n).$$

This P is called a *probability measure* on the measurable space $(\Omega, \mathcal{F})$, and $P(A)$ is called the *probability* of event A. The triplet $(\Omega, \mathcal{F}, P)$ is called a probability space.

Note that probabilities are not defined with respect to samples but rather with respect to events. So it is important to distinguish between samples and events. If Ω is countable, $\mathcal{F}$ is usually the family of all subsets of Ω, i.e., $\mathcal{F} = 2^{\Omega}$, and P can be defined with respect to each sample of Ω. In this case, we can define the probability of $A \in \mathcal{F}$, i.e., a subset of Ω, as

$$P(A) = \sum_{\omega \in A} P(\{\omega\}).$$

If Ω is not countable, then $\mathcal{F}$ is, in general, smaller than 2^{Ω}. For instance, if $\Omega = \mathbb{R}$, then $\mathcal{F}$ is chosen to be the smallest σ-field that contains all semi-open intervals $(a, b]$; this σ-field is called the *Borel field* on $\mathbb{R}$, and is denoted by $\mathcal{B}(\mathbb{R})$.

Let X be a function from Ω to $\mathbb{R}$ such that

$$\{\omega \in \Omega; X(\omega) \le a\} \in \mathcal{F}, \qquad \text{for all } a \in \mathbb{R}. \tag{1.1}$$

The function X is called a *random variable* on the measurable space $(\Omega, \mathcal{F})$. It can be shown that (1.1) holds if and only if

$$\{\omega \in \Omega; X(\omega) \in B\} \in \mathcal{F}, \qquad \text{for all } B \in \mathcal{B}(\mathbb{R}). \tag{1.2}$$

In what follows we simply denote the above event by $\{X \in B\}$. Let $\sigma(X)$ be the set of all events of (1.2), i.e., the events generated by X. Then it is easily seen that $\sigma(X)$ is a subset of $\mathcal{F}$ and it is a σ-field on Ω.

For a random variable X, its distribution function is defined as

$$F(t) = P(X \le t), \qquad t \in \mathbb{R}.$$

Clearly a distribution function $F(t)$ is nondecreasing in t.

Sometimes a vector is needed to describe a random phenomenon, e.g., the number of customers at different nodes of a network. A vector-valued function $\boldsymbol{X} = (X_1, X_2, \ldots, X_N)$ is called a random vector if all its components, the X_i's, are random variables.

A.2 STOCHASTIC PROCESSES

A stochastic process $\{X(t); t \in T\}$ is a collection of random variables. That is, for each fixed $t \in T$, $X(t)$ is a random variable. The index t is often referred to as time, and $X(t)$ is called the state of the process at time t. The state space of a stochastic process is defined as the set of all possible values that the random variables $X(t)$ can assume. For each fixed $\omega \in \Omega$, $X(t)(\omega)$ is a function of $t \in T$. This function is called a *sample path* of the stochastic process.

In this book, $\{X(t); t \in \mathbb{R}\}$ refers to a *continuous time stochastic process*, where $\mathbb{R}$ in certain cases may be replaced by $\mathbb{R}_+ \equiv [0, \infty)$. Similarly, $\{X_n; n = 0, \pm 1, \ldots\}$ is called a discrete time stochastic process. Note that even if $X(t)$ (or X_n) only takes a countable number of values, e.g., it is integer valued, the set of all sample paths may not be countable. Thus a sample space Ω for a stochastic process is in general not countable.

In what follows we focus on the continuous time stochastic process $\{X(t); t \in \mathbb{R}\}$. The discrete time case is analogous. We shall always assume that $X(t)$ is right-continuous, i.e., for each $\omega \in \Omega$,

$$\lim_{\epsilon \downarrow 0} X(t + \epsilon)(\omega) = X(t)(\omega), \qquad t \in \mathbb{R}.$$

When we observe a stochastic process from time $-\infty$ to time t, the cumulated information may be described by all events up to time t. We denote the set of such events by $\mathcal{F}_t(X)$, and call it the *history* of $\{X(t)\}$ up to time t. This history is formally defined by setting $\mathcal{F}_t(X)$ equal to the smallest σ-field that includes $\sigma(X(s))$ for all $s \le t$.

It can be shown that, if the state space $\mathcal{S}$ of $X(t)$ is countable, then $\mathcal{F}_t(X)$ is identical to the smallest σ-field that includes the events

$$\{X(t_1) = x_1, X(t_2) = x_2, \ldots, X(t_n) = x_n\},$$
$$n = 1, 2, \ldots, t_1 < t_2 < \cdots < t_n \le t, x_1, x_2, \ldots, x_n \in \mathcal{S}.$$

Clearly, the σ-field $\mathcal{F}_t(X)$ is increasing in t, i.e.,

$$\mathcal{F}_s(X) \subseteq \mathcal{F}_t(X), \qquad \text{for } s \le t.$$

An increasing family of σ-fields $\{\mathcal{F}_t; t \in \mathbb{R}\}$ is called a *filtration*. In particular, if $\mathcal{F}_t(X) \subset \mathcal{F}_t$ for all t, $\{\mathcal{F}_t; t \in \mathbb{R}\}$ is said to be a filtration of the stochastic process $\{X(t)\}$. Such a filtration in general contains more information than the history of $\{X(t)\}$, and describes the background for the process of interest. This is the case we considered in Section 3.1.

A.3 EXPECTATIONS

The expectation $E(X)$ of a random variable X can be defined in three steps. First, suppose that X takes finitely many values, say $x_1, x_2, \ldots, x_m$, then the expectation of X is defined as

$$E(X) = \sum_{i=1}^{m} x_i P(X = x_i)\,.$$

Thus the expectation $E(X)$ is the weighted sum of the possible values of X.

Next suppose that X is nonnegative. For each $n = 1, 2, \ldots$, divide the half line $[0, \infty)$ into the disjoint intervals

$$\left[0, \frac{1}{2^n}\right), \left[\frac{1}{2^n}, \frac{2}{2^n}\right), \ldots, \left[\frac{n2^n - 1}{2^n}, \frac{n2^n}{2^n}\right), [n, \infty)\,.$$

Define X_n by setting $X_n(\omega)$ equal to the minimum value of the interval to which $X(\omega)$ belongs. For instance, if $X(\omega) \in [(i-1)/2^n, i/2^n)$, then $X_n(\omega) = (i-1)/2^n$. We then have the following properties:

(a) X_n is a random variable, which takes finitely many values.

(b) $X_n(\omega) \leq X_{n+1}(\omega)$ for all $\omega \in \Omega$, and $n = 1, 2, \ldots$.

(c) If $X(\omega) \leq n$, then $0 \leq X(\omega) - X_n(\omega) \leq 1/2^n$.

From (b) and (c), $X_n(\omega)$ increasingly converges to $X(\omega)$ as n goes to infinity for each $\omega \in \Omega$. From (a), the expectation $E(X_n)$ can be defined, and, from (b), it follows that $E(X_n)$ increases in n. Hence, we can define the expectation $E(X)$ of X as

$$E(X) \equiv \lim_{n\to\infty} E(X_n).$$

Note that $E(X)$ may be infinity.

Finally, consider an arbitrary random variable X. Define nonnegative random variables X^+ and X^- by

$$X^+(\omega) = \max\{0, X(\omega)\}, \qquad X^-(\omega) = \max\{0, -X(\omega)\}.$$

Clearly,

$$X(\omega) = X^+(\omega) - X^-(\omega).$$

So we define the expectation $E(X)$ as

$$E(X) = E(X^+) - E(X^-),$$

provide one or both of $E(X^+)$ and $E(X^-)$ are finite. If not, it is said that the expectation $E(X)$ does not exist.

The expectation $E(X)$ can be expressed by the distribution function F of X as

$$E(X) = \int_{t \in \mathbb{R}} t dF(t).$$

In particular, if X is a discrete random variable taking values $x_0, x_1, \ldots$, we then have

$$E(X) = \sum_{i=0}^{\infty} x_i p(x_i), \tag{3.1}$$

where

$$p(x) = P(X = x) = F(x) - F(x-).$$

A.4 CONDITIONAL EXPECTATIONS

Let X and Y be random variables such that $E(X)$ is finite and Y only takes discrete values $y_0, y_1, \ldots$ with positive probabilities. Then, the conditional probability

$$P(A|Y = y_i) \equiv \frac{P(A \cap \{Y = y_i\})}{P(Y = y_i)}, \qquad A \in \mathcal{F},$$

defines a probability measure on $(\Omega, \mathcal{F})$ for each y_i. Note that $P(A|Y = y_i)$ is well defined because $P(Y = y_i) > 0$. Denote the expectation of X with respect to this probability measure by $E(X|Y = y_i)$, and call it the conditional expectation of X given the event $\{Y = y_i\}$. It can be shown that

$$E(X|Y = y_i) = \frac{E(X1[Y = y_i])}{P(Y = y_i)},$$

where $1[\cdot]$ is the indicator function.

We extend this notion of conditional expectation as follows. Define a function h on the set $\{y_i; i = 0, 1, \ldots\}$ as

$$h(y_i) = E(X|Y = y_i).$$

Define $E(X|Y)$ as

$$E(X|Y)(\omega) = h(Y(\omega)), \qquad \omega \in \Omega.$$

Note that $E(X|Y)$ is a random variable, and it takes the value $E(X|Y = y_i)$ on the set $\{\omega \in \Omega;\, Y(\omega) = y_i\}$. Hence, by (3.1),

$$\begin{aligned} E(E(X|Y)) &= \sum_{i=0}^{\infty} E(X|Y = y_i)P(Y = y_i) \\ &= \sum_{i=0}^{\infty} E(X1[Y = y_i]) \\ &= E(X)\,. \end{aligned}$$

The following result guarantees the existence of the conditional expectation $E(X|Y)$ in the general case. It is given without a proof.

Lemma A.1 (Radon–Nikodym Theorem) *There exists a random variable Z such that Z is a function of Y, and*

$$E\,(Z1_A) = E(X1_A), \qquad A \in \sigma(Y), \tag{4.1}$$

and such a Z is unique with probability 1, i.e., if Z' is also such a random variable, then $P(Z = Z') = 1$.

We denote this random variable Z by $E(X|Y)$, and call it the conditional expectation of X given Y. It is easily seen that, if Y assumes discrete values, then $Z = h(Y)$ satisfies (4.1).

The conditional expectation in Lemma A.1 can be further extended: if $\mathcal{G}$ is a σ-field on Ω, then there exists a random variable Z such that $\sigma(Z) \subset \mathcal{G}$ and

$$E\,(Z1_A) = E(X1_A), \qquad A \in \mathcal{G}. \tag{4.2}$$

This Z is also unique with probability 1. We denote this Z by $E(X|\mathcal{G})$, and call it the conditional expectation of X given $\mathcal{G}$.

As an example, if X is the indicator function of an event $D \in \mathcal{F}$, i.e., $X(\omega) = 1_D(\omega)$, then the conditional expectation $E(1_D|\mathcal{G})$ can be written as $P(D|\mathcal{G})$, and is called the conditional probability of D given $\mathcal{G}$.

A.5 REFERENCE NOTES

There are many excellent books on probability theory and measure theory. Karr (1992) is an introductory level textbook with rigorous mathematical proofs. More detailed results may be found, for example, in Billingsley (1990). The basic theory of stochastic processes can be found in Doob (1953).

Appendix B

Markov and Poisson Processes

We relegated the proofs of some of the basic results used in Chapters 2 and 3 to this appendix. They include the fact that the sojourn times of a Markov chain in any state are exponentially distributed, and the several characterizations of a Poisson process. We now provide the proofs of these results and discuss some additional properties of Markov and Poisson processes.

B.1 SOJOURN TIMES IN A MARKOV CHAIN

In Chapter 2, a continuous time Markov chain is defined without using the concept of a filtration. However, in Chapter 3 a filtration is required in order to consider a counting process driven by a Markov chain (see Appendix A for the definition of a filtration). In this section we consider Markov chains defined with respect to a given filtration.

Let $\{X(t)\}$ be a continuous time stochastic process with a countable state space $\mathcal{S}$, and let $\{\mathcal{F}_t\}$ be a filtration of $\{X(t)\}$. Let

$$\mathcal{F}_{t-} = \sigma\left(\cup_{s<t}\mathcal{F}_s\right), \qquad r \in \mathbb{R},$$

i.e., $\mathcal{F}_{t-}$ includes all the events before time t.

Definition B.1 (Markov chain with respect to filtration) The stochastic process $\{X(t)\}$ is a Markov chain with respect to $\{\mathcal{F}_t\}$ if, for all $A \in \mathcal{F}_{t-}$ such that $P(A, X(t) = x) > 0$,

$$P\big(X(t+s) = x' | A, X(t) = x\big) = P\big(X(t+s) = x' | X(t) = x\big), \quad x, x' \in \mathcal{S}. \quad (1.1)$$

In terms of the conditional probability defined in Appendix A, (1.1) can be expressed as

$$P(X(t+s) = x' \mid \mathcal{F}_t) = P(X(t+s) = x' \mid X(t)), \qquad x' \in \mathcal{S}. \tag{1.2}$$

If $\{\mathcal{F}_t\}$ is the history of $\{X(t)\}$, i.e., $\{\mathcal{F}_t(X)\}$, this definition of a Markov chain reduces to the definition in Chapter 2. Conditions (1.1) and (1.2) are called the *Markov property*.

Lemma B.2 *The Markov property (1.1) is equivalent to*

$$P(B|A, X(t) = x) = P(B|X(t) = x), \qquad x \in \mathcal{S}, \tag{1.3}$$

for each event B generated by $\{X(t+s); s > 0\}$.

PROOF. Expression (1.3) implies (1.1), since the latter is a special case of the former. We now prove the converse. Since events $\{X(u) = x\}$, for $u \geq t$ and $x \in \mathcal{S}$, generate B, it is sufficient to prove that, for each positive integer n and $t < t_1 < t_2 < \ldots < t_n$, (1.1) implies

$$\begin{aligned} &P(X(t_i) = x_i, i = 1, 2, \ldots, n|A, X(t) = x) \\ &\quad = P(X(t_i) = x_i, i = 1, 2, \ldots, n|X(t) = x), \quad x, x_1, \ldots, x_n \in \mathcal{S}, \end{aligned}$$

where $A \in \mathcal{F}_{t-}$. We only prove the case of $n = 2$.

$$\begin{aligned} &P(X(t_1) = x_1, X(t_2) = x_2|A, X(t) = x) \\ &\quad = P(X(t_2) = x_2|A, X(t) = x, X(t_1) = x_1)P(X(t_1) = x_1|A, X(t) = x) \\ &\quad = P(X(t_2) = x_2|X(t) = x, X(t_1) = x_1)P(X(t_1) = x_1|X(t) = x) \\ &\quad = P(X(t_1) = x_1, X(t_2) = x_2|X(t) = x), \end{aligned}$$

where the second equality follows from (1.1). Thus we obtain (1.3) for $n = 2$. The proof for the case of an arbitrary n is similar. □

Remark B.3 In the definition of a Markov chain, if the Markov property (1.1) holds when t is replaced by any random variable T that satisfies

$$\{T \leq u\} \in \mathcal{F}_u, \qquad \text{for all } u \in \mathbb{R}, \tag{1.4}$$

then the process $\{X(t)\}$ is called a strong Markov process with respect to $\{\mathcal{F}_t\}$. A random variable T satisfying (1.4) is called a Markov time or stopping time. It can be shown that any continuous time Markov chain with a right-continuous sample path is a strong Markov process.

As in Chapter 2, we assume that the Markov chain is time homogeneous, and denote its transition rate from state x to state x' by $q(x, x')$, which is defined by (1.5) of Chapter 2. The existence of the limit in this definition will be discussed in Remark B.7 (see Chung (1967) for its direct verification). As in Section 3.1, for each state x we add an extra jump process from x back to itself with rate $q(x, x)$. Thus we have two types of jumps at state x: those due to state changes with rate $q(x, x')$ for $x \neq x'$ and those due to the extra jumps with rate $q(x, x)$. For each time t, let $\tau(t)$ be the remaining sojourn time to the next jump of either type measured from time t. Then, the addition of the extra jumps is equivalent to assuming that

$$q(x, x) = \lim_{s \downarrow 0} \frac{1}{s} P(X(t + \tau(t)) = x, \tau(t) \leq s | X(t) = x). \tag{1.5}$$

Note that these extra jumps do not change the sample path of $X(t)$. Let $\{\mathcal{F}_t\}$ be the filtration generated by the history of $\{X(u); u \leq t\}$ and the history of the extra jumps. Then, the $\{X(t)\}$ is a Markov chain with respect to $\{\mathcal{F}_t\}$.

Remark B.4 The extra jumps with rate $q(x, x)$ can be generated using a sequence of i.i.d. exponential random variables with mean $q(x, x)^{-1}$ as their inter-occurrence times, as long as the process is in state x. Clearly, from the memoryless property of the exponential distribution, i.e.,

$$P(T > s + t | T > t) = P(T > s), \qquad s, t \geq 0, \tag{1.6}$$

where T is a random variable with exponential distribution, the Markov property is retained in the sense of (1.3), and (1.5) is verified.

Denote the total transition rate out of x by $q(x)$, which includes the transition $q(x, x)$, i.e.,

$$q(x) = \sum_{x' \in \mathcal{S}} q(x, x'), \qquad x \in \mathcal{S}.$$

Assume that

$$0 < q(x) < \infty, \qquad x \in \mathcal{S}.$$

This is a natural assumption, and it is satisfied in all our applications.

To prove (i) of Section 2.1, we need the following lemma.

Lemma B.5 *Let f be a function from $\mathbb{R}$ to $\mathbb{R}$. If $f(t)$ is right-continuous in t and satisfies*

$$f(s+t) = f(s) + f(t), \qquad s, t \geq 0, \tag{1.7}$$

then

$$f(t) = tf(1), \qquad t \geq 0. \tag{1.8}$$

PROOF. Replacing $s = (n-1)t$ in (1.7) for a positive integer n and performing substitutions repeatedly yields

$$\begin{aligned} f(nt) &= f((n-1)t) + f(t) \\ &= f((n-2)t) + 2f(t) \\ &\cdots \\ &= nf(t). \end{aligned}$$

Thus $f(n) = nf(1)$. Letting $t = m/n$ in the above equation for some positive integer m, we obtain

$$f\left(\frac{m}{n}\right) = \frac{1}{n} f(m) = \frac{m}{n} f(1).$$

Since for any real number t there exists a sequence of rational numbers m/n that converge to t from above and $f(t)$ is right-continuous in t, m/n and $f(m/n)$ converge to t and $f(t)$, respectively. This shows (1.8). □

Theorem B.6 *Let $\{X(t)\}$ be a continuous time Markov chain with respect to a filtration $\{\mathcal{F}_t\}$, and let $q(\cdot, \cdot)$ be its transition rate function. Then the sojourn time in state x is exponentially distributed with mean $1/q(x)$, and is independent of all events in the filtration before the current time.*

PROOF. Suppose $X(t) = x$, and let $s, h \geq 0$. Let $\tau(t)$ be the remaining sojourn time until the next jump at time t. Since

$$\{\tau(t) > s\} = \{\tau(t) \leq s\}^c \in \mathcal{F}_{s+t}$$

and

$$\{X(t) = x, \tau(t) > s\} \subset \{X(t+s) = x\},$$

the Markovian property (1.3) of Lemma B.2 implies

$$\begin{aligned} P(\tau(t) > s + h|X(t) = x, \tau(t) > s) \\ = P(\tau(t+s) > h|X(t+s) = x) \\ = P(\tau(t) > h|X(t) = x). \end{aligned}$$

Multiplying both sides by $P(\tau(t) > s|X(t) = x)$ yields

$$\begin{aligned} P(\tau(t) > s + h|X(t) = x) \\ = P(\tau(t) > s|X(t) = x)P(\tau(t) > h|X(t) = x). \end{aligned} \tag{1.9}$$

Fix $t > 0$ and $x \in \mathcal{S}$, and define

$$f(s) = \log(P(\tau(t) > s|X(t) = x)), \qquad s > 0.$$

Then, taking the natural logarithm of (1.9) yields (1.7) for $t = h$. Since f is right-continuous, it follows from Lemma B.5 that

$$P(\tau(t) > s|X(t) = x) = e^{f(1)s}, \qquad s > 0. \tag{1.10}$$

Since the Markov chain is time homogeneous, $f(1)$ is independent of t, but may depend on $x \in \mathcal{S}$. Noting that $f(1)$ is negative, we denote it by $-a(x)$.

We now show that $a(x) = q(x)$. From (1.5), (1.10) and the definition of the transition rate q, we have

$$P(X(t+s) = x|X(t) = x, \tau(t) = s) = \frac{q(x,x)}{a(x)}, \tag{1.11}$$

$$\lim_{h\downarrow 0} \frac{1}{h}(1 - P(X(t+h) = x|X(t) = x)) = \sum_{x'\neq x} q(x, x'). \tag{1.12}$$

Since $\tau(t)$, given $X(t) = x$, is exponentially distributed with mean $1/a(x)$, we have, by (1.10),

$$\begin{aligned} &P(X(t+h) = x|X(t) = x) \\ &= P(\tau(t) > h|X(t) = x) + P(X(t+h) = x, \tau(t) \le h|X(t) = x) \\ &= P(\tau(t) > h|X(t) = x) \\ &\quad + \int_0^h P(X(t+h) = x|X(t) = x, \tau(t) = s)a(x)e^{-a(x)s}ds \\ &= P(\tau(t) > h|X(t) = x) \\ &\quad + \int_0^h P(X(t+h) = x, X(t+s) \neq x|X(t) = x, \tau(t) = s)a(x)e^{-a(x)s}ds \\ &\quad + \int_0^h P(X(t+h) = x, X(t+s) = x|X(t) = x, \tau(t) = s)a(x)e^{-a(x)s}ds. \end{aligned}$$

Substituting this expression into (1.12) and noting that

$$\begin{aligned}
&\lim_{h \downarrow s} P(X(t+h) = x, X(t+s) \neq x | X(t) = x, \tau(t) = s) = 0, \\
&\lim_{h \downarrow s} P(X(t+h) = x, X(t+s) = x | X(t) = x, \tau(t) = s) \\
&\quad = \lim_{h \downarrow s} P(X(t+h) = x | X(t+s) = x) P(X(t+s) = x | X(t) = x, \tau(t) = s) \\
&\quad = \frac{q(x,x)}{a(x)},
\end{aligned}$$

where the last equality follows from (1.11), we obtain

$$\begin{aligned}
q(x) &= \sum_{x' \neq x} q(x, x') + q(x, x) \\
&= \lim_{h \downarrow 0} \frac{1}{h}(1 - P(X(t+h) = x | X(t) = x)) + q(x, x) \\
&= \lim_{h \downarrow 0} \frac{1}{h}(1 - P(\tau(t) > h | X(t) = x)) \\
&= a(x).
\end{aligned}$$

Hence, for each event $A \in \mathcal{F}_{t-}$ such that $P(A, X(t) = x) > 0$, the Markov property (1.3) implies

$$\begin{aligned}
P(\tau(t) > s | A, X(t) = x) &= P(\tau(t) > s | X(t) = x) \\
&= e^{-q(x)s}, \qquad s \geq 0.
\end{aligned}$$

Thus the remaining sojourn time is exponentially distributed with mean $1/q(x)$, and it is independent of $X(t)$ and of all events in the filtration before time t. We can repeat this argument by choosing the next jump epoch as t (to make the argument rigorous, we need the strong Markov property with respect to the jump times, see Remark B.3). This shows that the sojourn times are independent and have exponential distributions with mean $1/q(x)$ when the current state is x, and the sojourn times are independent of all events in the filtration before the current time. □

Remark B.7 In the proof of Theorem B.6, the existence of the limits in (1.5) and (1.12) is not an assumption but a conclusion, if we define $q(x, x)$ by (1.11) instead of assuming (1.5). Furthermore, under the same convention, the existence of the limit in (1.5) of Chapter 2 can be concluded in a similar way, and therefore $q(x, x')$ is well defined for $x \neq x'$.

B.2 POISSON PROCESSES

In Chapter 2 we defined the Poisson process with rate $\lambda > 0$ as a pure birth process with birth rate λ, i.e., the Markov chain with the state space $\mathcal{S} = \{0, 1, \ldots\}$ and the transition rate q given by

$$q(n, n+1) = \lambda, \qquad n = 0, 1, \ldots,$$

where the filtration is the history of the Markov chain. Since $q(n) = q(n, n+1)$, Theorem B.6 implies the following property.

(i) The time intervals between adjacent counting epochs in a Poisson process with rate λ are i.i.d. exponential random variables with mean $1/\lambda$.

Conversely, if a counting process satisfies (i), then the memoryless property (1.6) of the exponential distribution implies that, at any point in time t, the remaining sojourn time until the next counting epoch is independent of the counting epochs before time t, and exponentially distributed with mean $1/\lambda$. Hence, the counting epochs constitute a pure birth process with birth rate λ. Thus, Lemma 2.6 of Chapter 2 follows.

A stochastic process $\{N(t); t \geq 0\}$ is a counting process if $N(t)$ represents the number of events that have occurred up to time t (see Definition 2.3). Clearly, a counting process must satisfy the following properties.

(ii) $N(t) \geq 0$.

(iii) $N(t)$ is integer valued.

(iv) If $s < t$, then $N(s) \leq N(t)$.

(v) For $s < t$, $N(t) - N(s)$ equals the number of events that have occurred in time interval (s, t).

We assume that $N(t)$ is right-continuous in t. We shall need the following terminologies:

(a) **Independent Increments**: The numbers of events in disjoint time intervals are independent, i.e., for $0 < t_1 < t_2 < \cdots < t_n$,

$$N(t_1), N(t_2) - N(t_1), \ldots, N(t_n) - N(t_{n-1})$$

are independent random variables.

(b) **Stationary Increments**: The numbers of events in time intervals of the same length have a common distribution, i.e., for $s, t > 0$, $N(s)$ and $N(t+s) - N(t)$ have the same distribution.

(c) **Rarity**: No two events occur at once, i.e., $N(t)$ increases by at most 1 at any point in time.

A counting process that satisfies (c) is called simple. The following theorem shows that these conditions completely characterize a Poisson process.

Theorem B.8 *A counting process $\{N(t); t \geq 0\}$ with $\lambda = E(N(1)) < \infty$ is a Poisson process with rate λ if and only if assumptions (a), (b) and (c) hold. If this is the case, we have, for each $t > 0$,*

$$P(N(t) = n) = \frac{(\lambda t)^n}{n!} e^{-\lambda t}, \qquad n = 0, 1, \ldots, \tag{2.1}$$

i.e., $N(t)$ has a Poisson distribution with mean λt.

PROOF. The statement that $\{N(t); t \geq 0\}$ is a Poisson process with rate λ means that it is a pure birth process with rate λ. If this is the case, then property (i) and the memoryless property of the exponential distribution imply (a) and (b), and the fact that a pure birth process increases 1 at a time implies (c). On the other hand, it follows from (a) and (b) that $N(t)$ is a continuous time Markov chain with a constant rate, so (c) implies that $N(t)$ is a pure birth process. Hence, (a), (b), and (c) are equivalent to saying that $\{N(t)\}$ is a Poisson process.

We now show that the average number of arrivals in one unit of time $E(N(1))$ equals the rate λ of the pure birth process $N(t)$. From Theorem B.6, the state sojourn times of $N(t)$ are exponentially distributed with mean $1/\lambda$. If we let T_n denote the nth counting epoch and $T_0 = 0$, we then have

$$E(T_n - T_{n-1}) = \frac{1}{\lambda}, \qquad n = 1, 2, \ldots.$$

Thus, by the law of large numbers,

$$\begin{aligned} E(N(1)) &= \lim_{n\to\infty} \frac{N(n)}{n} \\ &= \lim_{n\to\infty} \frac{N(n)}{T_{N(n)}} \\ &= \lim_{n\to\infty} \frac{1}{T_n/n} = \lambda. \end{aligned}$$

It remains to prove (2.1). To this end, we use property (i). Since

$$\{N(t) = n\} = \{T_n \leq t < T_{n+1}\},$$

$$P(N(t) = n) = P(T_n \leq t) - P(T_{n+1} \leq t). \tag{2.2}$$

By property (i), T_n is the sum of n independent and exponentially distributed random variables with common mean $1/\lambda$. Using this fact, we show that

$$P(T_n \le t) = 1 - \sum_{i=0}^{n-1} \frac{(\lambda t)^i}{i!} e^{-\lambda t}, \qquad n = 1, 2, \ldots. \tag{2.3}$$

This distribution is said to be an *Erlang distribution with n phases and phase transition rate* λ, which is also known as a special case of the Gamma distribution. It is easily seen that the distribution has the following density function:

$$\frac{d}{dt} P(T_n \le t) = \frac{\lambda^n}{(n-1)!} t^{n-1} e^{-\lambda t}, \qquad t \ge 0.$$

Equation (2.3) obviously holds for $n = 1$, since T_1 has an exponential distribution with mean $1/\lambda$. Suppose it holds for n, then

$$\begin{aligned}
P(T_{n+1} \le t) &= P(T_n + (T_{n+1} - T_n) \le t) \\
&= \int_0^t P(T_n + y \le t) \lambda e^{-\lambda y} dy \\
&= \int_0^t \left(1 - \sum_{i=0}^{n-1} \frac{(\lambda(t-y))^i}{i!} e^{-\lambda(t-y)}\right) \lambda e^{-\lambda y} dy \\
&= \int_0^t \lambda e^{-\lambda y} dy - \sum_{i=0}^{n-1} \frac{\lambda^{i+1}}{i!} e^{-\lambda t} \int_0^t (t-y)^i dy \\
&= 1 - e^{-\lambda t} - \sum_{i=0}^{n-1} \frac{\lambda^{i+1}}{i!} e^{-\lambda t} \frac{t^{i+1}}{i+1} \\
&= 1 - \sum_{i=0}^{n} \frac{(\lambda t)^i}{i!} e^{-\lambda t}.
\end{aligned}$$

Hence, by induction, we obtain (2.3). Substituting this into (2.2) yields (2.1). This completes the proof. □

From (2.1) we obtain

$$E(N(t)) = \sum_{n=1}^{\infty} n \frac{(\lambda t)^n}{n!} e^{-\lambda t} = \lambda t \sum_{n=1}^{\infty} \frac{(\lambda t)^{n-1}}{(n-1)!} e^{-\lambda t} = \lambda t, \qquad t > 0. \tag{2.4}$$

Thus, the mean number of events in $[0, t]$ is a linear function of t. This linear property is not specific to a Poisson process, but it is implied by the stationary increments property (b).

In queueing applications, Poisson processes generate other processes such as queue length processes. In these situations the Poisson processes are usually assumed to be independent of the current state of the generated processes as well as the history up to the current time. Such Poisson processes can be defined as follows.

Definition B.9 (Poisson process with respect to filtration) Let $\{X(t); t \geq 0\}$ be a pure birth process, and let $\{\mathcal{F}_t\}$ be a filtration of $\{X(t)\}$. If $\{X(t)\}$ is a continuous time Markov chain with respect to the filtration $\{\mathcal{F}_t\}$, then the counting process $N(t) \equiv X(t)$ is called the *Poisson process with respect to* $\{\mathcal{F}_t\}$.

Similar to Theorem B.8, we have the following characterization.

Theorem B.10 *A counting process $\{N(t); t \geq 0\}$ with $\lambda = E(N(1)) < \infty$ is a Poisson process with respect to a filtration $\{\mathcal{F}_t\}$ with rate λ if and only if*

(a′) for any $t < t_1 < t_2 < \cdots < t_n$,

$$N(t_1), N(t_2) - N(t_1), \ldots, N(t_n) - N(t_{n-1})$$

are independent of all events in $\mathcal{F}_t$,

and assumptions (b) and (c) hold.

B.3 MARKOV DRIVEN POISSON PROCESSES

In this section we consider Poisson processes generated by a continuous time Markov chain. We first present the following lemma, which is a special case of a result in calculus known as the bounded convergence theorem.

Lemma B.11 *Let $\{p_n; n = 1, 2, \ldots\}$ be a set of nonnegative numbers such that*

$$\sum_{n=1}^{\infty} p_n < \infty, \tag{3.1}$$

and $\{f_n(t); t \geq 0, n = 1, 2, \ldots\}$ be a set of nonnegative numbers such that

$$\lim_{t \downarrow 0} f_n(t) = f_n(0), \qquad n = 1, 2, \ldots. \tag{3.2}$$

If there exists a set of nonnegative numbers $\{g_n; n = 1, 2, \ldots\}$ *such that*

$$\sum_{n=1}^{\infty} g_n p_n < \infty, \tag{3.3}$$

$$\sup_{t>0} f_n(t) \leq g_n, \qquad n = 1, 2, \ldots, \tag{3.4}$$

then

$$\lim_{t \downarrow 0} \sum_{n=1}^{\infty} f_n(t) p_n = \sum_{n=1}^{\infty} f_n(0) p_n. \tag{3.5}$$

PROOF. For each positive integer k,

$$\begin{aligned}\sum_{n=1}^{k} f_n(0) p_n &= \sum_{n=1}^{k} \lim_{t \downarrow 0} f_n(t) p_n \\ &= \lim_{t \downarrow 0} \sum_{n=1}^{k} f_n(t) p_n \leq \liminf_{t \downarrow 0} \sum_{n=1}^{\infty} f_n(0) p_n.\end{aligned}$$

Letting k go to infinity, we obtain

$$\sum_{n=1}^{\infty} f_n(0) p_n \leq \liminf_{t \downarrow 0} \sum_{n=1}^{\infty} f_n(t) p_n. \tag{3.6}$$

Since $g_n - f_n(t) \geq 0$ for all $t > 0$ and $n = 1, 2, \ldots$, a similar calculation shows

$$\sum_{n=1}^{\infty} (g_n - f_n(0)) p_n \leq \liminf_{t \downarrow 0} \sum_{n=1}^{\infty} (g_n - f_n(t)) p_n.$$

From (3.3), we cancel $\sum_{n=1}^{\infty} g_n p_n$ on both sides of the above equation and obtain

$$\limsup_{t \downarrow 0} \sum_{n=1}^{\infty} f_n(t) p_n \leq \sum_{n=1}^{\infty} f_n(0) p_n. \tag{3.7}$$

Thus, (3.5) follows from (3.6) and (3.7). □

Let $\{X(t)\}$ be a continuous time Markov chain with respect to a filtration $\{\mathcal{F}_t\}$ which has state space $\mathcal{S}$ and transition rate q such that

$$\sum_{x', x'' \in \mathcal{S}} q(x', x'') P(X(t) = x' | X(0) = x) < \infty, \qquad x \in \mathcal{S}. \tag{3.8}$$

Condition (3.8) implies that the sample path of $\{X(t)\}$ is well defined, which is implicitly assumed throughout the book.

Theorem B.12 *If, for the Markov chain $\{X(t)\}$, q_* is a thinning of q, i.e.,*

$$q_*(x, x') \leq q(x, x'), \qquad x, x' \in \mathcal{S},$$

and if the total rate going out of state x is positive and independent of $x \in \mathcal{S}$, i.e.,

$$q_*(x) \equiv \sum_{x' \in \mathcal{S}} q_*(x, x') = \alpha, \qquad x \in \mathcal{S}, \tag{3.9}$$

for some constant $\alpha > 0$, then the point process generated by q_* is a Poisson process with respect to the filtration $\{\mathcal{F}_t\}$.

PROOF. Let $\tau_*(t)$ be the time until the next jump due to q_* measured from time t. Then, by Definition 2.10 and (i), i.e., Lemma 2.6, the theorem is obtained if $\tau_*(t)$ is independent of $\mathcal{F}_t$ and exponentially distributed with mean $1/\alpha$. Note that $\tau_*(t)$ may not be the state sojourn time of the Markov chain $\{X(t)\}$. That is, $X(t)$ may change during the time interval $[t, \tau_*(t))$. Thus, the theorem is not a direct consequence of Theorem B.6, but we can apply a similar argument.

From the Markovian property (1.3), it follows that, for $x, x' \in \mathcal{S}$,

$$\begin{aligned}
&P(\tau_*(0) > s + t, X(s) = x' | \tau_*(0) > s, X(0) = x) \\
&\quad = P(\tau_*(0) > s + t | X(0) = x, \tau_*(0) > s, X(s) = x') \\
&\quad \times P(X(s) = x' | \tau_*(0) > s, X(0) = x) \\
&\quad = P(\tau_*(0) > s + t | \tau_*(0) > s, X(s) = x') P(X(s) = x' | \tau_*(0) > s, X(0) = x) \\
&\quad = P(\tau_*(s) > t | X(s) = x') P(X(s) = x' | \tau_*(0) > s, X(0) = x) \\
&\quad = P(\tau_*(0) > t | X(0) = x') P(X(s) = x' | \tau_*(0) > s, X(0) = x).
\end{aligned}$$

Multiplying both sides of this equation by $P(\tau_*(0) > s | X(t) = x)$ yields

$$\begin{aligned}
&P(\tau_*(0) > s + t, X(s) = x' | X(0) = x) \\
&\qquad = P(\tau_*(0) > t | X(0) = x') P(\tau_*(0) > s, X(s) = x' | X(0) = x).
\end{aligned} \tag{3.10}$$

Subtracting (3.10) with $t = 0$ from both sides of (3.10), we obtain

$$\begin{aligned}
&P(\tau_*(0) > s + t, X(s) = x' | X(0) = x) - P(\tau_*(0) > s, X(s) = x' | X(0) = x) \\
&\qquad = -P(\tau_*(0) \leq t | X(0) = x') P(\tau_*(0) > s, X(s) = x' | X(0) = x).
\end{aligned}$$

Summing this over all $x' \in \mathcal{S}$ and dividing both sides of the resulting expression by t yields

$$\begin{aligned}&\frac{1}{t}(P(\tau_*(0) > s + t|X(0) = x) - P(\tau_*(0) > s|X(0) = x))\\&\quad = -\sum_{x' \in \mathcal{S}} \frac{1}{t} P(\tau_*(0) \le t|X(0) = x') P(\tau_*(0) > s, X(s) = x'|X(0) = x).\end{aligned} \tag{3.11}$$

Since the state space $\mathcal{S}$ is countable, we can apply Lemma B.11 for

$$\begin{aligned}f_{x'}(t) &= \frac{1}{t} P(\tau_*(0) \le t|X(0) = x'),\\p_{x'} &= P(\tau_*(0) > s, X(s) = x'|X(0) = x),\\g_x &= q(x),\end{aligned}$$

provided conditions (3.3) and (3.4) are satisfied. Condition (3.3) is the regularity condition (3.8). To verify (3.4), we note that $\tau_*(t)$ is not less than the state sojourn time of $\{X(t)\}$. This fact and Theorem B.6 imply that

$$\begin{aligned}f_{x'}(t) &= \frac{1}{t} P(\tau_*(0) \le t|X(0) = x')\\&\le \frac{1 - e^{-q(x')t}}{t} \le q(x').\end{aligned}$$

Hence, (3.4) is satisfied. Furthermore, from the definition of q_* and Theorem B.6,

$$\lim_{t \downarrow 0} f_{x'}(t) = \lim_{t \downarrow 0} \frac{1}{t} P(\tau_*(0) \le t|X(0) = x') = q_*(x').$$

Thus it follows from (3.11) and the condition $q_*(x) = \alpha$ that

$$\frac{d}{ds} P(\tau_*(0) > s|X(0) = x) = -\alpha P(\tau_*(0) > s|X(0) = x).$$

Solving this differential equation with the boundary condition

$$P(\tau_*(0) > 0|X(0) = x) = 1,$$

we obtain

$$P(\tau_*(0) > s|X(0) = x) = e^{-\alpha s}, \qquad s \ge 0, x \in \mathcal{S},$$

and, equivalently,

$$P(\tau_*(t) > s | X(t) = x) = e^{-\alpha s}, \qquad s, t \geq 0, x \in \mathcal{S}.$$

Since $\{X(t)\}$ is Markov with respect to the filtration $\{\mathcal{F}_t\}$, the theorem is proved similar to Theorem B.6. □

We conclude this appendix with the following corollary.

Corollary B.13 *If, for the Markov chain $\{X(t)\}$ of Theorem B.12, q_u, $u = 1, 2, \ldots$, are thinnings of q such that*

$$\sum_{u=1}^{\infty} q_u(x, x') \leq q(x, x')$$

and

$$q_u(x) \equiv \sum_{x' \in \mathcal{S}} q_u(x, x') = \alpha_u > 0, \qquad u = 1, 2, \ldots, \tag{3.12}$$

then the point processes $\{N_u(t)\}$ generated by q_u are mutually independent Poisson with respect to the filtration $\{\mathcal{F}_t\}$.

PROOF. Let $\tau_u(t)$ be the time from t until the next jump due to q_u, for each $u = 1, 2, \ldots$. Let J be an arbitrary finite subset of $\{1, 2, \ldots\}$. Define

$$\tau_J(t) = \min_{u \in J} \tau_u(t), \qquad t \geq 0.$$

Clearly, $\tau_J(t)$ is the time from t until the next jump due to the transition function

$$q_J(x, x') \equiv \sum_{u \in J} q_u(x, x'), \qquad x, x' \in \mathcal{S}.$$

Thus, we can apply the same arguments as in the proof of Theorem B.12, and obtain

$$P(\tau_u(t) > s, u \in J | X(t) = x) = \prod_{u \in J} e^{-\alpha_u s}, \qquad s, t \geq 0, x \in \mathcal{S}.$$

Hence, $\tau_u(t)$ are mutually independent and exponentially distributed with mean α_u. The proof of the corollary is thus complete. □

B.4 REFERENCE NOTES

Formal treatments of Markov chains and processes are usually skipped in introductory level textbooks such as Ross (1983) and Çinlar (1975) as well as queueing books such as Kelly (1979), Walrand (1988) and Wolff (1989), while they are discussed in theoretical books such as Chung (1967) and Ethier and Kurtz (1986). This appendix intends to fill the gap between those two different treatments. A Poisson process with respect to a filtration is a special case of a point process with respect to a filtration. The filtration theory of point processes can be found, for example, in Bremaud (1981) and Daley and Vere-Jones (1988). A Markov driven Poisson process is a new terminology, and Theorem B.12 and Corollary B.13 are new. However, those results are straightforward from the filtration theory of point processes.

Appendix C

Brouwer's Fixed Point Theorem

In Chapter 11 we applied Brouwer's fixed point theorem to prove the existence of solutions to traffic equations. This theorem is one of the fundamental results in geometric topology, and has been extended in various ways. However, proving the theorem requires topological arguments that are highly technical and may not be of interest to the reader of this book. Thus in this section we only present the theorem and do not provide its proof. We do give a proof for the one-dimensional case because it is extremely simple, and yet it still sheds light on the structure of the proof for the general case.

C.1 EUCLIDEAN SPACE

We first introduce some notation. Let $\mathcal{X}$ be an N-dimensional vector space, i.e., $\mathbb{R}^N$, with the Euclidean metric $\|\cdot\|$, defined as

$$\|\boldsymbol{x} - \boldsymbol{y}\| = \sqrt{\sum_{\ell=1}^{N} (x_\ell - y_\ell)^2}, \qquad \boldsymbol{x}, \boldsymbol{y} \in \mathbb{R}^N,$$

where $\boldsymbol{x} = (x_1, \ldots, x_N)$ and $\boldsymbol{y} = (y_1, \ldots, y_N)$. For a positive number r and $\boldsymbol{x} \in \mathcal{X}$, $U_r(\boldsymbol{x})$, defined by

$$U_r(\boldsymbol{x}) = \{\boldsymbol{y} \in \mathcal{X};\ \|\boldsymbol{y} - \boldsymbol{x}\| < r\},$$

is called an open r-ball of $\mathcal{X}$ with center $\boldsymbol{x}$.

A subset V of $\mathcal{X}$ is said to be bounded if $V \subset U_r(\mathbf{0})$ for some $r > 0$. The subset V is said to be closed if, for any sequence $\{\boldsymbol{x}_n \in V; n = 1, 2, \ldots\}$ that converges to a point $\boldsymbol{x}_\infty \in \mathcal{X}$, i.e.,

$$\lim_{n\to\infty} \|\boldsymbol{x}_n - \boldsymbol{x}_\infty\| = 0,$$

we have $\boldsymbol{x}_\infty \in V$. The subset V is said to be convex if $\boldsymbol{x}, \boldsymbol{y} \in V$ implies

$$p\boldsymbol{x} + (1-p)\boldsymbol{y} \in V, \qquad 0 < p < 1.$$

For instance, the rectangle of the form

$$\{\boldsymbol{x} \in \mathcal{X}; 0 \le x_i \le r_i, i = 1, \ldots, N\}$$

for some given positive numbers r_i is clearly bounded, closed and convex. The fixed point theorem in Chapter 11 is applied to this type of subset.

For a subset V of $\mathcal{X}$, a function f from V to V is called continuous if, for any $\boldsymbol{x} \in V$ and any sequence $\{\boldsymbol{x}_n \in V\}$ that converges to $\boldsymbol{x}$ as n goes to infinity, $f(\boldsymbol{x}_n)$ converges to $f(\boldsymbol{x})$.

C.2 FIXED POINT

For a subset V of $\mathcal{X}$, let f be a function from V to V. An element $\boldsymbol{x} \in V$ is called a fixed point of f if

$$f(\boldsymbol{x}) = \boldsymbol{x}.$$

In Chapter 11 we considered iterative algorithms for finding fixed points. Such an algorithm is described by the sequence

$$\boldsymbol{x}_{n+1} = f(\boldsymbol{x}_n), \qquad n = 0, 1, \ldots.$$

Suppose that V is bounded and closed and that f is continuous. Then $\{f(\boldsymbol{x}_n); n = 0, 1, \ldots\}$ and $\{\boldsymbol{x}_n; n = 1, 2, \ldots\}$ are also bounded. By the well-known Bolzano–Weierstrass theorem in elementary calculus, there exist a point $\boldsymbol{x}_\infty \in V$ and a subsequence $\{\boldsymbol{x}_{n_i}; i = 1, 2, \ldots\}$ such that

$$\lim_{i\to\infty} \boldsymbol{x}_{n_i} = \boldsymbol{x}_\infty.$$

For this subsequence, we have

$$\lim_{i\to\infty} f(\boldsymbol{x}_{n_i-1}) = \lim_{i\to\infty} \boldsymbol{x}_{n_i} = \boldsymbol{x}_\infty. \tag{2.1}$$

However, the subsequence $\{\boldsymbol{x}_{n_i-1}; i = 1, 2, \ldots\}$, in general, may not converge to $\boldsymbol{x}_\infty$, so (2.1) does not imply that $\boldsymbol{x}_\infty$ is a fixed point of f. A sufficient condition for the convergence of $\{\boldsymbol{x}_{n_i-1}; i = 1, 2, \ldots\}$ is that f is a contraction mapping, i.e.,

$$\|f(\boldsymbol{x}) - f(\boldsymbol{y})\| \le r\|\boldsymbol{x} - \boldsymbol{y}\|, \qquad \boldsymbol{x}, \boldsymbol{y} \in V,$$

for a positive constant $r < 1$. It is not hard to show that a fixed point exists even for the case $r = 1$, for which f is called a non-expanding mapping. However, without these obvious conditions there are no easy ways to show the existence of a fixed point. Nevertheless, a fixed point does exist, as is shown in the following theorem, provided V is convex.

Theorem C.1 (Brouwer's fixed point theorem) *Let V be a subset of $\mathcal{X}$ that is bounded, closed and convex. If f is a continuous function from V to V, then f has a fixed point.*

As mentioned before, we omit the proof of this theorem except for the one-dimensional case (see the reference notes for a proof of the general case). The following proof for the one-dimensional case illustrates some of the ideas in the proof of the general case.

PROOF OF THE CASE $N = 1$ In this case, $\mathcal{X} = \mathbb{R}$. Without loss of generality we can assume $V = [0, 1]$. Let f be a continuous function from $[0, 1]$ to $[0, 1]$. Suppose that f does not have a fixed point. Define the function g by

$$g(x) = \begin{cases} 0, & x < f(x), \\ 1, & x \ge f(x). \end{cases} \tag{2.2}$$

Let $x_n \in [0, 1]$ be any sequence converging to $x \in [0, 1]$ as n goes to infinity. Since f has no fixed point, either $x < f(x)$ or $x > f(x)$. If $x < f(x)$, then $x_n < f(x_n)$ for sufficiently large n. Hence, $g(x_n)$ converges to 0, which equals $g(x)$. Similarly, $x > f(x)$ implies that $g(x_n)$ converges to $1 = g(x)$. Consequently, $g(x)$ is continuous in x. However, it is easily seen that a function g on [0,1] taking values 0 and 1 cannot be continuous. Hence, f must have a fixed point. □

The function g of (2.2) is a function from the closed interval $[0, 1]$ to its boundary $\{0, 1\}$. A similar function can be used to prove Theorem C.1 for the general case. To define it precisely, we first introduce a boundary of the closed set V. An element $\boldsymbol{x} \in V$ is called an interior point of V if there exists a number $r > 0$ such that

$$U_r(\boldsymbol{x}) \subset V,$$

and the set of all $\boldsymbol{x} \in V$ that are not interior points of V is called the boundary of V, and is denoted by ∂V. If f has no fixed point and if V is a convex set, then, for each $\boldsymbol{x} \in \mathcal{X}$, the set

$$\partial V \cap \{t(\boldsymbol{x} - f(\boldsymbol{x})) + \boldsymbol{x};\ t \geq 0\}$$

has a single element. Let $g(\boldsymbol{x})$ be this element for each $\boldsymbol{x}$. Then $g(\boldsymbol{x})$ is a mapping from V to ∂V, and unchanged on ∂V, i.e.,

$$g(\boldsymbol{x}) = \boldsymbol{x}, \qquad \boldsymbol{x} \in \partial V.$$

Furthermore the continuity of f implies that g is a continuous function. However, it is known from the theory of geometric topology that, if V is bounded, there are no continuous functions from V to ∂V such that they are unchanged on ∂V. Hence, f must have a fixed point for the convex set V. The non-existence of such a mapping from V to ∂V is trivial for the one-dimensional case, but there are no simple proofs for the general case.

C.3 REFERENCE NOTE

The proof of Brouwer's fixed point theorem can be found in most texts on geometric topology. The reader wishing to read a complete proof is referred to Todd (1976).

References

Asmussen, S. (1987), *Applied Probability and Queues*, John Wiley & Sons, New York.

Baccelli, F. and Bremaud, P. (1994), *Elements of Queueing Theory*, Springer-Verlag, Berlin.

Baccelli, F. and Foss, S. (1994), Stability of Jackson-type queueing networks, I, *Queueing Systems* **17**, 5–72.

Baskett, F., Chandy, K. M., Muntz, R. R. and Palacios, F. G. (1975), Open, closed and mixed networks of queues with different classes of customers, *Journal of the ACM* **22**, 248–260.

Baskett, F. and Palacios, F. G. (1972), Processor sharing in a central queueing model of multiprogramming with applications, *in* Proceedings of Sixth Annual Conference on Information Sciences and Systems, Princeton University, NJ, pp. 598–603.

Bharath and Kumar, K. (1980), Discrete time queueing systems and their networks, *IEEE Transactions on Communications* **COM-28**, 260–263.

Billingsley, P. (1990), *Probability and Measure*, 2nd edn, John Wiley & Sons, New York.

Boucherie, R. J. (1992), Product forms in queueing networks, Tinbergen Institute Research Series, Free University, The Netherlands.

Boucherie, R. J. and Van Dijk, N. M. (1991), Product forms for queueing networks with state-dependent multiple job transitions, *Advances in Applied Probability* **23**, 152–187.

Boucherie, R. J. and Van Dijk, N. M. (1993), A generalization of Norton's theorem for queueing networks, *Queueing Systems* **13**, 251–289.

Bremaud, P. (1981), *Point Processes and Queues: Martingale Dynamics*, Springer-Verlag, New York.

Burke, P. J. (1956), The output of a queueing system, *Operations Research* **4**, 699–704.

Burke, P. J. (1968), The output of a stationary $M/M/s$ queueing system, *Annals of Mathematical Statistics* **39**, 1144–1152.

Burman, D., Lehoczky, J. and Lim, Y. (1984), Insensitivity of blocking probabilities in a circuit switching network, *Journal of Applied Probability* **21**, 850–859.

Çinlar, E. (1975), *Introduction to Stochastic Processes*, Prentice-Hall, New Jersey.

Chandy, K. M. and Martin, J. (1983), A characterization of product form queueing networks, *Journal of the ACM* **24**, 250–263.

Chandy, K. M. and Palacios, F. G. (1972), The analysis and solutions for general queueing networks, *in* Proceedings of Sixth Annual Conference on Information Sciences and Systems, Princeton University, NJ, pp. 224–228.

Chang, C. S., Thomas, J. A. and Kiang, S. H. (1994), On the stability of open networks: an unified approach by stochastic dominance, *Queueing Systems* **15**, 239–260.

Chao, X. (1994), A note on queueing networks with signals and random triggering times, *Probability in the Engineering and Informational Sciences* **8**, 213–219.

Chao, X. (1995), Networks of queues with customers, signals and arbitrary service time distributions, *Operations Research* **43**, 537–544.

Chao, X. and Miyazawa, M. (1996*a*), A probabilistic decomposition approach to quasi-reversibility and its applications in coupling of queues. Preprint.

Chao, X. and Miyazawa, M. (1996*b*), Queueing networks with instantaneous movements: A coupling approach by quasi-reversibility. Submitted to *Advances in Applied Probability*.

Chao, X. and Miyazawa, M. (1998), On quasi-reversibility and local balance: An alternative derivation of the product-form results, *Operations Research* **46**, 927–933.

Chao, X., Miyazawa, M., Serfozo, R. and Takada, H. (1998), Markov network processes with product form stationary distributions, *Queueing Systems* **28**, 377–401.

Chao, X. and Pinedo, M. (1993), On generalized networks of queues with positive and negative arrivals, *Probability in the Engineering and Informational Sciences* **7**, 301–334.

Chao, X. and Pinedo, M. (1995*a*), Networks of queues with batch services, signals, and product form solutions, *Operations Research Letters* **17**, 237–242.

Chao, X. and Pinedo, M. (1995*b*), On queueing networks with signals and history dependent routing, *Probability in the Engineering and Informational Sciences* **9**, 341–354.

Chao, X., Pinedo, M. and Shaw, D. (1996), Networks of queues with batch services and customer coalescence, *Journal of Applied Probability* **33**, 858–869.

Chao, X. and Serfozo, R. F. (1997), String transition networks with mixed vector deletions and vector additions. Preprint.

Chao, X. and Zheng, S. (1998), A result on networks of queues with customer coalescence and state-dependent signaling, *Journal of Applied Probability* **35**, 151–164.

Chen, H. and Zhang, H. (1997), Stability of multiclass queueing networks under priority service disciplines. Preprint.

Chung, K. L. (1967), *Markov Chains with Stationary Transition Probabilities*, 2nd edn, Springer-Verlag, Berlin.

Dai, J. (1995), On positive Harris recurrence of multiclass queueing networks, *Annals of Applied Probability* **5**, 49–77.

Daley, D. J. and Vere-Jones, D. (1988), *An Introduction to the Theory of Point Processes*, Springer-Verlag, New York.

Doob, J. L. (1953), *Stochastic Processes*, John Wiley & Sons, New York.

Erlang, A. K. (1917), Solution of some problems in the theory of probabilities of significance in automatic telephone exchanges, *Post Office Electrical Engineer's Journal* **10**, 189–197.

Ethier, S. N. and Kurtz, T. G. (1986), *Markov Processes: Characterization and Convergence*, John Wiley & Sons, New York.

Flatto, L. and Hahn, S. (1984), Two parallel queues created by arrivals with two demands, *SIAM Journal of Applied Mathematics* **44**, 1041–1053.

Franken, P., König, D., Arndt, U. and Schmidt, V. (1982), *Queues and Point Processes*, John Wiley & Sons, Chichester.

Gelenbe, E. (1991), Product-form queueing networks with negative and positive customers, *Journal of Applied Probability* **28**, 656–663.

Gelenbe, E. (1993), G-networks with triggered customer movement, *Journal of Applied Probability* **30**, 742–748.

Gelenbe, E. and Pujolle, G. (1998), *Introduction to Queueing Networks*, John Wiley & Sons, Chichester.

Gelenbe, E. and Schassberger, R. (1992), Stability of product form G-network, *Probability in the Engineering and Informational Sciences* **6**, 271–276.

Glynn, W. P. (1990), Diffusion approximations in stochastic models, *in* D. Heyman and M. Sobel, eds, *Handbooks on OR & MS*, Vol. 2, North-Holland, pp. 145–198.

Gordon, W. J. and Newell, G. F. (1967*a*), Closed queueing systems with exponential servers, *Operations Research* **15**, 254–265.

Gordon, W. J. and Newell, G. F. (1967*b*), Cyclic queueing systems with restricted length queues, *Operations Research* **15**, 266–278.

Gross, D. and Harris, C. M. (1985), *Fundamentals of Queueing Theory*, 2nd edn, John Wiley & Sons, New York.

Harrison, M. and Williams, R. J. (1992), Brownian models of feedforward queueing networks: Quasireversibility and product form solutions, *Annals of Applied Probability* **2**, 263–293.

Henderson, W., Northcote, B. and Taylor, P. G. (1994*a*), Geometric equilibrium distributions for queues with interactive batch departures, *Annals of Operations Research* **48**, 493–511.

Henderson, W., Northcote, B. and Taylor, P. G. (1994*b*), State-dependent signaling in queueing networks, *Advances in Applied Probability* **26**, 436–455.

Henderson, W., Northcote, B. and Taylor, P. G. (1995), Triggered batch movement in queueing networks, *Queueing Systems* **21**, 125–141.

Henderson, W., Pearce, C., Pollett, P. and Taylor, P. (1992), Connecting internally balanced quasireversible Markov processes, *Advances in Applied Probability* **24**, 934–959.

Henderson, W., Pearce, C., Taylor, P. and Schassberger, R. (1995), Partial balance, product-form invariant measures and Markov processes: An integrated approach. Preprint.

Henderson, W. and Taylor, P. G. (1989), Insensitivity of processes with interruptions, *Journal of Applied Probability* **26**, 242–258.

Henderson, W. and Taylor, P. G. (1990), Product form in networks of queues with batch arrivals and batch services, *Queueing Systems* **6**, 71–88.

Henderson, W. and Taylor, P. G. (1997), State-dependent coupling of quasireversible nodes. Preprint.

Hsu, J. and Burke, P. J. (1976), Behavior of tandem buffers with geometric input and Markovian output, *IEEE Transactions on Communications* **COM-24**, 358–360.

Hunter, J. J. (1983), *Mathematical Techniques of Applied Probability*, Vol. 2, *Discrete Time Models: Techniques and Applications*, Academic Press, New York.

Jackson, J. R. (1957), Networks of waiting lines, *Operations Research* **5**, 518–521.

Jackson, J. R. (1963), Jobshop-like queueing systems, *Management Science* **10**, 131–142.

Jansen, U., König, D. and Nawrotzki, K. (1979), A criterion of insensitivity for a class of queueing systems with random marked point processes, *Math. Operationsforsch. u. Statist. Ser. Optimization* **10**, 379–403.

Karr, A. F. (1992), *Probability*, Springer-Verlag, New York.

Kelly, F. P. (1975), Network of queues with customers of different types, *Journal of Applied Probability* **12**, 542–554.

Kelly, F. P. (1976), Network of queues, *Advances in Applied Probability* **8**, 416–432.

Kelly, F. P. (1979), *Reversibility and Stochastic Networks*, John Wiley & Sons, New York.

Kelly, F. P. (1982), Networks of quasi-reversible nodes, *in* R.L.Disney and T. Ott, eds, *Applied Probability-Computer Science:* The *Interface*, Vol. I, Birkhäusel, pp. 3–26.

Kelly, F. P. (1991), Loss network, *Annals of Applied Probability* **1**, 319–378.

Kemeny, J. G., Snell, J. L. and Knapp, A. W. (1966), *Denumerable Markov Chains*, Van Nostrand, New York.

Kleinrock, L. (1975), *Queueing Systems*, Vol. 1, *Theory*, John Wiley & Sons, New York.

König, D., Matthes, K. and Nawrotzki, K. (1967), *Verallgemeinerungen der Erlangschen und Engsetschen Formeln (Eine Methode in der Bedienungs-theorie)*, Akademie-Verlag, Berlin.

Kumar, P. R. (1993), Re-entrant lines, *Queueing Systems* **13**, 87–110.

Loynes, R. M. (1962), The stability of a queue with non-independent inter-arrival and service times, *Proceedings of the Cambridge Philosophical Society* **58**, 497–520.

Malinkovsky, Y. (1990), A criterion for pointwise independence of states of units in an open stationary Markov queueing network with one class of customers, *Theory of Probability and Applications* **35**, 797–802.

Matthes, K. (1962), Zur Theorie der Bedienungsprozesse, *in* Trans. 3rd Prague Conference on Information Theor. Statist. Decision Function. Random Processes, Prague, pp. 308–326.

Miyazawa, M. (1991), The characterization of the steady state distributions of the supplemented self-clocking jump processes, *Mathematics of Operations Research* **16**, 547–565.

Miyazawa, M. (1993), Insensitivity and product-form decomposability of reallocatable *GSMP*, *Advances in Applied Probability* **25**, 415–437.

Miyazawa, M. (1994), On the characterization of departure rules for discrete-time queueing networks with batch movements and its applications, *Queueing Systems* **18**, 149–166.

Miyazawa, M. (1996), Stochastic bound and stability of discrete-time Jackson networks with batch movements, *in* P. Glasserman, K. Sigman and D. D. Yao, eds, *Stability and Rare Events in Queueing Networks*, Lecture Notes in Statistics 117, Springer-Verlag, pp. 75–94.

Miyazawa, M. (1997), Structure-reversibility and departure functions of queueing networks with batch movements and state dependent routing, *Queueing Systems* **25**, 45–75.

Miyazawa, M. (1998), Stochastic networks with countable state spaces: Modeling and tractable stationary distributions, *in* M. Miyazawa and H. Ōsawa, eds, *Memories on The Numazu Meeting on Applied Probability in 1995*, Numazu Institute of International Education, pp. 45–79.

Miyazawa, M. and Chao, X. (1998), Generating a queueing network with signals and state dependent local transitions from a quasi-reversible network. Preprint.

Miyazawa, M. and Taylor, P. G. (1997), A geometric product-form distribution for a queueing network with nonstandard batch arrivals and batch transfers, *Advances in Applied Probability* **29**, 523–544.

Miyazawa, M. and Wolff, R. W. (1996), Symmetric queues with batch departures and their networks, *Advances in Applied Probability* **28**, 308–326.

Muntz, R. R. (1972), Poisson departure processes and queueing networks, IBM Research Report RC 4145, IBM Thomas J. Watson Research Center.

Ōsawa, H. (1989), Discrete-time storage models with negative binomial inflow, *Journal of Operations Research Society of Japan* **32**, 218–232.

Ōsawa, H. (1994), Quasi-reversibility of a discrete-time queue and related models, *Queueing Systems* **18**, 133–148.

Pujolle, G., Claude, J. P. and Seret, D. (1985), A discrete tandem queueing system with a product form solution, *in* Seminar on Computer Networking and Performance Evaluation, North-Holland, Kyoto, pp. 139–147.

Reich, E. (1957), Waiting times when queues are in tandem, *Annals of Mathematical Statistics* **28**, 768–773.

Reich, E. (1963), Note on queues in tandem, *Annals of Mathematical Statistics* **34**, 338–341.

Ross, K. (1995), *Multiservice Loss Models for Broadband Telecommunication Networks*, Springer-Verlag, New York.

Ross, S. M. (1983), *Stochastic Processes*, John Wiley & Sons, New York.

Ross, S. M. (1997), *Introduction to Probability Models*, 6th edn, Academic Press, San Diego, CA.

Rumswicz, N. (1987), Insensitivity of $GSMP$s with age-dependent routing. Doctoral dissertation, University of Adelaide, Australia.

Schassberger, R. (1977), Insensitivity of steady state distributions of generalized semi-Markov processes. Part I, *Annals of Probability* **5**, 87–99.

Schassberger, R. (1986), Two remarks on insensitive stochastic models, *Advances in Applied Probability* **18**, 791–814.

Serfozo, R. F. (1989*a*), Markovian network processes: Congestion-dependent routing and processing, *Queueing Systems* **5**, 5–36.

Serfozo, R. F. (1989*b*), Poisson functionals of Markov processes and queueing networks, *Advances in Applied Probability* **21**, 595–611.

Serfozo, R. F. (1993), Queueing networks with dependent nodes and concurrent movements, *Queueing Systems* **13**, 143–182.

Serfozo, R. F. (1999), *Introduction to Stochastic Networks*, Springer-Verlag, New York. To appear.

Serfozo, R. F. and Yang, B. (1998), Markov network processes with string transitions, *Annals of Applied Probability* **8**, 793–821.

Shwartz, A. and Weiss, A. (1993), Induced rare events: Analysis via large deviations and time reversal, *Advances in Applied Probability* **25**, 667–689.

Sigman, K. (1990), The stability of open queueing networks, *Stochastic Processes and Their Applications* **35**, 11–25.

Song, J., Xu, S. and Liu, B. (1999), Order fulfillment performance measures in an assembly-to-order system with stochastic leadtime, *Operations Research* **47**, 131–149.

Stoyan, D. (1983), *Comparison Methods for Queues and Other Stochastic Models*, English edn, John Wiley & Sons, New York.

Takada, H. and Miyazawa, M. (1997), Necessary and sufficient conditions for product-form queueing networks. Preprint.

Takagi, H. (1993), *Queueing Analysis*, Vol. 3: *Discrete-Time Systems*, North-Holland, Amsterdam.

Todd, M. (1976), *The Computation of Fixed Points and Applications*, Vol. 124 of *Lecture Notes in Economics and Mathematical Systems*, Springer-Verlag, Berlin.

Towsley, D. E. (1980), Queueing network models with state-dependent routing, *Journal of the ACM* **27**, 323–337.

Van Dijk, N. M. (1993), *Queueing Networks and Product Forms: A Systematic Approach*, John Wiley & Sons, Chichester.

Walrand, J. (1983*a*), A discrete-time queueing network, *Journal of Applied Probability* **20**, 903–909.

Walrand, J. (1983*b*), A probabilistic look at networks of quasi-reversible queues, *IEEE Transactions on Information Theory* **IT-29**, 825–831.

Walrand, J. (1988), *An Introduction to Queueing Networks*, Prentice-Hall, New Jersey.

Walrand, J. and Varaiya, P. (1980), Interconnection of Markov chains and quasi-reversible queueing networks, *Stochastic Processes and Their Applications* **10**, 209–219.

Whitt, W. (1993), Large fluctuations in a deterministic multiclass network of queues, *Management Science* **39**, 1020–1028.

Whittle, P. (1967), Nonlinear migration processes, *Bulletin of the International Institute of Statistics* **42**, 642–647.

Whittle, P. (1968), Equilibrium distributions for an open migration process, *Journal of Applied Probability* **5**, 567–571.

Whittle, P. (1986), *Systems in Stochastic Equilibrium*, John Wiley & Sons, New York.

Wolff, R. W. (1982), Poisson arrivals see time averages, *Operations Reserach* **30**, 223–231.

Wolff, R. W. (1989), *Stochastic Modeling and the Theory of Queues*, Prentice Hall, New Jersey.

Woodward, M. E. (1994), *Communication and Computer Networks: Modeling with Discrete Time Queues*, IEEE Computer Society Press, Los Alamitos, CA.

Wright, P. E. (1992), Two parallel processors with coupled inputs, *Advances in Applied Probability* **24**, 986–1007.

Yamashita, H. and Miyazawa, M. (1998), Geometric product form queueing networks with concurrent batch movements, *Advances in Applied Probability* **30**, 1111–1129.

Index